Metals in Medicine (Inorganic Chemistry)

Metals in Medicine (Inorganic Chemistry)

Edited by
George Brown

WILLFORD PRESS

www.willfordpress.com

Published by Willford Press,
118-35 Queens Blvd., Suite 400,
Forest Hills, NY 11375, USA

ISBN: 978-1-64728-004-8

Cataloging-in-Publication Data

Metals in medicine (inorganic chemistry) / edited by George Brown.
    p. cm.
Includes bibliographical references and index.
ISBN 978-1-64728-004-8
1. Metals in medicine. 2. Chemistry, Inorganic. 3. Clinical chemistry. I. Brown, George.
QD152.5.M43 M48 2022
546--dc23

For information on all Willford Press publications
visit our website at www.willfordpress.com

WILLFORD PRESS

Contents

Permissions

List of Contributors

Index

Preface

Metals are used in medicine for diverse diagnostic and treatment purposes. It is noteworthy that metals are toxic for the human body when in high quantities. Mercury, lead, cadmium, plutonium, etc. are some toxic metals. The ingestion of metals or faulty metabolic pathways can lead to metal poisoning. Humans however require certain metals for normal functioning. Metals such as iron, sodium, zinc, magnesium, etc. are needed for their role as cofactors or prosthetics in enzymes and for the catalysis of specific reactions. Metal complexes can be used for radioisotope imaging or as contrast agents in magnetic resonance imaging. Technetium is the most common radioisotope agent that is used for imaging. Gallium-68 is used as a source of positrons in positron emission tomography. Metals are also used for various treatments. They can be used to treat cancer, by specifically attacking cancer cells and interacting with the DNA. Lithium compounds can be used for the management of manic-depressive disorder. Silver can be used to prevent infection, while gold salt complexes are used to treat rheumatoid arthritis. This book is a compilation of chapters that discuss the most vital concepts and emerging trends in the use of metals in medicine. Coherent flow of topics, student-friendly language and extensive use of examples make this book an invaluable source of knowledge.

This book is a result of research of several months to collate the most relevant data in the field.

When I was approached with the idea of this book and the proposal to edit it, I was overwhelmed. It gave me an opportunity to reach out to all those who share a common interest with me in this field. I had 3 main parameters for editing this text:

1. Accuracy – The data and information provided in this book should be up-to-date and valuable to the readers.

2. Structure – The data must be presented in a structured format for easy understanding and better grasping of the readers

3. Universal Approach – This book not only targets students but also experts and innovators in the field, thus my aim was to present topics which are of use to all

Thus, it took me a couple of months to finish the editing of this book.

I would like to make a special mention of my publisher who considered me worthy of this opportunity and also supported me throughout the editing process. I would also like to thank the editing team at the back-end who extended their help whenever required.

Editor

Pharmacological and Toxicological Threshold of Bisammonium Tetrakis 4-(*N, N*-Dimethylamino) pyridinium Decavanadate in a Rat Model of Metabolic Syndrome and Insulin Resistance

Samuel Treviño,[1] Alfonso Díaz,[1] Eduardo Sánchez-Lara,[2]
Víctor Enrique Sarmiento-Ortega,[1] José Ángel Flores-Hernández,[1] Eduardo Brambila ⓘ,[1]
Francisco J. Meléndez,[1] and Enrique González-Vergara ⓘ[2]

[1]*Facultad de Ciencias Químicas, Benemérita Universidad Autónoma de Puebla, 14 Sur y Av. San Claudio, Col. San Manuel, 72570 Puebla, PUE, Mexico*
[2]*Centro de Química, ICUAP, Benemérita Universidad Autónoma de Puebla, 14 Sur y Av. San Claudio, Col. San Manuel, 72570 Puebla, PUE, Mexico*

Correspondence should be addressed to Enrique González-Vergara; enrique.gonzalez@correo.buap.mx

Academic Editor: Zhe Liu

Vanadium(IV/V) compounds have been studied as possible metallopharmaceutical drugs against diabetes mellitus. However, mechanisms of action and toxicological threshold have been tackled poorly so far. In this paper, our purposes were to evaluate the metabolic activity on dyslipidemia and dysglycemia, insulin signaling in liver and adipose tissue, and toxicology of the title compound. To do so, the previously reported bisammonium tetrakis 4-(*N,N*-dimethylamino)pyridinium decavanadate, the formula of which is $[DMAPH]_4(NH_4)_2[V_{10}O_{28}] \cdot 8H_2O$ (where DMAPH is 4-dimethylaminopyridinium ion), was synthesized, and its dose-response curve on hyperglycemic rats was evaluated. A Long–Evans rat model showing dyslipidemia and dysglycemia with parameters that reproduce metabolic syndrome and severe insulin resistance was generated. Two different dosages, 5 μmol and 10 μmol twice a week of the title compound (equivalent to 2.43 mg·V/kg/day and 4.86 mg·V/kg/day, resp.), were administered intraperitoneal (i.p.) for two months. Then, an improvement on each of the following parameters was observed at a 5 μmol dose: weight reduction, abdominal perimeter, fatty index, body mass index, oral glucose tolerance test, lipid profile, and adipokine and insulin resistance indexes. Nevertheless, when the toxicological profile was evaluated at a 10 μmol dose, it did not show complete improvement, tested by the liver and adipose histology, as well as by insulin receptor phosphorylation and GLUT-4 expression. In conclusion, the title compound administration produces regulation on lipids and carbohydrates, regardless of dose, but the pharmacological and toxicological threshold for cell regulation are suggested to be up to 5 μmol (2.43 mg·V/kg/day) dose twice per week.

1. Introduction

Interest in vanadium actions on biological systems has gradually increased, and its physiological relevance has been established in recent years [1, 2]. Vanadium is the 18th most abundant element in the crust of our planet, even more than zinc (0.019% and 0.008%, resp.). Regarding this occurrence, vanadium is present in the soil, water, and air, in almost all types of ecosystems. It is absorbed by plants and travels along the food chain up to humans; thus, it is distributed ubiquitously in living organisms [1–3]. However, metabolism of vanadium in humans and animals has not been fully understood, and evidence is not conclusive as to its importance as a trace element [4]. The estimated daily consumption ranges from 10 to 60 μg depending on a mammal's diet, but its deficiency in mammals inhibits growth, impairs cellular regenerative functions, and affects thyroid metabolism and bone mineralization, and it causes disturbances in lipid and carbohydrate balance [3–6].

Lipid and carbohydrate disorders are strongly linked to obesity development, insulin resistance, type 2 diabetes, dyslipidemia, hepatic steatosis, and cardiovascular diseases. All of which have become known as metabolic syndrome. Metabolic syndrome is a major public health problem and an important clinical challenge worldwide because of its prevalence in 10 to 84% of the population depending on the world region. The International Diabetes Federation estimates that one-quarter of the world's adult population suffers from metabolic syndrome associated with being overweight and an increased body mass index that is reflected in a higher fatty body mass, distributed mainly in the visceral adipose tissue (abdominal circumference) [7–10]. Despite many years of research and partial success in the treatment of metabolic diseases, molecular mechanisms and their accompanying subcellular changes have not been fully elucidated. Therefore, research has focused on less invasive and more effective methods of treatment. Over the last twenty years of research, vanadium compounds have been shown to act in a similar way to insulin on selected diabetes and metabolic syndrome models, in tests both on animals and humans as far as physiological symptoms and biochemical parameters are concerned [11–13].

Depending on its speciation forms, physicochemical properties, and concentration, vanadium can be absorbed and distributed into the body. Most vanadium complexes are transformed into a cationic form (vanadyl) which binds to transferrin or albumin, where it can go through spontaneous oxidation to become vanadate [14]. Vanadate(V) and vanadyl(IV) are easily redox interconvertible, but vanadate is the main inorganic vanadium species available for interaction with cellular functions [2, 15–17]. The oxidation state of vanadium is a factor that determines its biological effects, which are present in the activity regulation of many enzymes and phosphorylation and dephosphorylation processes. In enzymes, regulatory properties of vanadate depend, at least in part, on its esterification with phosphate [18]. Therefore, enzymes can accept vanadate as an analogy to phosphate [19–21]. However, there is a subtle tight threshold between biological activity and toxicity that is correlated to its degree of oxidation (vanadyl < vanadate ion) and chemical form (organic < inorganic) [3, 21–23]. In this respect, the most often observed side effects include a loss of appetite and significant reduction of body weight (often leading to anorexia), as well as a low activity of hepatic damage enzymes, confirmed by histopathological studies [24]. However, reports are limited and divergent, but it has been suggested that vanadium toxicity is dependent on speciation and oxidation state via time and doses of administration. The therapeutic concentration limit of vanadium compounds is, as a general rule, below 0.01×10^{-3} M, which is a safe limit and maintains the biological activity, whereas several toxic effects associated with its threshold or accumulation in tissues may be expected to be above 1.0×10^{-3} M [25]. Regarding vanadium toxicity, it has been shown that it is dramatically reduced when oligomerized in the form of decavanadate [20–23]. The biochemical importance of decavanadate resides in protein-bound species because it is very stable to pH changes and is eventually

turned to labile oxovanadate species. Thus, its rate of disgregation can be very slow but effective, which allows it to exist under physiological conditions for some time [21]. Likewise, decavanadate is an anion that could get into the cell across proteins like channels or lipid interactions to trigger its biological effects on energy transduction systems [26]. Recently, our working team reported a decavanadate compound with DMAPH. The crystal structure of this complex has been completely characterized [27]. Thus, in this study, our aim was to evaluate the possible metabolic activity of the compound in a dyslipidemia and dysglycemia model, related to insulin signaling in liver and adipose tissue, as well as its toxicological and pharmacological effects.

2. Materials and Methods

2.1. Synthesis of [DMAPH]$_4$(NH$_4$)$_2$[V$_{10}$O$_{28}$]·8H$_2$O (V10-DMAP). All chemicals used were of reagent grade and were purchased from Sigma-Aldrich. The synthesis was performed with a 0.468 g (4 mmol) mixture of ammonium metavanadate in 18 mL of distilled water in an Erlenmeyer flask with magnetic stirring and heated at 70°C until it was dissolved. Then, four drops of concentrated hydrochloric acid (37%) at room temperature were added to allow decavanadate anion formation at pH 6. After obtaining an orange solution from this process, 0.122 g (1 mmol) of 4-dimethylaminopyridine dissolved in 2 mL of distilled water was added dropwise. Once 4-dimethylaminopyridine was added, the mixture was filtered off and allowed to stand at room temperature for one day to produce orange crystals of the title compound [27].

2.2. Dose-Response Curve of V10-DMAP. Long–Evans rats were used in this study because this strain is twice as susceptible to spontaneously develop obesity and hyperglycemia. Thirty male rats weighing 300 to 320 g underwent an intraperitoneal application of alloxan (150 mg/kg) to induce hyperglycemia. When hyperglycemia (HG) above 200 mg/dL was present, animals were split into six working groups (*n* = 5), all of which were given with dosages of the title compound abbreviated as V10-DMAP from now on, at 0 μmol, 2.5 μmol, 5 μmol, 7.5 μmol, 10 μmol, and 20 μmol, dissolved in 1 mL of injectable solution to be administered via intraperitoneal (i.p.). These dosages were administered twice a week for a month until glucose was normalized (≤130 mg/dL). Monitoring was carried out with serum collected in a BD Vacutainer® venous blood collection system, centrifuged at 2500 rpm for 5 min. Glucose quantification was carried out with a commercial kit, and the values obtained were used to determine a dose-response curve.

2.3. Dyslipidemia and Dysglycemia Model. Forty male Long–Evans rats (70–100 g) were provided by the vivarium "Claude Bernard" from Benemérita Universidad Autónoma de Puebla. The rats were housed in a climate-controlled environment and 12 h light-dark cycles with free access to food and water *ad libitum*. All procedures described in this study agreed with the Guide for the Care and Use of Laboratory Animals of the Mexican Council for Animal Care NOM-062-ZOO-1999. All applicable international, national, and institutional guidelines for the care and use of animals were followed. The rats were conditioned with a normal calorie diet during 15 days. The diet used was 5001 from LabDiet (Laboratory Rodent Diet), and its composition can be accessed on the manufacturer's website. Upon reaching the ideal weight (150 g), animals were randomly split into different groups. Two groups were formed: group 1: normal calorie or NC group fed with 5001 from LabDiet ($n = 10$); group 2: high calorie or HC group fed with 5008 from LabDiet ($n = 30$) Diet 5001 from LabDiet provides 58.0% carbohydrates, 13.5% fat, and 28.5% proteins. Diet 5008 from LabDiet provides 56.44% carbohydrates, 17.0% fat, and 26.85% proteins. The groups were fed for two months *ad libitum* with diet 5001 as a caloric control and diet 5008 until they developed the dyslipidemia and dysglycemia model. The model was validated by measuring body weight, abdominal perimeter, and tip of the nose to base of the tail length. BMI (body mass index) and body fat percentage were determined by the Lee index [28]. Glucose, fructosamine, triglycerides, total cholesterol, HDL cholesterol, insulin, and insulin resistance by a homeostasis model assessment insulin resistance (HOMA-IR) and insulin resistance adipocyte dysfunction (IDA-IR) were biochemically evaluated [28].

2.4. Effect of V10-DMAP on Dyslipidemia and Dysglycemia Model.

After appearance of the metabolic dysregulation validation, two groups of 10 rats from the 5008 group were randomly separated and divided into groups 3 and 4 as follows: group 3: HC-V10-DMAP-5 μmol group fed with diet 5008 and administered twice a week (i.p.) ($n = 10$); group 4: HC-V10-DMAP-10 μmol group was fed with diet 5008 and administered twice a week (i.p.) ($n = 10$). Both groups were fed for a period of two more months. The zoometry was monitored every two weeks for two months as described above. Toxicological and biochemical parameters in fasting conditions were assessed in serum at the end of the experimental period: total bilirubin, aspartate aminotransferase (AST), alanine aminotransferase (ALT), glutamyl transpeptidase (GT), lactate dehydrogenase (LDH), glucose, insulin, leptin, adiponectin, fructosamine, triglycerides, free fatty acids, total cholesterol, and HDL cholesterol were determined by using commercial kits. Additionally, an oral glucose tolerance test (OGTT, 1.75 g of glucose anhydrous/kg) was carried out after 4–6 h of fasting. A postload monitoring was carried out at 30, 60, and 90 min, and the area under the curve (AUC) was calculated. The samples were taken by cardiac puncture after anesthesia with ketamine and xylazine (20/137 mg/kg). Insulin resistance indexes (HOMA-IR and IDA-IR) were calculated.

2.5. Effects of V10-DMAP on the Liver and Adipose Tissue.

One week after the last biochemical evaluation, rats were anesthetized with sodium pentobarbital (40 mg/kg, i.p.) and then perfused with 200 mL of 4% paraformaldehyde. Biopsies from liver and visceral white adipose tissue were

removed and postfixed in the same solution for 48 h and then embedded in paraffin. Sections of 5 μm thick were taken from each tissue for subsequent staining. Histological evaluation was carried out with hematoxylin-eosin stain using standard procedures after paraffin removal and tissue rehydration. The degree of hepatic injury and adipose tissue remodeling (hypertrophy) were evaluated with light microscopy. Additionally, sections of tissue were rehydrated according to conventional histological techniques; then, nonspecific binding sites were blocked by incubation in 2% IgG-free bovine serum albumin (BSA, Sigma). After that, specimens were incubated with 0.2% Triton X-100. The sections were incubated overnight at 4 to 8°C with primary antibodies: *p*-insulin Rβ Antibody (Tyr 1162/1163) to the liver and adipose tissue and GLUT-4 to adipose tissue. The primary antibodies (Santa Cruz Biotechnology Inc., CA, USA) were 1 : 100 diluted. Fluorescein isothiocyanate (FITC) secondary antibodies (1 : 100, Jackson ImmunoResearch Laboratories Inc., PA, USA) revealed absence or presence of proteins. Slides were mounted with VectaShield containing 4′,6-diamidino-2-phenylindole (DAPI) (Vector Labs., CA, USA) for nucleus staining. Photomicrographs were taken using a fluorescence microscope (Leica Microsystems GmbH, Wetzlar, Germany) and projected with a Leica IM1000 version 1.20 release-9 computer-based program (Imagic Bildverarbeitung AG, Leica Microsystems, Heerbrugg, Switzerland).

2.6. Statistical Analysis.

Results were expressed as a mean ± standard error of the mean (SEM). A dietary comparison was performed by Student's *t*-test, considered significant $p \leq 0.05$. Meanwhile, a comparison of the groups studied after (V10-DMAP) treatments was performed by an analysis of variance and multiple comparisons using a two-way ANOVA analysis and a Bonferroni post hoc test, obtaining a significant $p < 0.05$. A GraphPad Prism 5.0 statistical program was used to perform this analysis.

3. Results

Hyperglycemic rats induced by alloxan initially showed an average of serum glucose of 300 mg/dL. The V10-DMAP dosages administrated twice a week for one month period produced a decrease in glucose levels depending on the vanadium compound concentration. According to the results, doses of 5 μmol and 10 μmol of the title compound (equivalent to 2.43 mg·V/kg/day and 4.86 mg·V/kg/day, resp.) twice a week, reduced glucose levels below 130 mg/dL in 50% of the subjects given with this compound (Figure 1). To determine the effective dose 50 (ED$_{50}$), we proceeded to perform a mathematical model to apply the exact concentration. Because the mathematical model did not discriminate between 5, 7.5, and 10 μmol doses, we decided to use only two different doses (5 μmol and 10 μmol, equivalent to 2.43 mg·V/kg/day and 4.86 mg·V/kg/day, resp.) in later experiments of metabolic regulation to investigate their toxicological effects.

On the contrary, in the model of dyslipidemia and dysglycemia produced by one-month consumption of diet

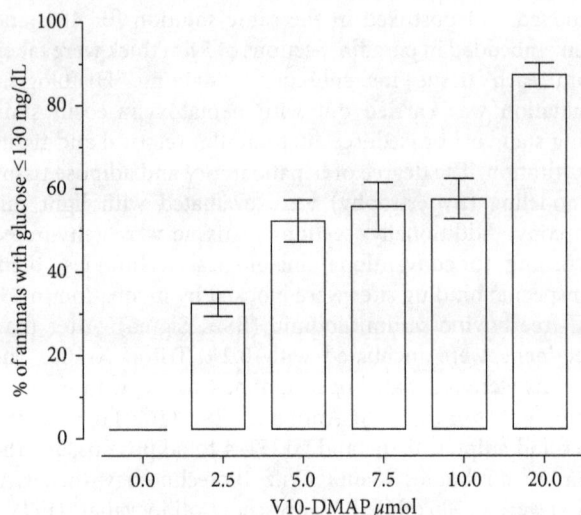

FIGURE 1: Effective dose 50 analysis (ED_{50}). The figure represents the percentage of glucose level reduction ≤ 130 mg/dL by administration of V10-DMAP. The hyperglycemic Long–Evans rats ($n = 5$/group) were given with dosages of 0.0, 2.5, 5.0, 7.5, 10.0, and 20 μmol of V10-DMAP for four weeks. The Long–Evans rats were alloxan-induced hyperglycemic (150 mg/kg). Mean glucose before V10-DMAP administration was ≈ 300 mg/dL.

5008 (which possesses 3.5% more fat than the normal-caloric diet 5001), there was a Long–Evans rat weight increase. Also, there was an elevation in the levels of glucose, fructosamine (protein glycation), and insulin levels with hepatic insulin resistance (HOMA-IR) and adipose resistance (IDA-IR) (Table 1). Moreover, hypertriglyceridemia and hypoalphalipoproteinemia (low level of HDL) were observed. These results strongly suggest the development of metabolic syndrome; however, the Long–Evans rats were fed with 5008 diet for one more month to confirm the model. After the second month, the Long–Evans rats showed a 20% weight gain in comparison to the diet 5001 group (Table 1). Similarly, glucose increased twice as much (206 mg/dL), fructosamine increased 40%, and insulin increased by 118%, which produced a noticeable insulin resistance level both in the liver and adipose tissue. Dyslipidemia showed a 71% triglyceride increase and a 15% HDL decrease (Table 1). These two-month-period results ensure the development of the dysglycemia and dyslipidemia model. Then, rats suffering from metabolic disorder were separated in individual cages, and V10-DMAP dosages of 5 μmol and 10 μmol (equivalent to 2.43 mg·V/kg/day and 4.86 mg·V/kg/day, resp.), twice per week, were administered intraperitoneally. Diet consumption was maintained *ad libitum*.

To justify the fact that the V10-DMAP dose may be used in the treatment of metabolic diseases, both zoometry and biochemical parameters were evaluated in all groups. Overweight was observed in groups which fed a HC diet (time dependent). Rats starting with an approximate 390 g weight showed a 36 g weight gain (426 g 8 weeks later). These results mean 5.5% more weight than the control group (Figure 2(a)). Nonetheless, BMI showed no change, and fat mass percentage showed a significant decrease in the HC diet

group. Remarkably, fat mass increased in the abdominal circumference area, suggesting impaired visceral white adipose tissue, because there was an increase of 8% in 8 weeks (Figure 2(d)).

Weight was reduced in both groups of rats given with V10-DMAP since the second week was in relation to the NC group. At the end of the V10-DMAP administration, group 3 (5 μmol) and group 4 (10 μmol) showed a significant weight reduction of 18% in relation to the NC group, while both were 22.5% lower about the HC group (Figure 2(a)). Likewise, the BMI of groups 3 and 4 improved after two weeks of V10-DMAP administration in 16.5% and 12%, diminishing up to 24.5% after eight weeks in comparison to the NC group and 17.5% in comparison to the HC group (Figure 2(b)). The body fat percentage had a similar pattern. The body fat percentage after two weeks had a similar pattern since the second week in groups 3 and 4, decreasing 7% and 6%, respectively. At the end of the experiment (8 weeks), the fat percentage was 10% lower in the NC group and 5% lower in the HC group (Figure 2(c)). Finally, abdominal perimeter showed a clear reduction. The maximum difference was 11% after eight weeks in groups 3 and 4 in comparison to the NC group, whereas the difference was 25.5% in comparison to the HC group for both groups (Figure 2(d)). Similarly, the oral glucose tolerance test showed significant differences in fasting (0 min) between the V10-DMAP groups in comparison to the NC group, but at 30 min, the V10-DMAP-10 μmol group was also slightly higher. Nevertheless, the area under curve (AUC) analysis showed no difference in both V10-DMAP groups. The HC group was evidently different in comparison to the NC group with a 26.5% increase at 0 min, a 48% increase at 30 min, a 45.5% increase at 60 min, a 69% increase at 90 min, and a 48.5% at AUC (Figures 3(a) and 3(b)).

Complete biochemical effects are shown in Table 2. In comparison to the NC group, the high-fat diet in the HC group maintains metabolic dysregulation. Meanwhile, the V10-DMAP administration shows substantial improvements in biochemical parameters. Both groups 3 and 4 regulate their fasting glucose and fructosamine levels, as well as insulin secretion until similar levels are reached as in the NC group. Therefore, hepatic and adipose insulin resistances disappear, and adipokine secretion was almost regulated. In the HC-V10-DMAP-10 μmol group, leptin remained slightly high, whereas adiponectin improved, but it did not recover the level shown by the NC group. In this regard, the HC-V10-DMAP-5 μmol group showed an adipokine and a lipid profile return to normal levels, but triglycerides did not diminish like the NC group. Complete lipid regulation was observed in the HC-V10-DMAP-10 μmol group. Notably, HDL level in both groups given with V10-DMAP showed an increase in its concentration, even higher than the NC group. All biochemical parameters in the V10-DMAP-administered groups presented an improvement in comparison to the HC group.

On the contrary, biochemical parameters related to hepatic toxicological effects were measured (Table 3). Total bilirubin showed no differences in either group. However, in both the HC group and group 4, the hepatic damage enzymes are increased to 65% and 26.5% in AST, increased to 38.5% in ALT (only in the HC group), 38% and 27% in γ-GT, 27% and

TABLE 1: Induction of the dyslipidemia and dysglycemia model.

	15 days fitting-out	1 month		2 months	
	5001 ($n = 40$)	5001 ($n = 10$)	5008 ($n = 30$)	5001 ($n = 10$)	5008 ($n = 30$)
Weight (g)	158 ± 6.3	230 ± 7.0	270 ± 5.0 ▲	323 ± 8.7	388 ± 11.4 ▲
Height (cm)	15.3 ± 0.6	20.3 ± 1.7	21.4 ± 1.4	23.2 ± 0.8	24.2 ± 1.1
Abdominal perimeter (cm)	13.9 ± 0.4	15.4 ± 0.6	16.8 ± 0.9	17.6 ± 0.7	18.9 ± 1.1
BMI (g/cm^2)	0.68 ± 0.07	0.57 ± 0.08	0.59 ± 0.08	0.59 ± 0.03	0.66 ± 0.04
% body fat (Lee index)	34.7 ± 0.9	29.8 ± 2.2	29.8 ± 1.8	29.2 ± 0.9	29.6 ± 1.1
Glucose (mg/dL)	106.0 ± 4.1	104.9 ± 4.5	174.3 ± 3.6 ▲	104.1 ± 6.8	206.3 ± 4.5 ▲
Fructosamine (mmol/L)	40.8 ± 0.8	42.8 ± 1.3	55.3 ± 2.1 ▲	43.2 ± 0.5	60.3 ± 2.3 ▲
Triglycerides (mg/dL)	58.4 ± 3.4	75.0 ± 1.5	104.7 ± 6.7 ▲	70.0 ± 8.2	119.6 ± 5.5 ▲
Total cholesterol (mg/dL)	86.8 ± 1.9	75.2 ± 1.9	72.5 ± 1.4	78.0 ± 2.3	80.0 ± 2.5
HDL (mg/dL)	45.7 ± 1.2	49.1 ± 1.7	41.3 ± 1.0 ▼	45.1 ± 2.0	38.5 ± 1.0 ▼
Insulin (μUI/mL)	15.3 ± 0.5	13.3 ± 0.8	25.9 ± 2.1 ▲	15.5 ± 1.1	33.8 ± 4.2 ▲
HOMA-IR	0.67 ± 0.04	0.58 ± 0.06	1.86 ± 0.19 ▲	0.67 ± 0.09	2.88 ± 0.42 ▲
IDA-IR	0.11 ± 0.01	0.18 ± 0.01	0.4 ± 0.02 ▲	0.19 ± 0.03	0.49 ± 0.01 ▲

Results shown are the average ± SEM. ▲ indicates a significant difference with values above the control group with a normal calorie diet. ▼ indicates a significant difference with values below the control group with a normal calorie diet. Comparisons between groups were performed by Student's t-test. BMI = body mass index; HDL = high density lipoprotein; HOMA-IR = homeostasis model assessment insulin resistance; IDA-IR = insulin resistance adipocyte dysfunction.

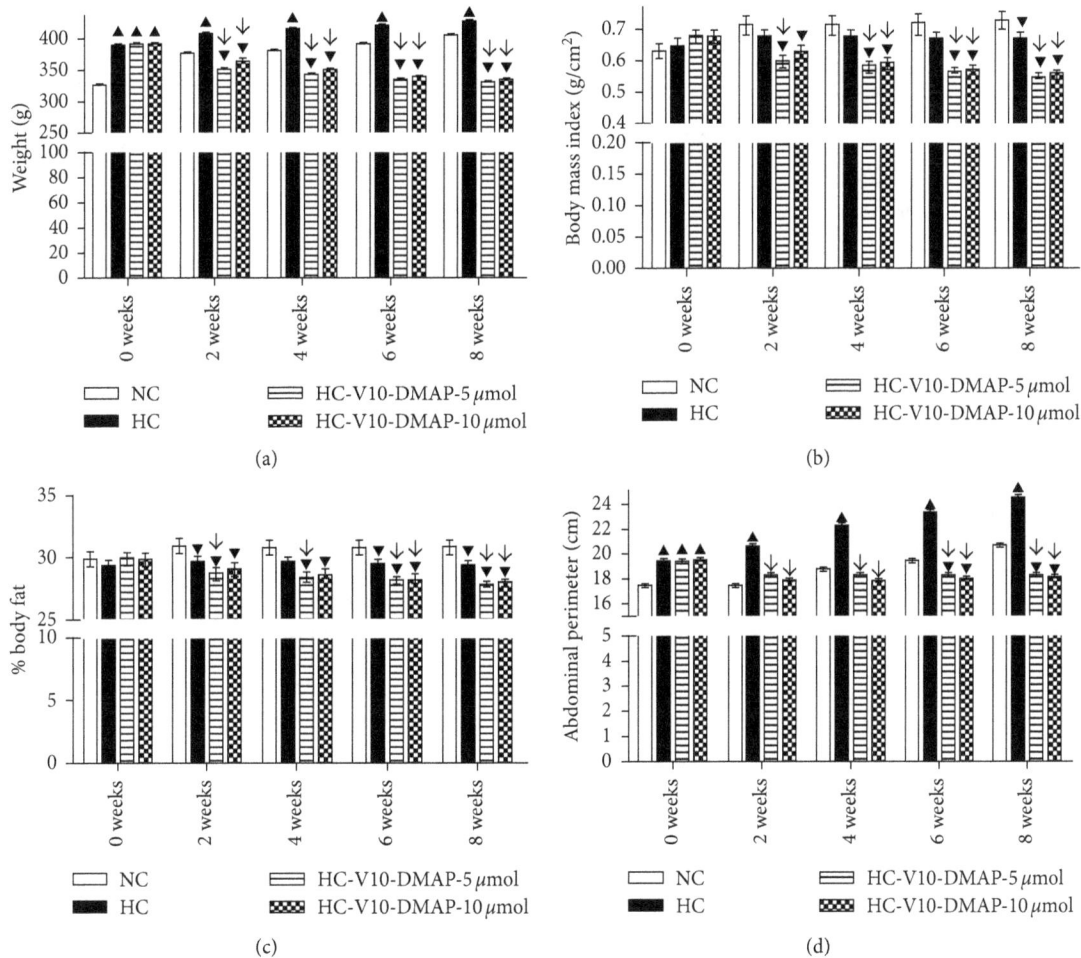

(a)

(b)

(c)

(d)

FIGURE 2: Zoometry after diets and V10-DMAP treatments: (a) body weight, (b) BMI, (c) Lee index, the percentage of the body fat mass index, and (d) abdominal perimeter or circumference. Results shown are the average ± SEM. ▲ indicates a significant difference between values above the control group with a normal calorie diet. ▼ indicates a significant difference between values below the control group with a normal calorie diet. ↓ indicates a significant difference between values below the HC group. Comparisons between groups were performed by two-way ANOVA and the Bonferroni post hoc test; $p < 0.05$.

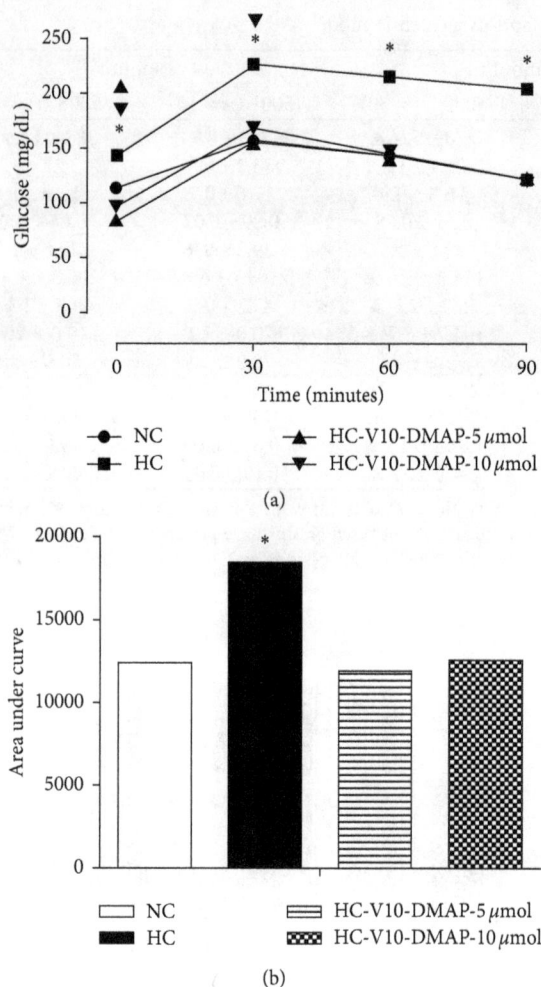

FIGURE 3: Oral glucose tolerance test and area under the curve after V10-DMAP administration: (a) OGTT (1.75 g/kg weight of anhydrous glucose) ($n = 10$ per group) and (b) graphical representation of AUC analysis. Results shown are the average ± SEM. *, ▲, and ▼ indicate a significant difference among the HC, HC-V10-DMAP-5 µM, and HC-V10-DMAP-10 µM groups, respectively, in comparison to the NC group. Comparisons between groups were performed by two-way ANOVA and Bonferroni's post hoc test; $p < 0.05$.

58% in LDH, respectively. Group 3, which belongs to HC-V10-DMAP-5 µmol, showed no differences in comparison to the NC group, and all serum biomarkers showed significant differences in comparison to the HC group.

Histological changes were evaluated in tissue sections stained with hematoxylin and eosin (H&E). These changes were viewed under standard bright field illumination. Eosinophilic cytoplasm and blue nuclei, which are characteristic of liver cells, were observed in the liver of the NC group with a distribution of the sinusoid spaces taking the central vein as a reference (Figure 4(a)). The HC group showed visible histology changes, including structure damage, hepatocellular necrosis (black arrow), leukocyte infiltration, recognized as portal inflammation (blue arrow), necrosis, and development of Mallory–Denk bodies, which are an inclusion found in the cytoplasm of hepatocytes. Also,

sinusoids between the hepatocyte plates were noticeably enlarged in the liver (Figure 4(b)). Despite that, injuries were of low grade. Liver injury notably diminishes with 5 and 10 µmol doses of V10-DMAP, but weak portal inflammation (blue arrow) and slight focal confluent necrosis were observed (Figures 4(c) and 4(d)).

On the contrary, normal histology of adipose tissue is shown in Figure 4(e). Regarding this type of histology, loose connective tissue composed mostly of adipocytes and their storage space (asterisk), and blood vessels can be observed (green arrow). Meanwhile, the HC group showed adipocyte hypertrophy which is characterized by fewer adipocytes (number per analyzed area) with a vascularization increase (Figure 4(f)). Conversely, adipocytes in both groups V10-DMAP reduced their storage area, increasing cell number and vascularization per analyzed area (Figures 4(g) and 4(h)).

To understand the effects caused by the administration of the V10-DMAP compound, the immunoreactivity in green for the insulin receptor phosphorylated in tyrosine (Tyr 1162/1163) was analyzed in the four study groups. Arbitrary units (pixels) were determined and normalized to the NC group using ImageJ program (data not shown). Regarding the control group, immunoreactivity to hepatic insulin receptor was higher in the HC group (120%), HC-V10-DMAP-5 µmol group (56%), and HC-V10-DMAP-10 µmol group (81%) (Figures 5(a)–5(d)). In visceral white adipose tissue, immunoreactivity in the HC group was 6% less than the NC group, while there was a 104% increase in the HC-V10-DMAP-5 µmol group but only a 20% increase in the HC-V10-DMAP-10 µmol group (Figures 5(e)–5(h)). Finally, GLUT-4 immunoreactivity in the HC group and HC-V10-DMAP-10 µmol group was a 54% and 28% less than the NC group, but there was some improvement in comparison to the HC group. Meanwhile, immunoreactivity for GLUT-4 in the HC-V10-DMAP-5 µmol group was higher (33%) to the NC group (Figures 5(i)–5(l)).

4. Discussion

In this paper, we evaluated the metabolic activity of V10-DMAP on the dyslipidemia and dysglycemia model. Previously, our working team reported the synthesis and characterization of V10-DMAP by infrared spectroscopy, thermal analysis, and single crystal X-ray diffraction [27]. Similarly, we have previously synthesized and reported another hybrid material based on metformin, a popular drug used to treat type 2 diabetes mellitus [29]. Metformin decavanadate has shown hypoglycemic and hypolipidemic effects on models of type 1 and type 2 diabetes mellitus. Despite vanadium compounds having insulin-mimetic properties, and the first report on their therapeutic properties about diabetes and dyslipidemia appeared in the 20th century [30, 31], its mode of action is still poorly understood at the molecular level [30, 32–39]. Although the general population is exposed to vanadium, normal blood concentration rarely exceeds 0.2 µM [3, 18, 40]; nonetheless, the therapeutic dose is controversial, because patients have a dependence of vanadium speciation and dose level.

TABLE 2: Effect of V10-DMAP on biochemical parameters.

	NC ($n = 10$)	HC ($n = 10$)	HC-V10-DMAP-5 μmol ($n = 10$)	HC-V10-DMAP-10 μmol ($n = 10$)
Glucose (mg/dL)	112.5 ± 3.5	142.3 ± 6.9 ▲	84.3 ± 2.1 ▼ ↓	94.8 ± 5.4 ▼ ↓
Fructosamine (mmol/L)	44.6 ± 0.8	63.8 ± 1.3 ▲	45.5 ± 2.1 ↓	38.2 ± 1.7 ▼ ↓
Triglycerides (mg/dL)	56.2 ± 3.4	140.0 ± 10 ▲	64.1 ± 2.8 ▲ ↓	60.0 ± 3.2 ↓
FFA (mg/dL)	2.0 ± 0.3	7.4 ± 1.1 ▲	2.2 ± 0.1 ↓	2.0 ± 0.1 ↓
Total cholesterol (mg/dL)	76.9 ± 4.3	105 ± 5.1 ▲	75.2 ± 3.1 ↓	79.0 ± 4.2 ↓
HDL (mg/dL)	43.3 ± 0.9	32.5 ± 1.1 ▼	51.1 ± 1.2 ▲ ↑	48.3 ± 1.5 ▲ ↑
Insulin (μUI/mL)	17.5 ± 0.8	63.4 ± 1.5 ▲	15.2 ± 1.5 ↓	20.2 ± 3.1 ↓
Leptin (ng/mL)	1.1 ± 0.02	5.7 ± 0.2 ▲	1.2 ± 0.09 ↓	2.1 ± 0.1 ↓
Adiponectin (μg/mL)	5.52 ± 0.3	3.21 ± 0.2 ▼	5.12 ± 0.2 ↑	4.32 ± 0.3 ▼ ↑
HOMA-IR	0.81 ± 0.06	3.84 ± 0.14 ▲	0.53 ± 0.06 ▼ ↓	0.81 ± 0.17 ↓
IDA-IR	0.11 ± 0.02	0.63 ± 0.02 ▲	0.09 ± 0.01 ▼ ↓	0.09 ± 0.01 ▼ ↓

Results shown are the average ± SEM. ▲ indicates a significant difference with values above the control group with a normal calorie diet. ▼ indicates a significant difference with values below the control group with a normal calorie diet. ↑ indicates a significant difference with values above the HC group. ↓ indicates a significant difference with values below the HC group. Comparisons between groups were performed by a two-way ANOVA and the Bonferroni post hoc test; $p < 0.05$. FFA = free fatty acids; HDL = high density lipoprotein; HOMA-IR = homeostasis model assessment insulin resistance; IDA-IR = insulin resistance adipocyte dysfunction.

TABLE 3: Toxicological effect of V10-DMAP administration.

	NC ($n = 10$)	HC ($n = 10$)	HC-V10-DMAP-5 μmol ($n = 10$)	HC-V10-DMAP-10 μmol ($n = 10$)
Total bilirubin (mg/dL)	0.65 ± 0.03	0.68 ± 0.02	0.62 ± 0.03	0.63 ± 0.04
AST (U/L)	117 ± 8	193 ± 11 ▲	105 ± 6 ↓	148 ± 7 ▲ ↓
ALT (U/L)	65 ± 4	90 ± 7 ▲	59 ± 4 ↓	78 ± 4 ↓
γ-GT (U/L)	55 ± 8	76 ± 6 ▲	63 ± 4 ↓	70 ± 5 ▲
LDH (U/L)	200 ± 12	455 ± 15 ▲	220 ± 14 ↓	316 ± 15 ▲ ↓

Results shown are the average ± SEM. ▲ indicates a significant difference with values above the control group with a normal calorie diet. ↓ indicates a significant difference with values below the HC group. Comparisons between groups were performed by a two-way ANOVA and the Bonferroni post hoc test; $p < 0.05$. AST = aspartate aminotransferase; ALT = alanine aminotransferase; γ-GT = gamma-glutamyl transpeptidase; LDH = lactate dehydrogenase.

In this regard, results showed that a micromole dose of decavanadate is sufficient to reach a hypoglycemic effect. Doses of 5 and 10 μmol (equivalent to 2.43 mg·V/kg/day and 4.86 mg·V/kg/day, resp.) showed no difference to discriminate between them. However, when toxicological effects were analyzed, the 10 μmol dose had a poor effect on AST, ALT, and LDH activities and did not improve γ-GT activity in comparison to the 5 μmol dose. Toxicological effects of decavanadate, usually in excess, are associated with the production of reactive oxygen species, lipid peroxidation, Fenton reaction increase, and mitochondrial antioxidant enzymes decrease, such as superoxide dismutase (SOD) and catalase activities, which lead to cell death and, releasing transaminases to plasma [1, 2, 22, 23, 41, 42]. Additionally, decavanadate in the range of 1 to 10 μmol produced 50% loss of cell viability in cardiomyocyte cultures [43]. Necrosis is involved in cell death by decavanadate excess, which is shown in Figure 4 where necrosis and inflammation were observed by infiltrating leukocytes, mainly in the V10-DMAP-10 μmol group, though there was a reduction of balloon cells or Mallory–Denk bodies and fatty deposits. Meanwhile, the 5 μmol (2.43 mg·V/kg/day) dose regulated the enzymatic activities and improved liver histology. In rodents, dosages that normalize glycemia were between 60 and 100 mg·V/kg/day, which produced vanadium levels in blood ranging from 10 to 20 μM. In humans, a dosage of 1.5 mg·V/kg/day has reported control of glycemia. On this condition, vanadium in blood is 1 to 5 μM [3]. Although in other studies the insulin-mimetic effects of multiple vanadium species (vanadate, peroxovanadates, vanadyl sulfate, etc.) and their powerful hypoglycemic effect have been reported, the ED$_{50}$ and toxic effects or damage to tissues by vanadium administration have not been shown. Our data suggest that vanadium concentration and the species administered are very important not only for glucose homeostasis but also for the toxicological threshold. Therefore, micromole quantities of decavanadate seem to be a highly druggable polyoxovanadate form.

Thus, it is possible to consider a new prodrug concept that implies dissociation of vanadium complexes before they reach the target biomolecules [44]. In this regard, decavanadate species are usually not considered in vanadium toxicological studies, mainly by speciation of vanadyl ($V^{IV}O$)$^{2+}$ or vanadate ($V^{V}O_4$)$^{3-}$ [45]. It has been proposed that the decomposition rate of decavanadate is slow and can promote either vanadyls or vanadates depending on the cell environment and its requirements, so in the body, its half-life expands hours (few hours in plasma, close to 26 hours in tissues, and after 10 days in urine) [18]. Therefore, the pharmacological activities can be exerted for long periods with a smaller dose that does not reach the toxicological threshold [46, 47]. During the time that it remains in the body, decavanadate can exert the biological actions.

The Long–Evans rats fed with lipid-rich diet 5008 developed a severe metabolic disorder in lipid and carbohydrates, as well as insulin resistance and hyperleptinemia with hypoadiponectinemia (Table 2 and Figure 3). Additionally, the Long–Evans rats observed overweight and visceral

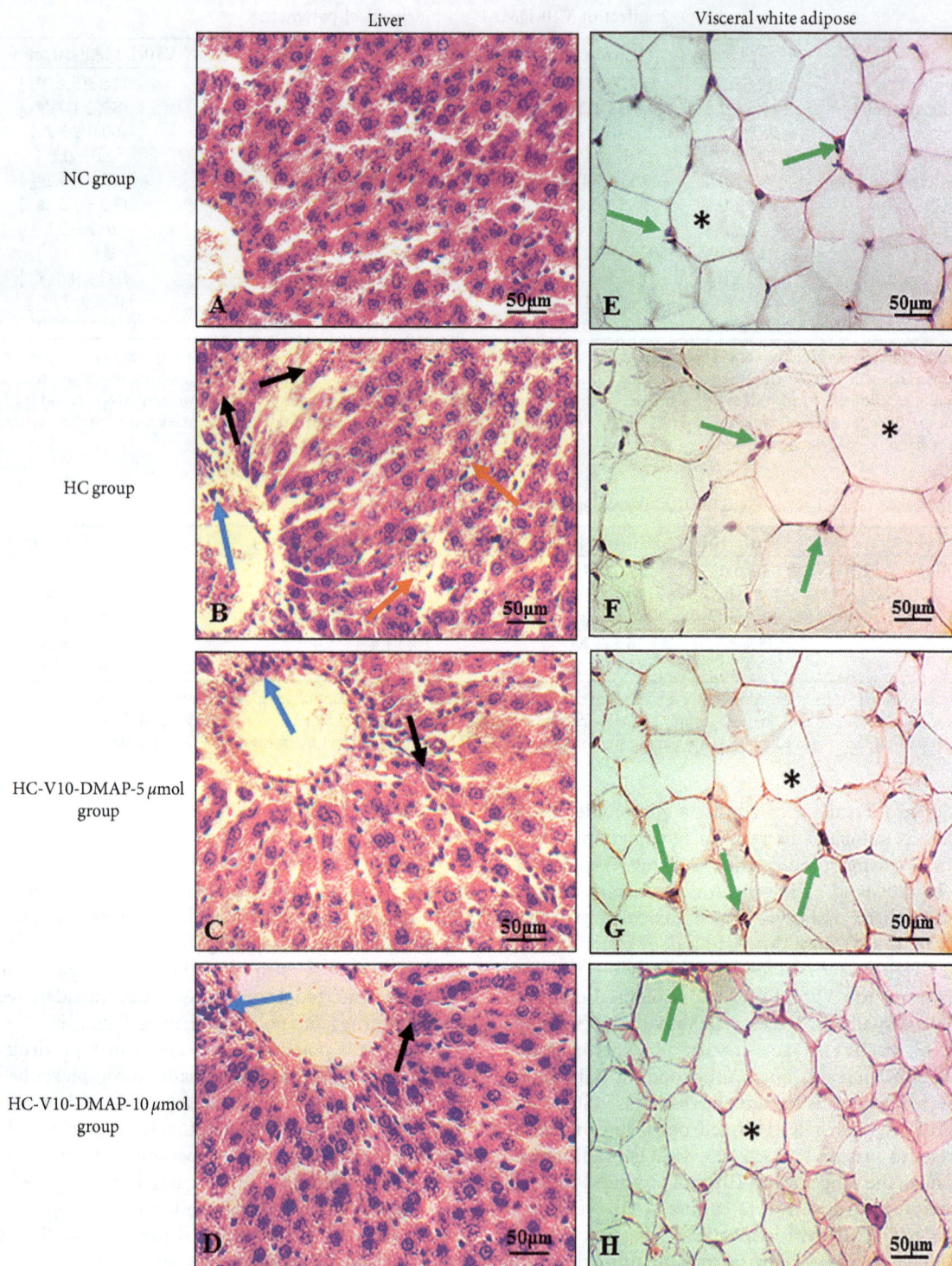

FIGURE 4: Histological features in the liver and visceral white adipose tissue after V10-DMAP administration. (a)–(d) Liver biopsies with H&E staining. The central vein was taken as a reference. (e)–(h) Visceral adipose tissue biopsies with H&E staining. Magnification images were acquired with a 40x objective. The blue arrow indicates inflammation by infiltrating leukocytes. The black arrow indicates hepatic cell necrosis. The red arrow indicates balloon cell or Mallory–Denk bodies. The asterisk indicates triglyceride storage area in adipocytes. The green arrow indicates blood vessels.

obesity (Figure 2). In adipocytes, hypertrophy and hyperplasia were evident with an increased area of stored triglycerides and impaired vascularization (Figure 4(f)). Likewise, a decrease in phosphorylation of insulin receptor

without GLUT-4 immunoreactivity was observed (Figures 5(f) and 5(j)). Meanwhile, hepatic tissue (Figure 4(b)) showed necrosis, fibrosis, inflammation, fatty droplets, Mallory–Denk bodies, and an increase in phosphorylation of insulin

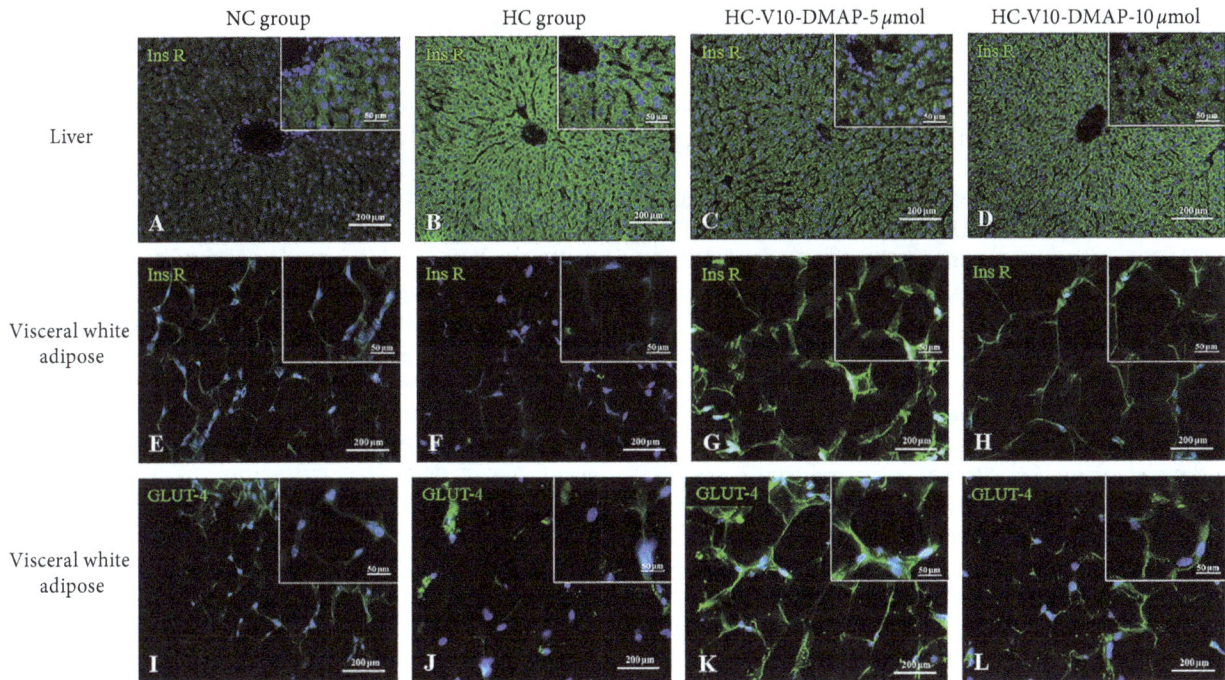

FIGURE 5: Insulin receptor and GLUT-4 in the liver and visceral white adipose tissue after V10-DMAP treatments. Photomicrographs show immunoreactivity to the insulin receptor and GLUT-4, both marked in green and counterstain with DAPI for nuclei, in the liver and adipose tissue. (a)–(d) Hepatic biopsies, insulin receptor phosphorylated in tyrosine (Tyr 1162/1163), and magnification 20x. (e)–(h) Visceral white adipose biopsies, insulin receptor phosphorylated in tyrosine (Tyr 1162/1163), and magnification 40x. (i)–(l) Visceral white adipose biopsies, glucose transporter type 4 insulin-dependent, and magnification 40x.

receptor (Figure 5(b)). These results are completely associated with insulin resistance, dyslipidemia, and metabolic syndrome. Some studies based on hyperinsulinemic-euglycemic clamp experiments have consistently demonstrated a whole-body insulin resistance in animal fat-fed rodents [48], thus matching with the biochemical parameters and the excess caloric intake reported herein. Insulin resistance is usually accompanied by hyperinsulinemia that promotes an ectopic lipid accumulation and a glucose transport decrease [49, 50]. Ectopic lipid accumulation is recognized as steatosis that leads to lipotoxicity and is strongly linked to insulin resistance, metabolic syndrome, and diabetes mellitus development [51, 52].

Particularly, fat mass accumulation during obesity development is characterized by adipocyte hyperplasia and hypertrophy, which increases the production of proinflammatory adipokines that lead to a state of chronic low-grade inflammation and may promote obesity linked to metabolic disorders, such as insulin resistance and metabolic syndrome [7]. Inflammation may impair insulin action systemically by increasing free fatty acids (FFA) or proinflammatory adipokines, such as leptin, which is a key hormone involved in the regulation of satiety, energy intake, and energy expenditure. Thus, hyperleptinemia does not suppress appetite, a phenomenon known as leptin resistance [53–57]. Leptin resistance is thought to be a fundamental pathology in nonalcoholic fatty liver development, insulin resistance, glucose tolerance impairment, and dyslipidemia as was observed in rats fed with diet 5008. On the contrary, adiponectin regulates lipid and glucose metabolism and

increases insulin sensitivity, regulates food intake and body weight, and protects against chronic inflammation. It also increases glucose transport in muscles and enhances fatty acid oxidation, so it is considered anti-inflammatory, antiatherogenic, anti-insulin resistance, and antimetabolic syndrome adipokine [7, 58–62]. Metabolically dysregulated rats showed severe hypoadiponectinemia and consequently insulin resistance in liver and adipose tissue. Insulin resistance implies changes upon insulin signaling, mainly in a transphosphorylation reaction that activates the insulin receptor substrate (IRS), an intrinsic kinase that activates the phosphatidylinositol 3-kinase- (PI3K-) AKT/protein kinase B (PKB) pathway to modulate most metabolic functions of insulin, such as glucose transport by GLUT-4 in adipocytes and myocytes. Meanwhile, this pathway modulates glycogen synthesis, gluconeogenesis, and protein synthesis in liver [63]. On the contrary, it also signals cell growth by the Ras-mitogen-activated protein kinase (MAPK) pathway [63–65]. Inflammatory adipokines enhance degradation of IRS1/2 by the protein tyrosine phosphatase 1B (PTP1B) activity. Therefore, GLUT-4 protein diminished in adipocytes, impairing glucose tolerance response in return to basal levels after a glucose load. In addition, it promoted triglyceride synthesis in the liver associated with insulin receptor hyperphosphorylation, as observed in rats which were fed a high-lipid diet.

It is remarkable that, since its early discovery, the insulin-like activity of vanadium has been considered to be an insulin-mimetic agent that reaches glucose regulation by similar pathways to the hormone. However, the metabolic model, where the defect is caused by hyperinsulinemia and insulin resistance, has not considered that the hormone is at high levels. Therefore, if vanadium has only affected, such as an insulin-mimetic agent, the defect would be promoted. In this respect, vanadium has proven to be a powerful inhibitor of protein tyrosine phosphatases (PTPs), improving tyrosine phosphorylation of IRS1/2 in rat adipocytes and cardiomyocytes, which lead to activation of its associated PI3K activity [66, 67] and PKB [68–70], stimulation of glucose uptake [71, 72], and GLUT-4 translocation [73, 74]. Also, vanadium in liver enhances phosphorylation of ERK1/2, PKB, and glycogen synthase kinase- (GSK-) 3β [71, 75]. As a result, vanadium acts efficiently on insulin phosphorylation cascade, probably because of its structural analogy between vanadate and phosphate, where the monomeric vanadate is slightly larger than the phosphate [18]. Moreover, several studies suggest that vanadium is involved in the regulation of phosphate-dependent processes, such as metabolic processes involving phosphatases and kinases. However, the toxicological threshold of vanadium compounds must always be taken into consideration because more metal into cells may cause damage or simply a loss of efficiency. As it can be seen, the V10-DMAP-10 μmol group does not reach the insulin receptor phosphorylation levels in both adipose tissue and liver compared with the V10-DMAP-5 μmol group, where even GLUT-4 protein is regulated more efficiently.

Although the vanadium mechanism on metabolic regulation has been described, decavanadate as an insulin-mimetic agent or enhancer is not fully understood, but its insulin-mimetic effects are probably related to the inhibition of tyrosine phosphatase-1B (PTP1B) [45]. Recently, it has been reported that the signal transduction pathway of PTP1B is influenced by reactive oxygen species (ROS) generated into metabolic cascades [44], so vanadium may be able to act as insulin mimetics due to changes in prooxidant and oxidant balance. In this respect, decavanadate has been observed to inhibit mitochondrial respiration of hepatocytes with 100-fold more power than vanadate, diminishing oxygen consumption and ROS production, which is related to mitochondrial complex III [43]. Inhibition of mitochondrial complexes make cells require more energy, so the burning of lipids and carbohydrates and the FFA flux from adipocytes to hepatocytes are continuously maintained for the adipose tissue to return to a morphology similar to the control group. Thus, the adiponectin level was regulated (decreasing leptin and increasing adiponectin). Meanwhile, decreased fatty content and glucose balance in the liver are improved, so glucose tolerance is also improved, and there are less fructosamine formation and insulin need, leading to insulin resistance reduction, both in adipocytes and hepatic tissue. Therefore, hepatocytes showed an injury reduction. Notably, the 5 μmol (equivalent to 2.43 mg·V/kg/day) dose was more efficient than the 10 μmol (4.86 mg·V/kg/day) dose. Supporting these facts, in previous studies, our working team has demonstrated that metforminium decavanadate produces metabolic regulation on carbohydrates and lipids in hyperinsulinemia and hypoinsulinemia models [76–79], regardless of the insulin-mimetic action by vanadium.

5. Conclusion

The V10-DMAP administration, only twice per week, produced regulation on lipids and carbohydrates, regardless of dose administered here; yet, the V10-DMAP-5 μmol (2.43 mg·V/kg/day) dose showed more effective effects than the V10-DMAP-10 μmol (4.86 mg·V/kg/day) dose in the liver, which was confirmed by enzymatic activity. In adipose tissue, the V10-DMAP-5 μmol (2.43 mg·V/kg/day) dose was also more beneficial on cells and adipokines. In this regard, a low decavanadate dose produced a clear regulation of the insulin signaling pathway. Furthermore, different experiments strongly suggest that vanadium and its species in the cell may act on proteins, such as phosphorylases and kinases, so it should not be considered only as an insulin-mimetic agent, and a deeper understanding of its mechanism of action needs to be further investigated. The pharmacological and toxicological threshold for cell regulation are suggested to be up to 5 μmol (2.43 mg·V/kg/day) for the title compound.

Acknowledgments

The authors wish to thank Vicerrectoria de Investigación y Estudios de Posgrado (VIEP) by Dr. Ygnacio Martínez Laguna and the Dirección de Innovación y Transferencia de Conocimiento (DITCO) through Dr. Pedro Hugo Hernández Tejeda and Dr. Martín Pérez Santos from Oficina de Comercialización de Tecnología (OCT) for financial support to carry out this research project (DITCO2016-8). The authors also thank Dr. Carlos Escamilla for the use of the Vivarium "Claude Bernard" and Yadira Rosas Bravo for technical assistance. E. Sánchez-Lara wishes to thank Consejo Nacional de Ciencia y Tecnología (CONACyT) for the PhD fellowship support granted with number 293256.

References

[1] D. Rehder, "The potentiality of vanadium in medicinal applications," *Future Medicinal Chemistry*, vol. 4, no. 14, pp. 1823–1837, 2012.

[2] D. Rehder, "The future of/for vanadium," *Dalton Transactions*, vol. 42, no. 33, p. 11749, 2013.

[3] A. Goc, "The biological activity of vanadium compounds," *Central European Journal of Biology*, vol. 1, no. 3, pp. 314–332, 2003.

[4] F. H. Nielsen and E. O. Uthus, "Essentiality and metabolism of vanadium," in *Vanadium in Biological Systems*, N. D. Chasteen, Ed., pp. 51–62, Kluwer Academic Publishers, Dordrecht, The Netherlands, 1990.

[5] R. J. French and P. J. H. Jones, "The role of vanadium in nutrition: metabolism, essentiality and dietary considerations," *Life Sciences*, vol. 52, no. 4, pp. 339–346, 1993.

[6] B. Mukherjee, B. Patra, S. Mahapatra, P. Banerjee, A. Tiwari, and M. Chatterjee, "Vanadium—an element of biological significance," *Toxicology Letters*, vol. 150, no. 2, pp. 135–143, 2004.

[7] J. Kaur, "A comprehensive review on metabolic syndrome," *Cardiology Research and Practice*, vol. 2014, Article ID 943162, 21 pages, 2014.

[8] S. Desroches and B. Lamarche, "Evolving definitions and increasing prevalence of metabolic syndrome," *Applied Physiology: Nutrition and Metabolism*, vol. 32, no. 1, pp. 23–32, 2007.

[9] G. D. Kolovou, K. K. Anagnostopoulou, K. D. Salpea, and D. P. Mikhailidis, "The prevalence of metabolic syndrome in various populations," *American Journal of the Medical Sciences*, vol. 333, no. 6, pp. 362–371, 2007.

[10] A. J. Cameron, J. E. Shaw, and P. Z. Zimmet, "Metabolic syndrome: prevalence in worldwide populations," *Endocrinology and Metabolism Clinics of North America*, vol. 33, no. 2, pp. 351–375, 2004.

[11] A. Nourparvar, A. Bullota, U. Di Mario, and R. Perfetti, "Novel strategies for the pharmacological management of type II diabetes," *Trends in Pharmacological Sciences*, vol. 25, no. 2, pp. 86–91, 2004.

[12] K. H. Thompson and C. Orvig, "Vanadium compounds in the treatment of diabetes," *Metal Ions in Biological Systems*, vol. 41, pp. 221–225, 2004.

[13] H. Sakurai, "Therapeutic potential of vanadium in treating diabetes mellitus," *Clinical Calcium*, vol. 15, pp. 49–57, 2005.

[14] X. G. Yang, K. Wang, J. Lu, and D. C. Crans, "Membrane transport of vanadium compounds and interaction with the erythrocyte membrane," *Coordination Chemistry Reviews*, vol. 273, no. 1-2, pp. 103–111, 2003.

[15] D. C. Crans, R. L. Bunch, and L. A. Theisen, "Interaction of trace levels of vanadium (IV) and vanadium (V) in biological systems," *Journal of the American Chemical Society*, vol. 111, no. 19, pp. 7597–7607, 1989.

[16] D. C. Crans, "Antidiabetic, chemical and physical properties of organic vanadate as presumed transition state inhibitors for phosphatases," *Journal of Organic Chemistry*, vol. 80, no. 24, pp. 11899–11915, 2015.

[17] D. C. Crans, "Fifteen years of dancing with vanadium," *Pure and Applied Chemistry*, vol. 77, no. 9, pp. 1497–1527, 2005.

[18] J. C. Pessoa, S. Etcheverry, and D. Gambino, "Vanadium compounds in medicine," *Coordination Chemistry Reviews*, vol. 301-302, pp. 24–48, 2015.

[19] D. C. Crans, "Vanadium and its role in life," in *Metal Ions in Biological Systems*, A. Sigel and H. Sigel, Eds., pp. 147–209, Marcel Dekker, New York, NY, USA, 1995.

[20] M. Aureliano, "Decavanadate: a journey in search of a role," *Dalton Transactions*, no. 42, pp. 9093–9100, 2009.

[21] M. Aureliano and D. C. Crans, "Decavanadate $V_{10}O_{28}^{6-}$ and oxovanadates: oxometalates with many biological activities," *Journal of Inorganic Biochemistry*, vol. 103, no. 4, pp. 536–546, 2009.

[22] J. L. Domingo, "Vanadium and tungsten derivatives as antidiabetic agents: a review of their toxic effects," *Biological Trace Element Research*, vol. 88, no. 2, pp. 97–112, 2002.

[23] J. L. Domingo, M. Gomez, D. J. Sanchez, J. M. Llobet, and C. L. Keen, "Toxicology of vanadium compounds in diabetic rats: action of chelating agents on vanadium accumulation," *Molecular and Cellular Biochemistry*, vol. 153, no. 1-2, pp. 233–240, 1995.

[24] M. C. Cam, R. A. Pederson, R. W. Brownsey, and J. H. McNeill, "Long-term effectiveness of vanadyl sulfate in streptozotocin-diabetic rats," *Diabetologia*, vol. 36, no. 3, pp. 218–224, 1993.

[25] T. Scior, J. A. Guevara-Garcia, P. Bernard, D. Quoc-Tuan, D. Domeyer, and S. Laufer, "Are vanadium compounds drugable? Structures and effects of antidiabetic vanadium compounds: a critical review," *Mini-Reviews in Medicinal Chemistry*, vol. 5, no. 11, pp. 995–1008, 2005.

[26] M. Aureliano and R. M. C. Gándara, "Decavanadate effects on biological systems," *Journal of Inorganic Biochemistry*, vol. 99, no. 5, pp. 979–985, 2005.

[27] E. Sánchez-Lara, I. Sánchez-Lombardo, A. Pérez-Benítez, A. Mendoza, M. Flores-Álamo, and E. González-Vergara, "A new dicationic ring [(water)$_6$–(ammonium)$_2$] acts as a building block for a supramolecular 3D assembly of decavanadate clusters and 4-(*N*, *N*-dimethylamino)pyridinium ions," *Journal of Cluster Science*, vol. 26, no. 3, pp. 901–912, 2014.

[28] S. Treviño, M. P. Waalkes, J. A. Flores Hernández, B. A. León-Chávez, P. Aguilar-Alonso, and E. Brambila, "Chronic cadmium exposure in rats produces pancreatic impairment and insulin resistance in multiple peripheral tissues," *Archives of Biochemistry and Biophysics*, vol. 583, pp. 27–35, 2015.

[29] I. Sánchez-Lombardo, E. Sánchez-Lara, A. Pérez-Benítez, A. Mendoza, S. Bernès, and E. González-Vergara, "Synthesis of metforminium (2+) decavanadates—crystal structures and solid-state characterization," *European Journal of Inorganic Chemistry*, vol. 2014, no. 27, pp. 4581–4588, 2014.

[30] J. Korbecki, I. Baranowska-Bosiacka, I. Gutowska, and D. Chlubek, "Biochemical and medical importance of vanadium compounds," *Acta Biochimica Polonica (ABP)*, vol. 59, no. 2, pp. 195–200, 2012.

[31] H. Sakurai, "A new concept: the use of vanadium complexes in the treatment of diabetes mellitus," *The Chemical Record*, vol. 2, no. 4, pp. 237–248, 2002.

[32] C. E. Heyliger, A. G. Tahiliani, and J. H. McNeill, "Effect of vanadate on elevated blood glucose and depressed cardiac performance of diabetic rats," *Science*, vol. 227, no. 4693, pp. 1474–1477, 1985.

[33] J. Meyerovitch, Z. Farfel, J. Sack, and Y. Shechter, "Oral administration of vanadate normalizes blood glucose levels in streptozotocin-treated rats. Characterization and mode of action," *Journal of Biological Chemistry*, vol. 262, pp. 6658–6662, 1987.

[34] M. Bendayan and D. Gingras, "Effect of vanadate administration on blood glucose and insulin levels as well as on the exocrine pancreatic function in streptozotocin-diabetic rats," *Diabetologia*, vol. 32, no. 8, pp. 561–567, 1989.

[35] S. M. Brichard, "Effects of vanadate on the expression of genes involved in fuel homeostasis in animal models of type I and type II diabetes," *Molecular and Cellular Biochemistry*, vol. 153, no. 1-2, pp. 121–124, 1995.

[36] S. M. Brichard, W. Okitolonda, and J. C. Henquin, "Long-term improvement of glucose homeostasis by vanadate treatment in diabetic rats," *Endocrinology*, vol. 123, no. 4, pp. 2048–2053, 1988.

[37] S. M. Brichard, J. Lederer, and J. C. Henquin, "Insulin-like properties of vanadium: a curiosity or a perspective for the treatment of diabetes?," *Diabetes & Metabolism*, vol. 17, pp. 435–440, 1991.

[38] R. A. Pederson, S. Ramanadham, A. M. Buchan, and J. H. McNeill, "Long-term effects of vanadyl treatment on streptozocin-induced diabetes in rats," *Diabetes*, vol. 38, no. 11, pp. 1390–1395, 1989.

[39] S. Ramanadham, J. J. Mongold, R. W. Brownsey, G. H. Cros, and J. H. McNeill, "Oral vanadyl sulfate in the treatment of diabetes mellitus in rats," *American Journal of Physiology-Heart and Circulatory Physiology*, vol. 257, no. 3, pp. H904–H911, 1989.

[40] Agency for Toxic Substance and Disease Registry (ATSDR), *U.S. Toxicological Profile for Vanadium*, Department of Healthand Human Services, Public Health Service, Centers for Disease Control, Atlanta, GA, USA, 2012, http://www.atsdr.cdc.gov/toxprofiles/tp58.pdf.

[41] M. Aureliano and C. A. Ohlin, "Decavanadate in vitro and in vivo effects: facts and opinions," *Journal of Inorganic Biochemistry*, vol. 137, pp. 123–130, 2014.

[42] J. Z. Byczkowski and A. P. Kulkarni, "Oxidative stress and prooxidant biological effects of vanadium," in *Vanadium in the Environment, Part 1: Chemistry and Biochemistry*, J. O. Nriagu, Ed., pp. 235–263, John Wiley & Sons, New York, NY, USA, 1998.

[43] S. S. Soares, F. Henao, M. Aureliano, and C. Gutierrez-Merino, "Vanadate induces necrotic death in neonatal rat cardiomyocytes through mitochondrial membrane depolarization," *Chemical Research in Toxicology*, vol. 21, no. 3, pp. 607–618, 2008.

[44] T. Scior, J. A. Guevara-García, Q. T. Do, P. Bernard, and S. Laufer, "Why antidiabetic vanadium complexes are not in the pipeline of "Big Pharma" drug research? A critical review," *Current Medicinal Chemistry*, vol. 23, no. 25, pp. 2874–2891, 2016.

[45] M. Aureliano, "Decavanadate toxicology and pharmacological activities: V10 or V1, both or none?," *Oxidative Medicine and Cellular Longevity*, vol. 2016, Article ID 6103457, 8 pages, 2016.

[46] G. Borges, P. Mendonça, N. Joaquim, J. Coucelo, and M. Aureliano, "Acute effects of vanadate oligomers on heart, kidney and liver histology in the Lusitanian toadfish (*Halobatrachus didactylous*)," *Archives of Environmental Contamination and Toxicology*, vol. 45, no. 3, pp. 415–422, 2003.

[47] S. Ramos, M. Manuel, T. Tiago et al., "Decavanadate interactions with actin: inhibition of G-actin polymerization and stabilization of decameric vanadate," *Journal of Inorganic Biochemistry*, vol. 100, no. 11, pp. 1734–1743, 2006.

[48] R. Buettner, J. Schölmerich, and L. C. Bollheimer, "High-fat diets: modeling the metabolic disorders of human obesity in rodents," *Obesity*, vol. 15, no. 4, pp. 798–808, 2007.

[49] G. Boden and F. Jadali, "Effects of lipid on basal carbohydrate metabolism in normal men," *Diabetes*, vol. 40, no. 6, pp. 686–692, 1991.

[50] G. Boden and X. Chen, "Effects of fat on glucose uptake and utilization in patients with non-insulin-dependent diabetes," *Journal of Clinical Investigation*, vol. 96, no. 3, pp. 1261–1268, 1995.

[51] A. Mitrakou, D. Kelley, M. Mokan et al., "The role of reduced suppression of glucose production and diminished early insulin release in impaired glucose tolerance," *New England Journal of Medicine*, vol. 326, no. 1, pp. 22–29, 1992.

[52] L. Rossetti, A. Giaccari, N. Barzilai, K. Howard, G. Sebel, and M. Hu, "The mechanism by which hyperglycemia inhibits hepatic glucose production in conscious rats: implications for the pathophysiology of fasting hyperglycemia in diabetes," *Journal of Clinical Investigation*, vol. 92, no. 3, pp. 1126–1134, 1993.

[53] J. A. Kim, M. Montagnani, K. K. Koh, and M. J. Quon, "Reciprocal relationships between insulin resistance and endothelial dysfunction: molecular and pathophysiological mechanisms," *Circulation*, vol. 113, no. 15, pp. 1888–1904, 2006.

[54] K. Makki, P. Froguel, and I. Wolowczuk, "Adipose tissue in obesity-related inflammation and insulin resistance: cells, cytokines, and chemokines," *ISRN Inflammation*, vol. 2013, Article ID 139239, 12 pages, 2013.

[55] I. Szanto and C. R. Kahn, "Selective interaction between leptin and insulin signaling pathways in a hepatic cell line," *Proceedings of the National Academy of Sciences*, vol. 97, no. 5, pp. 2355–2360, 2000.

[56] P. Z. Zimmet, V. R. Collins, M. P. de Courten et al., "Is there a relationship between leptin and insulin sensitivity independent of obesity? A population-based study in the Indian Ocean nation of Mauritius," *International Journal of Obesity*, vol. 22, no. 2, pp. 171–177, 1998.

[57] J. J. T. Liaw and P. V. Peplow, "Differential effect of electroacupuncture on inflammatory adipokines in two rat models of obesity," *Journal of Acupuncture and Meridian Studies*, vol. 9, no. 4, pp. 183–190, 2016.

[58] Y. Sharabi, M. Oron-Herman, Y. Kamari et al., "Effect of PPAR-gamma agonist on adiponectin levels in the metabolic syndrome: lessons from the high fructose fed rat model," *American Journal of Hypertension*, vol. 20, no. 2, pp. 206–210, 2007.

[59] Y. Matsuzawa, T. Funahashi, S. Kihara, and I. Shimomura, "Adiponectin and metabolic syndrome," *Arteriosclerosis, Thrombosis, and Vascular Biology*, vol. 24, no. 1, pp. 29–33, 2004.

[60] T. Kazumi, A. Kawaguchi, K. Sakai, T. Hirano, and G. Yoshino, "Young men with high-normal blood pressure have lower serum adiponectin, smaller LDL size and higher elevated heart rate than those with optimal blood pressure," *Diabetes Care*, vol. 25, no. 6, pp. 971–976, 2002.

[61] T. Pischon, C. J. Girman, G. S. Hotamisligil, N. Rifai, F. B. Hu, and E. B. Rimm, "Plasma adiponectin levels and risk of myocardial infarction in men," *Journal of the American Medical Association*, vol. 291, no. 14, pp. 1730–1737, 2004.

[62] L. Hutley and J. B. Prins, "Fat as an endocrine organ: relationship to the metabolic syndrome," *American Journal of the Medical Sciences*, vol. 330, no. 6, pp. 280–289, 2005.

[63] H. Kwon and J. E. Pessin, "Adipokines mediate inflammation and insulin resistance," *Frontiers in Endocrinology*, vol. 4, no. 71, pp. 1–13, 2013.

[64] S. Boura-Halfon and Y. Zick, "Phosphorylation of IRS proteins, insulin action, and insulin resistance," *American Journal of Physiology-Endocrinology and Metabolism*, vol. 296, no. 4, pp. E581–E591, 2009.

[65] L. Fontana, J. C. Eagon, M. E. Trujillo, P. E. Scherer, and S. Klein, "Visceral fat adipokine secretion is associated with systemic inflammation in obese humans," *Diabetes*, vol. 56, no. 4, pp. 1010–1013, 2007.

[66] S. K. Pandey, M. B. Anand-Srivastava, and A. K. Srivastava, "Vanadyl sulfate-stimulated glycogen synthesis is associated with activation of phosphatidylinositol 3 kinase and is independent of insulin receptor tyrosine phosphorylation," *Biochemistry*, vol. 37, no. 19, pp. 7006–7014, 1998.

[67] G. Elberg, Z. He, J. Li, N. Sekar, and Y. Shechter, "Vanadate activates membranous non-receptor protein tyrosine kinase in rat adipocytes," *Diabetes*, vol. 46, no. 11, pp. 1684–1690, 1997.

[68] A. Mohammad, S. Bhanot, and J. H. McNeill, "In vivo effects of vanadium in diabetic rats are independent of changes in PI-3 kinase activity in skeletal muscle," *Molecular and Cellular Biochemistry*, vol. 223, no. 1-2, pp. 103–108, 2001.

[69] L. Marzban, S. Bhanot, and J. H. McNeill, "In vivo effects of insulin and bis(maltolato)oxovanadium (IV) on PKB activity in the skeletal muscle and liver of diabetic rats," *Molecular and Cellular Biochemistry*, vol. 223, no. 1-2, pp. 147–157, 2001.

[70] S. Semiz and J. H. McNeill, "Oral treatment with vanadium of Zucker fatty rats activates muscle glycogen synthesis and insulin-stimulated protein phosphatase-1 activity," *Molecular*

and Cellular Biochemistry, vol. 236, no. 1-2, pp. 123–131, 2002.

[71] B. I. Posner, R. Faure, J. W. Burgess et al., "Peroxovanadium compounds: a new class of potent phosphotyrosine phosphatase inhibitors which are insulin mimetics," *Journal of Biological Chemistry*, vol. 269, pp. 4596–4604, 1994.

[72] J. Berger, N. Hayes, D. M. Szalkowski, and B. Zhang, "PI-3 kinase activation is required for insulin stimulation of glucose transport into L6 myotubes," *Biochemical and Biophysical Research Communications*, vol. 205, no. 1, pp. 570–576, 1994.

[73] R. V. Donthi, B. Huisamen, and A. Lochner, "Effect of vanadate and insulin on glucose transport in isolated adult rat cardiomyocytes," *Cardiovascular Drugs and Therapy*, vol. 14, no. 5, pp. 463–470, 2000.

[74] J. Li, G. Elberg, N. Sekar, H. Z. Bin, and Y. Shechter, "Antilipolytic actions of vanadate and insulin in rat adipocytes mediated by distinctly different mechanisms," *Endocrinology*, vol. 138, no. 6, pp. 2274–2279, 1997.

[75] G. Swarup, S. Cohen, and D. L. Garbers, "Inhibition of membrane phosphotyrosyl-protein phosphatase activity by vanadate," *Biochemical and Biophysical Research Communications*, vol. 107, no. 3, pp. 1104–1109, 1982.

[76] A. J. Davison, Q. Wu, J. Moon, and A. Stern, "Among a range of transition metals and ligands, vanadium desferroxamine excels in accelerating reactivity of ferrocytochrome C toward molecular oxygen," *Biochemistry and Cell Biology*, vol. 72, no. 5-6, pp. 169–174, 1994.

[77] M. J. Hosseini, F. Shaki, M. Ghazi-Khansari, and J. Pourahmad, "Toxicity of vanadium on isolated rat liver mitochondria: a new mechanistic approach," *Metallomics*, vol. 5, no. 2, pp. 152–166, 2013.

[78] S. Treviño, D. Velázquez-Vázquez, E. Sánchez-Lara et al., "Metforminium decavanadate as a potential metallopharmaceutical drug for the treatment of diabetes mellitus," *Oxidative Medicine and Cellular Longevity*, vol. 2016, Article ID 6058705, 14 pages, 2016.

[79] S. Treviño, E. Sánchez-Lara, V. E. Sarmiento-Ortega et al., "Hypoglycemic, lipid-lowering and metabolic regulation activities of metforminium decavanadate (H$_2$Metf)$_3$ [V$_{10}$O$_{28}$]· 8H$_2$O using hypercaloric-induced carbohydrate and lipid deregulation in Wistar rats as biological model," *Journal of Inorganic Biochemistry*, vol. 147, pp. 85–92, 2015.

2

The Structures, Spectroscopic Properties, and Photodynamic Reactions of Three [RuCl(QN)NO]⁻ Complexes (HQN = 8-Hydroxyquinoline and its Derivatives) as Potential NO-Donating Drugs

Leilei Xie,[1,2] Lifang Liu,[1,2] Wenming Wang,[1,2] Zhiou Ma,[1] Liqun Xu,[1] Xuan Zhao,[3] and Hongfei Wang ⓘ[1,2]

[1]Key Laboratory of Chemical Biology and Molecular Engineering of Education Ministry, Institute of Molecular Science, Shanxi University, Taiyuan 030006, China
[2]Key Laboratory of Energy Conversion and Storage Materials of Shanxi Province, Institute of Molecular Science, Shanxi University, Taiyuan 030006, China
[3]Department of Chemistry, University of Memphis, Memphis, TN 38152, USA

Correspondence should be addressed to Hongfei Wang; wanghf@sxu.edu.cn

Academic Editor: Concepción López

The structures and spectral properties of three ruthenium complexes with 8-hydroxyquinoline (Hhqn) and their derivatives 2-methyl-8-quinolinoline (H2mqn) and 2-chloro-8-quiolinoline (H2cqn) as ligands (QN = hqn, 2mqn, or 2cqn) were calculated with density functional theory (DFT) at the B3LYP level. The UV-Vis and IR spectra of the three [RuCl(QN)NO]⁻ complexes were theoretically assigned via DFT calculations. The calculated spectra reasonably correspond to the experimentally measured spectra. Photoinduced NO release was confirmed through spin trapping of the electron paramagnetic resonance spectroscopy (EPR), and the dynamic process of the NO dissociation upon photoirradiation was monitored using time-resolved infrared (IR) spectroscopy. Moreover, the energy levels and related components of frontier orbitals were further analyzed to understand the electronic effects of the substituent groups at the 2nd position of the ligands on their photochemical reactivity. This study provides the basis for the design of NO donors with potential applications in photodynamic therapy.

1. Introduction

The structure and reactivity of transition-metal-NO complexes have gained significant interest in recent years because of the important role of nitric oxide (NO) as a signaling molecule in biological systems [1–5]. NO plays important functions in various physiological processes [6–9]. Moreover, the active centers of several important biological enzymes contain metal ions bound with the NO ligand; therefore, studies of the structures and spectra of metal-NO complexes are important to understand the dynamic reactivity and their functions.

The utility of ruthenium (Ru) complexes to design potential anticancer drugs and cellular imaging agents has been extensively investigated [10–15]. Compared to iron-based nitrosyl complexes, Ru nitrosyls are promising candidates as potential NO-donating agents for targeted delivery of NO to physiological targets due to their inherent stability and modest photosensitivity [16–21]. Detailed structural and spectroscopic analyses of Ru complexes with different ligands are essential to investigate the kinetic process of the photoreaction. These studies provide a foundation to control NO release at the physiological target.

A series of nitrosylruthenium (Ru-NO) complexes with polypyridyl complexes have been reported. They are coordinated with (N,N) bidentate ligands forming cationing complexes. The cytotoxicity against tumor cell and their vasodilation effects have been studied [20–25]. Here, three

$[RuCl_3(QN)(NO)]^-$ anionic complexes were synthesized using 8-hydroxyquinoline and its derivatives (HQN) as ligands. These ligands are bidentate chelators that bind metal ions via O-N atoms. Figure 1 shows the structures of the $[RuCl_3(QN)(NO)]^-$ complexes and HQN ligands (HQN = 8-hydroxyquinoline (Hhqn), 2-methyl-8-quinolinol (H2mqn), and 2-chloro-8-quinolinol (H2cqn)). DFT calculations allowed the assignment of bands observed in the electronic and IR spectra of the complexes. Furthermore, the behavior of the three complexes upon photoirradiation was investigated using time-resolved spectroscopy technology. The electronic structures and molecular orbitals of these complexes were calculated to better understand the electronic effect of the substituted group at the 2nd position of the ligands. This study provides insight into the photodynamic properties and potential applications of the nitrosylruthenium (II) complexes.

2. Experimental

2.1. Synthesis. Chemical reagents and solvents were purchased from Sigma (St. Louis, MO, USA) and local vendors. The complexes were synthesized according to a previously described method with modifications [26, 27] and characterized by ^1H NMR spectroscopy using a Bruker 600 MHz spectrometer.

2.2. Spectra Measurements. After the complexes were dissolved in DMSO, the UV-visible spectra were recorded on a Thermo 220 spectrophotometer. The IR spectra were measured on an IS50R FT-IR spectrometer (Thermo Fisher) from 2000 to 1400 cm^{-1} at 1 cm^{-1} resolution. The sample solutions were added to an IR cell composed of two CaF_2 windows (25 mm in diameter and 2 mm thick), which were separated by an O-shaped 50 μm thick Teflon spacer.

The photoreaction kinetics was monitored via the IR spectra as a function of irradiation time. The IR spectra were recorded simultaneously for 30 min in the CaF_2 windows while being irradiated with a fiber connected to an Xe lamp with 420 nm band-pass filter (0.2 W/cm^2).

The electron paramagnetic resonance (EPR) spectra were obtained using a Bruker ESP-500E spectrometer at 9.8 GHz, X band, with 100 Hz field modulation. The three complexes (5 mM) mixed with 5 mM Fe(MGD)$_2$ were quantitatively injected into quartz capillaries, respectively. The sample was then illuminated in the cavity of the EPR spectrometer with an Hg lamp (LOT-QuantumDesign GmbH) at 365 nm. All experiments were performed at room temperature (20°C).

2.3. Quantum Chemical Calculations. Gaussian 09 and Gaussview 5 program packages were used for calculations and structure visualization, respectively [28, 29]. The original models for the three complexes were built based on the crystal structure of $[(CH_3)_4N][RuCl(2cqn)NO]$ complex [27]. All structures were fully optimized with Becke's three-parameter hybrid functional and the Lee–Yang–Parr correlation functional (B3LYP) [30–32] in the DMSO solvent. The basis sets aug-cc-pVDZ-PP and 6-311++G(d,p) were used to describe the Ru atom and the ligand atoms, respectively [33, 34]. The charge was set to −1, and both $S = 0$ and $S = 1$ states for the complexes were optimized.

The UV-visible spectra for the three complexes in DMSO solution were simulated with a time-dependent (TD-DFT) method, respectively, and the solvent effect was considered via the polarization continuum model [35, 36]. The natural atomic charges and Wiberg bond index of the complexes were obtained by natural population analysis (NPA) and natural bond orbital (NBO) analysis [37, 38].

3. Results and Discussion

3.1. Molecular Geometry. The selected and calculated bond lengths and the angles for three complexes are listed in Table 1. Most of the calculated bond lengths and angles of the optimized geometries (Table 1) deviate from crystal structural data by 0.03 Å and 2°, respectively. The theoretical bond lengths of Ru-N2, Ru-O1, and N2-O2 in $[RuCl_3(2cqn) NO]^-$ complex deviate the experimental data less than 0.01 Å, which is near to the uncertainty caused by the experiment measurements [39, 40].

The structures for both $S = 0$ and $S = 1$ as potential ground states were optimized, respectively. As shown in Table 2, the angles of the Ru-NO in the optimized structures of the lowest triplet excited states for three complexes are 143.2, 177.1, and 177.4 degrees, respectively. It is worth noting that the Ru-NO is in the bending model for $[RuCl_3(hqn)NO]^-$ in the triplet excited states. However, they are linear for both $[RuCl_3(2mqn)NO]^-$ and $[RuCl_3(2cqn)NO]^-$ in the singlet state and the lowest-triplet excited states. The calculated energy of the singlet state is the lowest one, suggesting the complexes with diamagnetic ground states.

3.2. Molecular Orbital Analyses. The HOMO-LUMO interactions were calculated to probe the reactivity of the various molecular systems [41–45]. The contour plots of the frontier orbitals for three complexes are shown in Figure 2, and the calculated HOMO and LUMO energy levels are shown in Table 2. The calculations were performed with the DMSO solvent. In the three complexes, the HOMO is described as a QN ligand-based orbital that contains some Ru (d) and NO (p) character, while the LUMO contains an antibonding overlap of the Ru (d) and π^* NO (p) orbitals. It suggests that the (Ru(II)-NO$^+$) group plays an important role in the photochemical reaction of nitrosylruthenium (II) complexes containing 8-quinoliolate and its derivatives.

For the $[RuCl_3(2mqn)NO]^-$ complex, the HOMO and LUMO relative orbital energies are higher and its LUMO-HOMO gap is smaller than those of $[RuCl_3(hqn)NO]^-$ complex. However, the HOMO and LUMO relative orbital energy is lower for $[RuCl_3(2cqn)NO]^-$ complex, while its LUMO-HOMO gap is larger than $[RuCl_3(hqn)NO]^-$. The variation of HOMO and LUMO energy orbitals suggests that different substituted groups in the 2nd ligand position could adjust the relative energies of the front orbitals and could affect the stabilities and reactivity of these complexes.

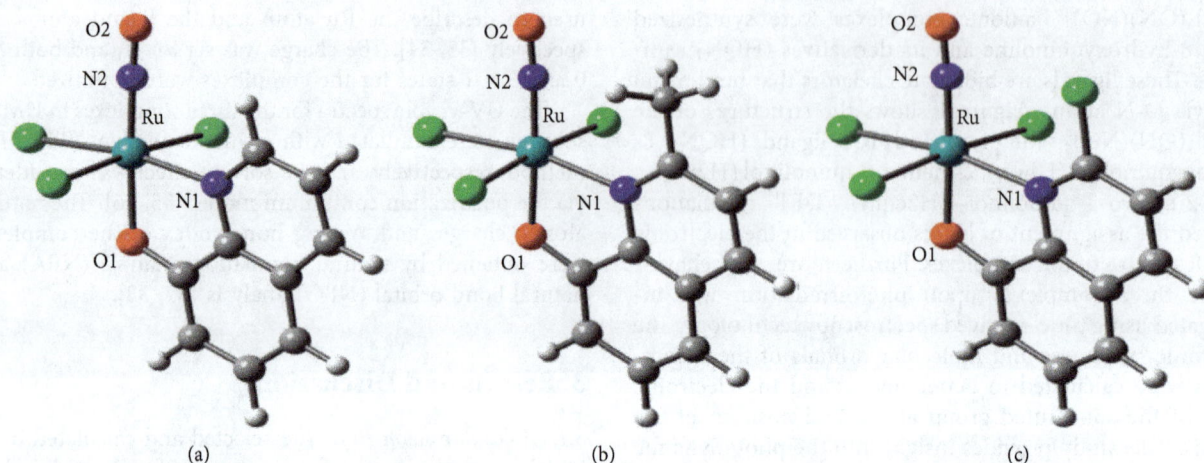

FIGURE 1: Optimized structures of [RuCl$_3$(hqn)NO]$^-$ (a), [RuCl$_3$(2mqn)NO]$^-$ (b), and [RuCl$_3$(2cqn)NO]$^-$ (c) complexes.

TABLE 1: Optimized vs. experimental geometries (in Å and °) with 6-311++G(d,p) and Aug-cc-pVDZ-PP as basis set.

| | [RuCl$_3$(hqn)NO]$^-$ | | [RuCl$_3$(2mqn)NO]$^-$ | | [RuCl$_3$(2cqn)NO]$^-$ | | X-ray data |
	Singlet state	Triplet state	Singlet state	Triplet state	Singlet state	Triplet state	
Ru-N1	2.099	2.087	2.139	2.227	2.154	2.543	2.088
Ru-N2	1.733	1.881	1.734	1.760	1.735	1.758	1.719
Ru-O1	2.011	2.092	1.999	1.981	2.003	1.991	1.993
N2-O2	1.149	1.168	1.148	1.144	1.145	1.144	1.149
∠Ru-N2-O2	177.3	143.2	176.1	177.1	176.0	177.4	174.1

TABLE 2: Relative energies (kcal/mol) and orbital energies (eV) of HOMOs and LUMOs for [RuCl$_3$(hqn)NO]$^-$, [RuCl$_3$(2mqn)NO]$^-$, and [RuCl$_3$(2cqn)NO]$^-$ complexes.

	[RuCl$_3$(hqn)NO]$^-$	[RuCl$_3$(2mqn)NO]$^-$	[RuCl$_3$(2cqn) NO]$^-$
Relative energies ($S = 0$)	−2082.3318613	−2121.6585791	−2541.9469066
Relative energies ($S = 1$)	−2082.291264	−2121.6052187	−2541.8973426
$\Delta_{(S1-S0)}$	25.475	33.484	31.102
LUMO	−2.728	−2.708	−2.762
LUMO-HOMO gap	3.207	3.164	3.222
HOMO	−5.935	−5.872	−5.984

3.3. Electronic Absorption Spectra. The UV-visible absorption spectra of the three [RuCl$_3$(QN)NO]$^-$ complexes in DMSO are shown in Figure 3. These three complexes have similar absorption curves in the ultraviolet and visible region with a 11 nm shift in the absorption peak and a 21 nm shift in the absorption peak between [RuCl$_3$(2mqn)NO]$^-$ and [RuCl$_3$(2cqn)NO]$^-$ in the ultraviolet region, respectively. In the visible region, the maximum absorption band is at 415 nm for [RuCl$_3$(hqn)NO]$^-$, 424 nm for [RuCl$_3$(2mqn) NO]$^-$, and 430 nm for [RuCl$_3$(2cqn)NO]$^-$.

In the UV region, the three complexes display absorption bands at 274 and 337 nm, 270 and 323 nm, 281 and 344 nm, respectively. The corresponding calculated values are 250 and 336 nm, 255 and 331 nm, 262 and 346 nm, respectively. The calculated wavelengths have an error of less than 24 nm compared to the experimental data from the TDDFT method while taking into account of the solvent effect.

The lowest-energy peak near 430 nm predominates the HOMOs-LUMOs excitation in the visible region. The absorption peaks for three complexes were calculated to be near 441, 467, and 461 nm, with deviation from the experimental value by about 30 nm. Analysis of the electronic structures and orbital components of the complexes indicates that these absorption bands mainly originate from the d(Ru)π + p(QN and Cl ligands) $\longrightarrow$ (d(Ru) + p^*(NO and QN ligands)) charge transfer processes, which were labeled as MLCT and LMCT processes (L stands for NO, Cl, and QN ligands).

3.4. Infrared Spectra. Figure 4 shows the infrared spectra of the three complexes recorded in DMSO. For comparison, the experimentally observed and calculated vibrational frequencies ranging from 2000 to 1400 cm^{-1} are presented in Table 3. The B3LYP functional tends to overestimate the fundamental normal modes of vibration, and thus the calculated frequencies were scaled with appropriate values to harmonize the theoretical and experimental wavenumbers [46]. In this study, the scale factor is about 0.97.

| [RuCl₃(hqn)NO]⁻ | [RuCl₃(2mqn)NO]⁻ | [RuCl₃(2cqn)NO]⁻ |

(a)

(b)

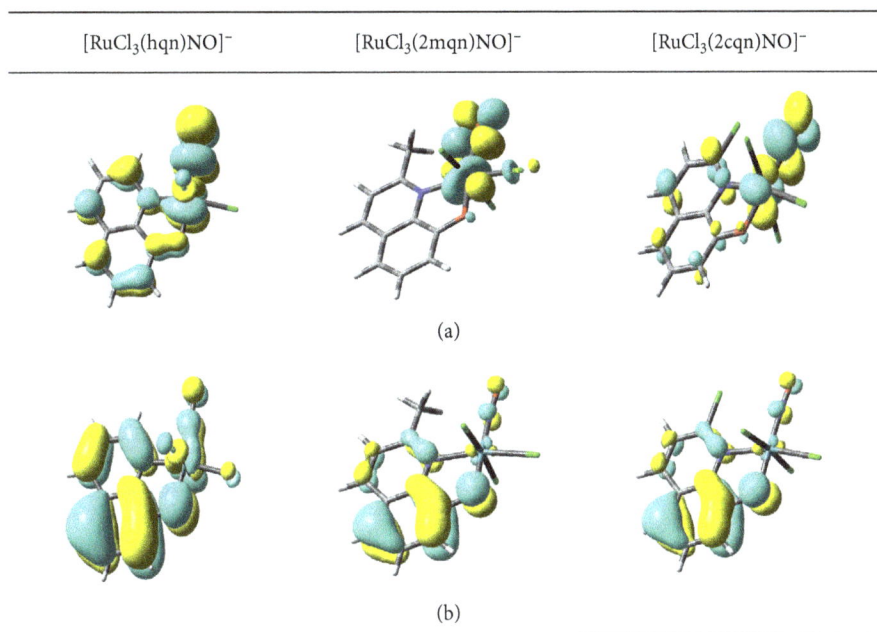

FIGURE 2: Contour diagrams of the calculated LUMO (a) and HOMO (b) of three complexes. Negative values of the wave function are represented in yellow.

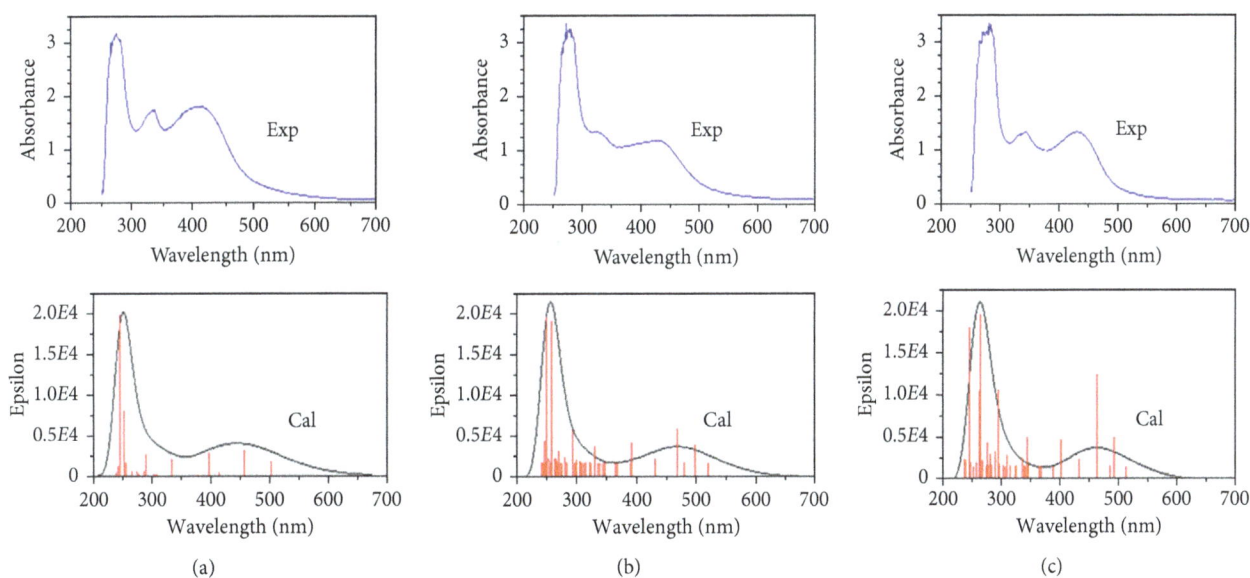

(a)   (b)   (c)

FIGURE 3: Recorded and calculated electronic absorption spectra of [RuCl₃(hqn)NO]⁻ (a), [RuCl₃(2mqn)NO]⁻ (b), and [RuCl₃(2cqn)NO]⁻ (c) complexes (blue: experimental; black: calculated; red: calculated oscillator strength).

The DFT calculation helps assigning vibrational modes to the observed frequencies. The three important vibrations correspond to the two ligands coordinated to the central Ru. There is a clear and strong vibration peak at ~1840 cm⁻¹ that is a stretching vibration for NO in the {Ru(II)-NO⁺} group. The vibration peaks at ~1560 and ~1500 cm⁻¹ correspond to the vibration of coordinated QN ligands. Monitoring the intensity variation of these peaks offers an important information to investigate the mechanism of the photoinduced reaction of ligand dissociation.

Table 3 lists a comparison of the NO stretching frequencies of the three complexes. Different substituted groups in the 2nd position of the ligand in the [RuCl(2mqn) NO] and [RuCl(hqn)NO] complexes cause a 5 cm⁻¹ red shift. This substitution in the [RuCl(2cqn)NO] and [RuCl (hqn)NO] complexes results in a 17 cm⁻¹ red shift in the IR absorption peak. Such a shift is clearly a ligand effect; the stretching frequency (ν_{NO}) of three complexes follows this order: ν_{NO} (2cqn) > ν_{NO} (2mqn) > ν_{NO} (hqn). It is clear that ligand substituents could tune the NO stretching frequency in the three nitrosylruthenium complexes.

3.5. Real-Time Measurement of NO Release. The photoinduced NO release from the three complexes was confirmed

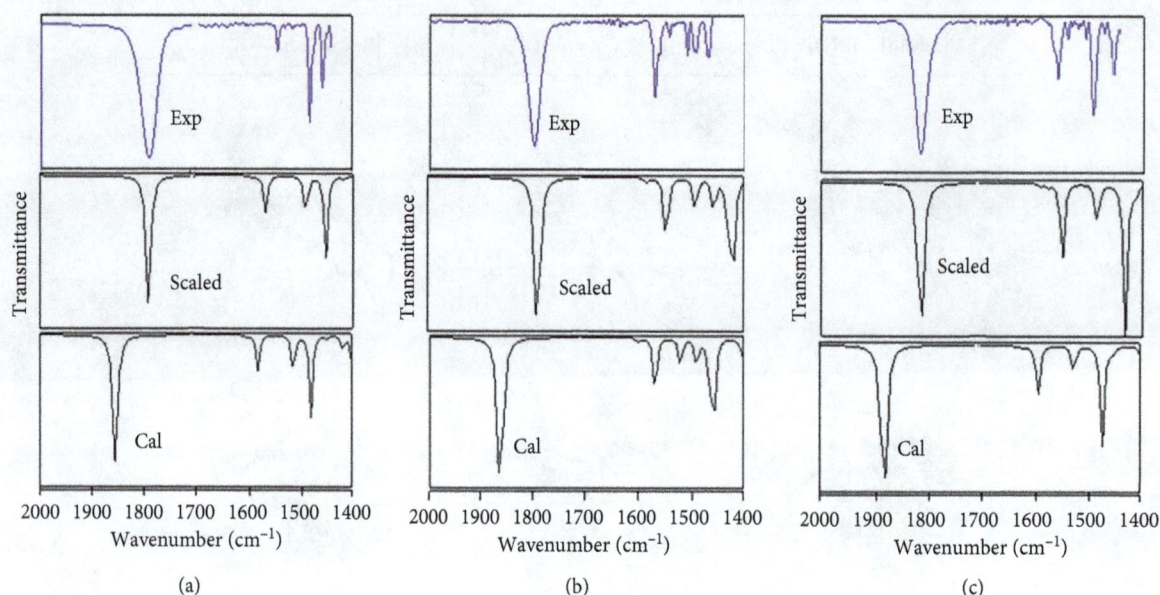

FIGURE 4: Recorded and calculated IR spectra of $[RuCl_3(hqn)NO]^-$ (a), $[RuCl_3(2mqn)NO]^-$ (b), and $[RuCl_3(2cqn)NO]^-$ (c) complexes in the 2000–1400 cm^{-1} region (blue: experimental; black: calculated).

TABLE 3: Observed and calculated vibrational frequencies (cm^{-1}) and intensities over 2000–1400 cm^{-1} region for $[RuCl_3(hqn)NO]^-$, $[RuCl_3(2mqn)NO]^-$, and $[RuCl_3(2cqn)NO]^-$ complexes.

$[RuCl_3(hqn)$ $NO]^-$		$[RuCl_3(2mqn)$ $NO]^-$		$[RuCl_3(2cqn)$ $NO]^-$		Assignment
Exp.	Cal.	Exp.	Cal.	Exp.	Cal.	
1839.40	1889.44	1844.01	1894.38	1856.55	1906.72	$\nu_{N=O}$: vs
1576.49	1609.36					δ_{hqn} : m
1500.01	1530.48					δ_{hqn} : m
1470.07	1491.36					δ_{hqn} : m
		1567.78	1593.19			δ_{2mqn} : m
		1540.36	1537.52			δ_{2mqn} : m
		1507.06	1502.23			δ_{2mqn} : m
		1468.27	1472.73			δ_{2mqn} : m
				1558.68	1590.61	δ_{2cqn} : m
				1493.57	1523.66	δ_{2cqn} : m
				1450.55	1470.57	δ_{2cqn} : m

with spin-trapping EPR spectroscopy via Fe(MGD)$_2$ for detecting NO· in real-time [47, 48]. Figure 5 shows the characteristic triplet signal with a hyperfine splitting constant (hfsc) value of 12.78 G and a g-factor of 2.039. These are consistent with published values for NO-Fe^{2+}-MGD adducts [49, 50]. It is obvious that free radicals were generated from the complexes with 365 nm photoirradiation, while almost no signal was observed in the dark. The intensity of resulting free radicals increased quickly upon photoirradiation, reaching a maximum at 30 seconds (Figure 5). It then decreased slowly over 5 minutes. Therefore, the NO release could be controlled with photoirradiation, providing the basis for further applications in photobiology and medicine.

3.6. NBO Analysis.

The natural atomic charges of the three complexes were obtained via natural population analysis (NPA) using the B3LYP method (Table 4). In the {Ru-NO} groups, all N atoms have a net positive charge from 0.451 to 0.467. The electronegative oxygen atoms have negative charges from −0.189 to −0.210, respectively. The calculated Wiberg bond index of NO increases from 1.8449 to 1.8501 and 1.8757 in the order of hqn, 2mqn, and 2cqn complexes. The NO stretching frequency (ν_{NO}) shifts from 1839.4 to 1844.01 and 1856.5 cm^{-1}, which is in agreement with the bond order analyses. The Wiberg bond index of Ru-N decreases from 1.6503 to 1.6408 and 1.6251 for the hqn, 2mqn, and 2cqn complexes, respectively, suggesting that NO is relatively easily released from the 2cqn complex. The results agree with the IR spectral measurements below.

3.7. Photoinduced NO Release.

Next, photoinduced NO release from the three $[RuCl_3(QN)(NO)]^-$ complexes was investigated using time-resolved IR spectra. A series of FT-IR spectra of the NO stretching mode were recorded as a function of photoirradiation. Figure 6 shows the change in the spectra over time. There is a significant decrease in the intensity of the NO vibrational peak around 1850 cm^{-1}, which dominated the photoinduced NO dissociation. The electronic transition from the metal and QN/Cl ligands to the antibonding orbitals of the {Ru(II)-NO$^+$} group upon photoirradiation weakens the bonding of Ru-NO and leads to dissociation of NO [51–53]. In addition, the decrease in the NO vibrational intensity for $[RuCl(2cqn)NO]^-$ complex is fast relative to the other two complexes, and its half-life of NO dissociation is shorter. Therefore, NO release could be adjusted by complexes using different ligands upon photoirradiation. This strategy can be applied for NO-donor design with potential applications in photobiology and clinical therapy.

Recently, we studied the cytotoxicity and photo-enhanced cytotoxicity of the three $[Ru(II)Cl_3(QN)(NO)]^-$

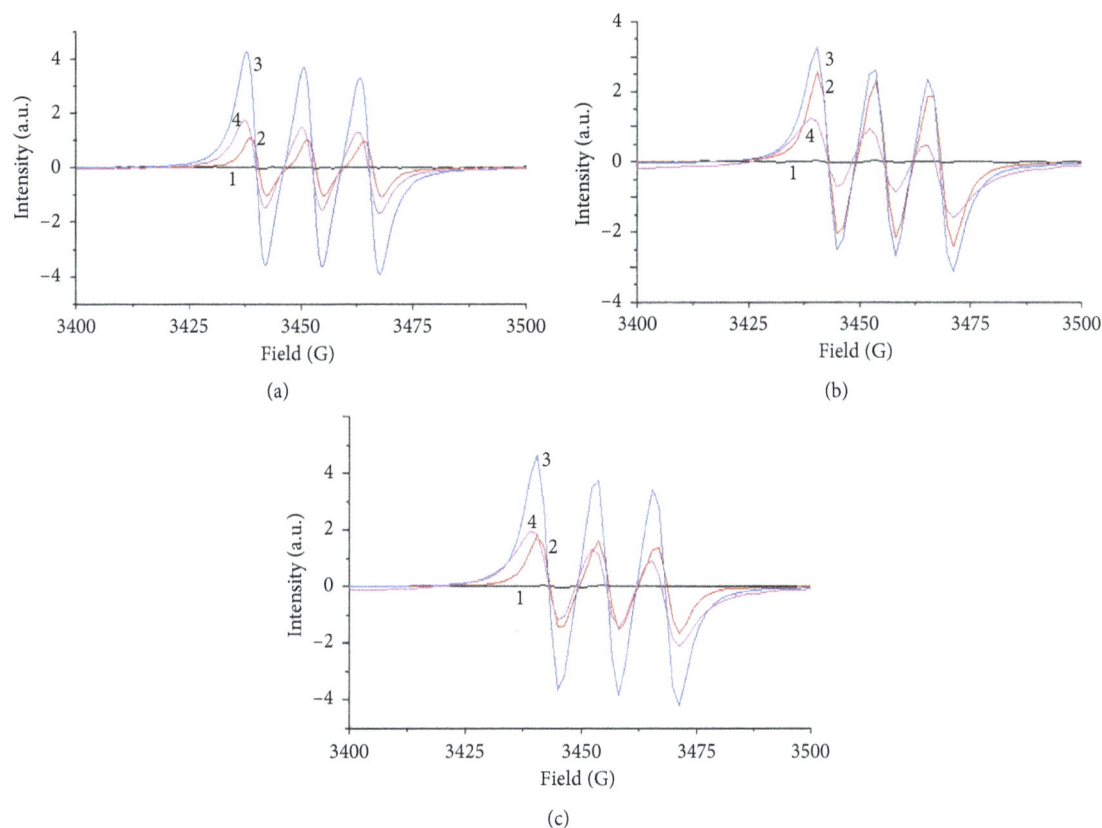

(a)

(b)

(c)

FIGURE 5: EPR spectra of $[RuCl_3(hqn)NO]^-$ (a), $[RuCl_3(2mqn)NO]^-$ (b), and $[RuCl_3(2cqn)NO]^-$ (c) complexes (1: control, without photoirradiation; 2: 15 s; 3: 30 s; 4: 5 min).

TABLE 4: Natural atomic charges and Wiberg bond index of NO in the three $[RuCl(QN)NO]^-$ complexes.

Molecule	Peak position/cm^{-1}(exp)	Atomic charge		Wiberg bond index	
		N	O	Ru-N	N-O
$[RuCl(hqn)NO]^-$	1839.40	0.454	−0.210	1.6503	1.8449
$[RuCl(2mqn)NO]^-$	1844.01	0.451	−0.208	1.6408	1.8501
$[RuCl(2cqn)NO]^-$	1856.55	0.467	−0.189	1.6251	1.8757

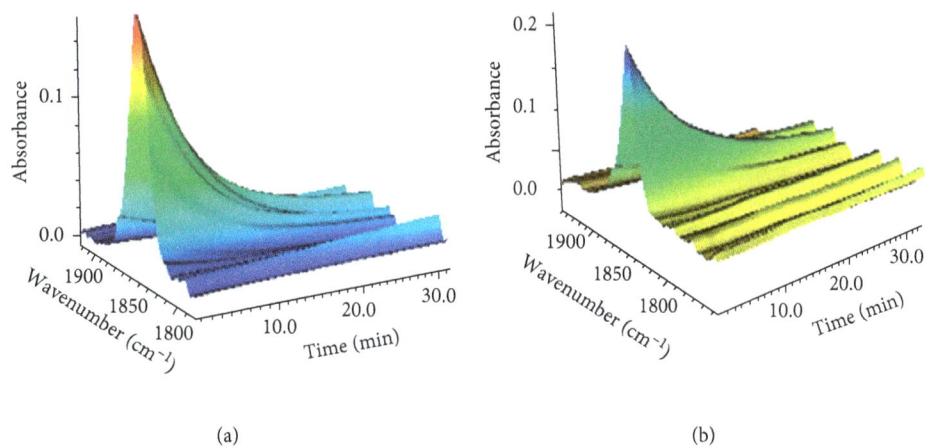

(a)

(b)

FIGURE 6: Continued.

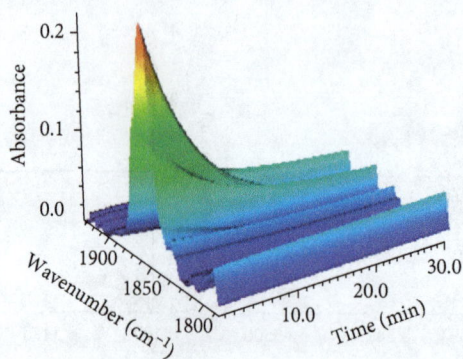

(c)

Figure 6: Time resolution of FT-IR spectra of [RuCl₃(hqn)NO]⁻ (a), [RuCl₃(2mqn)NO]⁻ (b), and [RuCl₃(2cqn)NO]⁻ (c) complexes upon photoirradiation.

complexes against HepG-2 cells [27]. The NO free radicals and [Ru(III)Cl₃(QN)]⁻ complexes resulting from photoirradiation of these complexes are bioactive and cytotoxic and can serve as the potential drugs with dual functions.

4. Conclusions

We have shown good agreement between the optimized structural parameters and their crystal structures via DFT calculation at the B3LYP level. The results provide valuable geometrical information and help to assign UV-visible spectra and FT-IR spectra. Meanwhile, DFT calculations for electronic structures and spectral characteristics of [RuCl₃(QN)(NO)]⁻ complexes provide a better understanding of the photophysical and photochemical properties of these complexes. Real-time NO release was studied via spin trapping of the EPR spectroscopy, and the time-resolved IR spectra showed that three complexes have slightly different half-lives of NO dissociation upon photoirradiation. Moreover, an energy level and component analysis of frontier orbitals shows that the different substituent groups at the 2nd position of the ligands affect their reactivities. This study provides the basis for the design of NO donors for their potential applications in photodynamic therapy.

Acknowledgments

The work was supported partially by the National Natural Science Foundation of China (21671125, 21601112, and 21543003) and by Talent Plan and Research Projects of Shanxi Province (2015081049 and 2015-021). The beam time provided by beamline 3W1A of Beijing Synchrotron Radiation Facility (BSRF) and BL17U and BL18U of the Shanghai Synchrotron Radiation Facility (SSRF) is specially acknowledged.

References

[1] E. G. Abucayon, D. R. Powell, and G. B. Richter-Addo, "Carbon-nitrogen and nitrogen-nitrogen bond formation from nucleophilic attack at coordinated nitrosyls in Fe and Ru Heme models," *Journal of the American Chemical Society*, vol. 139, no. 28, pp. 9495–9498, 2017.

[2] M. A. Herzik, R. Jonnalagadda, J. Kuriyan, and M. A. Marletta, "Structural insights into the role of iron-histidine bond cleavage in nitric oxide-induced activation of H-NOX gas sensor proteins," *Proceedings of the National Academy of Sciences of the United States of America*, vol. 111, no. 40, pp. 4156–4164, 2014.

[3] P. C. Ford, B. O. Fernandez, and M. D. Lim, "Mechanisms of reductive nitrosylation in iron and copper models relevant to biological systems," *Chemical Reviews*, vol. 105, no. 6, pp. 2439–2455, 2005.

[4] F. G. Doro, K. Q. Ferreira, Z. N. da Rocha, G. F. Caramori, A. J. Gomes, and E. Tfouni, "The versatile ruthenium(II/III) tetraazamacrocycle complexes and their nitrosyl derivatives," *Coordination Chemistry Reviews*, vol. 306, no. 2, pp. 652–677, 2016.

[5] M. J. Rose and P. K. Mascharak, "Photoactive ruthenium nitrosyls: effects of light and potential application as NO donors," *Coordination Chemistry Reviews*, vol. 252, no. 18–20, pp. 2093–2114, 2008.

[6] L. J. Ignarro, *Nitric Oxide: Biology and Pathobiology*, Academic Press, Burlington, MA, USA, 2010.

[7] D. Fukumura, S. Kashiwagi, and R. K. Jain, "The role of nitric oxide in tumour progression," *Nature Reviews Cancer*, vol. 6, no. 7, pp. 521–534, 2006.

[8] A. G. Tennyson and S. J. Lippard, "Generation, translocation, and action of nitric oxide in living systems," *Chemistry and Biology*, vol. 18, no. 10, pp. 1211–1220, 2011.

[9] P. T. Burks, J. V. Garcia, R. GonzalezIrias et al., "Nitric oxide releasing materials triggered by near-infrared excitation through tissue filters," *Journal of the American Chemical Society*, vol. 135, no. 48, pp. 18145–18152, 2013.

[10] M. R. Gill and J. A. Thomas, "Ruthenium(II) polypyridyl complexes and DNA—from structural probes to cellular imaging and therapeutics," *Chemical Society Reviews*, vol. 41, no. 8, pp. 3179–3192, 2012.

[11] H. Niyazi, J. P. Hall, K. O'Sullivan et al., "Crystal structures of Λ-[Ru(phen)₂dppz]²⁺with oligonucleotides containing TA/

TA and AT/AT steps show two intercalation modes," *Nature Chemistry*, vol. 4, no. 8, pp. 612–628, 2012.

[12] M. A. Sgambellone, A. David, R. N. Garner, K. R. Dunbar, and C. Turro, "Cellular toxicity induced by the photorelease of a caged bioactive molecule: design of a potential dual-action Ru(II) complex," *Journal of the American Chemical Society*, vol. 135, no. 30, pp. 11274–11282, 2013.

[13] L. X. Xue, T. T. Meng, W. Yang, and K. Z. Wang, "Recent advances in ruthenium complex-based light-driven water oxidation catalysts," *Journal of Photochemistry and Photobiology B: Biology*, vol. 152, pp. 95–105, 2015.

[14] N. Deepika, C. S. Devi, Y. P. Kumar et al., "DNA-binding, cytotoxicity, cellular uptake, apoptosis and photocleavage studies of Ru(II) complexes," *Journal of Photochemistry and Photobiology B-biology*, vol. 160, pp. 142–153, 2016.

[15] M. Schulze, V. Kunz, P. D. Frischmann, and F. Würthner, "A supramolecular ruthenium macrocycle with high catalytic activity for water oxidation that mechanistically mimics photosystem II," *Nature Chemistry*, vol. 8, no. 6, pp. 576–583, 2016.

[16] S. Sen, B. Kawahara, N. L. Fry et al., "A light-activated NO donor attenuates anchorage independent growth of cancer cells: important role of a cross talk between NO and other reactive oxygen species," *Archives of Biochemistry and Biophysics*, vol. 540, no. 1-2, pp. 33–40, 2013.

[17] A. P. de Lima Batista, A. G. S. de Oliveira-Filho, and S. E. Galembeck, "Photophysical properties and the NO photorelease mechanism of a ruthenium nitrosyl model complex investigated using the CASSCF-in-DFT embedding approach," *Physical Chemistry Chemical Physics*, vol. 19, no. 21, pp. 13860–13867, 2017.

[18] H. J. Xiang, Q. Deng, L. An, M. Guo, S. P. Yang, and J. G. Liu, "Tumor cell specific and lysosome-targeted delivery of nitric oxide for enhanced photodynamic therapy triggered by 808 nm near-infrared light," *Chemical Communications*, vol. 52, no. 1, pp. 148–151, 2016.

[19] A. Rathgeb, A. Böhm, M. S. Novak et al., "Ruthenium-nitrosyl complexes with glycine, L-alanine, L-valine, L-proline, D-proline, L-serine, L-threonine, and L-tyrosine: synthesis, X-ray diffraction structures, spectroscopic and electrochemical properties, and antiproliferative activity," *Inorganic Chemistry*, vol. 53, no. 5, pp. 2718–2729, 2014.

[20] M. G. de Oliveira, F. G. Doro, E. Tfouni, and M. H. Krieger, "Phenotypic switching prevention and proliferation/migration inhibition of vascular smooth muscle cells by the ruthenium nitrosyl complex trans-[Ru(NO)Cl(cyclam)](PF6)2," *Journal of Pharmacy and Pharmacology*, vol. 69, no. 9, pp. 1155–1165, 2017.

[21] A. J. Gomes, E. M. Espreafico, and E. Tfouni, "Trans-[Ru(NO)Cl(cyclam)](PF6)2 and [Ru(NO)(Hedta)] incorporated in PLGA nanoparticles for the delivery of nitric oxide to B16-F10 cells: cytotoxicity and phototoxicity," *Molecular Pharmacology*, vol. 10, no. 10, pp. 3544–3554, 2013.

[22] F. P. Rodrigues, C. R. Pestana, A. C. Polizello et al., "Release of NO from a nitrosyl ruthenium complex through oxidation of mitochondrial NADH and effects on mitochondria," *Nitric Oxide-Biology and Chemistry*, vol. 26, no. 3, pp. 174–181, 2012.

[23] A. C. Merkle, A. B. McQuarters, and N. Lehnert, "Synthesis, spectroscopic analysis and photolabilization of water-soluble ruthenium(III)-nitrosyl complexes," *Dalton Transactions*, vol. 41, no. 26, pp. 8047–8059, 2012.

[24] G. J. Rodrigues, A. C. Pereira, T. F. de Moraes, C. C. Wang, R. S. da Silva, and L. M. Bendhack, "Pharmacological characterization of the vasodilating effect induced by the ruthenium complex cis-[Ru(NO)(NO2)(bpy)2](PF6)2," *Journal of Cardiovascular Pharmacology*, vol. 65, no. 2, pp. 168–175, 2015.

[25] P. F. Castro, D. L. de Andrade, F. Reis Cde et al., "Relaxing effect of a new ruthenium complex nitric oxide donor on airway smooth muscle of an experimental model of asthma in rats," *Clinical and Experimental Pharmacology and Physiology*, vol. 43, no. 2, pp. 221–229, 2016.

[26] H. F. Wang, T. Hagihara, H. Ikezawa, H. Tomizawa, and E. Miki, "Electronic effects of the substituent group in 8-quinolinolatoligand on geometrical isomerism for nitrosylruthenium(II)complexes," *Inorganica Chimica Acta*, vol. 299, no. 1, pp. 80–90, 2000.

[27] L. Q. Xu, Z. O. Ma, W. M. Wang et al., "Photo-induced cytotoxicity, photo-controlled nitric oxide release, and DNA/human serum albumin binding of three water-soluble nitrosylruthenium complexes," *Polyhedron*, vol. 137, pp. 157–164, 2017.

[28] M. J. Frisch, G. W. Trucks, H. B. Schlegel et al., *Gaussian 09, revision C.01*, Gaussian, Inc, Wallingford, CT, USA, 2009.

[29] R. Dennington, T. Keith, and J. Millam, *GaussView Version 5*, Semichem Inc., Shawnee Mission KS, 2009.

[30] C. Lee, W. Yang, and R. G. Parr, "Development of the Colle-Salvetti correlation-energy formula into a functional of the electron density," *Physical Review B: Condensed Matter Matter Physics*, vol. 37, no. 2, pp. 785–789, 1988.

[31] A. D. Becke, "Density-functional exchange-energy approximation with correct asymptotic behavior," *Physical Review A General Physics*, vol. 38, no. 6, pp. 3098–3100, 1988.

[32] A. D. Becke, "Density-functional thermochemistry. III. The role of exact exchange," *Journal of Chemical Physics*, vol. 98, no. 7, pp. 5648–5652, 1993.

[33] K. A. Peterson, D. Figgen, M. Dolg, and H. Stoll, "Energy-consistent relativistic pseudopotentials and correlation consistent basis sets for the 4d elements Y-Pd," *Journal of Chemical Physics*, vol. 126, no. 12, pp. 124101–124112, 2007.

[34] W. J. Hehre, L. Radom, P. V. R. Schleyer, and J. Pople, *Ab Initio Molecular Orbital Theory*, Wiley and Sons, New York, NY, USA, 1986.

[35] S. Bhattacharya, T. K. Pradhan, A. De, S. R. Chaudhury, A. K. De, and T. Ganguly, "Photophysical processes involved within the anisole-thioindoxyl dyad system," *Journal of Physical Chemistry A*, vol. 110, no. 17, pp. 5665–5673, 2006.

[36] E. Cances, B. Mennucci, and J. Tomasi, "A new integral equation formalism for the polarizable continuum model: theoretical background and applications to isotropic and anisotropic dielectrics," *Journal of Chemical Physics*, vol. 107, no. 8, pp. 3032–3041, 1997.

[37] R. S. Mulliken, "Electronic population analysis on LCAO-MO molecular wave functions," *Journal of Chemical Physics*, vol. 23, no. 10, pp. 1833–1840, 1955.

[38] E. D. Glendening, A. E. Reed, J. E. Carpenter, and F. Weinhold, *NBO 3.0 Program Manual*, Theoretical Chemistry Institute, University of Wisconsin, Madison, WI, USA, 1990.

[39] R. Akesson, L. G. M. Pettersson, M. Sandstrom, and U. Wahlgren, "Ligand-field effects inthe hydrated divalent and trivalent metal-ions ofthefirst and 2nd transition periods," *Journal of the American Chemical Society*, vol. 116, no. 19, pp. 8691–8704, 1994.

[40] G. Frenking and U. J. Pidu, "Ab initio studies of transition-metal compounds: the nature of the chemical bond to a transition metal," *Journal of the Chemical Society, Dalton Transactions*, no. 10, pp. 1653–1662, 1997.

[41] M. K. Nazeeruddin, F. De Angelis, S. Fantacci et al., "Combined experimental and DFT-TDDFT computational study of photoelectrochemical cell ruthenium sensitizers," *Journal of the American Chemical Society*, vol. 127, no. 48, pp. 16835–16847, 2005.

[42] W. Grochala, A. Albrecht, C. Andreas, and R. Hoffmann, "Remarkably simple relationship connecting the calculated geometries of isomolecular states of three different multiplicities," *Journal of Physical Chemistry*, vol. 104, no. 11, pp. 2195–2203, 2000.

[43] L. Padmaja, M. Amalanathan, C. Ravikumar, and I. Hubert Joe, "NBO analysis and vibrational spectra of 2,6-bis(p-methyl benzylidene cyclohexanone) using density functional theory," *Spectrochimica Acta Part A*, vol. 74, no. 2, pp. 349–356, 2009.

[44] T. S. Sundar, R. Sen, and P. Johari, "Rationally designed donor-acceptor scheme based molecules for applications in opto-electronic devices," *Physical Chemistry Chemical Physics*, vol. 18, no. 13, pp. 9133–9147, 2016.

[45] H. F. Pan, W. M. Wang, Z. O. Ma et al., "Structures and spectroscopic properties of three [RuCl(2mqn)$_2$NO] (H2mqn = 2-methyl-8-quinolinol) isomers: an experimental and density functional theoretical study," *Polyhedron*, vol. 118, pp. 61–69, 2016.

[46] M. A. Palafox, M. Gill, N. J. Nunez, V. K. Rostogi, L. Mittal, and R. Sharm, "Scaling factors for the prediction of vibrational spectra. II. The aniline molecule and several derivatives," *International Journal of Quantum Chemistry*, vol. 103, no. 4, pp. 394–421, 2005.

[47] A. F. Vanin, A. P. Poltorakov, V. D. Mikoyan, L. N. Kubrina, and E. van Faassen, "Why iron-dithiocarbamates ensure detection of nitric oxide in cells and tissues," *Nitric Oxide*, vol. 15, no. 4, pp. 295–311, 2006.

[48] S. Porasuphatana, J. Weaver, T. A. Budzichowski, P. Tsai, and G. M. Rosen, "Differential effect of buffer on the spin trapping of nitric oxide by iron chelates," *Analytical Biochemistry*, vol. 298, no. 1, pp. 50–56, 2001.

[49] B. Gopalakrishnan, K. M. Nash, M. Velayutham, and F. A. Villamena, "Detection of nitric oxide and superoxide radical anion by electron paramagnetic resonance spectroscopy from cells using spin traps," *Journal of Visualized Experiments*, vol. 66, p. 2810, 2012.

[50] S. Pou, P. Tsai, S. Porasuphatana et al., "Spin trapping of nitric oxide by ferro-chelates: kinetic and in vivo pharmacokinetic studies," *Biochimica ET Biophysica Acta*, vol. 1427, no. 2, pp. 216–226, 1999.

[51] N. L. Fry and P. K. Mascharak, "Photolability of NO in designed metal nitrosyls with carboxamido-N donors: a theoretical attempt to unravel the mechanism," *Dalton Transactions*, vol. 41, no. 16, pp. 4726–4735, 2012.

[52] J. Wang, F. Yang, Y. Zhao et al., "Photoisomerization and structural dynamics of two nitrosylruthenium complexes: a joint study by NMR and nonlinear IR spectroscopies," *Physical Chemistry Chemical Physics*, vol. 16, no. 43, pp. 24045–24054, 2014.

[53] L. Freitag and L. González, "Theoretical spectroscopy and photodynamics of a ruthenium nitrosyl complex," *Inorganic Chemistry*, vol. 53, no. 13, pp. 6415–6426, 2014.

3

Gloriosa superba Mediated Synthesis of Platinum and Palladium Nanoparticles for Induction of Apoptosis in Breast Cancer

Shalaka S. Rokade,[1] **Komal A. Joshi,**[2] **Ketakee Mahajan,**[2] **Saniya Patil,**[2] **Geetanjali Tomar,**[2] **Dnyanesh S. Dubal,**[3] **Vijay Singh Parihar,**[4] **Rohini Kitture,**[5] **Jayesh R. Bellare,**[6] **and Sougata Ghosh** ⓘ[7]

[1]*Department of Microbiology, Modern College of Arts, Science and Commerce, Ganeshkhind, Pune 411016, India*
[2]*Institute of Bioinformatics and Biotechnology, Savitribai Phule Pune University, Pune 411007, India*
[3]*Indian Institute of Science, Education and Research, Pashan, Pune 411008, India*
[4]*Department of Biomedical Sciences and Engineering, BioMediTech, Tampere University of Technology, Korkeakoulunkatu 10, 33720 Tampere, Finland*
[5]*Department of Applied Physics, Defense Institute of Advanced Technology, Girinagar, Pune 411025, India*
[6]*Department of Chemical Engineering, Indian Institute of Technology Bombay, Powai, Mumbai 400076, India*
[7]*Department of Microbiology, School of Science, RK University, Kasturbadham, Rajkot 360020, India*

Correspondence should be addressed to Sougata Ghosh; ghoshsibb@gmail.com

Academic Editor: Konstantinos Tsipis

Green chemistry approaches for designing therapeutically significant nanomedicine have gained considerable attention in the past decade. Herein, we report for the first time on anticancer potential of phytogenic platinum nanoparticles (PtNPs) and palladium nanoparticles (PdNPs) using a medicinal plant *Gloriosa superba* tuber extract (GSTE). The synthesis of the nanoparticles was completed within 5 hours at 100°C which was confirmed by development of dark brown and black colour for PtNPs and PdNPs, respectively, along with enhancement of the peak intensity in the UV-visible spectra. High-resolution transmission electron microscopy (HRTEM) showed that the monodispersed spherical nanoparticles were within a size range below 10 nm. Energy dispersive spectra (EDS) confirmed the elemental composition, while dynamic light scattering (DLS) helped to evaluate the hydrodynamic size of the particles. Anticancer activity against MCF-7 (human breast adenocarcinoma) cell lines was evaluated using MTT assay, flow cytometry, and confocal microscopy. PtNPs and PdNPs showed $49.65 \pm 1.99\%$ and $36.26 \pm 0.91\%$ of anticancer activity. Induction of apoptosis was most predominant in the underlying mechanism which was rationalized by externalization of phosphatidyl serine and membrane blebbing. These findings support the efficiency of phytogenic fabrication of nanoscale platinum and palladium drugs for management and therapy against breast cancer.

1. Introduction

Spectacular development in the field of nanotechnology has led to the fabrication of exotic nanostructures with attractive physicochemical and optoelectronic properties. Nanomaterials have got broad-spectrum therapeutic applications which include carbon-based nanostructures, semiconductor quantum dots, polymeric particles, metallic nanoparticles, and magnetic nanoparticles. However, flexibility to vary the properties like shape, size, composition, assembly, and encapsulation has made metallic nanoparticles most preferred over others for biomedical applications [1]. Platinum-based therapeutic drugs, notably cisplatin and carboplatin, are exploited in chemotherapy against cancer, while platinum nanoparticles (PtNPs) have gained attention only recently [2]. Similarly, palladium nanoparticles (PdNPs) are also reported to exhibit anticancer activity against human leukemia (MOLT-4) cells [3]. Although there are so many physical and chemical methods for synthesis of PtNPs and PdNPs, biological methods are considered to be advantageous as they are more biocompatible and less toxic which is a prerequisite for an ideal candidate nanomedicine. Recently, we have shown the potential of medicinal plants like *Dioscorea bulbifera*, *Gnidia glauca*, *Plumbago zeylanica*, *Dioscorea oppositifolia*, *Barleria prionitis*, *Litchi chinensis*, and *Platanus orientalis* for synthesis of gold, silver, and bimetallic nanoparticles [4–15]. Medicinal plants are storehouses of variety of phytochemicals which may play a vital role in synthesis and stabilization of the bioreduced nanoparticles [16–23]. Hence, it is economical and efficient. Although we have reported its potential for synthesis of gold nanoparticles (AuNPs) and silver nanoparticles (AgNPs) earlier, there are no reports on synthesis of PtNPs and PdNPs till date by *Gloriosa superba* tuber extract (GSTE) [24]. *G. superba* is reported to harbour several groups of secondary metabolites such as alkaloids, flavonoids, glycosides, phenols, saponins, steroids, tannins, and terpenoids [25]. The roots are widely used as germicide, to cure ulcers, piles, haemorrhoids, inflammation, scrofula, leprosy, dyspepsia, worm's infestation, flatulence, intermittent fevers, debility, arthritis, and against snake poison [26]. But no extensive studies have been carried out till date on its nanobiotechnological applications.

In view of the background, herein we report synthesis of PtNPs and PdNPs using GSTE which was further characterized using UV-visible spectroscopy, high-resolution transmission electron microscopy (HRTEM), energy dispersive spectroscopy (EDS), dynamic light scattering (DLS), and X-ray diffraction (XRD) analysis. Furthermore, the bioreduced nanoparticles were checked for anticancer activity against MCF-7 cell lines.

2. Materials and Methods

2.1. Plant Material and Extract Preparation. GSTE was prepared by collecting *G. superba* fresh tubers from the Western Ghats of Maharashtra, India, which were thoroughly washed, chopped into small pieces, and shade-dried for 2 days. The dried tubers were reduced to fine powder in an electric blender, 5 g of which was added to 100 mL of distilled water in a 300 mL Erlenmeyer flask and boiled for 5 minutes and eventually collected by decantation followed by filtration through a Whatman number 1 filter paper. The resulting filtrate was used for synthesis of nanoparticles [14].

2.2. Synthesis and UV-Vis Spectroscopy. Reduction of $PtCl_6^{2-}$ ions was initiated by addition of 5 mL of GSTE to 95 mL of 10^{-3} M aqueous $H_2PtCl_6 \cdot 6H_2O$ solution, while for synthesis of PdNPs, 5 mL of GSTE was mixed with 95 mL of 10^{-3} M aqueous $PdCl_2$. The resulting mixtures were incubated at 100°C for 5 hours with constant stirring for synthesis of PtNPs and PdNPs which was monitored at regular intervals using UV-Vis spectroscopy on a spectrophotometer (SpectraMax M5, Molecular Devices Corp, USA) operated at resolution of 1 nm [18, 27].

2.3. High-Resolution Transmission Electron Microscopy (HRTEM), Energy Dispersive Spectroscopy (EDS), Dynamic Light Scattering (DLS), and X-Ray Diffraction (XRD). Morphological features like size and shape of bioreduced PtNPs and PdNPs were determined using JEOL-JEM-2100 high-resolution transmission electron microscope (HRTEM) equipped with a energy dispersive spectrometer (EDS) at an energy range of 0–20 keV. Particle size was analyzed using the dynamic light scattering equipment (Zetasizer Nano-2590, Malvern Instruments Ltd., Worcestershire, UK) in polystyrene cuvette [14, 15]. The diffraction data for the dry powder were recorded on a Bruker X-ray diffractometer using a Cu Kα (1.54 Å) source [28].

2.4. Fourier-Transform Infrared (FTIR) Spectroscopy. After 5 hours of synthesis of PtNPs and PdNPs using GSTE, the resulting mixture was centrifuged at 10,000 rpm for 15 minutes. The supernatant was collected which was added on KBr and dried. Similarly, GSTE before bioreduction was also used to compare the alteration of the phytochemistry. The KBr pellet containing GSTE before and after bioreduction was subjected to FTIR (IRAffinity-1, Shimadzu Corp, Tokyo, Japan) spectroscopy measurement in the diffused reflection mode at a resolution of $4 \, cm^{-1}$ subjected to the IR source 500–4000 cm^{-1} [8].

2.5. Anticancer Activity. Anticancer activities of PtNPs and PdNPs were compared using MTT (3-(4,5-dimethyl-thiazol-2-yl)-2,5-diphenyl-tetrazolium bromide) assay. MCF-7 cells were seeded (4×10^4 cells/well) in a 96-well plate and incubated for adherence for 24 hours, at 37°C with 5% CO_2 concentration followed by which nanoparticles were added at a final concentration of 200 μg/mL and incubated for 48 hours. Medium was removed thereafter, and PBS was used to wash the cells. In each well, MTT (0.5 mg/mL) was added and incubated for 3 hours. The resulting formazan crystals were solubilised in acidified isopropanol, and the absorbance was measured at 570 nm. The statistical analysis was done by using one-way ANOVA.

2.6. Flow Cytometric Analysis. The mechanism underlying the anticancer activity of the PtNPs and PdNPs against MCF-7 cells was studied using flow cytometric analysis of cells treated with respective nanoparticles. 5×10^5 cells were initially seeded in a T-25 flask and incubated for 24 hours followed by addition of PtNPs and PdNPs nanoparticles at a concentration of 200 μg/mL. After 48 hours of incubation, the cells were harvested and stained with Annexin V-FITC (dilution 1 : 20) and propidium iodide (dilution 1 : 20) for 15 minutes at 4°C and were acquired using BD FACSVerse and analyzed by BD FACSuit software as reported earlier [8, 14].

2.7. Confocal Microscopy. In order to support flow cytometric analysis, immunofluorescence staining was performed to find out the mechanism of cell death in MCF-7 cells on treatment with PtNPs and PdNPs. Cells were seeded at a density of 5×10^4 cells on to glass coverslips followed by

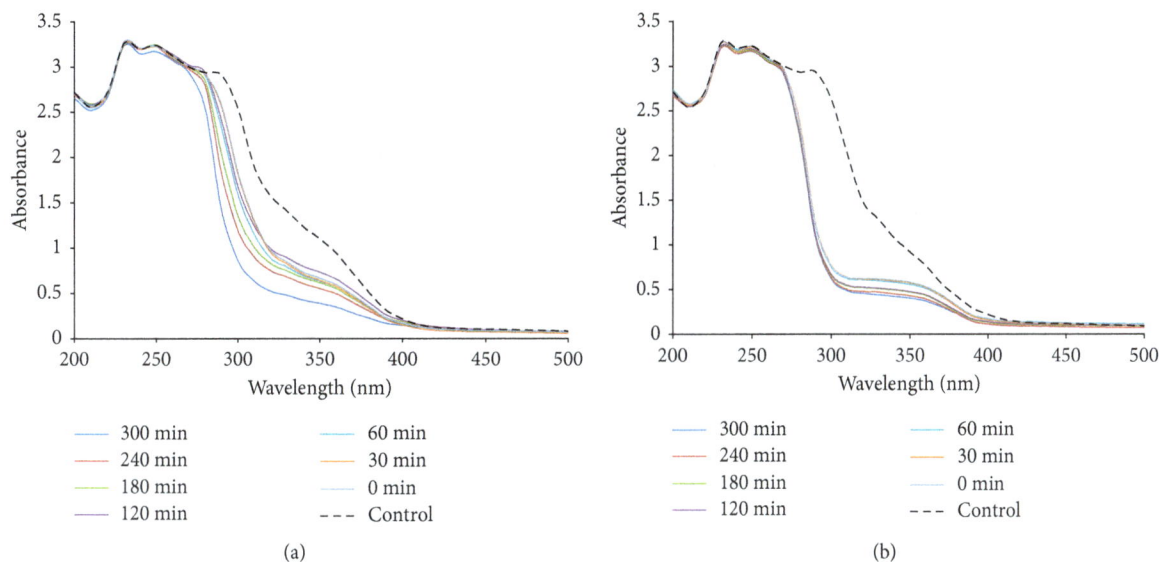

FIGURE 1: UV-Vis spectra recorded at different time intervals for nanoparticle formation using GSTE at 100°C with (a) 1 mM $H_2PtCl_6 \cdot 6H_2O$ solution and (b) 1 mM $PdCl_2$ solution. Control represents corresponding salt solution without GSTE.

incubation for 24 hours for adherence and then treated thereafter with 200 μg/mL of PtNPs and PdNPs for 48 hours. The treated cells were stained with Annexin V(AV)-FITC and PI, both at a dilution of 1 : 20 for 15 minutes at 4°C followed by observation under the LSM 780 confocal laser scanning microscope, Carl Zeiss [8, 14, 24].

3. Results and Discussion

3.1. UV-Visible Spectra. GSTE served as source of the phytomolecules which could efficiently synthesize and stabilize PtNPs and PdNPs that were further studied for anticancer activity. Development of brown colour on addition of GSTE in $H_2PtCl_6 \cdot 6H_2O$ salt solution on incubation at 100°C indicated the synthesis of PtNPs. UV-visible spectra showed the decrease in the intensity specific to the $H_2PtCl_6 \cdot 6H_2O$ salt solution till 5 hours, beyond which no significant decrease was observed which confirmed the completion of the synthesis (Figure 1(a)). Similarly, initially, dark brown colour was developed which eventually turned into black on reaction of GSTE with $PdCl_2$ solution under same conditions. Decrease in the intensity of the UV-spectrum corresponding to $PdCl_2$ solution confirmed the synthesis of PdNPs within 5 hours (Figure 1(b)). This result is well in agreement with the previous reports where nanoscale PtNPs and PdNPs were synthesized using medicinal plants like *D. bulbifera* and *B. prionitis* [8, 14]. The synthesis was found to be faster as compared to synthesis using *Glycine max* and *Cinnamomum camphora*, both of which took 48 hours for complete synthesis of PdNPs [29, 30]. As displayed in Figure 1, the absorption spectra of platinum and palladium colloidal suspensions after 5 hours of bioreduction by GSTE were compared with the absorption spectra of their respective salt solution. Previous reports confirm that the absorption bands appearing in the contrast spectrum of corresponding salt solution were ascribed to the

ligand-to-metal charge-transfer transition of the ions. The absence of the absorption peaks above 300 nm in all the samples after 5 hours indicated complete reduction of the metal ions. Similar accreditation was made during thermally induced reduction of Pd(Fod)₂ in o-xylene and sonochemical reduction of Pd(NO₃)₂ in aqueous solution, respectively. Absence of absorption peaks was consistent with the theoretical study of the surface plasmon resonance absorption of PdNPs. The spectra of colloidal suspensions of PtNPs and PdNPs presented broad absorption continua extending throughout the visible-near-ultraviolet region, which were also observed earlier for the platinum group of metals [31–35].

3.2. HRTEM Analysis. Morphological analysis of the as-synthesized PtNPs and PdNPs was performed using high-resolution transmission electron microscopy (HRTEM). Figures 2(a) and 2(b) reveal the size and shape of the bioreduced PtNPs. The synthesized PtNPs were very small that were majorly of spherical shape, while the diameter was in a range from 0.8 nm to 3 nm. In the magnified overview of the image, the particles were seen to be embedded in a biological matrix may be derived from the GSTE which can play a critical role in the stabilization process. *Diospyros kaki* was reported to synthesize PtNPs of larger size, the diameter was found to be in a range between 2 and 12 nm [36]. At 90°C, *Cacumen platycladi* is reported to synthesize very small PtNPs varying in a range of 2.4 ± 0.8 nm [37, 38]. Figures 2(c) and 2(d) showed the morphological characteristics of the PdNPs which were also predominantly spherical in shape, and the diameter of the particles was found to vary in a narrow range between 5 and 8 nm. It is very rare to get such monodispersed uniform nanoparticles using a biological route. Similarly, previous study reports that PdNPs synthesized using *Glycine max* were found to be

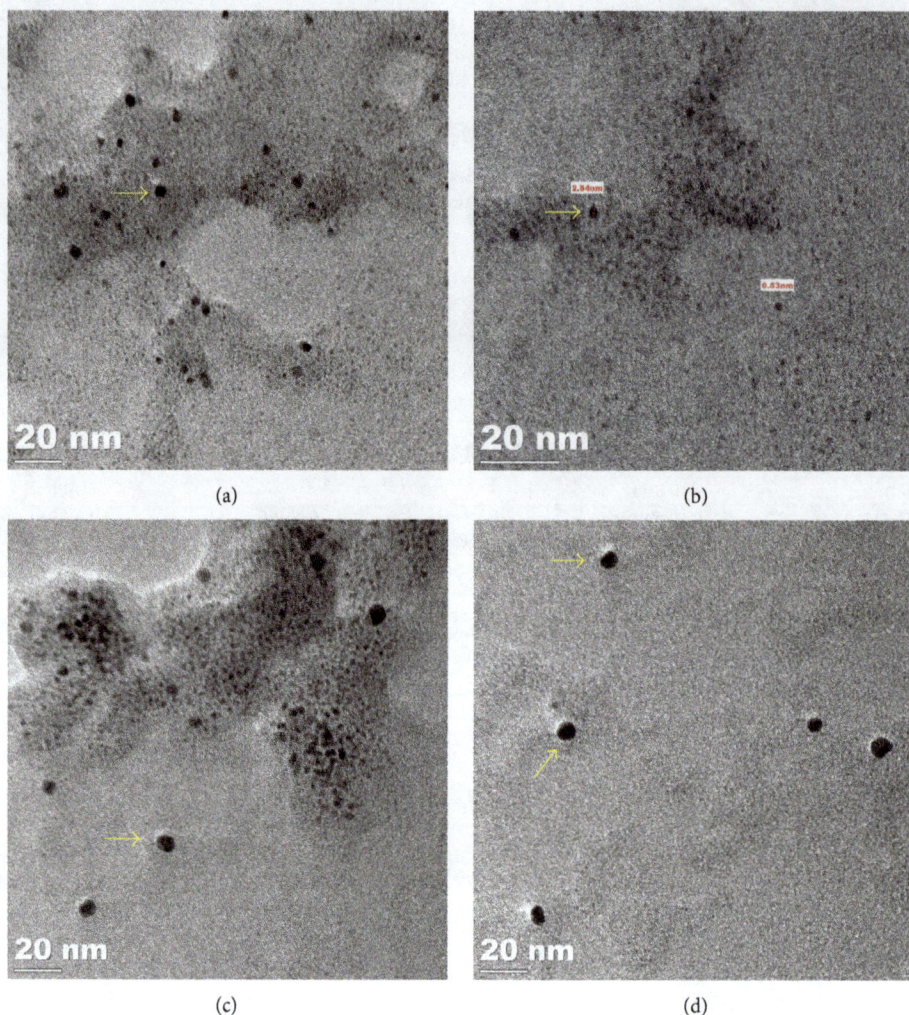

Figure 2: HRTEM images of nanoparticles synthesized by GSTE: PtNPs with inset scale bar showing (a) 100 nm and (b) 20 nm; PdNPs with inset scale bar showing (c) 20 nm and (d) 20 nm.

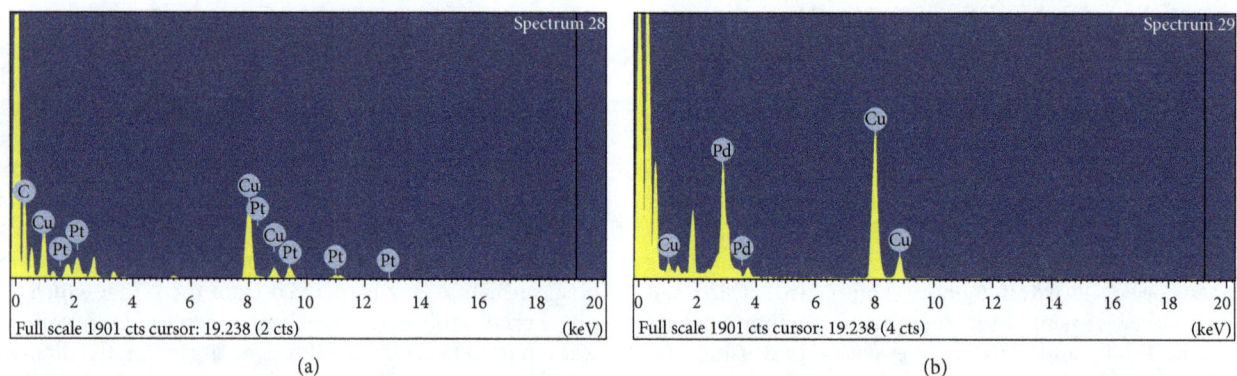

Figure 3: Representative spot EDS of nanoparticles synthesized by GSTE: (a) presence of platinum in PtNPs; (b) presence of palladium in PdNPs.

bigger in size which was 15 nm in diameter [29]. The energy dispersive spectra profile confirmed the presence of elemental platinum and palladium in PtNPs and PdNPs, respectively (Figure 3). Hydrodynamic size recorded for the bioreduced nanoparticles was also in agreement with the observed HRTEM data. However, larger dimensions were also visualized in DLS spectra which may be due to the nanoparticles trapped in the phytochemical entities from GSTE (Figure 4) [7]. Table 1 gives a comprehensive account of various medicinal plants like *Anacardium occidentale*,

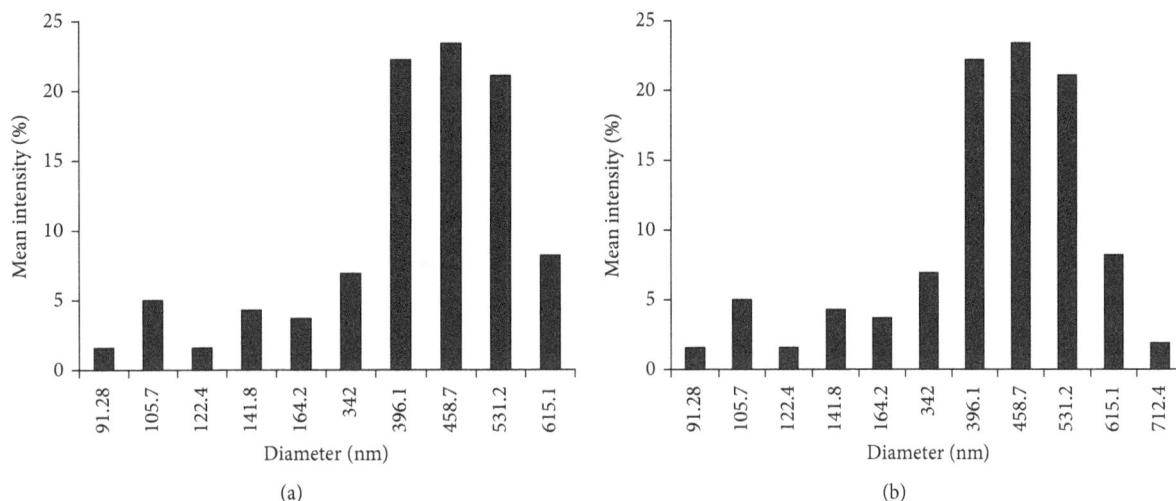

FIGURE 4: Dynamic light scattering measurement showing size distribution of nanoparticles synthesized by GSTE: (a) PtNPs; (b) PdNPs.

TABLE 1: Phytogenic PtNPs and PdNPs.

Serial number	Plant	Extract used	NPs	Shape	Size (nm)	Reference
1	*Cacumen platycladi*	Whole biomass	PtNPs	Spherical	2.4 ± 0.8	[38]
2	*Anacardium occidentale*	Leaf	PtNPs	Irregular and rod shaped	—	[39]
3	*Diospyros kaki*	Leaf	PtNPs	Spheres and plates	2–20	[36]
4	*Ocimum sanctum*	Leaf	PtNPs	Irregular	23	[42]
5	*Fumariae herba*	Whole herb	PtNPs	Hexagonal and pentagonal	30	[43]
6	*Curcuma longa*	Tuber	PdNPs	Spherical	15–20	[44]
7	*Gardenia jasminoides* Ellis	Fruit	PdNPs	Spherical, rod, and three-dimensional polyhedra	3–5	[45]
8	*Glycine max*	Leaf	PdNPs	Spherical	15	[29]
9	*Punica granatum*	Peel	PtNPs	Spherical	16–23	[46]
10	*Cinnamomum camphora*	Leaf	PdNPs	Irregular	6	[30]
11	*Annona squamosa* L.	Peel	PdNPs	Spherical	100	[41]
12	*Pulicaria glutinosa*	Whole plant	PdNPs	Spherical	20–25	[46, 47]
13	*Delonix regia*	Leaf	PdNPs	Spherical	2–4	[48]
14	*Piper betle* L.	Leaf	PtNPs	Spherical	2.1 ± 0.4	[40]
			PdNPs	Spherical	3.8 ± 0.2	
15	*Dioscorea bulbifera*	Tuber	PtNPs	Spherical	2–5	[8]
			PdNPs	Spherical and blunt ended cubes	10–25	
16	*Barleria prionitis*	Leaf	PtNPs	Spherical	1–2	[14]
			PdNPs	Spherical and irregular	5–7	

Piper betle, *Annona squamosa*, *Terminalia chebula*, and *Pulicaria glutinosa*, which are reported to synthesize either PtNPs, PdNPs, or both [37, 39–41].

3.3. X-Ray Diffraction (XRD) Analysis. The as-synthesized nanoparticles were characterized for their phase with the help of XRD. The powder diffraction data of the dried powder was recorded on a Bruker X-ray diffractometer with Cu Kα (1.54 Å) source. Figure 5 shows the XRD data of the PtNPs and PdNPs. The sharp peaks in case of PtNPs and PdNPs represent the crystalline nature of both the nanoparticles. The phase formation has also been confirmed from the data [8]. The characteristic peaks, as seen in Figure 5, correspond to the lattice planes (111), (200), and (220) in case of PtNPs; however, (111) plane was not seen in case of PdNPs. The reason for absence (or no growth) of the (111) plane in case of PdNPs needs to be explored, but at the preliminary stage, we feel that the plant extract might have some crucial role in such restricted growth.

3.4. FTIR Analysis. FTIR spectral analysis showed various functional groups in GSTE before bioreduction and their alteration after synthesis of PtNPs and PdNPs (Figure 6). GSTE showed a prominent peak of the hydroxyl group specific to alcoholic and phenolic compounds at ~3300 cm^{-1}, which remain unaltered even after nanoparticles synthesis.

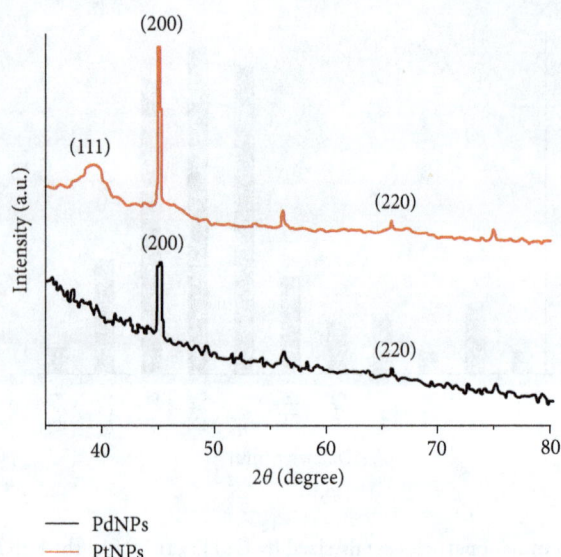

FIGURE 5: Representative X-ray diffraction profile of thin film PtNPs and PdNPs, synthesized by GSTE.

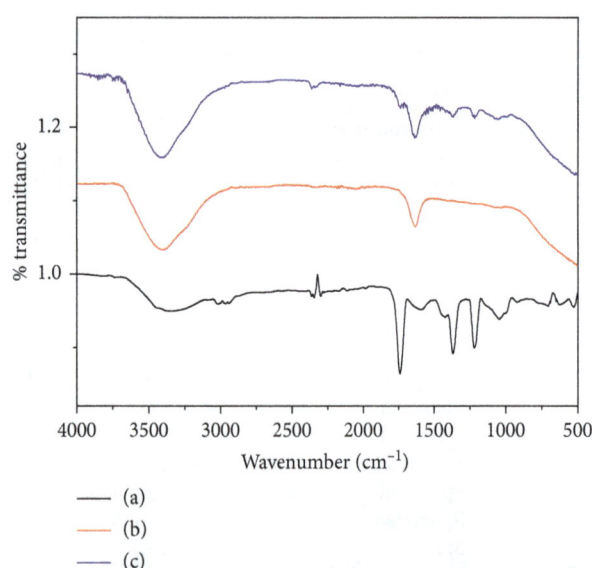

FIGURE 6: FTIR spectra of GSTE. (a) Before synthesis of nanoparticles, (b) after synthesis of PtNPs, and (c) after synthesis of PdNPs.

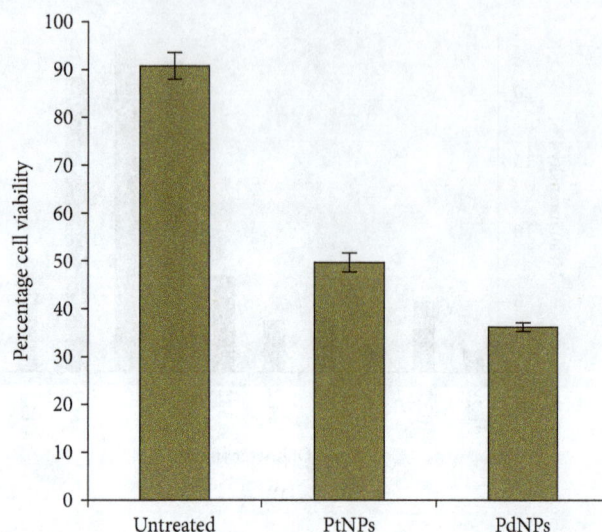

FIGURE 7: Anticancer activity against MCF-7 cells using MTT reduction assay. The data are indicated as the mean ± SEM ($n = 5$).

Similarly, peaks observed at 1049, 1218, 1369, and 1737 cm^{-1} can be attributed to the C-O-C bond in ether, unassigned amide mode, CH$_3$ bend, and stretching of C=O bond, respectively, which disappeared after synthesis of nanoparticles. This indicates that phytochemicals with abovementioned functional groups are responsible for reduction of the metal ions salts leading to synthesis of corresponding nanoparticles. However, a significant feature of the amide bond at 1627 cm^{-1} seen in GSTE is recovered after synthesis, suggesting the replacement of carboxylic group by amines, which in turn again supports the hypothesis of role of carboxylic and similar groups in reduction of the metal salts into the corresponding metal nanoparticles [49].

3.5. Anticancer Activity. Apoptosis is considered as programmed cell death orchestrated by cascade of interdependent synchronised cellular events. It is the most critical process for maintenance of homeostasis, where an efficient balance between cell proliferation and cell death is maintained [50]. Fabrication of apoptotic nanoinducers is of prime importance to develop novel nanomedicine against cancer. Platinum drugs like cisplatin, oxaliplatin, and carboplatin are considered as candidates for treatment and management of cancer, although they pose a threat of potential adverse effects. However, there are very less studies on the anticancer activity of biologically synthesized PtNPs and PdNPs. In our study, both PtNPs and PdNPs showed superior anticancer activity by reducing the viability of MCF-7 cells on treatment till 48 hours. PtNPs showed an anticancer activity up to 49.65 ± 1.99%, while PdNPs showed an activity up to 36.26 ± 0.91% (Figure 7). PtNPs and PdNPs are reported to exhibit high cytotoxicity owing to their physicochemical interactions with the functional groups of cellular proteins, nitrogen bases, and phosphate groups of the DNA leading to cell death. Earlier reports confirm that Pd leads to formation of free radicals, leakage of lactate dehydrogenase, and cell-cycle disturbances which can be the key underlying mechanism behind the anticancer activity [3]. Cellular deaths are mainly due to either apoptosis, autophagy, or necrosis. In order to determine the percentage of apoptotic and necrotic cells, MCF-7 cells were treated with 200 µg/mL of both PtNPs and PdNPs for 48 hours and stained with Annexin V and PI followed by flow cytometric analysis (Figure 8). Both PtNPs and PdNPs were capable of inducing apoptosis in MCF-7 cells up to 12.32% and 31.3%, respectively, which was found to be higher compared to previous reports on human lung adenocarcinoma (A549), ovarian teratocarcinoma (PA-1), pancreatic cancer (Mia-Pa-Ca-2) cells, and normal peripheral blood mononucleocyte (PBMC) cells [2]. Our results were comparable to anticancer activity of PdNPs synthesized using *Camellia sinensis* against

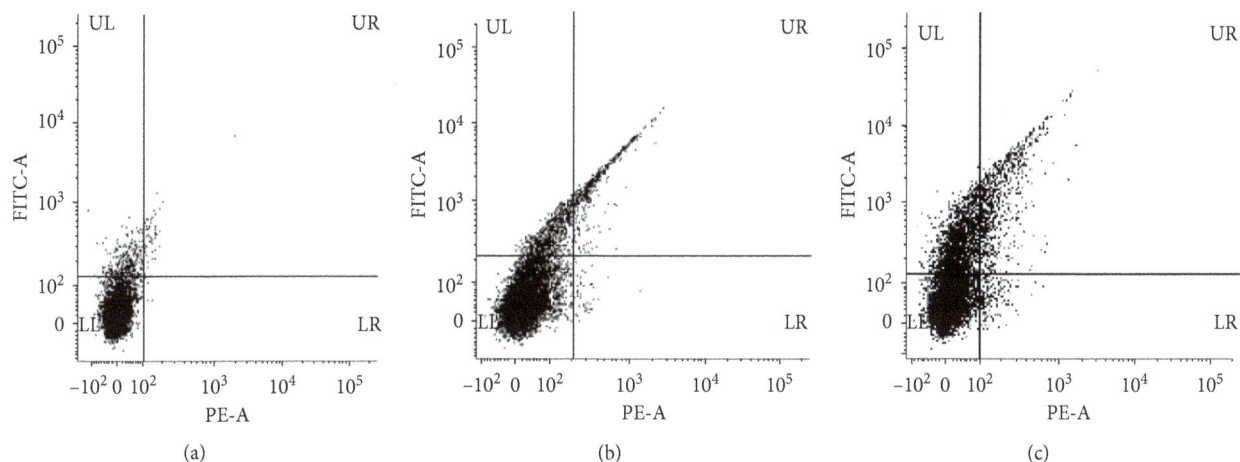

(a)   (b)   (c)

FIGURE 8: Flow cytometric analysis for MCF-7 cells treated with PtNPs and PdNPs for 48 hours confirming phosphatidyl serine externalization (Annexin V-FITC binding) and cell membrane disintegration (PI staining) The dual parametric dot plots combining Annexin V-FITC and PI fluorescence show the viable cell population (lower left quadrant, Annexin V-FITC⁻ PI⁻), the early apoptotic cells (lower right quadrant, Annexin V-FITC⁺PI⁻), and the late apoptotic cells (upper right quadrant, Annexin V-FITC⁺PI⁺). (a) Untreated cells; (b) treatment with PtNPs; (c) treatment with PdNPs.

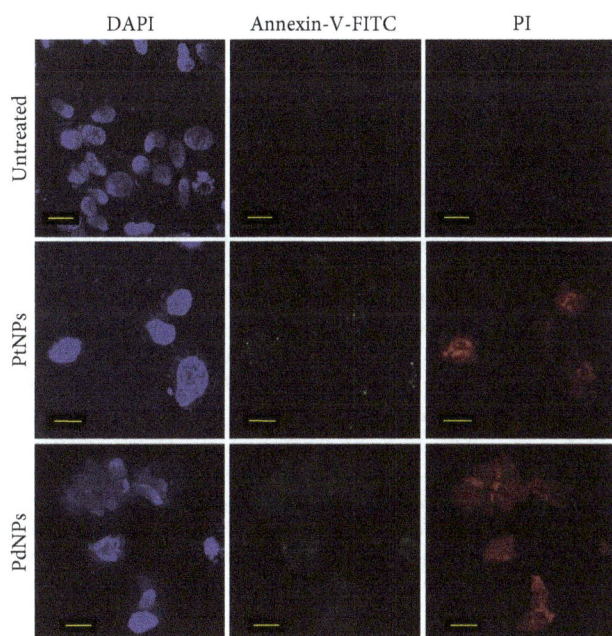

FIGURE 9: Confocal imaging of apoptosis induction by PtNPs and PdNPs in MCF-7 cells seeded on coverslips and stained with Annexin V-FITC and PI.

human leukemia (MOLT-4) [3]. Recently, such unconventional platinum anticancer agents and associated nanomedicines have got more attention as clinically successful platinum drugs like cisplatin, carboplatin, and oxaliplatin have exhibited tremendous deleterious side effects that include nephrotoxicity, fatigue, emesis, alopecia, ototoxicity, peripheral neuropathy, and myelosupression [51, 52]. Confocal images also confirmed the induction of apoptosis (Figure 9). Externalization of phosphatidyl serine and membrane disintegration was evident from Annexin V-FITC⁺PI⁺ MCF-7 cells. Similarly, membrane blebbing

and chromosome condensation were also observed in PtNPs treated cells, which is a critical hallmark of apoptosis [8].

4. Conclusion

Monodispersed PtNPs and PdNPs were synthesized using *G. superba* tuber extract which were found to be uniformly spherical and almost isodiametric. The synthesis was found to be rapid, efficient, and environmentally benign. Both PtNPs and PdNPs showed potent anticancer activity against MCF-7 (human breast adenocarcinoma) cells. The mechanism of cell death was confirmed to be induction of apoptosis characterized by phosphatidyl serine externalization, membrane disintegration, and blebbing with chromosome condensation. Further studies on these phytogenic nanoparticles might help to establish their potential as candidate drugs against breast cancer.

Acknowledgments

The authors acknowledge the help extended for the use of TEM and HRTEM facilities in Chemical Engineering and CRNTS funded by the DST through Nanomission and IRPHA schemes. Dr. Geetanjali Tomar thanks DST INSPIRE for the faculty position and the research grants (nos. IFA13 LSBM73 and GOI-E-161(2), resp.).

References

[1] D. G. Sant, T. R. Gujarathi, S. R. Harne et al., "*Adiantum philippense* L. frond assisted rapid green synthesis of gold and silver nanoparticles," *Journal of Nanoparticles*, vol. 2013, Article ID 182320, 9 pages, 2013.

[2] Y. Bendale, V. Bendale, and S. Paul, "Evaluation of cytotoxic activity of platinum nanoparticles against normal and cancer

cells and its anticancer potential through induction of apoptosis," *Integrative Medicine Research*, vol. 6, no. 2, pp. 141–148, 2017.

[3] S. Azizi, M. M. Shahri, H. S. Rahman, R. A. Rahim, A. Rasedee, and R. Mohamad, "Green synthesis palladium nanoparticles mediated by white tea (*Camellia sinensis*) extract with antioxidant, antibacterial, and antiproliferative activities toward the human leukemia (MOLT-4) cell line," *International Journal of Nanomedicine*, vol. 12, pp. 8841–8853, 2017.

[4] S. Ghosh, S. Patil, M. Ahire et al., "Synthesis of silver nanoparticles using *Dioscorea bulbifera* tuber extract and evaluation of its synergistic potential in combination with antimicrobial agents," *International Journal of Nanomedicine*, vol. 7, pp. 483–496, 2012.

[5] A. K. Mittal, Y. Chisti, and U. C. Banerjee, "Synthesis of metallic nanoparticles using plant extracts," *Biotechnology Advances*, vol. 31, no. 2, pp. 346–356, 2013.

[6] M. R. Bindhu and M. Umadevi, "Synthesis of monodispersed silver nanoparticles using *Hibiscus cannabinus* leaf extract and its antimicrobial activity," *Spectrochimica Acta Part A: Molecular and Biomolecular Spectroscopy*, vol. 101, pp. 184–190, 2013.

[7] A. Schröfel, G. Kratošová, I. Šafařík, M. Šafaříková, I. Raška, and L. M. Shor, "Applications of biosynthesized metallic nanoparticles–a review," *Acta Biomaterialia*, vol. 10, no. 10, pp. 4023–4042, 2014.

[8] S. Ghosh, R. Nitnavare, A. Dewle et al., "Novel platinum-palladium bimetallic nanoparticles synthesized by *Dioscorea bulbifera*: anticancer and antioxidant activities," *International Journal of Nanomedicine*, vol. 10, no. 1, pp. 7477–7490, 2015.

[9] M. Prathap, A. Alagesan, and B. D. R. Kumari, "Anti-bacterial activities of silver nanoparticles synthesized from plant leaf extract of *Abutilon indicum* (L.) Sweet," *Journal of Nanostructure in Chemistry*, vol. 4, p. 106, 2014.

[10] R. Majumdar, B. G. Bag, and N. Maity, "*Acacia nilotica* (Babool) leaf extract mediated size-controlled rapid synthesis of gold nanoparticles and study of its catalytic activity," *International Nano Letters*, vol. 3, p. 53, 2013.

[11] C. Krishnaraj, R. Ramachandran, K. Mohan, and P. T. Kalaichelvan, "Optimization for rapid synthesis of silver nanoparticles and its effect on phytopathogenic fungi," *Spectrochimica Acta Part A: Molecular and Biomolecular Spectroscopy*, vol. 93, pp. 95–99, 2012.

[12] J. K. Andeani, H. Kazemi, S. Mohsenzadeh, and A. Safavi, "Biosynthesis of gold nanoparticles using dried flowers extract of *Achillea wilhelmsii* plant," *Digest Journal of Nanomaterials and Biostructures*, vol. 6, no. 3, pp. 1011–1017, 2011.

[13] G. R. Salunke, S. Ghosh, R. J. S. Kumar et al., "Rapid efficient synthesis and characterization of AgNPs, AuNPs and AgAuNPs from a medicinal plant *Plumbago zeylanica* and their application in biofilm control," *International Journal of Nanomedicine*, vol. 9, no. 1, pp. 2635–2653, 2014.

[14] S. S. Rokade, K. A. Joshi, K. Mahajan et al., "Novel anticancer platinum and palladium nanoparticles from *Barleria prionitis*," *Global Journal of Nanomedicine*, vol. 2, no. 5, article 555600, 2017.

[15] S. Shende, K. A. Joshi, A. S. Kulkarni et al., "*Litchi chinensis* peel: a novel source for synthesis of gold and silver nano-catalysts," *Global Journal of Nanomedicine*, vol. 3, no. 1, article 555603, 2017.

[16] R. Kitture, S. Ghosh, P. Kulkarni et al., "Fe$_3$O$_4$-citrate curcumin: promising conjugates for superoxide scavenging, tumor suppression and cancer hyperthermia," *Journal of Applied Physics*, vol. 111, no. 6, article 064702, 2012.

[17] J. L. Gardea-Torresdey, E. Gomez, J. R. Peralta-Videa, J. G. Parsons, H. Troiani, and M. Jose-Yacaman, "Alfalfa sprouts: a natural source for the synthesis of silver nanoparticles," *Langmuir*, vol. 19, no. 4, pp. 1357–1361, 2003.

[18] J. L. Gardea-Torresdey, J. G. Parsons, E. Gomez et al., "Formation and growth of Au nanoparticles inside live Alfalfa plants," *Nano Letters*, vol. 2, no. 4, pp. 397–401, 2002.

[19] R. Rajan, K. Chandran, S. L. Harper, S. I. Yun, and P. Thangavel Kalaichelvan, "Plant extract synthesized silver nanoparticles: an ongoing source of novel biocompatible materials," *Industrial Crops and Products*, vol. 70, pp. 356–373, 2015.

[20] K. S. Kavitha, S. Baker, D. Rakshith et al., "Plants as green source towards synthesis of nanoparticles," *International Research Journal of Biological Sciences*, vol. 2, no. 6, pp. 66–76, 2013.

[21] M. Khan, M. Khan, S. F. Adil et al., "Green synthesis of silver nanoparticles mediated by *Pulicaria glutinosa* extract," *International Journal of Nanomedicine*, vol. 8, no. 1, pp. 1507–1516, 2013.

[22] P. Kuppusamy, M. M. Yusoff, G. P. Maniam, and N. Govindan, "Biosynthesis of metallic nanoparticles using plant derivatives and their new avenues in pharmacological applications–an updated report," *Saudi Pharmaceutical Journal*, vol. 24, no. 4, pp. 473–484, 2016.

[23] S. Arokiyaraj, M. V. Arasu, S. Vincent et al., "Rapid green synthesis of silver nanoparticles from *Chrysanthemum indicum* L and its antibacterial and cytotoxic effects: an in vitro study," *International Journal of Nanomedicine*, vol. 9, no. 1, pp. 379–388, 2014.

[24] K. Jyoti, M. Baunthiyal, and A. Singh, "Characterization of silver nanoparticles synthesized using *Urtica dioica* Linn. leaves and their synergistic effects with antibiotics," *Journal of Radiation Research and Applied Sciences*, vol. 9, no. 3, pp. 217–227, 2016.

[25] M. Senthilkumar, "Phytochemical screening of *Gloriosa superba* L. from different geographical positions," *International Journal of Scientific and Research Publications*, vol. 3, pp. 1–5, 2013.

[26] K. J. R. Gopi and R. Panneerselvam, "Quantification of colchicine in seed and tuber samples of *Gloriosa superba* by high performance liquid chromatography method," *Journal of Applied Pharmaceutical Science*, vol. 1, no. 7, pp. 116–119, 2011.

[27] S. S. Shankar, A. Rai, B. Ankamwar, A. Singh, A. Ahmad, and M. Sastry, "Biological synthesis of triangular gold nano-prisms," *Nature Materials*, vol. 3, no. 7, pp. 482–488, 2004.

[28] S. S. Shankar, A. Rai, A. Ahmad, and M. Sastry, "Controlling the optical properties of lemongrass extract synthesized gold nanotriangles and potential application in infrared-absorbing optical coatings," *Chemistry of Materials*, vol. 17, no. 3, pp. 566–572, 2005.

[29] R. K. Petla, S. Vivekanandhan, M. Misra, A. K. Mohanty, and N. Satyanarayana, "Soybean (*Glycine max*) leaf extract based green synthesis of palladium nanoparticles," *Journal of Biomaterials and Nanobiotechnology*, vol. 3, no. 1, pp. 14–19, 2012.

[30] X. Yang, Q. Li, H. Wang et al., "Green synthesis of palladium nanoparticles using broth of *Cinnamomum camphora* leaf," *Journal of Nanoparticle Research*, vol. 12, no. 5, pp. 1589–1598, 2010.

[31] T. Teranishi and M. Miyake, "Size control of palladium nanoparticles and their crystal structures," *Chemistry of Materials*, vol. 10, no. 2, pp. 594–600, 1998.

[32] T. Yonezawa, K. Imamura, and N. Kimizuka, "Direct prep-

aration and size control of palladium nanoparticle hydrosols by water-soluble isocyanide ligands," *Langmuir*, vol. 17, no. 16, pp. 4701–4703, 2001.

[33] C. Luo, Y. Zhang, and Y. Wang, "Palladium nanoparticles in poly(ethyleneglycol): the efficient and recyclable catalyst for Heck reaction," *Journal of Molecular Catalysis A: Chemical*, vol. 229, no. 1-2, pp. 7–12, 2005.

[34] P. F. Ho and K. M. Chi, "Size-controlled synthesis of Pd nanoparticles from β-diketonato complexes of palladium," *Nanotechnology*, vol. 15, no. 8, pp. 1059–1064, 2004.

[35] A. Nemamcha, J. Rehspringer, and D. Khatmi, "Synthesis of palladium nanoparticles by sonochemical reduction of palladium(II) nitrate in aqueous solution," *Journal of Physical Chemistry B*, vol. 110, no. 1, pp. 383–387, 2006.

[36] J. Y. Song, E. Y. Kwon, and B. S. Kim, "Biological synthesis of platinum nanoparticles using *Diopyros kaki* leaf extract," *Bioprocess and Biosystems Engineering*, vol. 33, no. 1, pp. 159–164, 2010.

[37] S. F. Adil, M. E. Assal, M. Khan, A. Al-Warthan, M. R. H. Siddiqui, and L. M. Liz-Marzán, "Biogenic synthesis of metallic nanoparticles and prospects toward green chemistry," *Dalton Transactions*, vol. 44, no. 21, pp. 9709–9717, 2015.

[38] B. Zheng, T. Kong, X. Jing et al., "Plant-mediated synthesis of platinum nanoparticles and its bioreductive mechanism," *Journal of Colloid and Interface Science*, vol. 396, pp. 138–145, 2013.

[39] D. S. Sheny, D. Philip, and J. Mathew, "Synthesis of platinum nanoparticles using dried *Anacardium occidentale* leaf and its catalytic and thermal applications," *Spectrochimica Acta Part A: Molecular and Biomolecular Spectroscopy*, vol. 114, pp. 267–271, 2013.

[40] P. Rajasekharreddy and P. U. Rani, "Biosynthesis and characterization of Pd and Pt nanoparticles using *Piper betle* L. plant in a photoreduction method," *Journal of Cluster Science*, vol. 25, no. 5, pp. 1377–1388, 2014.

[41] S. M. Roopan, A. Bharathi, R. Kumar, V. G. Khanna, and A. Prabhakarn, "Acaricidal, insecticidal, and larvicidal efficacy of aqueous extract of *Annona squamosa* L peel as biomaterial for the reduction of palladium salts into nanoparticles," *Colloids and Surfaces B: Biointerfaces*, vol. 92, pp. 209–212, 2012.

[42] C. Soundarrajan, A. Sankari, P. Dhandapani et al., "Rapid biological synthesis of platinum nanoparticles using *Ocimum sanctum* for water electrolysis applications," *Bioprocess Biosystem Engineering*, vol. 35, no. 5, pp. 827–833, 2012.

[43] R. Dobrucka, "Synthesis and structural characteristic of platinum nanoparticles using herbal bidens tripartitus extract," *Journal of Inorganic and Organometallic Polymers and Materials*, vol. 26, no. 1, pp. 219–225, 2015.

[44] M. Sathishkumar, K. Sneha, and Y. S. Yun, "Palladium nanocrystal synthesis using *Curcuma longa* tuber extract," *International Journal of Materials Sciences*, vol. 4, no. 1, pp. 11–17, 2009.

[45] L. Jia, Q. Zhang, Q. Li, and H. Song, "The biosynthesis of palladium nanoparticles by antioxidants in *Gardenia jasminoides Ellis*: long life time nanocatalysts for *p*-nitrotoluene hydrogenation," *Nanotechnology*, vol. 20, no. 38, article 385601, 2009.

[46] P. Dauthal and M. Mukhopadhyay, "Biofabrication, characterization, and possible bio-reduction mechanism of platinum nanoparticles mediated by agro-industrial waste and their catalytic activity," *Journal of Industrial and Engineering Chemistry*, vol. 22, pp. 185–191, 2015.

[47] M. Khan, M. Khan, M. Kuniyil et al., "Biogenic synthesis of palladium nanoparticles using *Pulicaria glutinosa* extract and their catalytic activity towards the Suzuki coupling reaction," *Dalton Transactions*, vol. 43, no. 24, pp. 9026–9031, 2014.

[48] P. Dauthal and M. Mukhopadhyay, "Biosynthesis of palladium nanoparticles using *Delonix regia* leaf extract and its catalytic activity for nitro-aromatics hydrogenation," *Industrial and Engineering Chemistry Research*, vol. 52, no. 51, pp. 18131–18139, 2013.

[49] G. Socrates, *Infrared and Raman Characteristic Group Frequencies*, John Wiley & Sons, Hoboken, NJ, USA, 3rd edition, 2001.

[50] M. Khan, M. Khan, A. H. Al-Marri et al., "Apoptosis inducing ability of silver decorated highly reduced graphene oxide nanocomposites in A549 lung cancer," *International Journal of Nanomedicine*, vol. 11, pp. 873–883, 2016.

[51] T. C. Johnstone, G. Y. Park, and S. J. Lippard, "Understanding and improving platinum anticancer drugs–phenanthriplatin," *Anticancer Research*, vol. 34, no. 1, pp. 471–476, 2014.

[52] I. DeAlba-Montero, J. Guajardo-Pacheco, E. Morales-Sánchez et al., "Antimicrobial properties of copper nanoparticles and amino acid chelated copper nanoparticles produced by using a soya extract," *Bioinorganic Chemistry and Applications*, vol. 2017, Article ID 1064918, 6 pages, 2017.

4

Cytotoxic and Bactericidal Effect of Silver Nanoparticles Obtained by Green Synthesis Method using *Annona muricata* Aqueous Extract and Functionalized with 5-Fluorouracil

María del Carmen Sánchez-Navarro,[1] Claudio Adrian Ruiz-Torres ⓘ,[2]
Nereyda Niño-Martínez,[2] Roberto Sánchez-Sánchez,[3]
Gabriel Alejandro Martínez-Castañón ⓘ,[2] I. DeAlba-Montero,[2] and Facundo Ruiz ⓘ[2]

[1]*Facultad de Estomatología, Universidad Autónoma de San Luis Potosí (UASLP), Avenida Manuel Nava 2, Zona Universitaria, 78290 San Luis Potosí, Mexico*
[2]*Facultad de Ciencias, Universidad Autónoma de San Luis Potosí (UASLP), Avenida Manuel Nava 6, Zona Universitaria, 78290 San Luis Potosí, Mexico*
[3]*Instituto Nacional de Rehabilitación LGII, CENIAQ, Calzada México Xochimilco No. 289, Colonia Arenal de Guadalupe, Delegación Tlalpan, 14389 Ciudad de México, Mexico*

Correspondence should be addressed to Claudio Adrian Ruiz-Torres; ruiztorresclaudio@gmail.com and Facundo Ruiz; ruizfacundo1@gmail.com

Guest Editor: Aurel Tabacaru

Nanomaterials obtained by green synthesis technologies have been widely studied in recent years owing to constitute cost-effective and environmental-friendly methods. In addition, there are several works that report the simultaneous performance of the reducer agent as a functionalizing agent, modifying the properties of the nanomaterial. As a simple and economical synthesis methodology, this work presents a method to synthesize silver nanoparticles (AgNPs) using *Annona muricata* aqueous extract and functionalized with 5-fluorouracil (5-FU). The processes of reduction, nucleation, and functionalization of the nanoparticles were analyzed by UV-Vis absorption spectroscopy, and it was found that they are the function of the contact time of the metal ions with the extract. The structural characterization was carried out by transmission electron microscopy (TEM) and X-ray diffraction patterns (XRD). The antibacterial properties of the synthetized nanomaterials were tested using minimum inhibitory concentration (MIC) and minimum bactericidal concentration (MBC) against *Enterococcus faecalis*, *Staphylococcus aureus*, and *Escherichia coli* growth.

1. Introduction

The use of nanoparticles as nanodelivery vehicles has received quite great interest in the medical sector in recent years owing to the fact that, by biochemical engineering, it is possible to design multifunctional nanostructured biomaterials to deliver specific drugs to target tumor or cancer cells [1]. The best performance of nanobiomaterials with respect to other types of biomaterials is due to quite high compatibility and adaptability to biological systems, which additionally represents nonviral systems, constituting promising tools in biomedicine research. A clear example of this fact is silver nanoparticles; due to the application of these types of materials to biological systems, there has been development of numerous nanodelivery vehicles, in view of their intrinsic properties, biocompatibility, and antimicrobial capacity [2]. It is necessary to improve material properties and biocompatibility for a more efficient yield in drug delivery to a specific target, avoiding a wide distribution of the medicine. In light of this, the material functionalization and organometallic science by the development of covalent nets or polymeric functionalization have modified the interaction of inorganic surface on metallic NPs with the surrounding media, improving their performance in biomedicine [3, 4].

5-Fluorouracil (5-FU) is a potent broadly used antimetabolite for cancer treatments such as advanced oral cancer [5]. Nevertheless, several limitations exist on its use related to the short half-life, lack of control on selective delivery, and ample diffusion on body, limiting its antitumor applicability. Therefore, through the functionalization of nanostructured materials with 5-FU, it has remarkably improved the utility of the drug and produced novel and proficient tools for oncologic research [6].

Currently, ample spectrum of different methodologies for silver nanobiomaterials obtained and functionalization for biomedical purposes exists [7]. However, in the previous years, a remarkable interest in the green synthesis methods of nanobiomaterials has been rising in response which represents environmental-friendly methods, low-toxic methodologies, cost-effective alternatives, and one-step NP synthesis-functionalization method [8]. In general, nanomaterial synthesis by "green methods" considers three main characteristics according to Raveendran et al. [9] as follows: (1) solvent friendly as a reaction medium; (2) environmentally beneficial reducing agents; and (3) use of nontoxic material as capping agents [9]. In addition, different reports had reported not only the achievement of obtaining nanoparticles by green synthesis but also discussed the medicinal properties of the materials associated with the active ingredients present in the natural extracts used, which has promoted a scientific focus on the biological activities of these kinds of substances [10, 11]. Furthermore, in relation to the information described above, the synthesis of silver nanomaterials through an efficient, economically cheap, and environmentally safe method has become an important research area in nanobiotechnology. Therefore, currently, diverse plant extracts have been used as excellent bioreducing agents in the synthesis of silver nanoparticles (AgNPs) [12, 13].

The extract Annona muricata, generally known as guanabana, is largely distributed in tropical areas of South America and North America; all fractions of the A. muricata tree are widely used as traditional medicines against human diseases, including cancer and infections. The antiinflammatory, hypoglycemic, sedative, smooth muscle relaxant, hypotensive, and antispasmodic effects are attributed to the leaves, barks, and roots of Annona muricata. The leaves of this plant are also employed against tumors and cancer in South America and tropical Africa [14]. Phytochemical evaluations of the Annona muricata plant have shown the presence of alkaloids, megastigmanes, flavonol triglycosides, phenolics, cyclopeptides, essential oils, and some minerals such as K, Ca, Na, Cu, Fe, and Mg [14].

In the present work, the synthesis of silver nanoparticles using Annona muricata aqueous extracts as a bioreducing agent is reported. The materials were characterized in terms of their optical properties, crystallinity, morphology, hydrodynamic radius, and surface charge. The antibacterial capacities of the materials were evaluated by their bactericidal effect against oral microorganisms such as E. Faecalis, S. Mutans, S. Oralis, S. Aureus, and E. Coli by minimum inhibitory concentration (MIC) and minimum bactericidal concentration (MBC). Moreover, the evaluation of the cytotoxicity of the samples was done by the cellular viability of fibroblast cells using MTT assay and fluorescent microscopy.

2. Materials and Methods

2.1. Reagents. A. muricata leaves were purchased from a local supermarket; silver nitrate ($AgNO_3$) and 5-FU were obtained from Sigma-Aldrich; serological pipettes (5, 10, and 25 mL) and 50 mL centrifuge tubes were purchased from Santa Cruz Biotechnology, Inc.; 25 cm^3 cell-culture flask were purchased from Corning®; and 48-well cell-culture cluster were purchased from Costar®."

Dulbecco's Modified Eagle Medium (DMEM), phosphate-buffered saline (PBS) of pH 7.4, fetal bovine serum (FBS), penicillin/streptomycin, and doxorubicin were purchased from Gibco®. The following bacteria were used: *Enterococcus faecalis* (ATCC 29212), *Staphylococcus aureus* (ATCC 29213), and *Escherichia coli* (ATCC 25922). Human fibroblasts were donated by Dr. Roberto Sánchez-Sánchez, biotechnology laboratory of Instituto Nacional de Rehabilitación LGII, CENIAQ (Ciudad de México).

2.2. AgNP Synthesis. The synthesis of the materials was carried out by the use of *Annona muricata* extract as a bioreducing agent based on the previously reported method by Santhosh et al. [15]. Initially, the leaves were washed with deionized water to remove impurities. After being crushed in a blender, 5 g of the powder was previously ground in 125 mL of deionized water and was boiled to the boiling point. After obtaining the infusion of the extract, in a separate vessel, Ag salt was added and the process of reduction and formation of the nanoparticles began, which was evidenced by the immediate color change, indicating the formation of the same ones. The color of the mixture changed from pale brown to dark brown for AgNPs, and no synthetic reagents were required for this synthesis. Finally, the NP sedimentation was induced by centrifugation and washed with ethanol three times.

2.3. Silver NP Functionalization with 5-Fluorouracil. Prior to Ag nanoparticle synthesis and washing, 50 ml of NP solution was taken in a vessel; subsequently, 0.5 g of 5-fluorouracil was added, and the solution was sonicated for 20 min.

2.4. Physical Characterization Methods. UV-Vis absorption spectra were obtained using the S2000 UV-Vis spectrometer (OceanOptics, Inc.). Transmission electron microscopy (TEM) images were obtained at 100 kV using a JEOL-1230. The hydrodynamic radius and Z-potential of the samples were measured with a Nanosizer DLS. Furthermore, the X-ray diffraction (XRD) patterns were obtained using a GBC-Difftech MMA diffractometer with filtered CuKα ($\lambda = 1.54$ Å) radiation.

2.5. Antibacterial Activity of AgNPs. The antimicrobial activities of the AgNPs were confirmed via minimum inhibitory concentration (MIC) and minimum bactericidal concentration (MBC) against strains of *E. faecalis* (ATCC 29212), *S. aureus* (ATCC 29213), and *E. coli* (ATCC 25922); they were studied at 0.5 of the McFarland Scale determined with a colorimeter (LaMotte Smart3) according to the standard microdilution method (CLSI M100-S25, January 2015) [16].

2.6. Cytotoxicity Assay. With respect to cytotoxicity assays, human fibroblasts isolated from the dermis were cultured and stored in liquid nitrogen at −196°C, in order to retain the viability of the cells. A cell culture was carried out in 25 cm^3 plates (Costar®), until the surface of the culture vessel was covered, waiting for the formation of a monolayer (layer thickness of one cell). Previously, subcultures were carried out for three weeks using the DMEM culture medium (Gibco®), which contains specific proteins essential for cell survival, development, and proliferation. Subsequently, the incubation was performed at 37°C under a CO_2 atmosphere using a NUAIRE Autoflow Ir Water-Jacketed CO_2 incubator; medium changes were made and observed on an inverted microscope (Axio-Zeiss). Once the desired cell confluence was obtained, the cultures were treated with trypsin (a proteolytic enzyme that degrades the extracellular matrix and sequesters the calcium ion, which is essential for cell adhesion). In addition, by gentle agitation after 5–7 min, the cells detached, and the cell suspension required to calculate the number of cells present in a certain volume were obtained and analyzed in a Neubauer chamber. Furthermore, 20,000 cells per cm^2 were seeded in 48-well plates and cultured for 24 hrs. With the purpose of material cytotoxicity assays performed with calcein and ethidium homodimer (Thermo®), different dilutions of silver nanoparticles synthesized by a green method were made using an extract of *Annona muricata*. The treatment groups used to perform the cytotoxicity test are shown in Table 1:

The viability of human fibroblasts after exposure to AgNPs was evaluated by the amount of viable cells stained by MTT assay. The human fibroblasts were plated in 96-well plates and exposed to AgNPs, AgNPs + 5-FU, 5-FU, and *Annona muricata*. Cells were added into the medium at concentrations of 30 μg/ml (ppm) maintained in a humidified atmosphere at 37°C and 5% CO_2. After 24 and 48 h, the medium was removed from each well, replaced with a new medium with MTT solution in an amount of 10% of culture volume, and incubated for 4 h at 37°C until a blue-colored formazan product developed. The resulting formazan product was dissolved in DMSO, and the absorbance was measured at 570–690 nm by using a Synergy HTX Multi-Mode Microplate reader (BioTek Instrument, Inc.).

3. Results and Discussion

3.1. Structural Characterization of AgNPs

3.1.1. Ultraviolet-Visible (UV-Vis) Absorption Spectroscopy. The UV-Vis spectra of the materials were made in order to follow the nucleation process of the particles as a function of the contact time with the extract. In Figure 1, a band at 425 nm can be observed, which is associated with the surface

TABLE 1: Treatments and different concentrations used in the cytotoxicity tests carried out in fibroblasts.

Concentrations	C1	C2	C3	C4
Annona muricata	100 μg/mL	75 μg/mL	50 μg/mL	25 μg/mL
5-FU	20 μg/mL	15 μg/mL	10 μg/mL	5 μg/mL
AgNPs	32 μg/mL	24 μg/mL	16 μg/mL	8 μg/mL
AgNPs + 5-FU	23 μg/mL	18 μg/mL	13 μg/mL	8 μg/mL

plasmon of silver nanoparticles. The intensity of the band increases as the contact time of the metal ions with the extract is increased, having its maximum intensity at 60 min. Additionally, it is possible to observe a thinning of the band of the final solution with respect to the initial solution. This thinning can be associated to two phenomena; firstly, the influence of the symmetry of the particles and their size distribution on their optical properties has been previously reported, where irregular particles (nonspherical) will show two or more surface plasmon bands, which will result in the widening of the band; this is due to the fact that as the reduction process begins, the nucleation and the generation of the primary particles will take place, which will have irregular shapes and smaller diameters, explaining the blue shift of the band in the initial solutions and a red shift in the final solutions where the process of nucleation and generation of the particles is concluded [17, 18].

On the other hand, due to the fact that the nanoparticle formation process is interrupted, there will possibly be several particle sizes influencing the excitation surface plasmon peak. Finally, another phenomenon involved to this fact is the band of absorption of the extract, which is at 435 nm, affecting the band position and possibly influencing the red shift. In relation to this, the preservation of the organic agent on the surface of silver nanoparticles evidenced its role as a functionalizing agent.

3.1.2. Morphological Characterization: TEM. Figure 2(a) presents the TEM images and histograms of the synthetized materials. AgNPs present a quasi-spherical morphology arranged in isolated clusters. The differential distribution corresponding to this sample displays a particle size range between 4.54 and 16.48 nm with an asymmetrical geometry and a bimodal distribution (Figure 2(b)). The average particle size and coefficient of variation (CV%) calculated for this sample were 10.87 nm and 22.94%, respectively. The statistical parameters obtained are shown in Table 2.

3.1.3. Dynamic Light Scattering (DLS) and Zeta Potential. Table 3 displays the values of particle size and Z-potential obtained for the particles. The acquired value for particle size was 16.46 ± 0.46, slightly differing from the obtained diameter mean of 10.87 from TEM images analysis, indicating the association of the value corresponding to DLS analysis to the hydrodynamic radius of the particles and corroborating the results shown in size statistical analysis. The measured Z-potential of the sample displays a value of −27.3 ± 1.22 mV, which represent an electrostatic repulsion between the

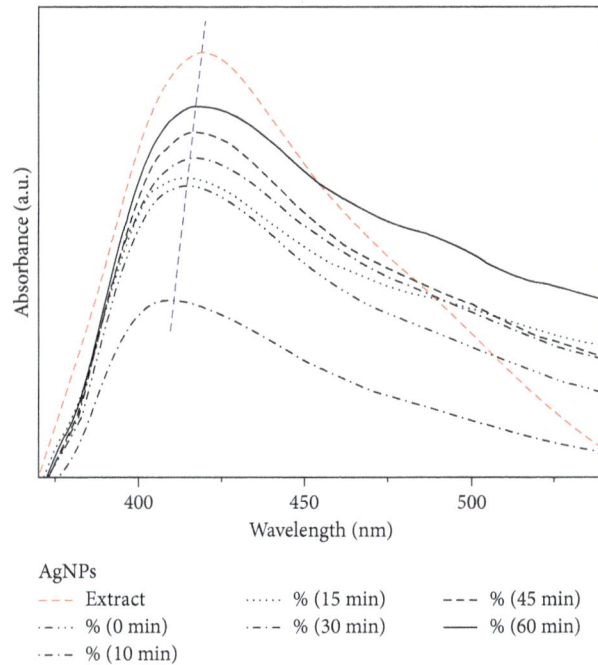

AgNPs
- - - Extract   ······ % (15 min)   - - - % (45 min)
·-··· % (0 min)   ·-··· % (30 min)   —— % (60 min)
·-·- % (10 min)

Figure 1: UV-Vis absorption spectra of silver nanoparticles at different time periods and their reaction with *A. muricata* extract at 1000 μL.

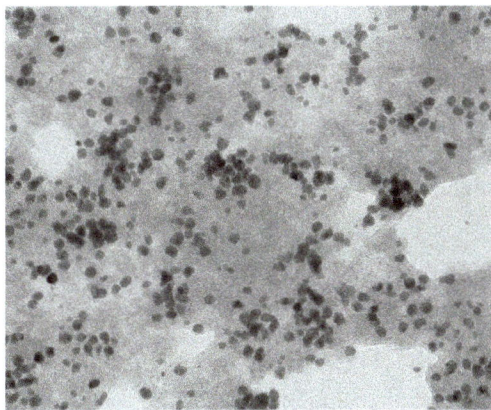

Ag 9.tif
Print mag: 293000x @ 7,0 in   HV = 100 kV
11:21 02/01/17   Direct mag: 40000x
TEM mode: imaging   AMT camera system

(a)

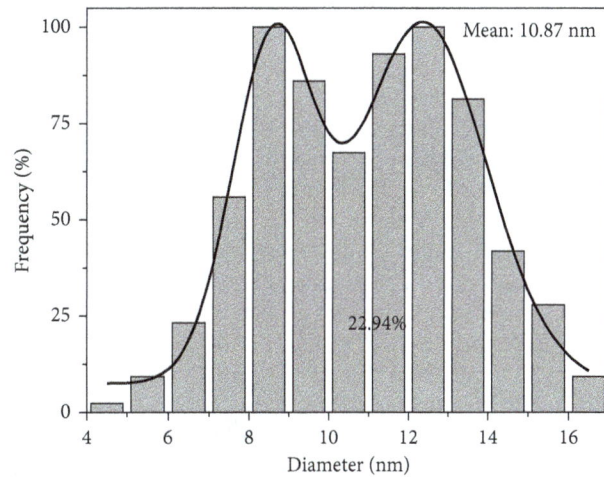

(b)

Figure 2: (a) TEM photomicrography and (b) differential size distribution of silver nanoparticles biosynthesized using *A. muricata* extract.

Table 2: Size distribution statistical parameters of ZVI materials in nm.

Sample	Mean	CV (%)	D10	D50	D90	D50-D10
AgNPs	10.87	22.94	7.12	10.54	13.72	3.42

Table 3: Dynamic light scattering (DLS) and zeta potential of Ag sample.

Sample	Particle size (nm)	Z-potential (mV)
AgNPs	16.46 ± 0.46	-27.3 ± 1.22

particles as shown in the TEM image corresponding to this sample, where most of the NPs are well dispersed (Table 3).

3.1.4. X-Ray Diffraction. The X-ray diffractogram corresponding to AgNPs presents the peaks associated with the

Ag cubic phase at 38.23, 44.18, 64.74, 75.51, and 81.74°, which can be indexed as the (111), (200), (220), (311), and (222) planes (JCPDS File No.: 04–0783), respectively, illustrating with this information the presence of crystalline silver nanoparticles (Figure 3).

3.2. Antimicrobial Activity of AgNPs. In this study, MIC values were obtained for the AgNPs tested against *E. coli* (ATCC 25922), *S. aureus* (ATCC 29213), and *E. faecalis* (ATCC 292129). The results are presented in Table 4, where the MIC of the AgNPs synthesized by *Annona muricata* presents a lower antibacterial activity in comparison with 5-fluoracil, but the functionalization/combination of the silver nanoparticles with 5-fluoracil showed a synergism because all the strains were inhibited at a concentration of 1.95 Dg/ml or less. With respect to the tree strains tested with AgNPs and 5-fluoracil, *Enterococcus faecalis,* which is a facultative anaerobic Gram-positive coccus, shows the lowest sensitivity in comparison with *E. coli* and *S. aureus* [19]; A. Manten and J. I. Terra were the first to report the antibacterial activity of antineoplastic drugs as well as the combination effects between the antibacterial and antineoplastic agents [19]. The results corresponding to 5-fluoracil alone displayed similar results compared to the data reported by Gieringer et al. [20], who found that concentrations of 0.8 μg/ml or less inhibit all strains of *Staphylococcus aureus* [20].

3.3. Cytotoxicity Assay

3.3.1. MTT Assay. The cell viability was evaluated by the cytotoxicity test and fluorescent microscopy in order to compare the effects produced on oral fibroblasts by green AgNPs and the presence and absence of 5-FU as the functionalization agent on the nanomaterials.

The results corresponding to the cytotoxicity test show slight differences in cell viability between 24 and 48 hrs of cell exposition to AgNPs. It is possible to observe toxicity for silver nanoparticles at the highest concentrations used (Figure 4). AshaRani et al. [21] report similar results for lung fibroblasts in response to silver NPs exposition, arguing a cytotoxicity dependence on NPs concentration as well as cell-cycle detection [21]. Additionally, Ahmad et al. [22] evaluated the effect of silver and gold nanoparticles obtained by the green synthesis method in murine macrophages using concentrations in the range of 10–1000 μg/mL, obtaining less cytotoxicity at a concentration lower than 80 μg/mL [22]. Furthermore, the information presented illustrates a cytotoxicity-effect reduction of the nanomaterials by 5-FU functionalization, evidencing the biocompatibility improvement of the nanobiomaterials.

3.3.2. Fluorescence Microscopy. Corresponding to the fluorescence microscopy analysis, it was observed that, at a lower concentration of treatments, there was more confluence and cell density compared to the control group (Figures 5 and 6). On the other hand, as concentrations increased, alterations in cells morphology and few cell extensions were observed. Moreover, an important characteristic observed at 24 is the presence of few dead cells for AgNPs exposition, indicating the role of the bioextract as a possible cytoprotective agent since it encapsulates the nanoparticles and 5-FU, preventing direct contact with cells, and additionally, the cell cultures remained viable with only

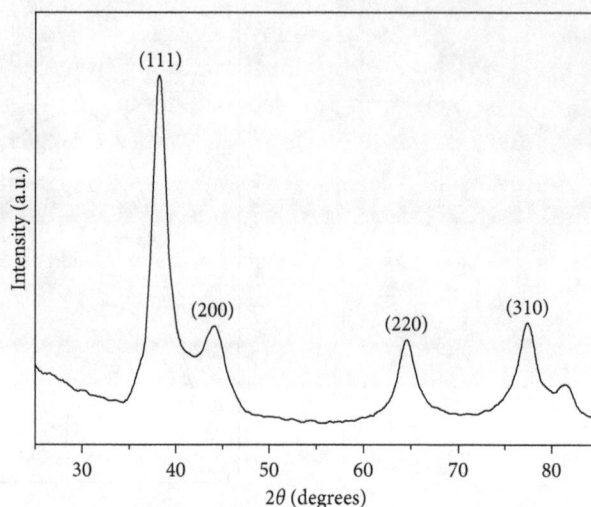

FIGURE 3: X-ray diffraction patterns of the Ag sample.

TABLE 4: Minimum inhibitory concentrations of Ag nanomaterials.

Sample	MIC of silver nanoparticles (mg/ml)		
	E. coli (ATCC 25922)	Bacterial strains *S. aureus* (ATCC 29213)	*E. faecalis* (ATCC 29212)
AgNPs	6.68 ± 0.0	13.36 ± 0.0	26.72 ± 0.0
5-Fluorouracil	7.8 ± 0.0	7.8 ± 0.0	15.62 ± 0.0
AgNPs + 5-FU	1.95 ± 0.0	0.97 ± 0.0	0.97 ± 0.0
Annona muricata extract	–a	–a	–a
Amikacin	1 ± 0.0	2 ± 0.0	128 ± 0.0

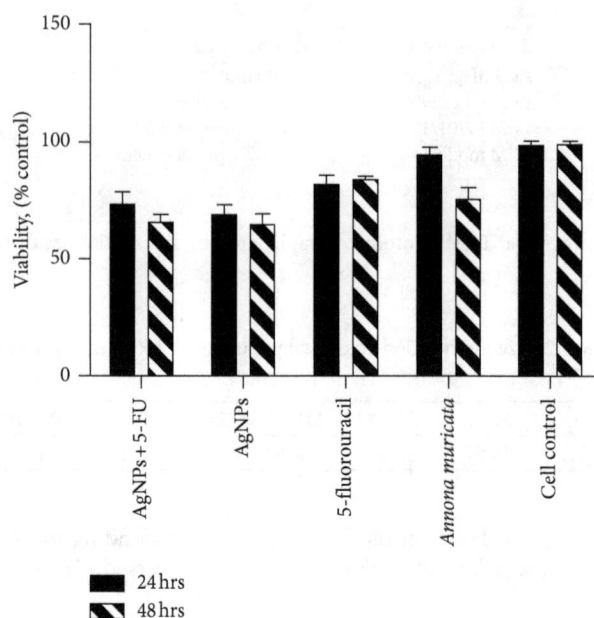

FIGURE 4: Graph of cytotoxicity of fibroblasts exposed to different concentrations of Ag nanoparticles for 24 and 48 hours.

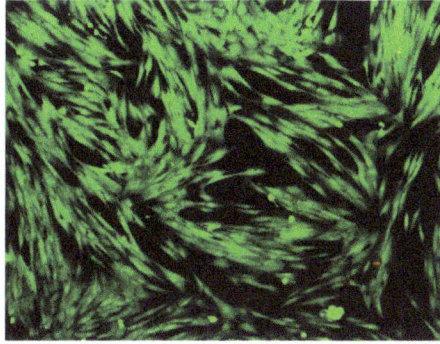

FIGURE 5: Fluorescent microscopy image of the control group of fibroblasts at 24 hrs.

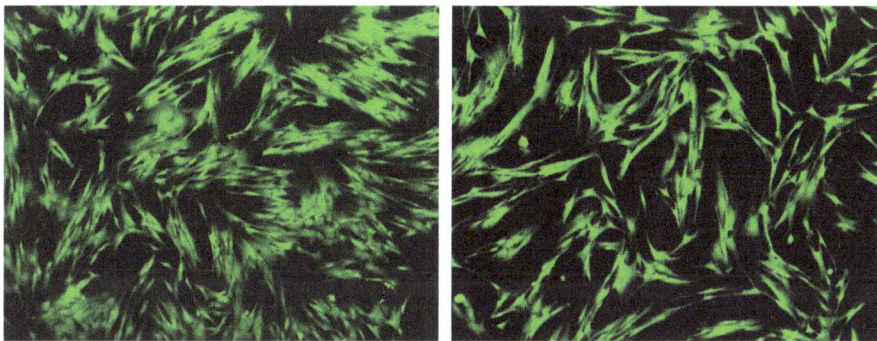

(a)

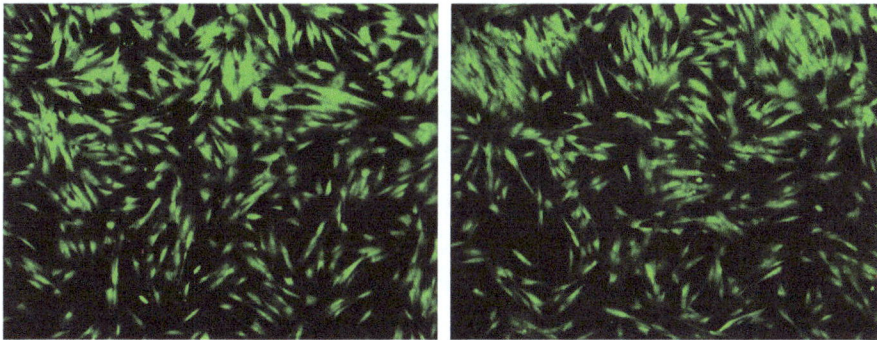

(b)

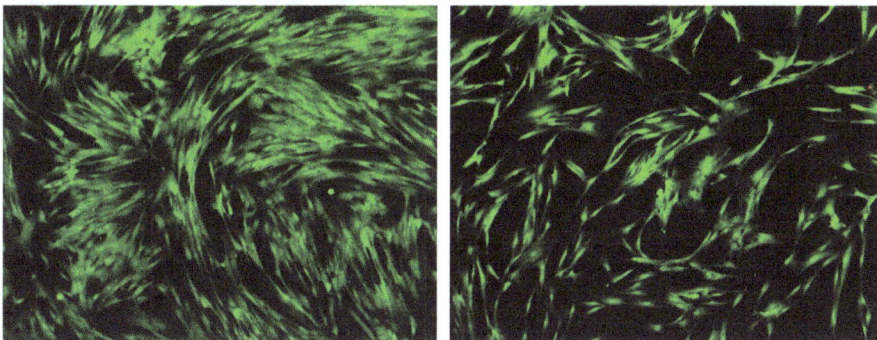

(c)

FIGURE 6: Continued.

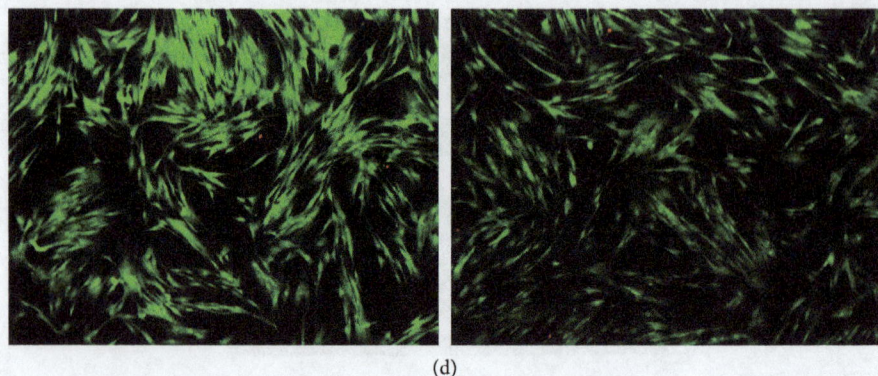

(d)

FIGURE 6: Fluorescent microscopy images of cytotoxicity assays in fibroblasts at 24 hrs and with respective concentrations at which it was evaluated (lowest and highest). (a) *A. muricata*. (b) 5-FU group. (c) AgNPs group. (d) Group of AgNPs and 5-FU.

the presence of the extract (Figures 6(b)–6(d)). Similarly, AshaRani et al. [21] previously reported different morphological changes, indicating unhealthy cells, few cell extensions, and restricted propagation patterns with respect to control [21].

4. Conclusions

In this work, it was demonstrated that, in the green synthesis of silver nanoparticles using *Annona muricata* as a bioreducer, the obtained NPs were characterized in terms of their optical properties, crystallinity, morphology, hydrodynamic radius, and surface charge. The UV-Vis monitoring of the AgNPs formation displayed the nucleation process of the particles and the increasing intensity of the silver characteristic band at 435 nm depending on time reaction. The nanoparticles presented a quasi-spherical shape with an average particle size of 10.87 nm and a hydrodynamic radius of 16.46 ± 0.46 nm. The Z-potential obtained had a value of -27.3 ± 1.22 mV, demonstrating repulsion between the particles and good colloidal stability of the material. The antimicrobial properties of the materials showed a great inhibition against Gram-positive and Gram-negative bacteria. The cytotoxicity of the NPs at 24 and 48 hrs displayed an increment in cell viability associated with the particles functionalization by 5-FU, and only a few dead cells at 24 hrs were observed in the fluorescence microscopy images.

Acknowledgments

This study was supported by Consejo Nacional de Ciencia y Tecnología (CONACyT). DeAlba-Montero would like to thank CONACyT for Scholarship no. 358436, and Claudio Adrian Ruiz-Torres would like to thank to CONACyT for its support with Scholarship no. 412287.

References

[1] Y.-Y. Tseng, Y.-C. Kau, and S.-J. Liu, "Nanomedicine-the future of cancer treatment: a review," *Expert Opinion on Drug Delivery*, vol. 13, no. 11, pp. 1533–1544, 2016.

[2] K. Rani, "Biomedical applications of silver and gold nanoparticles: effective and safe non-viral delivery vehicles," *Journal of Applied Biotechnology and Bioengineering*, vol. 3, no. 2, article 00059, 2017.

[3] D. Dhanyalayam, L. Scrivano, O. Ilaria Parisi et al., "Biopolymeric self-assembled nanoparticles for enhanced antibacterial activity of Ag-based compounds," *International Journal of Pharmaceutics*, vol. 517, no. 1, pp. 395–402, 2017.

[4] M.-A. Neouze and U. Schubert, "Surface modification and functionalization of metal and metal oxide nanoparticles by organic ligands," *Monatshefte Für Chemie-Chemical Monthly*, vol. 139, no. 3, pp. 183–195, 2008.

[5] C. Andreadis, K. Vahtsevanos, T. Sidiras, I. Thomaidis, K. Antoniadis, and D. Mouratidou, "5-Fluorouracil and cisplatin in the treatment of advanced oral cancer," *Oral Oncology*, vol. 39, no. 4, pp. 380–385, 2003.

[6] L. Nair, J. Sankar, S. A. Nair, and G. V. Kumar, "Biological evaluation of 5-fluorouracil nanoparticles for cancer chemotherapy and its dependence on the carrier, PLGA," *International Journal of Nanomedicine*, vol. 6, pp. 1685–1697, 2011.

[7] D. Dhanya, P. Giuseppe, A. R. Cappello et al., "Phosphonium salt displays cytotoxic effects against human cancer cell lines," *Anti-Cancer Agents in Medicinal Chemistry*, vol. 17, no. 13, 2017.

[8] D. MubarakAli, N. Thajuddin, K. Jeganathan, and M. Gunasekaran, "Plant extract mediated synthesis of silver and gold nanoparticles and its antibacterial activity against clinically isolated pathogens," *Colloids and Surfaces B: Biointerfaces*, vol. 85, no. 2, pp. 360–365, 2011.

[9] P. Raveendran, J. Fu, and S. L. Wallen, "Completely "green" synthesis and stabilization of metal nanoparticles," *Journal of the American Chemical Society*, vol. 125, no. 46, pp. 13940-13941, 2003.

[10] V. Ganesh Kumar, S. Dinesh Gokavarapu, A. Rajeswari et al., "Facile green synthesis of gold nanoparticles using leaf extract of antidiabetic potent cassia auriculata," *Colloids and Surfaces B: Biointerfaces*, vol. 87, no. 1, pp. 159–163, 2011.

[11] P. Tippayawat, N. Phromviyo, P. Boueroy, and A. Chompoosor, "Green synthesis of silver nanoparticles in

aloe vera plant extract prepared by a hydrothermal method and their synergistic antibacterial activity," *PeerJ*, vol. 4, article e2589, 2016.

[12] S. Patra, S. Mukherjee, A. Kumar Barui, A. Ganguly, B. Sreedhar, and C. R. Patra, "Green synthesis, characterization of gold and silver nanoparticles and their potential application for cancer therapeutics," *Materials Science and Engineering C*, vol. 53, pp. 298–309, 2015.

[13] M. V. Sujitha and S. Kannan, "Green synthesis of gold nanoparticles using citrus fruits (citrus limon, citrus reticulata and citrus sinensis) aqueous extract and its characterization," *Spectrochimica Acta-Part A: Molecular and Biomolecular Spectroscopy*, vol. 102, pp. 15–23, 2013.

[14] S. Z. Moghadamtousi, M. Fadaeinasab, S. Nikzad, G. Mohan, H. M. Ali, and H. Abdul Kadir, "*Annona muricata* (Annonaceae): a review of its traditional uses, isolated acetogenins and biological activities," *International Journal of Molecular Sciences*, vol. 16, no. 7, pp. 15625–15628, 2015.

[15] S. B. Santhosh, R. Yuvarajan, and D. Natarajan, "*Annona muricata* leaf extract- mediated silver nanoparticles synthesis and its larvicidal potential against dengue, malaria and filariasis vector," *Parasitology Research*, vol. 114, no. 8, pp. 3087–3096, 2015.

[16] Testing Susceptibility, *M100-S24 Performance Standards for Antimicrobial*, SAI Global, Sydney, NSW, Australia, 2013.

[17] G. A. Martinez-Castanon, N. Niño-Martínez, F. Martínez-Gutierrez, J. R. Martínez-Mendoza, and F. Ruiz, "Synthesis and antibacterial activity of silver nanoparticles with different sizes," *Journal of Nanoparticle Research*, vol. 10, no. 8, pp. 1343–1348, 2008.

[18] S. Pal, K. T. Yu, and J. M. Song, "Does the antibacterial activity of silver nanoparticles depend on the shape of the nanoparticle? A study of the gram-negative bacterium *Escherichia coli*," *Journal of Biological Chemistry*, vol. 290, no. 42, pp. 1712–1720, 2015.

[19] A. Manten and J. I. Terra, "Some observations on antagonism between penicillin and antineoplastic antibiotics," *Acta Physiologica et Pharmacologica Neerlandica*, vol. 14, no. 3, pp. 250–258, 1967.

[20] J. H. Gieringer, A. F. Wenz, H. M. Just, and F. D. Daschner, "Effect of 5-fluorouracil, mitoxantrone, methotrexate, and vincristine on the antibacterial activity of ceftriaxone, ceftazidime, cefotiam, piperacillin, and netilmicin," *Chemotherapy*, vol. 32, no. 5, pp. 418–424, 1986.

[21] P. V. AshaRani, G. L. Kah Mun, M. P. Hande, and S. Valiyaveettil, "Cytotoxicity and genotoxicity of silver nanomaterials," in *Proceedings of Technical Proceedings of the 2009 NSTI Nanotechnology Conference and Expo NSTI-Nanotech*, pp. 383–386, Houston, TX, USA, May 2009.

[22] A. Ahmad, F. Syed, A. Shah et al., "Silver and gold nanoparticles from *Sargentodoxa cuneata*: synthesis, characterization and antileishmanial activity," *RSC Advances*, vol. 5, no. 90, pp. 73793–73806, 2015.

5

A Review on Platelet Activating Factor Inhibitors: Could a New Class of Potent Metal-Based Anti-Inflammatory Drugs Induce Anticancer Properties?

Vasiliki D. Papakonstantinou,[1] Nefeli Lagopati,[2] Effie C. Tsilibary,[2]
Constantinos A. Demopoulos,[1] and Athanassios I. Philippopoulos[3]

[1]Laboratory of Biochemistry, Faculty of Chemistry, National and Kapodistrian University of Athens, 15771 Athens, Greece
[2]Institute of Biosciences & Applications, NCSR Demokritos, 15310 Agia Paraskevi, Greece
[3]Laboratory of Inorganic Chemistry, Department of Chemistry, National and Kapodistrian University of Athens,
 Panepistimiopolis Zografou, 15771 Athens, Greece

Correspondence should be addressed to Athanassios I. Philippopoulos; atphilip@chem.uoa.gr

Academic Editor: Concepción López

In this minireview, we refer to recent results as far as the Platelet Activating Factor (PAF) inhibitors are concerned. At first, results of organic compounds (natural and synthetic ones and specific and nonspecific) as inhibitors of PAF are reported. Emphasis is given on recent results about a new class of the so-called metal-based inhibitors of PAF. A small library of 30 metal complexes has been thus created; their anti-inflammatory activity has been further evaluated owing to their inhibitory effect against PAF in washed rabbit platelets (WRPs). In addition, emphasis has also been placed on the identification of preliminary *structure-activity* relationships for the different classes of metal-based inhibitors.

1. Introduction

This work was motivated by the fact that a comprehensive survey on metal-based potent inhibitors of PAF as active anti-inflammatory drugs has not been previously described in the literature. It is the aim of this microreview to reveal the critical role of the metal center and of the molecular structure (different coordination geometries) of the relevant metal complexes within this series of new *metal-based* potent inhibitors of PAF. Biological results of these compounds are reviewed and added to the dataset base of inorganic metal-based anti-inflammatory drugs. The review is divided into two general parts. At first, the general characteristics of PAF are described, followed by selected, known organic inhibitors of PAF. In the second part, the structural characteristics and the biological activity against PAF, of different classes of metal-based inhibitors, are presented.

2. General Characteristics of the Platelet Activating Factor

2.1. Platelet Activating Factor, Structure, and Activity. Platelet Activating Factor (PAF) has been characterized as a new, ubiquitous, potent, and unique class of lipid chemical mediators that share similar biological activities, namely, PAF-like activity molecules [1]. Originally, the term PAF was meant to be one phosphoglycerylether lipid, identified as 1-O-alkyl-2-acetyl-*sn*-glycero-3-phosphocholine with 16–18 carbons at the ether-bonded fatty chain at the sn-1 position of the glycerol backbone, as shown in Figure 1 [2]. Interestingly, while most of the ether lipids replaced their ether bond with esterified analogues through time, PAF conserved its ether bond in sn-1 position because of its important biological role [3].

PAF is produced by a plethora of cells as platelets, neutrophils, monocytes/macrophages, lymphocytes, basophiles,

TABLE 1: IC$_{50}$ (μM, final concentration) of various PAF inhibitors that induces 50% inhibition against PAF aggregation in washed rabbit platelets (WRPs). IC$_{50}$ (μL) values are expressed in microliters of initial volume of oil that induces 50% inhibition against 1×10^{-11} M PAF final concentration in the aggregometer cuvette.

PAF inhibitor	IC$_{50}$	Reference
Ginkgolides		
BN 52021	3.6	[13]
BN 52020	9.7	[13]
BN 52022	38	[13]
Alpha-bulnesene	0.024	[17]
Cedrol	13	[29]
Paricalcitol	0.3	[100]
Di-hydroxy-chimyl-alcohol (pine pollen)	4.5	[27]
Olive oil	6.5 μL	[74]
Sunflower	21 μL	[88]
Sesame oil	150 μL	[88]

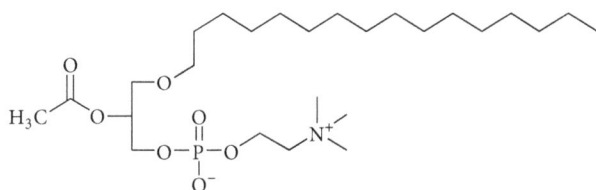

FIGURE 1: PAF molecular structure.

eosinophils, mast cells, and endothelial cells and has many biological roles, both physiological and pathological depending mainly on the extent of PAF production and enzymatic regulation [4]. Under physiological conditions, the production of PAF is regulated by its enzymes [5] and the produced PAF participates in physiological processes as reproduction, memory formation, vascular tone, apoptosis, and angiogenesis. However, in pathological conditions, excess amounts of PAF can cause inflammation and lead to inflammatory conditions or diseases, such as allergy, asthma, atherosclerosis, diabetes, renal diseases, cancer, HIV pathogenesis, and periodontitis [6].

2.2. PAF Receptor. PAF exerts its autocrine and paracrine actions through binding to a well-characterized G-protein coupled receptor (GPCR) located on the plasma membrane of a wide variety of mammalian cells such as endothelial cells, neutrophils, monocytes, dendritic cells, platelets, and leukocytes [7]. There are few studies that prove the existence of PAFR in endomembranes as well. The first one revealed three distinct classes of binding sites for PAF. Two of them were found on the microsomes and the third one was on the synaptic plasma membranes of rat cerebral cortex [8]. Later, other studies certified the intracellular existence of PAFR as well as the presence of PAF binding sites in the nucleus, both at the nuclear envelope and at nuclear matrix [9–11].

Finally, it seems that cytoplasmic PAFR is implicated in the acute effects of PAF, while the intracellular PAFR intermediates the long-standing effects of PAF via the modulation of gene expression.

The interaction of PAF with its receptor (PAFR) initiates a cascade of diverse intracellular signaling pathways, which translate the mediator's message to the final cell response [12]. The signal transduction pathways activated by PAF are cell and species dependent and are regulated by several mechanisms. These include covalent modification and the internalization of PAFR, the modulation of PAFR gene expression, and binding potential along with the fine-tuning of the signal transduction pathways by a plethora of intracellular and extracellular modulators [6].

2.3. PAF Inhibitors. There are hundreds of molecules that are able to inhibit biological actions, mediated by PAF. Most of them have been screened in vitro in cell systems and animal models and the ones with the best properties, in terms of potency, bioavailability, and safety, have been also tested in clinical trials. There are many ways to classify PAF inhibitors such as (i) their source, isolated from natural sources or chemically synthesized; (ii) their chemical structure, nitrogen heterocyclic compounds, PAF analogues (PAF-like molecules), dihydropyridines, natural medicines, and others; or (iii) the way they interact with PAF receptor, specific or nonspecific inhibitors. IC$_{50}$, of various PAF inhibitors against PAF aggregation in washed rabbit platelets, are presented in Table 1. In the present minireview, PAF inhibitors will be divided into a new classification method as organic and inorganic.

3. Organic PAF Inhibitors

3.1. Phytochemical Products. There are several hundreds of phytochemical products that can inhibit PAF biological activities and therefore act as anti-inflammatory agents. The purpose of this review is not to list all of them, so below are some of the most promising ones for therapeutic use or dietary supplements (Figure 2). Ginkgolides are diterpenes with a structure of twenty carbon cage molecules that come from the leaves and the roots of an ancient Chinese tree called *Ginkgo biloba* and belong to the natural specific inhibitors which antagonize the binding to PAF's membrane receptor by

Ginkgolide	R_1	R_2	R_3
A	H	H	OH
B	OH	H	OH
C	OH	OH	OH
J	H	OH	OH
M	OH	OH	H

Ginkgolide	R
P	H
Q	OH

Ginkgolide	R
K	OH
L	H

(a)

α-Bulnesene

(b)

Andrographolide

(c)

Kadsurenone

(d)

L-652,469/tussilagone

(e)

Yangambin

(f)

Cedrol

(g)

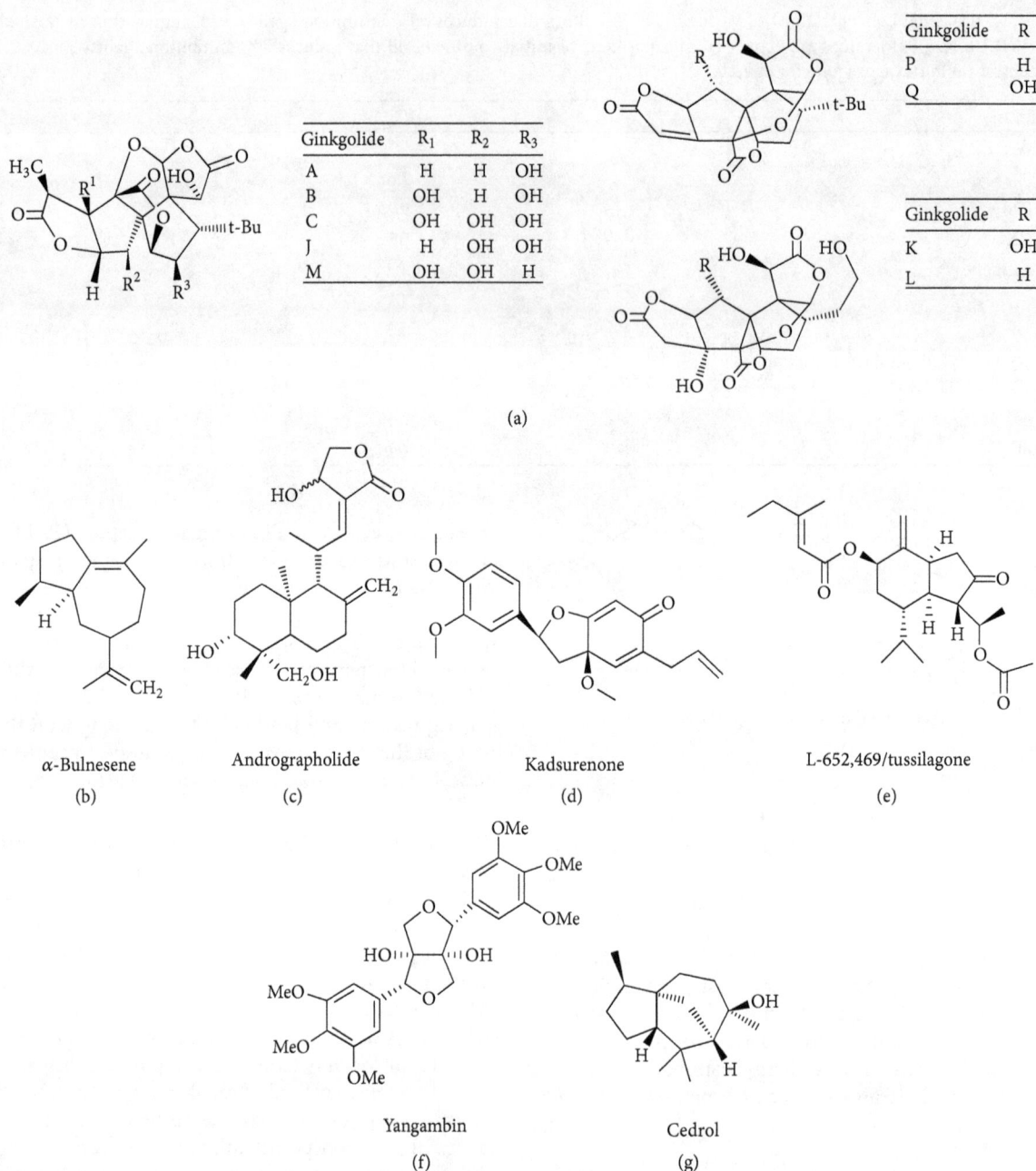

FIGURE 2: Phytochemical products with anti-PAF activity.

a competitive way. Based on the molecular dynamics simulations, the binding of PAF to PAFR leads to its activated state, while the binding of *Ginkgo biloba* locks PAFR in its inactive state. There is a plethora of pathological conditions, where PAF is implicated, and *Ginkgo biloba* extracts have managed to ameliorate, as cognitive disorders, HIV infection, ischemia, tissue injuries, cancer, and airway diseases as asthma and allergy [13].

Andrographolide has a labdane diterpenoid structure and it is obtained from the stems and the leaves of a Asian plant called *Andrographis paniculata* which was traditionally used

against viral infections and inflammatory diseases [14]. The studies around its inhibition against PAF have revealed that it is a nonspecific PAF inhibitor [15] with a plethora of promising effects as anti-inflammatory, anticancer, immunomodulatory, antiviral, and cardioprotective [16].

Alpha-bulnesene is a sesquiterpenoid molecule that comes from the essential oil of an Asian plant called *Pogostemon cablin* and acts as a PAF inhibitor exerting its inhibitory effect by antagonizing competitively PAF binding to its receptor and thus inhibits intracellular Ca^{2+} increase [17]. As it is a potent anti-inflammatory agent, it may be

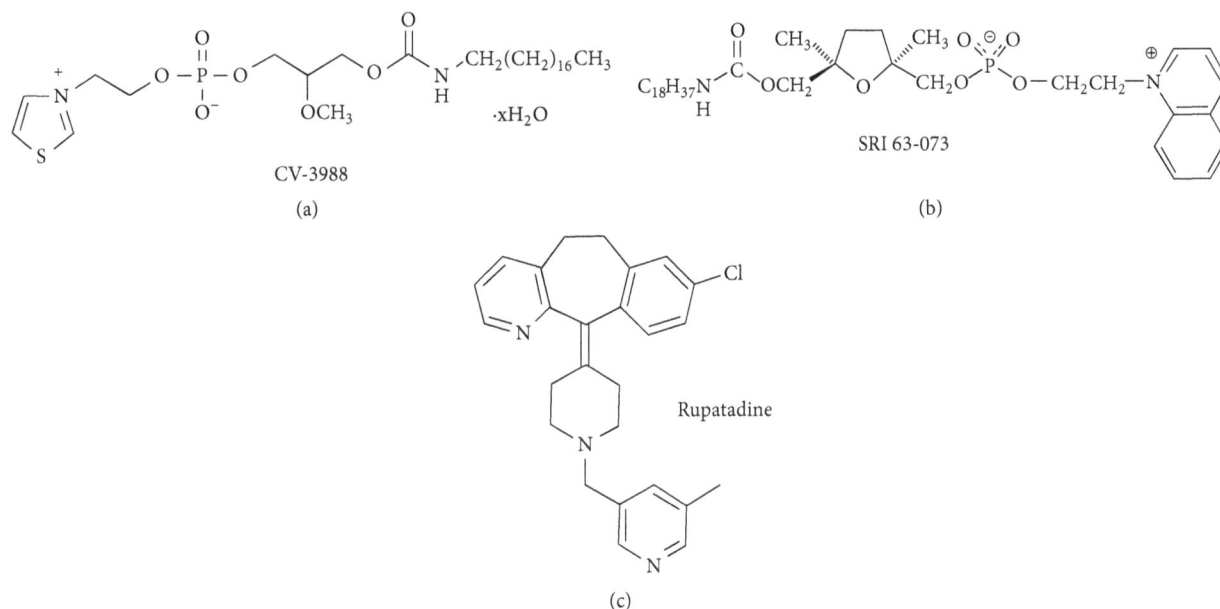

FIGURE 3: Synthetic PAF inhibitors.

used for the inhibition or prevention of allergic syndromes, providing new antiallergic medicine that exhibits fewer side effects [18].

Kadsurenone is the first potent PAF antagonist that was discovered and comes from the Chinese herb *Piper futokadsurae*. It is a benzofuranoid neolignan whose structure was used as a template for the development of synthetic PAF antagonists as well as for the molecular modeling of PAFR [19].

Tussilagone (or L-652,469) is a terpene that comes from the buds of the plant *Tussilago farfara* and acts as a nonspecific PAF inhibitor by blocking the calcium channels [20]. It is extensively used in inflammatory respiratory diseases including cough, asthma, and acute/chronic bronchitis because of its anti-inflammatory properties [21].

Yangambin is a natural furofuran lignan, isolated from the Brazilian plant *Ocotea duckei Vattimo*. Studies have revealed that it is a specific PAF inhibitor which antagonizes competitively PAF binding to its receptor [22]. It displays beneficial actions against allergy and headache [23, 24] along with its cardioprotective role [25].

Lipid extracts from the plant *Urtica dioica*, the common nettle, were found to contain both PAF and PAF-like activity molecules and glycolipid derivatives which act as PAF inhibitors and hence explain the sense of urticaria and the beneficial effects of nettle as traditional allergy relief and diuretic remedy [26].

The lipid molecule, di-hydroxy-chimyl-alcohol, derived from pine pollen, was also found to exert anti-PAF action [27].

Cedrol is a sesquiterpene alcohol which is found as an essential oil in many plants as *Biota orientalis*, *Pterocarpus indicusirginia*, in various types of conifers as *Cupressus* and *Juniperus*, and in *Origanum onites* [28]. It is one of the well-known PAF antagonists [29].

3.2. Synthetic Products. The promising results of natural PAF inhibitors evoke the need for the development of synthetic PAF inhibitors. The first molecules that were synthesized had similar chemical structure with PAF, meaning a glycerol backbone such as CV-3988 [30], CV-6209 [31], ONO-6240 [32], and Ro 19-3704 [33]. The idea that followed was to replace the glycerol backbone with a cyclic structure such as SRI 63-073 [34], SRI 63-441 [35], UR-11353 [36], and CL-184,005 [37] (Figure 3).

Later, the PAF inhibitors that were synthesized had no similar structure to PAF. These molecules contain heterocyclic structures which are characterized by sp^2 nitrogen atom that is able to interact with PAFR as a hydrogen bond acceptor. These kinds of inhibitors are pyrrolothiazole-related antagonists as tulopafant [38], thiazolidine derivatives as SM-10661 [39], imidazolyl derivatives as modipafant [40], and lexipafant [41] and hetrazepine derivatives as WEB-2086 and WEB-2170 [42]. All the above synthetic antagonists display a great variability in their chemical structure which might have importance in their different pharmacological profile.

The success of the in vitro studies of PAF inhibitors in combination with the use of many natural PAF inhibitors as traditional remedies for inflammatory diseases had been very promising for the use of PAF inhibitors in the clinical practice. Many PAF inhibitors were tested in clinical trials displaying tolerability and safety but with no effectiveness. Modipafant [40], UK-74,505 [43], WEB 2086 [44], and SR27417A [45, 46] showed no effect against asthma disease. Lexipafant, one of the most potent PAF inhibitors, was tested against cognitive impairment [47], myocardial infraction [48], sepsis [49], and organ failure in severe pancreatitis [50] with no significant results. Moreover, SR27417A did not decrease the symptoms of acute ulcerative colitis [51]; BN 50730 showed no significant amelioration of rheumatoid arthritis [52] as well as Ro 24-238 against psoriasis [53] and TCV-309 against septic shock [54]. However, there are also trials with positive results such as UK,74505 on airway and systemic responses [55], levocetirizine in chronic idiopathic urticaria [56], WEB 2086 gel against UVB-induced dermatitis [57], BN 52021 on pulmonary function in the early postischemic period [58], and Y-24180 in bronchial hyper-responsiveness in patients with asthma [59]. Last but not least comes the use of rupatadine against allergic rhinitis and several other allergic disorders [60], the positive outcomes of which resulted in the production of the first circulating drug based on a PAF inhibitor, named Rupafin.

In addition, molecules demonstrating dual antagonistic actions against PAF as well as against another inflammatory mediator have been synthesized and their therapeutic properties have been studied in vitro. These molecules inhibit both PAF actions and 5-lipoxygenase, as LDP-392 [61], or thromboxane synthase [62], or iNOS induction [63]. Rupatadine, as well, is both an oral PAFR antagonist and a histamine H(1)-receptor antagonist [64].

As far as it concerns already used pharmaceutical products, there are many of them that were found to exhibit potent inhibitory effect against PAF actions, which is expected as Platelet Activating Factor is implicated in inflammatory processes and therefore in a plethora of pathological conditions with inflammatory background. PAF is known to be implicated in atherosclerosis [65] and many statins as well as digoxin, which are used for their cardioprotective properties, are also PAF inhibitors [66, 67]. Many of the antiretrovirals used for HIV infection are also PAF inhibitors, both in vitro and in vivo [68–72].

3.3. Diet Products. The regulation of PAF actions is also succeeded by the daily ingestion of PAF inhibitors which are consumed through our diet (see Figures S1 and S2, in the Supplementary Material available online at https://doi.org/10.1155/2017/6947034). Among all diets, the Mediterranean diet consists of numerous products that exert potent anti-PAF activities in the washed platelet aggregation assay, such as virgin or regular olive oil [73, 74], wine and grapes [75–78], honey [79], fish [80–82], milk and yoghurt [83], and traditional Greek meals [84], the isolation and structural characterization of which demonstrated the presence of complex polar lipids as glycolipids and ganglioside derivatives, phenolics, and phenol glucosides [65]. Furthermore, from the acetylation of poplar lipids and phenolic compounds derived acetylated products with anti-PAF action [85, 86], especially, olive oil was found to contain high levels of PAF antagonists among other vegetable oils and the structure of its most active fraction was found to be a glycerolglycolipid [74]. In another study, olive pomace, crude olive pomace oil, and waste byproducts from olive pomace oil production were found to contain polar lipids as bioactive compounds that inhibit or antagonize PAF. The most bioactive compound came from olive pomace and has been chemically characterized as a glycerylether-sn-2-acetyl glycolipid [87]. Moreover, microconstituents of several seed oils, such as sesame, corn, and sunflower, were found to have either inhibitors or agonists of PAF. The most active agonist came from corn oil and was found to be almost five orders of

magnitude less active than PAF, meaning that its action, in several cells and/or tissues, through PAF receptors, would minimize the biological effects of PAF [88]. In vivo, polar extracts from olive oil and olive pomace have reduced the atheromatic lesion in hypercholesterolemic rabbits [89, 90] and mellitus patients reduced platelet sensitivity after one month of traditional Mediterranean diet [84], while wine consumption improved platelet sensitivity independently of alcohol [91]. Considering the expanding role of PAF and other lipid mediators in human pathophysiology, the study of their dietary modulation is of increasing scientific interest. Today, a plethora of studies report molecules that inhibit the cellular effects of PAF in vitro and in experimental animal models or human studies in vivo, which are well presented in the review of Nomikos et al. [92].

Apart from Mediterranean products, PAF inhibitors can also be found in many other products as garlic [93–95], soy sauce [96], and tea [97]. Last but not least comes curcumin which exerts potent antiplatelet activity through the inhibition of COX and the blockade of calcium signaling. Additionally, curcumin managed to reduce colonic mucosal and tumor PLA_2 [98, 99].

Referring to vitamins, paracitol (Vitamin D) is known to improve the inflammatory status of hemodialysis patients, an ability possibly due to its anti-PAF activities [100] and also tocopherol (Vitamin E) inhibits PAF-induced aggregation [101, 102]. Besides, there are endogenous PAF inhibitors as blood cardiolipin which was shown to control PAF actions [103].

The involvement of PAF in cancer is already known and especially in carcinogenesis, melanoma, or metastasis [104–107]. Also, various compounds with anti-PAF activities, as *Ginkgo biloba* extracts [108, 109], andrographolide [110, 111], tussilagone [112], WEB 2170 [113], and WEB 2086 [107], have been shown to exert anticancer properties in vitro. In addition, both cisplatin and WEB 2086 administered to animal models decreased dramatically tumor growth [114]. This manuscript studies whether a new class of potent metal-based anti-inflammatory drugs could induce indirect anticancer properties.

The hypothesis claims that the tumor cells express PAF receptor and therefore there are high PAF levels all around the microenvironment of the tumor. The use of anti-PAF compounds decreases PAF levels ergo inflammation and therefore enhances the anticancer properties of the chemotherapy. Moreover, the use of PAF inhibitors, such as metal compounds, which may have additional direct anticancer properties, will amplify the effectiveness of the common anticancer treatments.

4. Inorganic Metal-Based Inhibitors

As mentioned in the previous part, organic compounds represent well-known antagonists in the area, while application of metal-based compounds as PAF inhibitors has been systematically ignored, although the advantages of metal-based therapeutics over the organic analogues have been well addressed recently [115]. In fact, transition metal coordination compounds display a number of different coordination numbers and geometries (square planar, tetrahedral, and octahedral) as opposed to the typical tetrahedral geometry for the organic congeners. Moreover, metal complexes present a number of special characteristics such as wide structural diversity, the possibility of tuning thermodynamic and kinetic ligand substitution, and different oxidation states [116–118]. In addition, a number of metal ions, playing a vital role for life, are involved in many natural biological processes [119–122]. As a result, our current research interest has been focused on the synthesis of transition metal coordination compounds and their applications mainly as potent inhibitors of PAF, which may possess similar or significantly higher activity and/or selectivity in regard to the common organic-based inhibitors. This is a vast research area and work on this field has not been referred to previously. A recent review by Leung et al. [123] has highlighted some examples of transition metal complexes that have been investigated for their anti-inflammatory activity. However, a significant number of coordination compounds with potential anti-inflammatory action have not been included. In this respect, it is the aim of the present minireview to present existing *metal-based* coordination compounds that exert their inflammatory activity by blocking the expression of PAF. Preliminary *structure-activity* relationships have been established, studying diverse parameters such as (i) the coordination geometry of the metal complex (square planar versus octahedral, etc.), (ii) the nature of the bidentate ligands and occasionally of the moieties in its periphery, (iii) the effect of the counteranions, and (iv) the total charge and size of the complex.

Below, we describe the existing different categories of metal-based inhibitors of PAF which have been systematically studied. Other classes of anti-inflammatory compounds do not constitute the aim of this minireview.

4.1. Rhodium Complexes with Bidentate Nitrogen Containing Ligands

4.1.1. Rhodium(III) Compounds. In 2009, we have successfully reported a preliminary work [124] on the anti-PAF activity of a Rh(III) coordination compound, namely, cis-[Rh(L)$_2$Cl$_2$]Cl (**Rh-1**), where L stands for the bidentate ligand 2,2′-pyridylquinoxaline with an IC$_{50}$ value of 250 nM (Figure 4). The corresponding bidentate ligand has been chosen owing to the remarkable biological properties reported [125, 126], while, on the other hand, investigation of the biological activity of inert metal complexes, such as rhodium(III) compounds, shows gradually increasing interest [127–130]. These reasons led us to design the target molecule **Rh-1** (Figure 4) with octahedral coordination geometry and evaluate in vitro its inhibitory effect against PAF-induced platelet aggregation in washed rabbit platelets (WRPs). The inhibitory effect of **Rh-1** and of all other new compounds tested (vide infra) was expressed by their IC$_{50}$ value in μM (Table 2). The IC$_{50}$ values reflect the inhibition strength of each compound, since a low IC$_{50}$ value reveals stronger inhibition of the PAF-induced aggregation for a given metal complex concentration. **Rh-1** shows a strong inhibitory effect towards the PAF-induced WRPs aggregation with IC$_{50}$ = 125 nM. This is an improved value as compared to the first

TABLE 2: IC$_{50}$ data (μM) final concentration in the aggregometer cuvette for rhodium PAF inhibitors towards WRPs.

Number	Rhodium complexes	IC$_{50}$
Rh-1	cis-[Rh(L)$_2$Cl$_2$]Cl	0.12 ± 0.11
Rh-2	cis-[Rh(L1)$_2$Cl$_2$]Cl	0.51 ± 0.23
Rh-3	cis-[Rh(L2)$_2$Cl$_2$]Cl	0.35 ± 0.20
Rh-4	mer-[Rh(L)Cl$_3$(MeOH)]	2.6 ± 2.0
Rh-5	[Rh(L)(cod)]Cl	0.016 ± 0.015
Rh-6	[Rh(L)(cod)]NO$_3$	0.015 ± 0.015
—	cisplatin	0.55 ± 0.22

measurement performed (IC$_{50}$ = 250 nM) because of slight changes in the experimental protocol during the biochemical experiment [131]. Since PAF is participating in inflammation, these compounds are potential anti-inflammatory drugs. After that, following a systematic approach [132], we have successfully developed a series of analogous rhodium(III)-based inhibitors of PAF of the general type cis-[Rh(L1)$_2$Cl$_2$]Cl (**Rh-2**) and cis-[Rh(L2)$_2$Cl$_2$]Cl (**Rh-3**) incorporating the bidentate ligands L1 = 4-carboxy-2-(2′-pyridyl)quinoline and L2 = 2,2′-bipyridine-4,4′-dicarboxylic acid that bear one and two carboxylic acid groups (-COOH), respectively, in the ligand sphere (Figure 4).

The idea, behind this, was to examine the possible effect of the carboxylic acid groups on the inhibition of the PAF, since, recently, an apparent dependence of biological activities (cytotoxicity and antioxidant efficiency) on incorporation of -COOH groups in the bipyridine ring has been reported [133]. The importance of the carboxylate moiety as a pharmacophore to the medicinal chemistry has been reported [134], while, interestingly, it has been proposed that approximately 25% of all drugs contain a COOH moiety.

The results of the biological assay performed reveal that indeed both substances (**Rh-2** and **Rh-3**), containing different number of carboxylic acid groups in the ligand periphery, display submicromolar activity (IC$_{50}$ = 0.35 μM and 0.51 μM, resp.) that is in the same order of magnitude as that of the **Rh-1** inhibitor. Evaluation of the anticancer activity of these substances is currently underway so as to check the relationship between PAF and cancer (vide infra). On the other hand, the implication of PAF in carcinogenesis and cancer metastasis has been well documented, whereas the known BN 5202 inhibitor has showed antitumor effects [135]. Cytotoxicity tests of the **Rh-1** compound against HEK-293 and cancer cell lines MCF-7 (breast cancer) were performed using cisplatin as a reference. Interestingly, preliminary results [131] reveal that **Rh-1**, a potent anti-PAF inhibitor (IC$_{50}$ = 0.12 μM), is less cytotoxic (54% viability on HEK-293 cells) compared to the cisplatin analogue (10.3% viability) with an IC$_{50}$ value of 0.55 μM against PAF. On the other hand, **Rh-1** was proved to be less potent (77% viability) on the MCF-7 cancer cells. For cisplatin, a 16% viability was observed in this cell line, evidence of the lack of harsh toxicity of the **Rh-1** accompanied by anticancer activity. Perhaps these compounds of the chemical formulas cis-[Rh(L)$_2$Cl$_2$]Cl, that inhibit PAF action, may also be used as a potential anticancer agent.

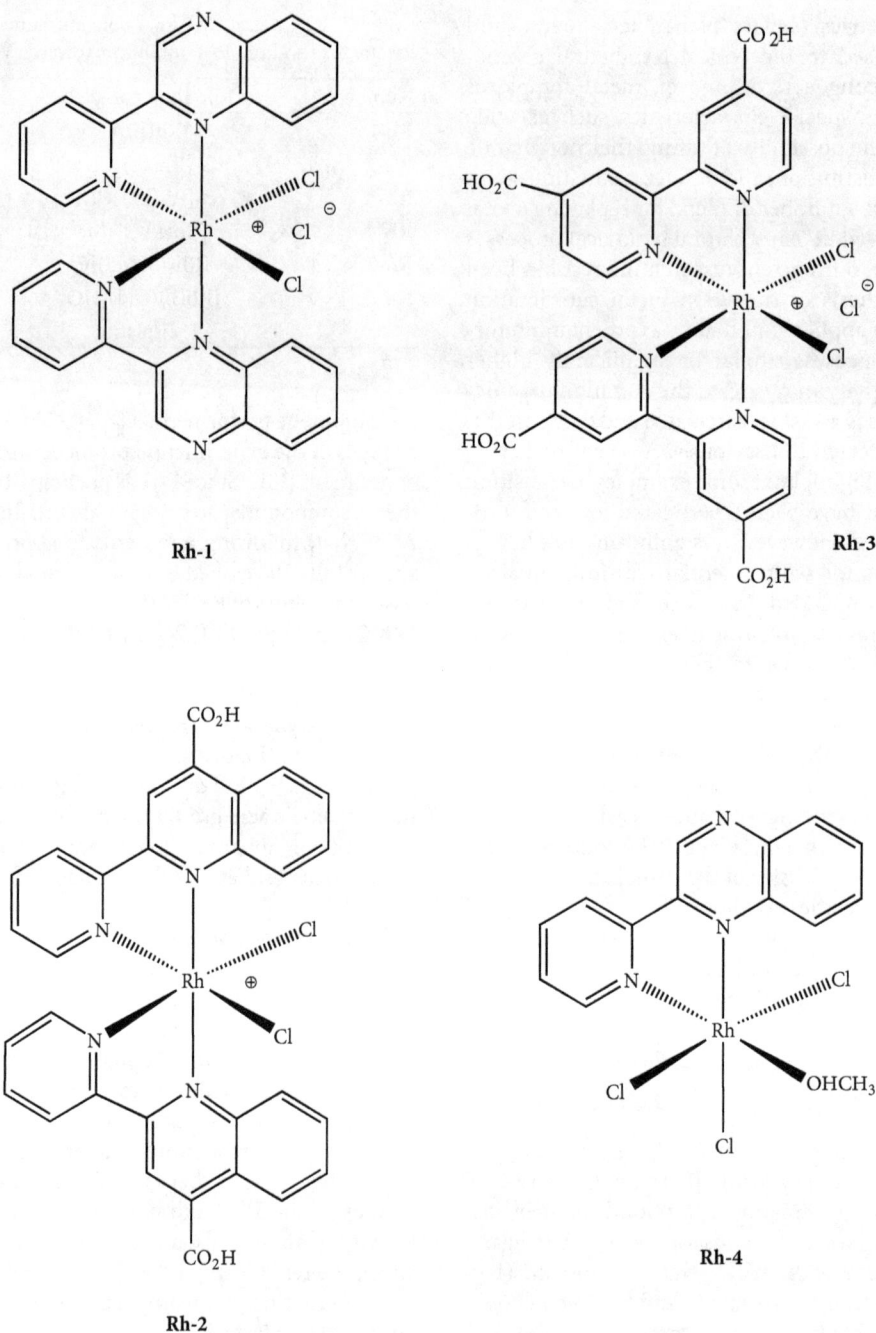

FIGURE 4: Molecular structures of rhodium(III) inhibitors of PAF.

In an effort to get an insight about the possible mechanism of action of these substances related to the PAFR, we performed radioactivity experiments (Scatchard analysis) involving the specific binding of [^{3}H]-PAF to washed rabbit platelets and its inhibition by complex **Rh-1**, that is the most potent of these Rh(III) derivatives. The specific PAF antagonist, BN 52021, was used as a reference compound and the radioactivity was measured by scintillation counting. This compound induced 18% inhibition of PAF binding onto platelet PAFR, whereas the highest % I (% I = [(total

binding − total binding with tested compound)/specific binding] × 100) was observed at 50 nM (34% inhibition). These data suggest that **Rh-1** inhibits PAF action on platelets and this effect is attributed only in part to the inhibition of PAF binding onto PAFR (nonspecific binding). It seems that, at low concentrations, **1** affects PAFR much more effectively and inhibits PAF action by another pathway [124]. Molecular docking theoretical calculations were in accord with the previous reported finding, denoting that the octahedral Rh(III) complex **Rh-1** (and **Rh-2** and **Rh-3**) with two aromatic

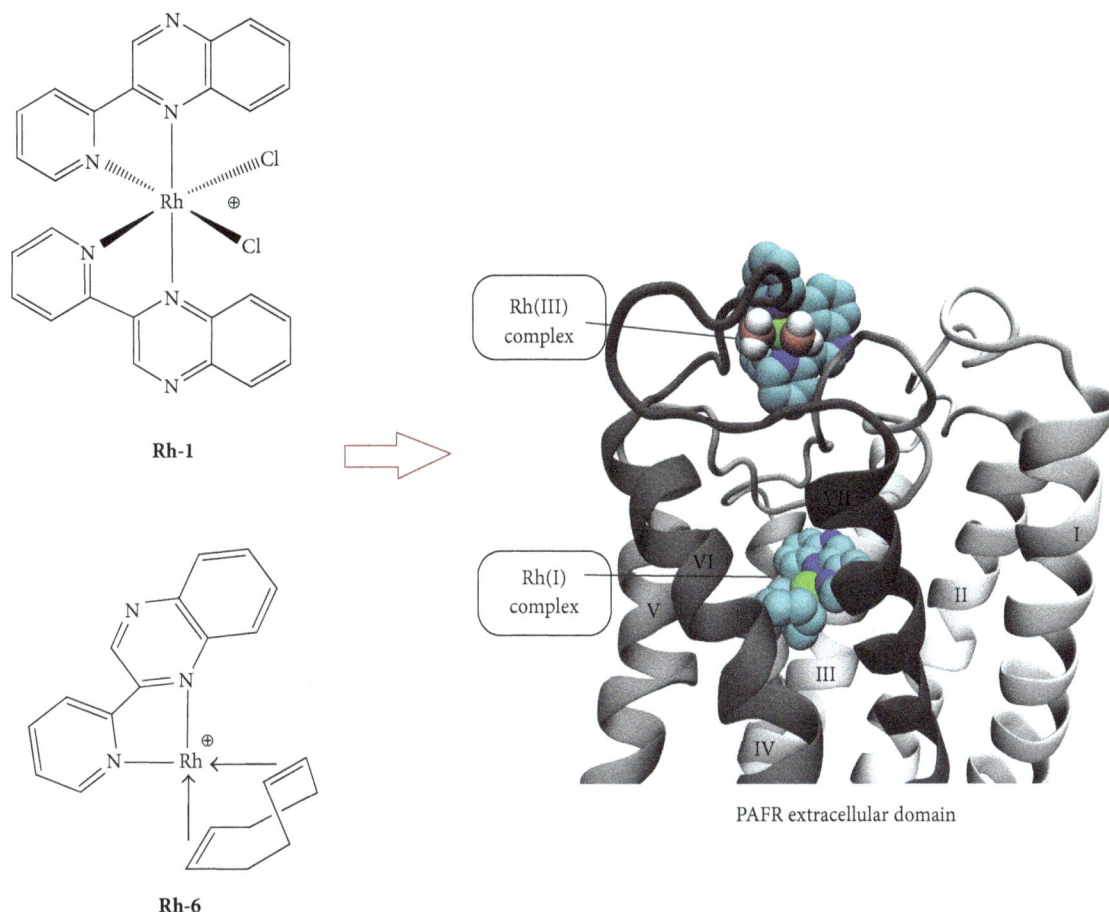

FIGURE 5: Molecular model of PAFR with the two predicted binding sites of the octahedral Rh(III) complex, **Rh-1**, and the square planar Rh(I) complexes, **Rh-5** and **Rh-6** (cartoon representation).

ligands cannot fit into the ligand-binding site of the PAFR model. Instead, they could bind to the extracellular domain of the receptor (Figure 5) and, therefore, antagonize the substrate's entrance to PAFR.

From the reaction of L with one equivalent of $RhCl_3 \times 3H_2O$, the $[Rh(L)Cl_3(MeOH)]$ complex has been prepared as a mixture of the *mer* and *fac* isomers. Upon purification the *mer*-$[Rh(L)Cl_3(MeOH)]$ isomer (**Rh-4**) crystallizes selectively from a mixture of both isomers in solution, while the *fac*-isomer was not isolated in a pure form. **Rh-4** is less effective ($IC_{50} = 2.6\,\mu M$) compared to the more bulky compounds **Rh-1**, **Rh-2**, and **Rh-3** (Figure 4). The calculated binding affinity of $3.20\,\mu M$ is in very good agreement with the in vitro results. As far as the rhodium(III) compounds are concerned, we conclude that the number of bidentate ligands and, as a result, the size of the target molecule (metal complex) affect dramatically the biological activity expressed. Thus, **Rh-4** which comprises a single aromatic ligand is less effective compared to the sterically demanding **Rh-2** and **Rh-3** complexes with two ligand molecules. At the same time, the role of total charge on the metal complex and ion size cannot be ruled out as recently reported for metal complexes as inhibitors of the 26S proteasome in tumor cells [136]. The rhodium precursor $RhCl_3 \times 3H_2O$ and the organic ligands

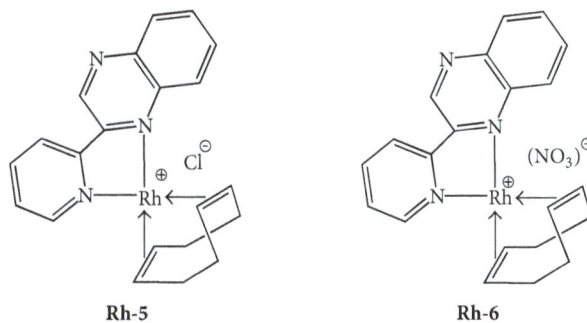

FIGURE 6: Molecular structure of Rh(I) inhibitors.

L, L1, and L2 have also been tested, exhibiting a weak inhibitory effect, ranging from $20\,\mu M$ to $67\,\mu M$. From this comparison, it becomes evident the crucial role of the metal center in the final structure of these coordination compounds towards these inflammatory mediators.

4.1.2. Rhodium(I) Compounds. The square planar Rh(I) complexes $[Rh(L)(cod)]Cl$ (**Rh-5**) and $[Rh(L)(cod)](NO_3)$ (**Rh-6**) (Figure 6) were prepared from the $[Rh(cod)Cl]_2$ dimer (cod = C_8H_{12}) precursor and the corresponding ligand L [132].

The biological evaluation showed that **Rh-5** and **Rh-6** exhibit an IC_{50} value within the range of 15 to 16 nM, indicating that these organometallic compounds are *selective* and potent inhibitors of PAF in the nanomolar scale. Docking theoretical calculations support the experimental results reported, predicting a high affinity of **Rh-5** and **Rh-6** for PAFR (estimated at $K_i = 20$ nM) that is mainly attributed to the favourable van der Waals and desolvation energy terms of its hydrophobic ligand (cyclooctadiene moiety). Finally, a clear *structure-activity* relationship is deduced, denoting that the square planar organometallic rhodium(I) complexes **Rh-5**, **Rh-6** are better inhibitors of the PAF-induced aggregation in comparison to the Rh(III) octahedral complexes **Rh-1/Rh-2/Rh-3/Rh-4**. The enhanced inhibitory activity of **Rh-5** and **Rh-6** could be further attributed to the presence of the hydrophobic cyclooctadiene moiety (vide infra the docking experiments, Figure 6). Interestingly, the biologic activity (observed IC_{50} values) of **Rh-5** and **Rh-6** against the PAF-induced aggregation is comparable to the corresponding action of some of the most potent PAF receptor antagonists, namely, WEB 2170, BN 52021, and rupatadine (IC_{50} = 0.02, IC_{50} = 0.03 and IC_{50} = 0.26 μM, resp.) [137–139]. In any case, it has to be pointed out that, upon coordination of the organic ligands (L, L1, and L2) to the rhodium center a dramatic increase *(synergetic effect)* of the inhibitory activity towards, the PAF-induced rabbit PRP aggregation is taking place in a dose-dependent manner. Therefore, the effects of rhodium metal complexes towards this inflammatory mediator can be clearly attributed to the final structure (square planar and/or octahedral) of these coordination compounds.

4.2. Copper(II), Zinc(II), Ni(II), and Ga(III) with Chalcogenated Imidodiphosphinato Ligands.

Tsoupras and coworkers [140, 141] have described a series of metal complexes (M = Cu, Co, Ni, Zn, Ga) with chalcogenated imidodiphosphinato ligands showing an inhibitory effect towards PAF-induced aggregation in the nanomolar to micromolar range (vide infra). The metal complexes examined are as follows: [M{(OPPh$_2$)(OPPh$_2$)N}$_2$], M = Cu (E = O,O; **Cu-1**); M = Zn ((E = O,O; **Zn-1**); (E = O,S); (E = S,S; **Zn-2**); (E = O,Se; **Zn-3**)); [M{(OPPh$_2$)(OPPh$_2$)N}$_3$], M = Ga (E = O,O; **Ga-1**); [M{(Ph$_2$P)$_2$N-S-CHMe}X$_2$], M = Ni (E = P,P; X = Cl, **Ni-1**); (E = P,P; X = Br, **Ni-2**); (E = O,Se; **Ni-3**); M = Co, Ni (E = O,Se; **Co-1**). These complexes display diverse coordination geometries, which range from square planar for **Cu-1**, **Ni-1** and **Ni-2** to tetrahedral for **Zn-1** and **Zn-2**, while a pseudotetrahedral environment has been found for the corresponding **Zn-3**, **Co-1**, and **Ni-3** analogues. Their molecular structures are depicted in Figure 7 and the corresponding IC_{50} values are included in Table 3. Remarkably, **Co-1** and **Ni-3** (E = O,Se) exhibit the strongest inhibitory effect against the PAF-induced aggregation in WRPs with an IC_{50} value of 18 to 19 nM. The increased biological activity of both complexes has been mainly attributed to the typical *open shell* structure they possess. It seems also that, in these ligands, the combination of O and Se, as donor atoms, is beneficial for Co(II) and Ni(II) ions over the Zn(II) ions which possess

TABLE 3: IC_{50} data (μM) final concentration in the aggregometer cuvette for metal complexes bearing PNP ligands as PAF inhibitors.

Number	PNP complexes	IC_{50}
Cu-1	Cu{(OPPh$_2$)(OPPh$_2$)N-O,O}$_2$	~1.0
Zn-1	Zn{(OPPh$_2$)(OPPh$_2$)N-O,O}$_2$	0.54
Zn-2	Zn{(SPPh$_2$)(SPPh$_2$)N-S,S}$_2$	0.36
Zn-3	Zn{(OPPh$_2$)(SePPh$_2$)N-O,Se}$_2$	1.11 ± 0.22
Ga-1	Ga{(OPPh$_2$)(OPPh$_2$)N-O,O}$_3$	0.062 ± 0.045
Ni-1	Ni{(Ph$_2$P)$_2$N-S-CHMePh-P,P}Cl$_2$	~16
Ni-2	Ni{(Ph$_2$P)$_2$N-S-CHMePh-P,P}Br$_2$	~3.0
Ni-3	Ni{(OPPh$_2$)(SePPh$_2$)N-O,Se}$_2$	0.019 ± 0.006
Co-1	Co{(OPPh$_2$)(SePPh$_2$)N-O,Se}$_2$	0.018 ± 0.005

a *closed shell* structure. In fact, the biological activity of **Zn-3** drops almost two orders of magnitude (IC_{50} = 1.11 μM) than that of **Co-1** and **Ni-3**. For the structurally related diamagnetic Zn(II) complexes **Zn-1** (E = O,O) and **Zn-2** (E = S,S), the presence of O or S donor atoms increases slightly the inhibitory capacity against PAF (IC_{50} = 0.54 μM and 0.36 μM, resp.). Moreover, the square planar **Cu-1** compound exhibited medium inhibitory effect against PAF. On the other hand, the more bulky **Ga-1** complex, consisting of a main group element and three bidentate PNP ligands with O,O as the donor atoms, shows an appreciable anti-PAF activity (IC_{50} = 62 nM). The authors have commented that the higher potency may be due to the octahedral structure of this compound. Finally, the square planar compounds **Ni-1** and **Ni-2** are the less active, owing to a partial degradation in DMSO. At this point, it should be stated that for biological purposes chemical stability in solution is a prerequisite. However, a large number of biological activity studies have been carried out so as to systematically ignore the possible effects of the corresponding medium that is used to dissolve a new substance under investigation. In particular, the presence of easily dissociating groups in a metal complex (halides, etc.) and the use of organic solvents with strong donor abilities (DMSO, DMF, and MeCN) constitute it susceptible to a possible decomposition in solution. The *nature of the active species* in solution remains unclear affecting the results of the biological experiment. In summary, the authors of this study have concluded that the stereochemical and electronic characteristics of the metal complexes previously described determine their inhibitory effect. Among the inhibitors studied, **Co-1**, **Ni-3**, and **Ga-1** could be potentially examined for their anti-inflammatory activity. It may be that these compounds express their biological activity via a selective interaction with the PAF receptor, although this has not been examined. Docking theoretical calculations might be helpful towards this aim as previously shown for the rhodium(I)/(III) series [124, 132].

4.3. Re(I) Complex with a Phenanthroline-Dione Ligand.

Kaplanis and coworkers [142] reported on a Re(I) derivative (Figure 8) of the general type *fac*-[Re(phendione)(CO)$_3$Cl] (**Re-1**) (where phendione = 1,10-phenanthroline-5,6-dione)

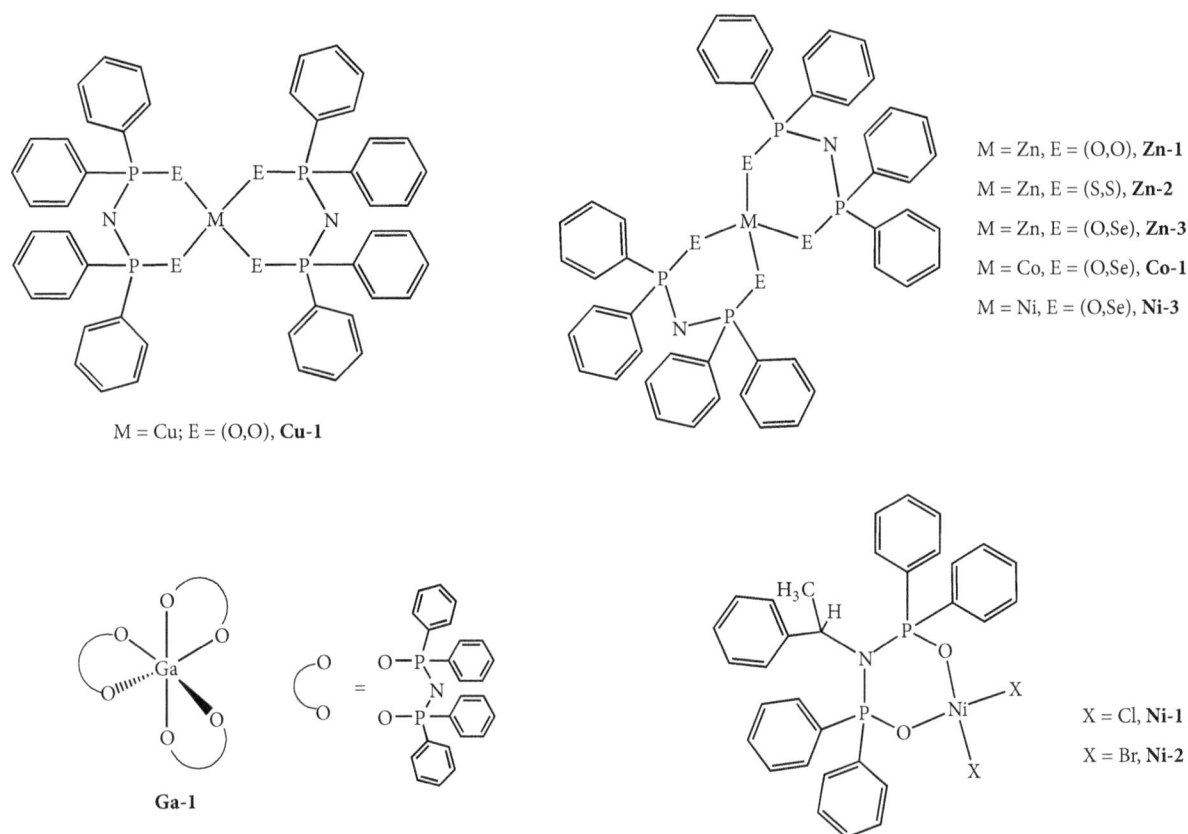

M = Cu; E = (O,O), **Cu-1**

M = Zn, E = (O,O), **Zn-1**

M = Zn, E = (S,S), **Zn-2**

M = Zn, E = (O,Se), **Zn-3**

M = Co, E = (O,Se), **Co-1**

M = Ni, E = (O,Se), **Ni-3**

Ga-1

X = Cl, **Ni-1**

X = Br, **Ni-2**

FIGURE 7: Molecular structures of metal complexes bearing PNP ligands.

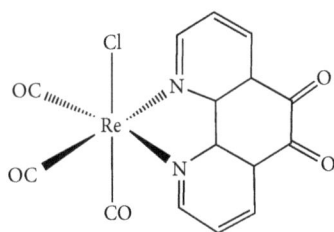

FIGURE 8: Molecular structure of a Re(I) inhibitor.

that has been evaluated as inhibitor of PAF-induced aggregation of washed rabbit platelets (IC$_{50}$ = 0.86 μM).

It is to be noted that the starting material [Re(CO)$_5$Cl] displayed a better biological activity (IC$_{50}$ = 0.17 μM) as compared to the free ligand phendione (IC$_{50}$ = 0.93 μM) and the Re(I) complex mentioned above. This is in contrast to previous data where coordination of a ligand to a metal ion enhances the inhibitory activity towards PAF aggregation of the metal complex [132, 141]. It may be that **Re-1** complex does not fit into the binding site of PAFR; presumably it dissociates at the extracellular domain and finally the observed activity could be attributed to the activity of the phendione only. In fact, **Re-1** and the free ligand exhibit the same inhibitory effect, while [Re(CO)$_5$Cl] does not seem to contribute to the biological activity of the synthesized complex **Re-1**. In this respect, analogous observations have been reported for two new Rh(III) complexes

of the formula *mer*-[Rh(L)Cl$_3$(MeOH)], where L stands for substituted pyridylquinoline ligands [143]. Coordination of the pyridylquinoline bidentate ligands to the Rh(III) source (RhCl$_3$×3H$_2$O, IC$_{50}$ = 67 μM) improves slightly the biological profile of the metal-based inhibitors, although the Rh(III) metal precursor is not that potent inhibitor as the relevant [Re(CO)$_5$Cl] analogue. A plausible answer to the questions raised may be derived with the help of molecular docking theoretical calculations. Finally, **Re-1** and [Re(CO)$_5$Cl] showed activity against PAF-basic metabolic enzyme activities in rabbit leukocyte homogenates.

4.4. Ru(II) Complexes with Bidentate (N$^\wedge$N) and Tridentate (N$^\wedge$N$^\wedge$N) Nitrogen Based Ligands. Inspired by the previously described results on the field, we wanted to further extend our work to ruthenium(II) coordination compounds and further explore the activity of a variety of coordination compounds containing this metal, as PAF inhibitors [144]. The advantages of ruthenium in a number of biological actions have been well documented [145–147]. Thus a series of octahedral ruthenium(II)/(III) complexes have been tested towards this goal. These substances that have been synthesized by our group incorporate heterocyclic bidentate (N$^\wedge$N) and tridentate ligands (N$^\wedge$N$^\wedge$N) with or without carboxylic acid ancillary functionalities displaying typical octahedral coordination geometries as the majority of ruthenium(II) coordination compounds *(structurally related compounds)*. Among them, those with the -COOH moieties have been

(a)

(b)

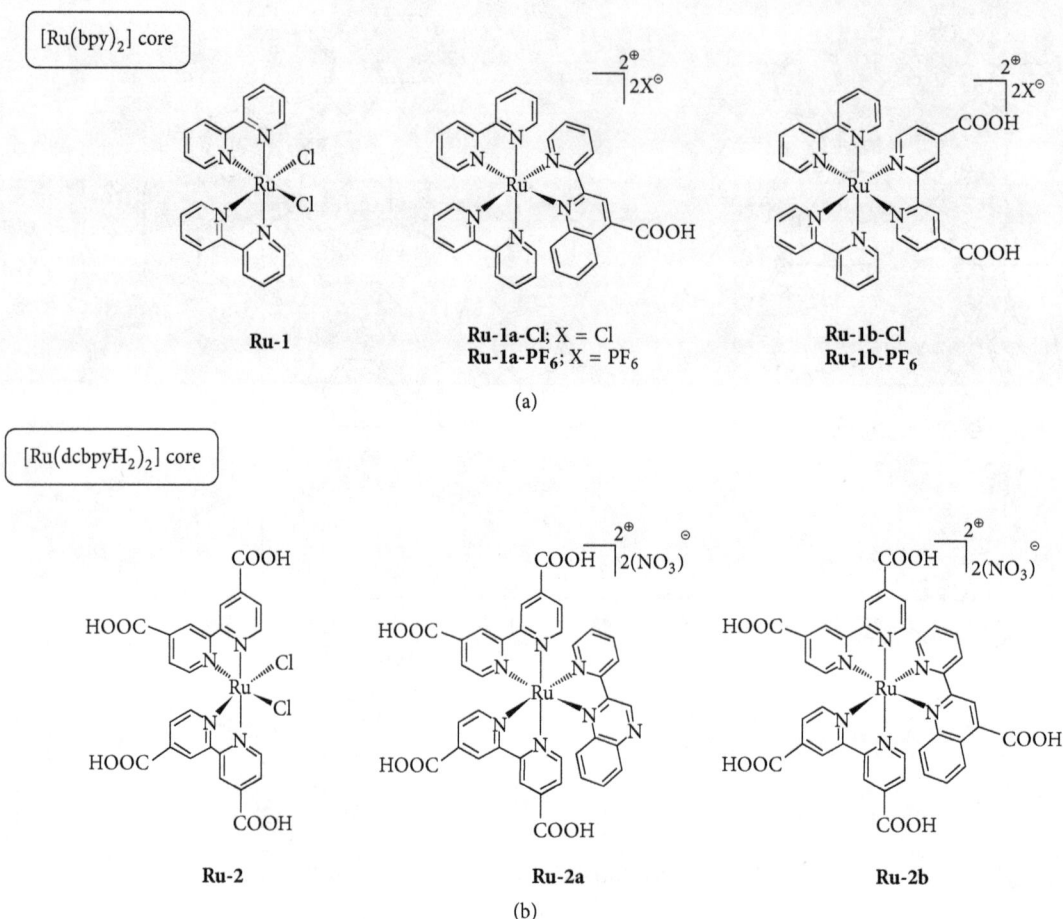

FIGURE 9: Molecular structures of Ru(II) inhibitors with bidentate ligands.

mainly tested as sensitizers in third-generation photovoltaic solar cells. Depending on the ruthenium(II) source, these substances can be divided into three categories. The first two comprise the cis-[Ru(bpy)$_2$(L)]$^{2+}$ [148] (Figure 9(a)) and cis-[Ru(dcbpyH$_2$)$_2$(L)]$^{2+}$ [149] units (where bpy = 2,2'-bipyridine; L = 2-(2'-pyridyl)quinoxaline (L^1), 4-carboxy-2-(2'-pyridyl)quinoline (L^2), 2,2'-dipyridine-4,4'-dicarboxylic acid (dcbpyH$_2$) (Figure 9(b))), while the third one displays the [Ru(bpp)]$^{2+}$ (bpp is the tridentate ligand 2,6-Bis(1-pyrazolyl)pyridine) [150] and [Ru(bdmpp)]$^{2+}$ (bdmpp = 2,6-bis(3,5-dimethyl-1-pyrazolyl)pyridine) core [144] (Figure 10).

In total, the formulas of the ruthenium complexes examined (studied) are as follows: cis-[Ru(bpy)$_2$(L)]X$_2$ (Ru-1a-Cl; Ru-1a-PF$_6$, L = L^2 = 4-carboxy-2-(2'-pyridyl)quinoline; Ru-1b-Cl, Ru-1b-PF$_6$, L = dcbpyH$_2$ = 2,2'-Bipyridine-4,4'-dicarboxylic acid), cis-[Ru(dcbpyH$_2$)$_2$(L)]X$_2$ (Ru-2a, X = NO$_3$, L = L^1 = 2-(2'-pyridyl)quinoxaline; Ru-2b, X = NO$_3$, L = L^2), [Ru(bpp)Cl(dcbpyH)] (Ru-3a), [Ru(bpp)Cl(dcbpyH$_2$)]Cl (Ru-3b) (bpp = 2,6-Bis(1-pyrazolyl)pyridine), and [Ru(bdmpp)Cl(dcbpyH$_2$)]PF$_6$ (Ru-4a) (bdmpp = 2,6-bis(3,5-dimethyl-1-pyrazolyl)pyridine), where cis-[Ru(bpy)$_2$Cl$_2$] (Ru-1), cis-[Ru(dcbpyH$_2$)$_2$Cl$_2$] (Ru-2), [Ru(bpp)Cl$_3$] (Ru-3) and [Ru(bdmpp)Cl$_3$] (Ru-4). Results of the biological experiment show that most of the

ruthenium complexes are potent inhibitors of PAF with IC$_{50}$ values ranging from micromolar to submicromolar concentrations, extending therefore the range of their applications (Table 4). Inhibition is taking place in a dose-dependent manner. In addition, it has been demonstrated that coordination of the organic ligands to the ruthenium metal center enhances significantly the inhibitory potency against the PAF-induced aggregation. This remark corroborates the results of our previous work based on rhodium(I) and rhodium(III) inhibitors of PAF [124, 132].

Next, we studied the possible effect of the counterion upon ligand exchange reactions. Remarkably, we noticed that, upon substitution reaction of the chloride counteranion of the bpy derivatives Ru-1a-Cl and Ru-1b-Cl into the hexafluorophosphate salts, Ru-1a-PF6 (0.48 µM), and Ru-1b-PF6 (0.5 µM), a twentyfold and fourfold increase, respectively, of the PAF inhibitory effect is observed, rendering them more potent against PAF inhibition. This can be attributed in part to the higher inhibitory effect towards PAF of the trifluoroacetyl-analogues, fluoride containing substances, compared to the trichloroacetyl ones [151].

In the second class of complexes, we observed that the inhibitory activity of the ionic compounds Ru-2a, Ru-2b (IC$_{50}$ values of approximately 0.2 µM) is considerably higher than that of the neutral precursor Ru-2 (4.5 µM), while

[Ru(bpp)] core

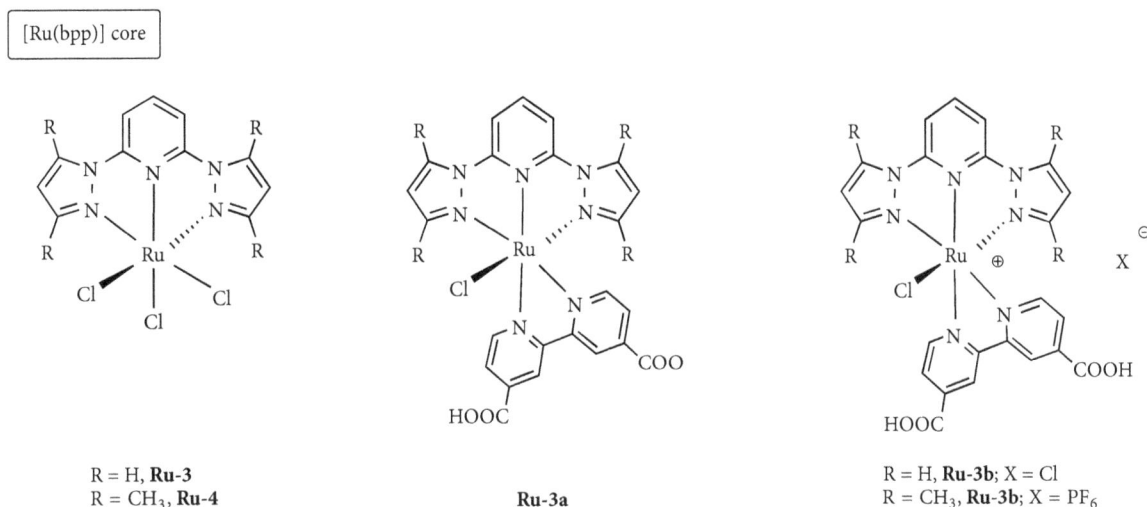

R = H, **Ru-3**
R = CH$_3$, **Ru-4**

Ru-3a

R = H, **Ru-3b**; X = Cl
R = CH$_3$, **Ru-3b**; X = PF$_6$

FIGURE 10: Molecular structures of Ru(III)/(II) inhibitors with tridentate ligands.

TABLE 4: IC$_{50}$ data (μM) of various Ru(II)/(III) PAF inhibitors.

Number	Complexes	IC$_{50}$
Ru-1	[Ru(bpy)$_2$Cl$_2$]	7.0 ± 0.7
Ru-1a-Cl	[Ru(bpy)$_2$(L^2)]Cl$_2$	11 ± 1
Ru-1a-PF$_6$	[Ru(bpy)$_2$(L^2)](PF$_6$)$_2$	0.48 ± 0.06
Ru-1b-Cl	[Ru(bpy)$_2$(dcbpyH$_2$)]Cl$_2$	1.2 ± 0.1
Ru-1b-PF$_6$	[Ru(bpy)$_2$(dcbpyH$_2$)](PF$_6$)$_2$	0.50 ± 0.05
Ru-2	[Ru(dcbpyH$_2$)$_2$Cl$_2$]	4.5 ± 0.5
Ru-2a	[Ru(dcbpyH$_2$)$_2$(L^1)](NO$_3$)$_2$	0.18 ± 0.01
Ru-2b	[Ru(dcbpyH$_2$)$_2$(L^2)](NO$_3$)$_2$	0.24 ± 0.03
Ru-3	[Ru(bpp)Cl$_3$]	2.1 ± 0.2
Ru-3a	[Ru(bpp)(dcbpyH)Cl]	3.1 ± 0.3
Ru-3b	[Ru(bpp)(dcbpyH$_2$)Cl]Cl	11.8 ± 0.1
Ru-4	[Ru(bdmpp)Cl$_3$]	2.6 ± 0.3
Ru-4b	[Ru(bdmpp)(L^4)Cl](PF$_6$)	6.4 ± 1.1

an almost 3-fold increase over the bpy derivatives **Ru-1a-PF$_6$** and **Ru-1b-PF$_6$** (0.5 μM) is shown. We can conclude therefore that the presence of the cis-[Ru(dcbpyH$_2$)$_2$]$^{2+}$ core containing -COOH acid groups as auxiliary substituents is more favourable over the classical cis-[Ru(bpy)$_2$]$^{2+}$ core. The inhibitory effect of the most potent ruthenium(II) compounds reported previously is comparable to that of the well-established rhodium(III) inhibitors, **Rh-1**, **Rh-2**, and **Rh-3** (0.12 μM, 0.51 μM, and 0.35 μM, resp.) displaying octahedral coordination geometries as well. Presumably this could be attributed to the influence of the electronic configurations and the size of Ru(II) and Rh(III) ions. Remarkably, **Ru-2a** and **Ru-2b** compounds display comparable biological activity (IC$_{50}$ values of 0.18 and 0.24 μM) with rupatadine fumarate, a potent PAF receptor antagonist (0.26 μM) that is used currently (Rupafin) in clinical practice [152]. Within this series, it seems that the ionic compounds are more potent as compared to the neutral precursors (**Ru-1**, **Ru-2**),

a result which correlates well with our previous notes on the rhodium-based inhibitors [124, 132]. However, the neutral ruthenium(III) precursor **Ru-3** (IC$_{50}$ = 2.1 μM) displays a slightly higher potency compared to the internal salt **Ru-3a** (3.1 μM) and a more intense inhibitory effect than that of the salt-like complex **Ru-3b** (11.8 μM). In addition, the neutral ruthenium(III) precursor **Ru-4** (IC$_{50}$ = 2.6 μM) is less potent than the ionic complex **Ru-4b** (IC$_{50}$ = 6.4 μM). Moreover, **Ru-3** and **Ru-4** are more active than the ruthenium(II) compounds Ru-1 (7 μM) and **Ru-2** (4.5 μM), respectively. Obviously, these discrepancies indicate that the inhibitory effect of Ru(III) based inhibitors against PAF is an issue that merits further investigation.

4.5. Synopsis of the Metal-Based Inhibitors Reported in This Work. According to the experimental findings, it seems that the nature of the metal center and the nature of the organic ligand attached to it alter substantially the biological action expressed (vide infra). Although the biological probe studied is a complicated system, following simple coordination chemistry principles, we have managed to get some quite helpful structure-activity relationships as mentioned in the conclusions part that follows.

In general, the square planar Rh(I) complexes incorporating a 2-(2'-pyridyl)quinoxaline ligand (**Rh-1**, **Rh-2**; IC$_{50}$ = 15-16 nM), the pseudotetrahedral Co(II) and Ni(II) complexes (**Co-1**, **Ni-3**, IC$_{50}$ = 18-19 nM), and the octahedral Ga(III) complex bearing chalcogenated imidodiphosphinato ligands (**Ga-1**, IC$_{50}$ = 62 nM) inhibited PAF in the nanomolar scale. Theoretical docking calculations (for the rhodium complexes) are in accord with the experimental findings denoting that **Rh-1**, **Rh-2** inhibitors could fit in the ligand-binding site of PAF receptor (PAFR). **Rh-1**, **Rh-2**, and **Rh-3** complexes as well as **Ru-2a**, **Ru-2b** and **Zn-1**, **Zn-2**, and **Re-1** were potent inhibitors in the submicromolar range. A schematic representation of the more potent metal-based inhibitors is given in Figure 11.

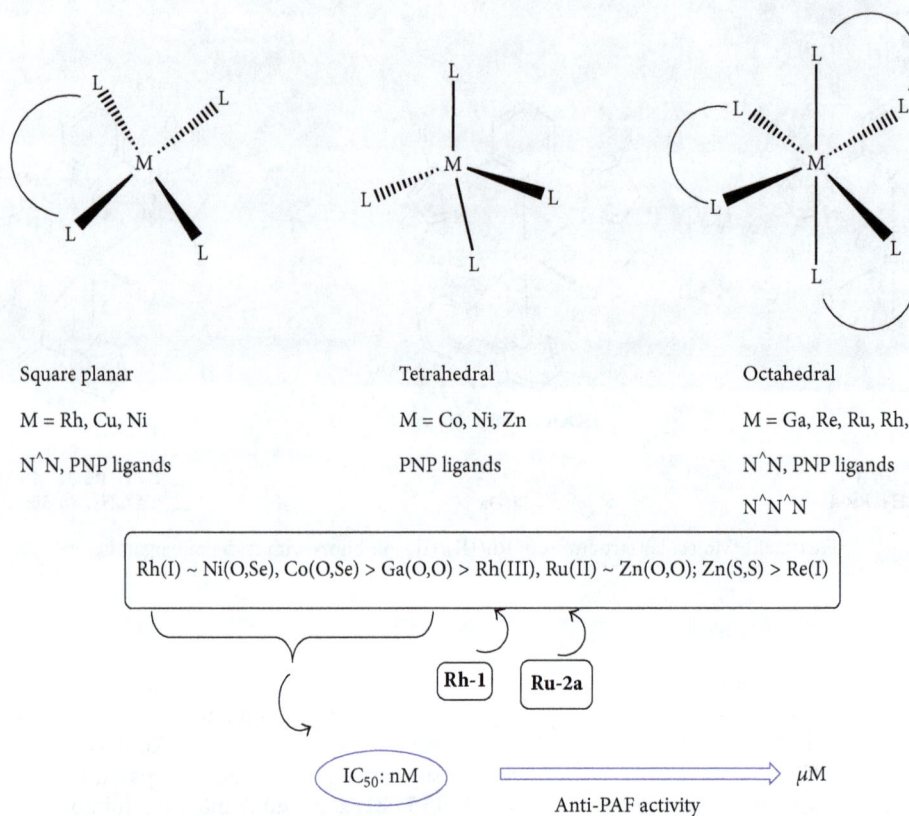

FIGURE 11: The most potent inhibitors of PAF described in this work.

5. Conclusions

The present minireview describes the latest results on the application of a series of metal complexes with diverse structures and different metal cores as PAF inhibitors in WPRs, a new class of the so-called metal-based inhibitors of PAF. From this work, some preliminary structure-activity relationships have been established based on simplified structural comparison among the library of 30 small molecules investigated. A number of factors such as the counteranion, the total charge of the inhibitor, and the size of the target molecule determine, at least in part, the extent of their inhibitory action. Interestingly, the results show the important role of the metal center and the coordination geometries adopted upon coordination of the organic ligand. Thus, for the rhodium complexes, a straightforward relationship has been addressed, denoting that the square planar organometallic Rh(I) complexes display an inhibitory effect at the nanomolar level than the Rh(III) congeners. The effect of the cyclooctadiene moiety has been analyzed on the basis of theoretical docking calculations, and presumably the improved anti-PAF action has been attributed to the hydrophobic interactions of the cyclooctadiene within the special domains of the PAFR. In addition, the effect of the total charge has been examined, showing that the positively charged Rh compounds and the carboxylic acid containing Ru(II) analogous ($[Ru(dcbpyH_2)_2]^{2+}$ core) seem to be the more potent. It may be that, in vitro, these substances display a better permeability across the negatively charged lipid membranes. However, the trend is not followed

for the ruthenium bipyridine class ($[Ru(bpy)]^{2+}$ core). On the other hand, on moving to the relevant PNP bifunctional entities, the biological motif alters. The neutral Ni(II)-(O,Se) and Co(I)-(O,Se) analogues now present a strong anti-PAF activity and possibly this may be an effect of the ligand properties.

Upon a thorough study of the molecular structures proposed, we realized that, in most of the cases (not the example of the Re(I) complex), the obtained metal complexes are characterized by an improved biological activity as compared to both the organic precursor and the metal source (synergetic effect). To obtain reasonable *structure-activity* relationships for the other categories tested, as for the rhodium metal compounds, theoretical docking calculations are required.

From the results of this study, one may conclude that indeed metal-based inorganic compounds are a very promising class of anti-PAF and anti-inflammatory drugs. For the rhodium(III) PAF inhibitor **Rh-1**, the moderate cytotoxicity observed in HEK 293 cell lines corroborates the increased anti-inflammatory action observed. Cisplatin that is also active against inflammation is less potent in HEK 293 cell lines. We suggest that this biological effect is due to the overall structural characteristics of **Rh-1** that shows a combined anti-PAF and anticancer activity.

This is of interest, denoting that current organic inhibitors existing could be potentially replaced with a new class of metal-based inhibitors of PAF in the near future. Finally, the results of this study could help us to create a database set of coordination compounds with potential anti-inflammatory activity. In such a way, we could roughly predict or estimate the potency of other new metal-based substances against PAF, displaying similar molecular structures. Thus the preliminary results reported could provide us with the required knowledge so as to rationally design more potent anti-inflammatory drugs in the future.

Acknowledgments

Athanassios I. Philippopoulos would like to thank Professor Dr. A. C. Filippou of the Chemistry Department of the University of Bonn for his support on the spectroscopic characterization of the compounds presented in this work. Also, Nefeli Lagopati would like to thank Dr. Paraskevi Kitsiou of the Institute of Biosciences & Applications, NCSR Demokritos, for her precious help with the cytotoxicity tests performed.

References

[1] N. J. Cusack, "Platelet-activating factor," *Nature*, vol. 285, no. 5762, p. 193, 1980.

[2] C. A. Demopoulos, R. N. Pinckard, and D. J. Hanahan, "Platelet-activating factor. Evidence for 1-O-alkyl-2-acetyl-sn-glyceryl-3-phosphorylcholine as the active component (a new class of lipid chemical mediators)," *The Journal of Biological Chemistry*, vol. 254, no. 19, pp. 9355–9358, 1979.

[3] V. I. Kulikov and G. I. Muzya, "Ether lipids and platelet-activating factor: evolution and cellular function," *Biochemistry*, vol. 62, no. 10, pp. 1103–1108, 1997.

[4] M. L. Balestrieri, D. Castaldo, C. Balestrieri, L. Quagliuolo, A. Giovane, and L. Servillo, "Modulation by flavonoids of PAF and related phospholipids in endothelial cells during oxidative stress," *Journal of Lipid Research*, vol. 44, no. 2, pp. 380–387, 2003.

[5] F. Snyder, "Platelet-activating factor: the biosynthetic and catabolic enzymes," *Biochemical Journal*, vol. 305, no. 3, pp. 689–705, 1995.

[6] S. Antonopoulou, T. Nomikos, H. C. Karantonis, E. Fragopoulou, and C. A. Demopoulos, "PAF, a potent lipid mediator," in *Bioactive Phospholipids. Role in Inflammation and Atherosclerosis*, A. D. Tselepis, Ed., pp. 85–134, Transworld Research Network, Trivandrum, India, 2008.

[7] Z.-I. Honda, S. Ishii, and T. Shimizu, "Platelet-activating factor receptor," *Journal of Biochemistry*, vol. 131, no. 6, pp. 773–779, 2002.

[8] V. L. Marcheselli, M. J. Rossowska, M.-T. Domingo, P. Braquet, and N. G. Bazan, "Distinct platelet-activating factor binding sites in synaptic endings and in intracellular membranes of rat cerebral cortex," *The Journal of Biological Chemistry*, vol. 265, no. 16, pp. 9140–9145, 1990.

[9] F. Gobeil Jr., A. Fortier, T. Zhu et al., "G-protein-coupled receptors signalling at the cell nucleus: an emerging paradigm," *Canadian Journal of Physiology and Pharmacology*, vol. 84, no. 3-4, pp. 287–297, 2006.

[10] A. M. Marrache, F. Gobeil Jr., T. Zhu, and S. Chemtob, "Intracellular signaling of lipid mediators via cognate nuclear G protein-coupled receptors," *Endothelium: Journal of Endothelial Cell Research*, vol. 12, no. 1-2, pp. 63–72, 2005.

[11] T. Zhu, F. Gobeil Jr., A. Vazquez-Tello et al., "Intracrine signaling through lipid mediators and their cognate nuclear G-protein-coupled receptors: a paradigm based on PGE2, PAF, and LPA1 receptors," *Canadian Journal of Physiology and Pharmacology*, vol. 84, no. 3-4, pp. 377–391, 2006.

[12] G. A. Zimerman, M. R. Elstad, D. E. Lorant et al., "Platelet-activating factor (PAF): signalling and adhesion in cell-cell interactions," in *Platelet-Activating Factor and Related Lipid Mediators 2*, vol. 416 of *Advances in Experimental Medicine and Biology*, pp. 297–304, Springer, 1996.

[13] V. D. Papakonstantinou, "Ginkgo biloba and its anti-inflammatory value as a medical tool," *Hellenic Journal of Atherosclerosis*, vol. 4, no. 2, pp. 109–115, 2013.

[14] Y.-C. Shen, C.-F. Chen, and W.-F. Chiou, "Suppression of rat neutrophil reactive oxygen species production and adhesion by the diterpenoid lactone andrographolide," *Planta Medica*, vol. 66, no. 4, pp. 314–317, 2000.

[15] E. Amroyan, E. Gabrielian, A. Panossian, G. Wikman, and H. Wagner, "Inhibitory effect of andrographolide from *Andrographis paniculata* on PAF-induced platelet aggregation," *Phytomedicine*, vol. 6, no. 1, pp. 27–31, 1999.

[16] T. Jayakumar, C.-Y. Hsieh, J.-J. Lee, and J.-R. Sheu, "Experimental and clinical pharmacology of andrographis paniculata and its major bioactive phytoconstituent andrographolide," *Evidence-Based Complementary and Alternative Medicine*, vol. 2013, Article ID 846740, 16 pages, 2013.

[17] H.-C. Hsu, W.-C. Yang, W.-J. Tsai, C.-C. Chen, H.-Y. Huang, and Y.-C. Tsai, "α-Bulnesene, a novel PAF receptor antagonist isolated from *Pogostemon cablin*," *Biochemical and Biophysical Research Communications*, vol. 345, no. 3, pp. 1033–1038, 2006.

[18] W.-H. Peng, T.-C. Lu, J.-C. Liao et al., "Analgesic and anti-inflammatory activities of the methanol extract from pogostemon cablin," *Evidence-based Complementary and Alternative Medicine*, vol. 2011, Article ID 671741, 2011.

[19] T. Y. Shen, S. B. Hwang, M. N. Chang et al., "Characterization of a platelet-activating factor receptor antagonist isolated from haifenteng (Piper futokadsura): specific inhibition of in vitro and in vivo platelet-activating factor-induced effects," *Proceedings of the National Academy of Sciences of the United States of America*, vol. 82, no. 3, pp. 672–676, 1985.

[20] S.-B. Hwang, M. N. Chang, M. L. Garcia et al., "L-652,469—a dual receptor antagonist of platelet activating factor and dihydropyridines from Tussilago farfara L," *European Journal of Pharmacology*, vol. 141, no. 2, pp. 269–281, 1987.

[21] C. Hwangbo, H. S. Lee, J. Park, J. Choe, and J.-H. Lee, "The anti-inflammatory effect of tussilagone, from Tussilago farfara, is mediated by the induction of heme oxygenase-1 in murine macrophages," *International Immunopharmacology*, vol. 9, no. 13-14, pp. 1578–1584, 2009.

[22] H. C. Castro-Faria-Neto, P. T. Bozza, H. N. Cruz et al., "Yangambin: a new naturally-occurring platelet-activating factor receptor antagonist: binding and in vitro functional studies," *Planta Medica*, vol. 61, no. 2, pp. 101–105, 1995.

[23] R. C. P. Marques, S. R. B. De Medeiros, C. Da Silva Dias, J. M. Barbosa-Filho, and L. F. Agnez-Lima, "Evaluation of the mutagenic potential of yangambin and of the hydroalcoholic extract of Ocotea duckei by the Ames test," *Mutation Research*, vol. 536, no. 1-2, pp. 117–120, 2003.

[24] F. C. F. de Sousa, B. A. Pereira, V. T. M. Lima et al., "Central nervous system activity of yangambin from *Ocotea duckei* vattimo (Lauraceae) in mice," *Phytotherapy Research*, vol. 19, no. 4, pp. 282–286, 2005.

[25] E. Tibiriçá, "Cardiovascular properties of yangambin, a lignan isolated from Brazilian plants," *Cardiovascular Drug Reviews*, vol. 19, no. 4, pp. 313–328, 2001.

[26] S. Antonopoulou, C. A. Demopoulos, and N. K. Andrikopoulos, "Lipid separation from urtica dioica: existence of platelet-activating factor," *Journal of Agricultural and Food Chemistry*, vol. 44, no. 10, pp. 3052–3056, 1996.

[27] A. Siafaka-Kapadai, C. A. Demopoulos, and N. K. Andrikopoulos, "Biological activity of lipids of pine pollen on platelet aggregation in correlation with the platelet activating factor," *Biochemistry International*, vol. 12, no. 1, pp. 33–41, 1986.

[28] S. P. Bhatia, D. McGinty, C. S. Letizia, and A. M. Api, "Fragrance material review on cedrol," *Food and Chemical Toxicology*, vol. 46, no. 11, pp. S100–S102, 2008.

[29] H. O. Yang, D.-Y. Suh, and B. H. Han, "Isolation and characterization of platelet-activating factor receptor binding antagonists from Biota orientalis," *Planta Medica*, vol. 61, no. 1, pp. 37–40, 1995.

[30] Z. Terashita, S. Tsushima, Y. Yoshioka, H. Nomura, Y. Inada, and K. Nishikawa, "CV-3988—a specific antagonist of platelet activating factor (PAF)," *Journal of Agricultural and Food Chemistry*, vol. 32, no. 17, pp. 1975–1982, 1983.

[31] Z.-I. Terashita, Y. Imura, M. Takatani, S. Tsushima, and K. Nishikawa, "CV-6209, a highly potent antagonist of platelet activating factor in vitro and in vivo," *Journal of Pharmacology and Experimental Therapeutics*, vol. 242, no. 1, pp. 263–268, 1987.

[32] T. Toyofuku, K. Kubo, T. Kobayashi, and S. Kusama, "Effects of ONO-6240, a platelet-activating factor antagonist, on endotoxin shock in unanesthetized sheep," *Prostaglandins*, vol. 31, no. 2, pp. 271–281, 1986.

[33] S. D'Humières, F. Russo-Marie, and B. Boris Vargaftig, "PAF-acether-induced synthesis of prostacyclin by human endothelial cells," *European Journal of Pharmacology*, vol. 131, no. 1, pp. 13–19, 1986.

[34] R. N. Saunders and D. A. Handley, "Platelet-activating factor antagonists," *Annual Review of Pharmacology and Toxicology*, vol. 27, pp. 237–255, 1987.

[35] D. A. Handley, J. C. Tomesch, and R. N. Saunders, "Inhibition of PAF-induced systemic responses in the rat, guinea pig, dog and primate by the receptor antagonist SRI 63-441," *Thrombosis and Haemostasis*, vol. 56, no. 1, pp. 40–44, 1986.

[36] M. Merlos, L. A. Gómez, M. Giral, M. L. Vericat, J. García-Rafanell, and J. Forn, "Effects of PAF-antagonists in mouse ear oedema induced by several inflammatory agents," *British Journal of Pharmacology*, vol. 104, no. 4, pp. 990–994, 1991.

[37] A. Wissner, M. L. Carroll, K. E. Green et al., "Analogues of platelet activating factor. 6. Mono- and bis-aryl phosphate antagonists of platelet activating factor," *Journal of Medicinal Chemistry*, vol. 35, no. 9, pp. 1650–1662, 1992.

[38] D. P. O'Hair, A. M. Roza, R. Komorowski et al., "Tulopafant, a PAF receptor antagonist, increases capillary patency and prolongs survival in discordant cardiac xenotransplants," *Journal of Lipid Mediators*, vol. 7, no. 1, pp. 79–84, 1993.

[39] Y. Komuro, N. Imanishi, M. Uchida, and S. Morooka, "Biological effect of orally active platelet-activating factor receptor antagonist SM-10661," *Molecular Pharmacology*, vol. 38, no. 3, pp. 378–384, 1990.

[40] L. M. Kuitert, R. M. Angus, N. C. Barnes et al., "Effect of a novel potent platelet-activating factor antagonist, modipafant, in clinical asthma," *American Journal of Respiratory and Critical Care Medicine*, vol. 151, no. 5, pp. 1331–1335, 1995.

[41] A. N. Kingsnorth, S. W. Galloway, and L. J. Formela, "Randomized, double-blind phase II trial of Lexipafant, a platelet-activating factor antagonist, in human acute pancreatitis," *British Journal of Surgery*, vol. 82, no. 10, pp. 1414–1420, 1995.

[42] J. Casals-Stenzel and H. O. Heuer, "Use of WEB 2086 and WEB 2170 as platelet-activating factor antagonists," *Methods in Enzymology*, vol. 187, pp. 455–465, 1990.

[43] L. M. Kuitert, K. P. Hui, S. Uthayarkumar et al., "Effect of the platelet-activating factor antagonist UK-74,505 on the early and late response to allergen," *The American Review of Respiratory Disease*, vol. 147, no. 1, pp. 82–86, 1993.

[44] A. Freitag, R. M. Watson, G. Matsos, C. Eastwood, and P. M. O'Byrne, "Effect of a platelet activating factor antagonist, WEB 2086, on allergen induced asthmatic responses," *Thorax*, vol. 48, no. 6, pp. 594–598, 1993.

[45] D. J. Evans, P. J. Barnes, M. Cluzel, and B. J. O'Connor, "Effects of a potent platelet-activating factor antagonist, SR27417A, on allergen-induced asthmatic responses," *American Journal of Respiratory and Critical Care Medicine*, vol. 156, no. 1, pp. 11–16, 1997.

[46] F. P. Gómez, J. Roca, J. A. Barberà, K. F. Chung, V. I. Peinado, and R. Rodriguez-Roisin, "Effect of a platelet-activating factor (PAF) antagonist, SR 27417A, on PAF-induced gas exchange abnormalities in mild asthma," *The European Respiratory Journal*, vol. 11, no. 4, pp. 835–839, 1998.

[47] D. P. Taggart, S. M. Browne, D. T. Wade, and P. W. Halligan, "Neuroprotection during cardiac surgery: a randomised trial of a platelet activating factor antagonist," *Heart*, vol. 89, no. 8, pp. 897–900, 2003.

[48] R. Taylor, D. Fatovich, T. Hitchcock, C. Morrison, and L. Curtis, "Platelet-activating factor antagonism and streptokinase-induced hypotension in clinical acute myocardial infarction," *Clinical Science*, vol. 100, no. 6, pp. 601–607, 2001.

[49] Y. Suputtamongkol, S. Intaranongpai, M. D. Smith et al., "A double-blind placebo-controlled study of an infusion of lexipafant (platelet-activating factor receptor antagonist) in patients with severe sepsis," *Antimicrobial Agents and Chemotherapy*, vol. 44, no. 3, pp. 693–696, 2000.

[50] C. D. Johnson, A. N. Kingsnorth, C. W. Imrie et al., "Double blind, randomised, placebo controlled study of a platelet activating factor antagonist, lexipafant, in the treatment and prevention of organ failure in predicted severe acute pancreatitis," *Gut*, vol. 48, no. 1, pp. 62–69, 2001.

[51] W. A. Stack, D. Jenkins, P. Vivet, and C. J. Hawkey, "Lack of effectiveness of the platelet-activating factor antagonist SR27417A in patients with active ulcerative colitis: a randomized controlled trial," *Gastroenterology*, vol. 115, no. 6, pp. 1340–1345, 1998.

[52] P. Hilliquin, V. Chermat-Izard, and C.-J. Menkes, "A double blind, placebo controlled study of a platelet activating factor antagonist in patients with rheumatoid arthritis," *The Journal of Rheumatology*, vol. 25, no. 8, pp. 1502–1507, 1998.

[53] M. E. Elders, M. J. P. Gerritsen, and P. C. M. van de Kerkhof, "The effect of topical application of the platelet-activating factor-antagonist, Ro 24-0238, in psoriasis vulgaris—a clinical and immunohistochemical study," *Clinical and Experimental Dermatology*, vol. 19, no. 6, pp. 453–457, 1994.

[54] M. Poeze, A. H. M. Froon, G. Ramsay et al., "Decreased organ failure in patients with severe sirs and septic shock

treated with the platelet-activating factor antagonist TCV-309: a prospective, multicenter, double-blind, randomized phase II trial," *Shock*, vol. 14, no. 4, pp. 421–428, 2000.

[55] B. J. O'Connor, S. Uden, T. J. Carty, J. D. Eskra, P. J. Barnes, and K. F. Chung, "Inhibitory effect of UK,74505, a potent and specific oral platelet activating factor (PAF) receptor antagonist, on airway and systemic responses to inhaled PAF in humans," *American Journal of Respiratory and Critical Care Medicine*, vol. 150, no. 1, pp. 35–40, 1994.

[56] M. Johnson, G. Kwatra, D. K. Badyal, and E. A. Thomas, "Levocetirizine and rupatadine in chronic idiopathic urticaria," *International Journal of Dermatology*, vol. 54, no. 10, pp. 1199–1204, 2015.

[57] E. Baltás, V. Trach, A. Dobozy, and L. Kemény, "Platelet-activating factor antagonist WEB 2086 inhibits ultraviolet-B radiation-induced dermatitis in the human skin," *Skin Pharmacology and Applied Skin Physiology*, vol. 16, no. 4, pp. 259–262, 2003.

[58] T. Wittwer, M. Grote, P. Oppelt, U. Franke, H.-J. Schaefers, and T. Wahlers, "Impact of PAF antagonist BN 52021 (Ginkolide B) on post-ischemic graft function in clinical lung transplantation," *The Journal of Heart and Lung Transplantation*, vol. 20, no. 3, pp. 358–363, 2001.

[59] S. Hozawa, Y. Haruta, S. Ishioka, and M. Yamakido, "Effects of a PAF antagonist, Y-24180, on bronchial hyperresponsiveness in patients with asthma," *American Journal of Respiratory and Critical Care Medicine*, vol. 152, no. 4 I, pp. 1198–1202, 1995.

[60] J. Mullol, J. Bousquet, C. Bachert et al., "Update on rupatadine in the management of allergic disorders," *Allergy: European Journal of Allergy and Clinical Immunology*, vol. 70, no. 100, pp. 1–24, 2015.

[61] C. Qian, S.-B. Hwang, L. Libertine-Garahan et al., "Anti-inflammatory activities of LDP-392, a dual PAF receptor antagonist and 5-lipoxygenase inhibitor," *Pharmacological Research*, vol. 44, no. 3, pp. 213–220, 2001.

[62] M. Fujita, T. Seki, H. Inada, K. Shimizu, A. Takahama, and T. Sano, "Novel agents combining platelet activating factor (PAF) receptor antagonist with thromboxane synthase inhibitor (TxSI)," *Bioorganic and Medicinal Chemistry Letters*, vol. 12, no. 5, pp. 771–774, 2002.

[63] I. M. Hussaini, Y.-H. Zhang, J. J. Lysiak, and T. Y. Shen, "Dithiolane analogs of lignans inhibit interferon-γ and lipopolysaccharide-induced nitric oxide production in macrophages," *Acta Pharmacologica Sinica*, vol. 21, no. 10, pp. 897–904, 2000.

[64] S. J. Keam and G. L. Plosker, "Rupatadine: a review of its use in the management of allergic disorders," *Drugs*, vol. 67, no. 3, pp. 457–474, 2007.

[65] C. A. Demopoulos, H. C. Karantonis, and S. Antonopoulou, "Platelet activating factor—a molecular link between atherosclerosis theories," *European Journal of Lipid Science and Technology*, vol. 105, no. 11, pp. 705–716, 2003.

[66] N. Tsantila, A. B. Tsoupras, E. Fragopoulou, S. Antonopoulou, C. Iatrou, and C. A. Demopoulos, "In vitro and in vivo effects of statins on platelet-activating factor and its metabolism," *Angiology*, vol. 62, no. 3, pp. 209–218, 2011.

[67] D. Kelefiotis, E. Lanara, C. Vakirtzi-Lemonias et al., "Study of Digoxin as inhibitor of the in vivo effects of acetyl glyceryl ether phosphorycholine (AGEPC) in mice," *Life Sciences*, vol. 42, no. 6, pp. 623–633, 1988.

[68] M. Chini, A. B. Tsoupras, N. Mangafas et al., "Effects of HAART on platelet-activating factor metabolism in naive HIV-infected patients I: study of the tenofovir-DF/emtricitabine/efavirenz

HAART regimen," *AIDS Research and Human Retroviruses*, vol. 28, no. 8, pp. 766–775, 2012.

[69] M. Chini, A. B. Tsoupras, N. Mangafas et al., "Effects of highly active antiretroviral therapy on platelet activating factor metabolism in naïve HIV-infected patients: II study of the abacavir/lamivudine/efavirenz haart regimen," *International Journal of Immunopathology and Pharmacology*, vol. 25, no. 1, pp. 247–258, 2012.

[70] V. D. Papakonstantinou, M. Chini, N. Mangafas et al., "In vivo effect of two first-line ART regimens on inflammatory mediators in male HIV patients," *Lipids in Health and Disease*, vol. 13, no. 1, article 90, 2014.

[71] A. B. Tsoupras, M. Chini, N. Tsogas et al., "Anti-platelet-activating factor effects of highly active antiretroviral therapy (HAART): a new insight in the drug therapy of HIV infection?" *AIDS Research and Human Retroviruses*, vol. 24, no. 8, pp. 1079–1086, 2008.

[72] T. Kelesidis, V. Papakonstantinou, P. Detopoulou et al., "The role of platelet-activating factor in chronic inflammation, immune activation, and comorbidities associated with HIV infection," *AIDS Reviews*, vol. 17, no. 4, pp. 191–201, 2015.

[73] S. Antonopoulou, C. E. Semidalas, S. Koussissis, and C. A. Demopoulos, "Platelet-Activating Factor (PAF) antagonists in foods: a study of lipids with PAF or anti-PAF-like activity in cow's milk and yogurt," *Journal of Agricultural and Food Chemistry*, vol. 44, no. 10, pp. 3047–3051, 1996.

[74] H. C. Karantonis, S. Antonopoulou, and C. A. Demopoulos, "Antithrombotic lipid minor constituents from vegetable oils. Comparison between olive oils and others," *Journal of Agricultural and Food Chemistry*, vol. 50, no. 5, pp. 1150–1160, 2002.

[75] E. Fragopoulou, S. Antonopoulou, and C. A. Demopoulos, "Biologically active lipids with antiatherogenic properties from white wine and must," *Journal of Agricultural and Food Chemistry*, vol. 50, no. 9, pp. 2684–2694, 2002.

[76] E. Fragopoulou, S. Antonopoulou, T. Nomikos, and C. A. Demopoulos, "Structure elucidation of phenolic compounds from red/white wine with antiatherogenic properties," *Biochimica et Biophysica Acta (BBA)—Molecular and Cell Biology of Lipids*, vol. 1632, no. 1–3, pp. 90–99, 2003.

[77] E. Fragopoulou, T. Nomikos, S. Antonopoulou, C. A. Mitsopoulou, and C. A. Demopoulos, "Separation of biologically active lipids from red wine," *Journal of Agricultural and Food Chemistry*, vol. 48, no. 4, pp. 1234–1238, 2000.

[78] E. Fragopoulou, T. Nomikos, N. Tsantila, A. Mitropoulou, I. Zabetakis, and C. A. Demopoulos, "Biological activity of total lipids from red and white wine/must," *Journal of Agricultural and Food Chemistry*, vol. 49, no. 11, pp. 5186–5193, 2001.

[79] S. Koussissis, C. E. Semidalas, E. C. Hadzistavrou, V. Kalyvas, S. Antonopoulou, and C. A. Demopoulos, "PAF antagonists in food: isolation and identification of PAF antagonists in honey and wax," *Revue Francaise des Corps Gras*, vol. 5, no. 6, p. 127, 1994.

[80] J. Rementzis, S. Antonopoulou, D. Argyropoulos, and C. A. Demopoulos, "Biologically active lipids from *S. scombrus*," in *Platelet-Activating Factor and Related Lipid Mediators 2*, vol. 416 of *Advances in Experimental Medicine and Biology*, pp. 65–72, Springer, 1996.

[81] A. Panayiotou, D. Samartzis, T. Nomikos et al., "Lipid fractions with aggregatory and antiaggregatory activity toward platelets in fresh and fried cod (Gadus morhua): correlation with platelet-activating factor and atherogenesis," *Journal of*

Agricultural and Food Chemistry, vol. 48, no. 12, pp. 6372–6379, 2000.

[82] T. Nomikos, H. C. Karantonis, C. Skarvelis, C. A. Demopoulos, and I. Zabetakis, "Antiatherogenic properties of lipid fractions of raw and fried fish," *Food Chemistry*, vol. 96, no. 1, pp. 29–35, 2006.

[83] S. Antonopoulou, C. E. Semidalas, S. Koussissis, and C. A. Demopoulos, "Platelet Activating Factor (PAF) antagonists in foods: a study of lipids with PAF or anti-PAF-like activity in cow's milk and yogurt," *Journal of Agricultural and Food Chemistry*, vol. 44, no. 10, pp. 3047–3051, 1996.

[84] S. Antonopoulou, E. Fragopoulou, H. C. Karantonis et al., "Effect of traditional Greek Mediterranean meals on platelet aggregation in normal subjects and in patients with type 2 diabetes mellitus," *Journal of Medicinal Food*, vol. 9, no. 3, pp. 356–362, 2006.

[85] C. A. Demopoulos and S. Antonopoulou, "A discovery trip to compounds with PAF-like activity," *Advances in Experimental Medicine and Biology*, vol. 416, pp. 59–63, 1997.

[86] E. Fragopoulou, T. Nomikos, H. C. Karantonis et al., "Biological activity of acetylated phenolic compounds," *Journal of Agricultural and Food Chemistry*, vol. 55, no. 1, pp. 80–89, 2007.

[87] H. C. Karantonis, N. Tsantila, G. Stamatakis et al., "Bioactive polar lipids in olive oil, pomace and waste byproducts," *Journal of Food Biochemistry*, vol. 32, no. 4, pp. 443–459, 2008.

[88] H. C. Karantonis, I. Zabetakis, T. Nomikos, and C. A. Demopoulos, "Antiatherogenic properties of lipid minor constituents from seed oils," *Journal of the Science of Food and Agriculture*, vol. 83, no. 12, pp. 1192–1204, 2003.

[89] H. C. Karantonis, S. Antonopoulou, D. N. Perrea et al., "In vivo antiatherogenic properties of olive oil and its constituent lipid classes in hyperlipidemic rabbits," *Nutrition, Metabolism and Cardiovascular Diseases*, vol. 16, no. 3, pp. 174–185, 2006.

[90] N. Tsantila, H. C. Karantonis, D. N. Perrea et al., "Antithrombotic and antiatherosclerotic properties of olive oil and olive pomace polar extracts in rabbits," *Mediators of Inflammation*, vol. 2007, Article ID 36204, 11 pages, 2007.

[91] M. N. Xanthopoulou, K. Kalathara, S. Melachroinou et al., "Wine consumption reduced postprandial platelet sensitivity against platelet activating factor in healthy men," *European Journal of Nutrition*, 2016.

[92] T. Nomikos, E. Fragopoulou, and S. Antonopoulou, "Food ingredients and lipid mediators," *Current Nutrition and Food Science*, vol. 3, no. 4, pp. 255–276, 2007.

[93] R. Apitz-Castro, S. Cabrera, M. R. Cruz, E. Ledezma, and M. K. Jain, "Effects of garlic extract and of three pure components isolated from it on human platelet aggregation, arachidonate metabolism, release reaction and platelet ultrastructure," *Thrombosis Research*, vol. 32, no. 2, pp. 155–169, 1983.

[94] H. Lim, K. Kubota, A. Kobayashi, T. Seki, and T. Ariga, "Inhibitory effect of sulfur-containing compounds in Scorodocarpus borneensis Becc. on the aggregation of rabbit platelets," *Bioscience, Biotechnology and Biochemistry*, vol. 63, no. 2, pp. 298–301, 1999.

[95] J. Morris, V. Burke, T. A. Mori, R. Vandongen, and L. J. Beilin, "Effects of garlic extract on platelet aggregation: a randomized placebo-controlled double-blind study," *Clinical and Experimental Pharmacology and Physiology*, vol. 22, no. 6-7, pp. 414–417, 1995.

[96] H. Tsuchiya, M. Sato, and I. Watanabe, "Antiplatelet activity of soy sauce as functional seasoning," *Journal of Agricultural and Food Chemistry*, vol. 47, no. 10, pp. 4167–4174, 1999.

[97] Y. M. Sagesaka, T. Uemura, N. Watanabe, K. Sakata, and J. Uzawa, "A new glucuronide saponin from tea leaves *Camellia sinensis* var. *sinensis*," *Bioscience, Biotechnology and Biochemistry*, vol. 58, no. 11, pp. 2036–2040, 1994.

[98] C. V. Rao, A. Rivenson, B. Simi, and B. S. Reddy, "Chemoprevention of colon carcinogenesis by dietary curcumin, a naturally occurring plant phenolic compound," *Cancer Research*, vol. 55, no. 2, pp. 259–266, 1995.

[99] B. H. Shah, Z. Nawaz, S. A. Pertani et al., "Inhibitory effect of curcumin, a food spice from turmeric, on platelet-activating factor- and arachidonic acid-mediated platelet aggregation through inhibition of thromboxane formation and Ca^{2+} signaling," *Biochemical Pharmacology*, vol. 58, no. 7, pp. 1167–1172, 1999.

[100] S. N. Verouti, A. B. Tsoupras, F. Alevizopoulou, C. A. Demopoulos, and C. Iatrou, "Paricalcitol effects on activities and metabolism of platelet activating factor and on inflammatory cytokines in hemodialysis patients," *The International Journal of Artificial Organs*, vol. 36, no. 2, pp. 87–96, 2013.

[101] F. Violi, D. Pratico, A. Ghiselli et al., "Inhibition of cyclooxygenase-independent platelet aggregation by low vitamin E concentration," *Atherosclerosis*, vol. 82, no. 3, pp. 247–252, 1990.

[102] E. Kakishita, A. Suehiro, Y. Oura, and K. Nagai, "Inhibitory effect of vitamin E (α-tocopherol) on spontaneous platelet aggregation in whole blood," *Thrombosis Research*, vol. 60, no. 6, pp. 489–499, 1990.

[103] D. Tsoukatos, C. A. Demopoulos, A. D. Tselepis et al., "Inhibition by cardiolipins of platelet-activating factor-induced rabbit platelet activation," *Lipids*, vol. 28, no. 12, pp. 1119–1124, 1993.

[104] S. Ishii and T. Shimizu, "Platelet-activating factor (PAF) receptor and genetically engineered PAF receptor mutant mice," *Progress in Lipid Research*, vol. 39, no. 1, pp. 41–82, 2000.

[105] V. O. Melnikova, K. Balasubramanian, G. J. Villares et al., "Crosstalk between protease-activated receptor 1 and platelet-activating factor receptor regulates melanoma cell adhesion molecule (MCAM/MUC18) expression and melanoma metastasis," *The Journal of Biological Chemistry*, vol. 284, no. 42, pp. 28845–28855, 2009.

[106] V. O. Melnikova, G. J. Villares, and M. Bar-Eli, "Emerging roles of PAR-1 and PAFR in melanoma metastasis," *Cancer Microenvironment*, vol. 1, no. 1, pp. 103–111, 2008.

[107] A. B. Tsoupras, C. Iatrou, C. Frangia, and C. A. Demopoulos, "The implication of platelet activating factor in cancer growth and metastasis: potent beneficial role of PAF-inhibitors and antioxidants," *Infectious Disorders—Drug Targets*, vol. 9, no. 4, pp. 390–399, 2009.

[108] S. Yeu, C.-M. Sun, H.-H. Chuang, and P.-T. Chang, "Studies on the cytotoxic mechanisms of ginkgetin in a human ovarian adenocarcinoma cell line," *Naunyn-Schmiedeberg's Archives of Pharmacology*, vol. 362, no. 1, pp. 82–90, 2000.

[109] Y. Zhang, A. Y. Chen, M. Li, C. Chen, and Q. Yao, "*Ginkgo biloba* extract kaempferol inhibits cell proliferation and induces apoptosis in pancreatic cancer cells," *Journal of Surgical Research*, vol. 148, no. 1, pp. 17–23, 2008.

[110] T. Matsuda, M. Kuroyanagi, S. Sugiyama, K. Umehara, A. Ueno, and K. Nishi, "Cell differentiation-inducing diterpenes from Andrographis paniculata NEES," *Chemical and Pharmaceutical Bulletin*, vol. 42, no. 6, pp. 1216–1225, 1994.

[111] S. Nanduri, V. K. Nyavanandi, S. Sanjeeva Rao Thunuguntla et al., "Synthesis and structure-activity relationships of andrographolide analogues as novel cytotoxic agents," *Bioorganic and Medicinal Chemistry Letters*, vol. 14, no. 18, pp. 4711–4717, 2004.

[112] H. Li, H. J. Lee, Y. H. Ahn et al., "Tussilagone suppresses colon cancer cell proliferation by promoting the degradation of β-catenin," *Biochemical and Biophysical Research Communications*, vol. 443, no. 1, pp. 132–137, 2014.

[113] S. I. de Oliveira, L. N. S. Andrade, A. C. Onuchic et al., "Platelet-activating factor receptor (PAF-R)-dependent pathways control tumour growth and tumour response to chemotherapy," *BMC Cancer*, vol. 10, article 200, 2010.

[114] A. C. Onuchic, C. M. L. Machado, R. F. Saito, F. J. Rios, S. Jancar, and R. Chammas, "Expression of PAFR as part of a prosurvival response to chemotherapy: a novel target for combination therapy in melanoma," *Mediators of Inflammation*, vol. 2012, Article ID 175408, 6 pages, 2012.

[115] T. W. Hambley, "Metal-based therapeutics," *Science*, vol. 318, no. 5855, pp. 1392–1393, 2007.

[116] S. H. van Rijt and P. J. Sadler, "Current applications and future potential for bioinorganic chemistry in the development of anticancer drugs," *Drug Discovery Today*, vol. 14, no. 23-24, pp. 1089–1097, 2009.

[117] E. Meggers, "Exploring biologically relevant chemical space with metal complexes," *Current Opinion in Chemical Biology*, vol. 11, no. 3, pp. 287–292, 2007.

[118] G. Gasser, I. Ott, and N. Metzler-Nolte, "Organometallic anticancer compounds," *Journal of Medicinal Chemistry*, vol. 54, no. 1, pp. 3–25, 2011.

[119] N. Farrell, "Biomedical uses and applications of inorganic chemistry. An overview," *Coordination Chemistry Reviews*, vol. 232, no. 1-2, pp. 1–4, 2002.

[120] H. B. Gray, "Biological inorganic chemistry at the beginning of the 21st century," *Proceedings of the National Academy of Sciences of the United States of America*, vol. 100, no. 7, pp. 3563–3568, 2003.

[121] Y. K. Yan, M. Melchart, A. Habtemariam, and P. J. Sadler, "Organometallic chemistry, biology and medicine: ruthenium arene anticancer complexes," *Chemical Communications*, no. 38, pp. 4764–4776, 2005.

[122] S. J. Lippard and J. M. Berg, "Bio-inorganic chemistry: newly charted waters," *Current Opinion in Chemical Biology*, vol. 4, no. 2, pp. 137–139, 2000.

[123] C.-H. Leung, S. Lin, H.-J. Zhong, and D.-L. Ma, "Metal complexes as potential modulators of inflammatory and autoimmune responses," *Chemical Science*, vol. 6, no. 2, pp. 871–884, 2015.

[124] A. I. Philippopoulos, N. Tsantila, C. A. Demopoulos, C. P. Raptopoulou, V. Likodimos, and P. Falaras, "Synthesis, characterization and crystal structure of the *cis*-[RhL$_2$Cl$_2$]Cl complex with the bifunctional ligand (L) 2-(2*'*-pyridyl)quinoxaline. Biological activity towards PAF (Platelet Activating Factor) induced platelet aggregation," *Polyhedron*, vol. 28, no. 15, pp. 3310–3316, 2009.

[125] A. E. A. Porter, "The structure, reactions, synthesis, and uses of heterocyclic compounds," in *Comprehensive Heterocyclic Chemistry*, vol. 3, pp. 157–197, Pergamon Press, New York, NY, USA, 1984.

[126] C. A. Mitsopoulou and C. Dagas, "Synthesis, characterization, DNA binding, and photocleavage activity of oxorhenium (V)

complexes with α-diimine and quinoxaline ligands," *Bioinorganic Chemistry and Applications*, vol. 2010, Article ID 973742, 9 pages, 2010.

[127] M. Harlos, I. Ott, R. Gust et al., "Synthesis, biological activity, and structure-activity relationships for potent cytotoxic rhodium(III) polypyridyl complexes," *Journal of Medicinal Chemistry*, vol. 51, no. 13, pp. 3924–3933, 2008.

[128] S. Thangavel, M. Paulpandi, H. B. Friedrich, K. Murugan, S. Kalva, and A. A. Skelton, "Synthesis, characterization, antiproliferative and molecular docking study of new half sandwich Ir(III), Rh(III) and Ru(II) complexes," *Journal of Inorganic Biochemistry*, vol. 159, pp. 50–61, 2016.

[129] D. A. Medvetz, K. D. Stakleff, T. Schreiber et al., "Ovarian cancer activity of cyclic amine and thiaether metal complexes," *Journal of Medicinal Chemistry*, vol. 50, no. 7, pp. 1703–1706, 2007.

[130] F. P. Pruchnik, P. Jakimowicz, Z. Ciunik et al., "Rhodium(III) complexes with polypyridyls and pyrazole and their antitumor activity," *Inorganica Chimica Acta*, vol. 334, pp. 59–66, 2002.

[131] P. Falaras and A. I. Philippopoulos, "Platelet Activating Factor (PAF) inhibitors with possible antitumor activity," patent number 1006959, Industrial Property Organization of Greece, 2009.

[132] A. B. Tsoupras, A. Papakyriakou, C. A. Demopoulos, and A. I. Philippopoulos, "Synthesis, biochemical evaluation and molecular modeling studies of novel rhodium complexes with nanomolar activity against platelet activating factor," *Journal of Inorganic Biochemistry*, vol. 120, pp. 63–73, 2013.

[133] T. S. Kamatchi, N. Chitrapriya, S. K. Kim, F. R. Fronczek, and K. Natarajan, "Influence of carboxylic acid functionalities in ruthenium (II) polypyridyl complexes on DNA binding, cytotoxicity and antioxidant activity: synthesis, structure and *in vitro* anticancer activity," *European Journal of Medicinal Chemistry*, vol. 59, pp. 253–264, 2013.

[134] A. Böcker, P. R. Bonneau, O. Hucke, A. Jakalian, and P. J. Edwards, "Development of specific "drug-like property" rules for carboxylate-containing oral drug candidates," *ChemMedChem*, vol. 5, no. 12, pp. 2102–2113, 2010.

[135] S.-H. Xia and D.-C. Fang, "Pharmacological action and mechanisms of ginkgolide B," *Chinese Medical Journal*, vol. 120, no. 10, pp. 922–928, 2007.

[136] C. N. Verani, "Metal complexes as inhibitors of the 26S proteasome in tumor cells," *Journal of Inorganic Biochemistry*, vol. 106, no. 1, pp. 59–67, 2012.

[137] H. O. Heuer, "Current status of PAF antagonists," *Clinical and Experimental Allergy*, vol. 22, no. 11, pp. 980–983, 1992.

[138] H. O. Heuer, "Involvement of platelet-activating factor (PAF) in septic shock and priming as indicated by the effect of hetrazepinoic PAF antagonists," *Lipids*, vol. 26, no. 12, pp. 1369–1373, 1991.

[139] I. Izquierdo, M. Merlos, and J. García-Rafanell, "Rupatadine: A new selective histamine H1 receptor and platelet-activating factor (PAF) antagonist. A review of pharmacological profile and clinical management of allergic rhinitis," *Drugs of Today*, vol. 39, no. 6, pp. 451–468, 2003.

[140] A. B. Tsoupras, M. Roulia, E. Ferentinos, I. Stamatopoulos, C. A. Demopoulos, and P. Kyritsis, "Structurally diverse metal coordination compounds, bearing imidodiphosphinate and diphosphinoamine ligands, as potential inhibitors of the platelet activating factor," *Bioinorganic Chemistry and Applications*, vol. 2010, Article ID 731202, 9 pages, 2010.

[141] E. Ferentinos, A. B. Tsoupras, M. Roulia, S. D. Chatziefthimiou, C. A. Demopoulos, and P. Kyritsis, "Inhibitory activity of the

novel $Zn[(OPPh_2)(SePPh_2)N]_2$ complex towards the Platelet Activating Factor (PAF) and thrombin: comparison with its isomorphous Co(II) and Ni(II) analogues," *Inorganica Chimica Acta*, vol. 378, no. 1, pp. 102–108, 2011.

[142] M. Kaplanis, G. Stamatakis, V. D. Papakonstantinou, M. Paravatou-Petsotas, C. A. Demopoulos, and C. A. Mitsopoulou, "Re(I) tricarbonyl complex of 1,10-phenanthroline-5,6-dione: DNA binding, cytotoxicity, anti-inflammatory and anticoagulant effects towards platelet activating factor," *Journal of Inorganic Biochemistry*, vol. 135, pp. 1–9, 2014.

[143] A. Margariti, *Rh(III) complexes and study of their biological activity against the Platelet Acivating Factor [M.S. thesis]*, National and Kapodistrian University of Athens, Athens, Greece, 2014.

[144] A.-B. Tsoupras, V. Papakonstantinou, G. M. Stamatakis, C. A. Demopoulos, P. Falaras, and A. I. Philippopoulos, "Biochemical evaluation of ruthenium-based complexes towards PAF (Platelet Activating Factor) and Thrombin. Potent anti-inflammatory agents," *Science Letters Journal*, vol. 4, article 208, 2015.

[145] W. H. Ang and P. J. Dyson, "Classical and non-classical ruthenium-based anticancer drugs: towards targeted chemotherapy," *European Journal of Inorganic Chemistry*, vol. 2006, no. 20, pp. 4003–4018, 2006.

[146] A. Bergamo, C. Gaiddon, J. H. M. Schellens, J. H. Beijnen, and G. Sava, "Approaching tumour therapy beyond platinum drugs: status of the art and perspectives of ruthenium drug candidates," *Journal of Inorganic Biochemistry*, vol. 106, no. 1, pp. 90–99, 2012.

[147] M. J. Clarke, "Ruthenium metallopharmaceuticals," *Coordination Chemistry Reviews*, vol. 232, no. 1-2, pp. 69–93, 2002.

[148] G. Konti, E. Chatzivasiloglou, V. Likodimos et al., "Influence of pyridine ligand nature and the corresponding ruthenium(II) dye molecular structure on the performance of dye-sensitized solar cells," *Photochemical and Photobiological Sciences*, vol. 8, no. 5, pp. 726–732, 2009.

[149] A. Philippopoulos, P. Falaras, E. Chatzivasiloglou, O. Igglessi-Markopoulou, V. Likodimos, and G.-C. Konti, "Synthesis and spectroscopic characterization of new heteroleptic ruthenium(II) complexes incorporating 2-(2ʹ-pyridyl) quinoxaline and 4-carboxy-2-(2ʹ-pyridyl)quinoline," *Journal of Coordination Chemistry*, vol. 65, no. 14, pp. 2535–2548, 2012.

[150] A. I. Philippopoulos, A. Terzis, C. P. Raptopoulou, V. J. Catalano, and P. Falaras, "Synthesis, characterization, and sensitizing properties of heteroleptic Ru^{II} complexes based on 2,6-bis(1-pyrazolyl)pyridine and 2,2ʹ-bipyridine-4,4ʹ-dicarboxylic acid ligands," *European Journal of Inorganic Chemistry*, vol. 2007, no. 36, pp. 5633–5644, 2007.

[151] L. G. Kritikou, M. C. Moschidis, A. Siafaka, and A. C. Demopoulos, "Biological activities of phosphono-, 2-O-trichloroacetyl, 2-O-trifluoroacetyl- analogs of 1-O- Alkyl-2-O-acetyl-SN-glyceryl-3-phosphoryl-choline (AGEPC) and of acetylated glyceryl ether derivatives: platelet-Activating Factor," in *Proceedings of the INSERM Symposium no. 23*, pp. 65–72, Elsevier Science Publishers B.V., Amsterdam, The Netherlands, 1983.

[152] M. Merlos, M. Giral, D. Balsa et al., "Rupatadine, a new potent, orally active dual antagonist of histamine and platelet-activating factor (PAF)," *Journal of Pharmacology and Experimental Therapeutics*, vol. 280, no. 1, pp. 114–121, 1997.

6

Structure-Based Drug Design for Cytochrome P450 Family 1 Inhibitors

Zbigniew Dutkiewicz ⓘ[1] and Renata Mikstacka[2]

[1]Department of Chemical Technology of Drugs, Poznań University of Medical Sciences,
 Grunwaldzka 6, 60-780 Poznań, Poland
[2]Department of Inorganic and Analytical Chemistry, Ludwik Rydygier Collegium Medicum,
 Nicolaus Copernicus University in Toruń, Dr A. Jurasza 2, 85-089 Bydgoszcz, Poland

Correspondence should be addressed to Zbigniew Dutkiewicz; zdutkie@ump.edu.pl

Academic Editor: Albrecht Messerschmidt

Cytochromes P450 are a class of metalloproteins which are responsible for electron transfer in a wide spectrum of reactions including metabolic biotransformation of endogenous and exogenous substrates. The superfamily of cytochromes P450 consists of families and subfamilies which are characterized by a specific structure and substrate specificity. Cytochromes P450 family 1 (CYP1s) play a distinctive role in the metabolism of drugs and chemical procarcinogens. In recent decades, these hemoproteins have been intensively studied with the use of computational methods which have been recently developed remarkably to be used in the process of drug design by the virtual screening of compounds in order to find agents with desired properties. Moreover, the molecular modeling of proteins and ligand docking to their active sites provide an insight into the mechanism of enzyme action and enable us to predict the sites of drug metabolism. The review presents the current status of knowledge about the use of the computational approach in studies of ligand-enzyme interactions for CYP1s. Research on the metabolism of substrates and inhibitors of CYP1s and on the selectivity of their action is particularly valuable from the viewpoint of cancer chemoprevention, chemotherapy, and drug-drug interactions.

1. Introduction

In the past decade, an interest in the use of computational methods in preclinical drug discovery has been continuously growing. Structure-based drug design (SBDD) became possible due to the availability of X-ray structures of receptors and the development of molecular modeling methods. The combined techniques employed in a drug discovery with respect to numerous receptors are still being improved and are becoming of better and better quality. There are thorough reviews of the actual possibilities of and prospects for using computational methods in drug design [1–5]. The present review is devoted to studies on cytochrome P450 family 1 (CYP1), an important family of enzymes responsible for drug metabolism and procarcinogen activation, with the use of molecular docking and molecular dynamics simulations. In some respects, family 1 of cytochromes P450 is exceptional. It comprises two isozymes, CYP1A1 and CYP1A2, of a great

similarity and a third, more distinctive CYP1B1, which displays overlapping substrate specificity with the other members of the family. These three enzymes possessing relatively small binding cavities are valuable objects for comparative studies of enzymatic catalysis and ligand-enzyme interactions with the use of computational methods.

Cytochromes P450 (CYPs) are a superfamily of constitutive and inducible enzymes—hemoproteins—responsible for the oxidative metabolism of various xenobiotics and bioactive endogenous compounds. At present, 57 human genes of cytochromes P450 are known; they demonstrate a significant interindividual genetic variability [6]. On the basis of structural homology, CYPs may be assigned to the same family if they share not less than 40% of the amino acid sequence identity. Isoforms showing more than 55% sequence identity belong to the same subfamily. Family 1 (CYP1) comprises three isoforms: CYP1A1, CYP1A2, and CYP1B1. The amino acid sequence of CYP1A2 is 72% identical to that of CYP1A1, while CYP1B1 has lower amino acid sequence identity with both CYP1A1 (38%) and CYP1A2 (37%). Despite that, CYP1B1 is qualified as a CYP1 member on the basis of similar substrate specificity and the common induction of CYP1s by the aryl hydrocarbon receptor (AHR) [7].

CYP superfamily enzymes contain a heme prosthetic group catalyzing oxidation reactions and N- and O-dealkylations of substrates. CYPs are diversified in respect to substrate specificity and inhibitor susceptibility. Family 1 of these enzymes/CYP1s is responsible for the phase I metabolism of endogenous and exogenous substrates. CYP1s participate in the oxidative metabolism of endogenous substances, such as bile acids, steroid hormones, and lipids, exogenous compounds, and numerous pharmaceuticals and compounds derived from environmental pollution. CYP1s metabolize potential carcinogens: aryl hydrocarbons, aromatic amines, heterocyclic aromatic amines, and heterocyclic amines. CYP1s play a pivotal role in procarcinogen activation catalyzing metabolism of 66% of potential carcinogens [8]. The biotransformation of procarcinogens leads to the formation of mutagenic compounds, which form adducts with nucleic bases, responsible for the initiation of carcinogenesis.

CYP1B1 plays a particularly important role in pathogenesis of hormone-induced cancers, being responsible for metabolism of 17-alpha-estradiol (E2) to highly mutagenic and carcinogenic 4-hydroxy-E2 [9]. 4-Hydroxy-E2 and other products of E2 metabolism—quinones and semiquinones—exhibit genotoxic activity by forming adducts with nucleic acids [10]. Recently, CYP1B1's role in cancer progression and metastasis was reported [11]. CYP1A1 and CYP1B1 are targets of anticancer agents because of their overexpression in tumor cells compared to their normal counterparts [12, 13]. They may be used as a marker/tumor antigen in therapeutic strategies [14]. Overexpressed CYP1A1 or CYP1B1 in target tissues may play a double-sided role: they may activate prodrugs to their therapeutic forms, and on the contrary, they metabolize chemotherapeutics to their inactive forms. Inhibitors of CYP1B1 activity are used in mechanistic studies of drug metabolism [15].

Multidrug resistance caused by an efficient metabolism of chemotherapeutics catalyzed by CYP1A1 and CYP1B1 is a crucial problem in cancer chemotherapy. Moreover, inter- and intraspecies variability of CYP structures results in unique profiles of enzyme activities, which influence the therapeutic action of drugs metabolized by CYPs. The modeling of unique CYP structures may help to explain individual variation in response to drugs. To stop the inactivation of chemotherapeutics and to avoid drug resistance, the use of CYP1s inhibitors is proposed [9, 14–16].

The expression of CYP1 isoenzymes is tissue specific [17]. CYP1A1 is an inducible enzyme that occurs in the lungs and the trachea. Its induction is dependent on the presence of pollutants in the environment. CYP1A2 is a hepatic constitutive form of the enzyme that is responsible for drug metabolism in the liver. CYP1B1 occurs in the majority of extrahepatic tissues. A high level of CYP1B1 is observed in the bone marrow, kidney, spleen, thyroid gland, and reproductive tissues such as the uterus or prostate and mammary glands [17].

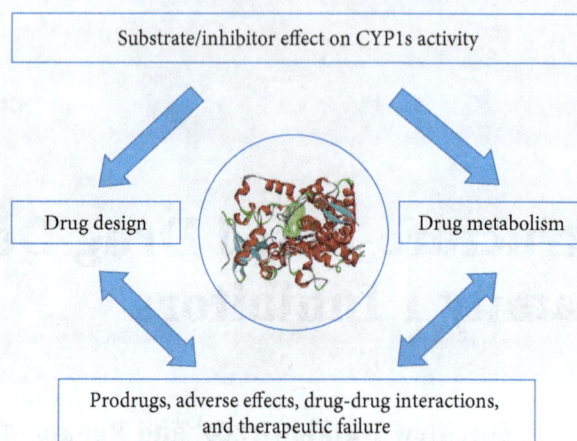

FIGURE 1: Relevance of CYP1 structure-activity relationship studies.

A focus on CYPs, in particular on CYP1s, arises from the role they play in the activation of procarcinogens. CYPs involved in the initiation of carcinogenesis have become the targets for anticancer strategy [18, 19]. One of the prophylactic actions is chemoprevention, which was defined as a prevention, inhibition, or reversal of carcinogenesis with the use of compounds of natural origin, their derivatives, or synthetic compounds [20]. The inhibition of CYP1A1 and CYP1B1 activities by natural compounds present in human diet constitutes one of the chemopreventive strategies [21].

In the past, the interactions between cytochromes P450 (CYPs) and their ligands have been investigated on the basis of the homology models of cytochrome P450 isozymes [22–24]. The homology models of CYPs had been used until the crystallographic structures of CYP1A1, CYP1A2, and CYP1B1 were determined and described [25–27]. Now, the crystal structures of CYP1s are available in the Protein Data Bank (PDB, http://www.rcsb.org/pdb/home/home.do), which greatly facilitates the progress of hit identifications.

The computational approach in structure-activity relationship studies is conducted with the aim of determining the site of metabolism of drugs and prodrugs and explaining the mechanism of therapeutic failure, prodrug toxicity, and adverse effects (Figure 1). In computational-aided drug design at the "hit-to-lead" stage, two different approaches are used: the structure-based (or receptor-based) drug design (SBDD) and the ligand-based drug design (LBDD) in relation to the known structure of a receptor or a ligand.

This review summarizes in silico studies and the studies that combine investigations in vitro with a computational approach to CYP family 1 enzymes as targets for the inhibitory activity of ligands. In our review, we focused on the SBDD concerning CYP1 family enzymes. We surveyed the studies of interactions of lead compounds with CYP1A1, CYP1A2, and CYP1B1. We also discussed the studies focused on the search of compounds: substrates and inhibitors, which display selective and potent molecular interactions with individual isozymes.

2. Structures of CYP1 Family Members

2.1. CYP1A1. Before the structure of human CYP1A1 was determined, most reports concerning CYP1A1 were

TABLE 1: The summary of studies on CYP1-ligand interactions.

Cytochrome P450	Ligand	Methods	Notes	References
CYP1A1 and other cytochromes				
CYP1A1 and CYP1A2	Aromatic amines, heterocyclic amines, aromatic hydrocarbons (benzo(a) pyrenemethylcholantrene), phenacetin, furafylline, and 7-methoxyresorufin	Homology modeling based on the CYP102 crystal structure	Human, mouse, rabbit, and trout CYP sequences	[22]
CYP1A1	7-Ethoxyresorufin, 7-methoxyresorufin, and benzo[a] pyrene	Homology modeling		[28]
CYP1A1, CYP1A2, and CYP1B1	Rutaecarpine and its derivatives	Homology modeling		[29]
CYP1A1	Arachidonic acid and eicosopentaenoic acid	Homology modeling	Molecular docking explains regiospecificity of metabolism	[30]
CYP1A1 and CYP1A2	7-Methoxyresorufin and 7-ethoxyresorufin	Homology modeling	Active site mutations in human CYP1A1 and CYP1A2	[31]
CYP1A1 and CYP1A2	Dietary flavonoids	Homology modeling		[32]
CYP1A1	B[a]P, B[a]P-7R, 8R-dihydrodiol, B[a]P-7S, 8S-dihydrodiol, eicosapentaenoate, and arachidonate	Homology modeling	Regioselectivity	[33]
CYP1A1	Ethoxyresorufin	Homology modeling		[34]
CYP1A1	B[a]P	Wild-type and exon 6 del CYP1A1 homology models		[35]
CYP1A1, CYP1A2, and CYP1B1	Alkoxyl derivatives of 7,8-dehydrorutaecarpine	Homology models based on the crystal structure of rabbit CYP2C5		[36]
CYP1A1	B[a]P, TCB, and TCDD	Rat, human, scup, and killifish homology models		[37]
CYP1A1	Representative ligands: α-naphthoflavone and benzothiazole	Homology modeling		[38]
CYP1A1 and CYP1A2 (CYP2A6 and CYP2B1)	Arylacetylenes	CYP1A2 crystal structure (PDB: 2HI4) and homology model of CYP1A1	Distances of ligands to heme, Fe, and Phe residues were analyzed	[39]
CYP1A1	Benzoxazoles and benzothiazoles	CoMFA, homology modeling, and molecular docking		[40]
CYP1A1, CYP1A2, and CYP1B1 (CYP2C9 and CYP3A4)	33 flavonoid derivatives	PDB: 2HI4 and homology models of CYP1A1 and CYP1B1	Hydroxyl and methoxy derivatives of flavone more potent as CYP inhibitors	[41]
CYP1A1, CYP1A2, and CYP1B1	Methoxyflavonoids	PDB: 2HI4 and homology models of CYP1A1 and CYP1B1	Important amino acid residues	[42]
CYP1A1 and CYP2B1	p-Aminophenol-succinic acid derivatives (acetylcholinesterase inhibitors)	Homology modeling of rat CYPs based on structures of CYP1A2 and CYP3A4 and molecular dynamics	Biological experiments on rat microsomes induced with 5,6-benzoflavone and phenobarbital	[43]
CYP1A1, CYP1A2, and CYP1B1	17-β-Estradiol	PDB: 2HI4 and homology models of CYP1A1 and CYP1B1	Important amino acid residues	[44]
CYP1A1	3,3′,4,4′, 5-Pentachlorobiphenyl	Homology modeling	Rat and human recombinant microsomes	[45]
CYP1A1 and CYP1B1	Resveratrol and its derivatives	Homology modeling based on CYP1A2 crystal structure		[46]
CYP1A1 and CYP1B1	Dietary flavonoids	Homology models based on the structure of CYP1A2 (PDB: 2HI4)		[47]

TABLE 1: Continued.

Cytochrome P450	Ligand	Methods	Notes	References
CYP1A1 and CYP1A2 (CYP1A6 and CYP2B1)	Flavone propargyl ethers	CYP1A2 crystal structure (PDB: 2HI4) and homology model of CYP1A1	Flavone propargyl ethers are more potent inhibitors of CYP1A1 and CYP1A2 than the parent hydroxy flavones	[48]
CYP1A1 and CYP1A2	Phenacetin and acetaminophen	CYP1A2 crystal structure (PDB: 2HI4) and homology model of CYP1A1	Isoform-selective metabolism	[49]
CYP1A1 and CYP1B1	Polycyclic aromatic hydrocarbons	Homology modeling		[50]
CYP1A1	Sulforaphane	The tertiary structure of CYP1A1 was generated with the combination methods of threading, ab initio modeling, and structural refinement	Sulforaphane failed to reduce the genotoxic effect of TCDD in yeast cells	[51]
CYP1A1	Pyrimidobenzothiazole (NSC745689)	Homology modeling and molecular dynamics		[52]
CYP1A1, CYP1A2, and CYP1B1 (CYP2A6 and CYP2B1)	Pyranoflavones		Molecular surface images generated from UCSF Chimera	[53]
CYP1A1 and CYP1A2	Ethynylflavones	PDB: 4I8V and PDB: 2HI4	Selective inhibitory activity toward CYP1A1	[54]
CYP1A1	Polychlorinated dibenzo-p-dioxins and coplanar polychlorinated biphenyls	Homology modeling	Rat and human CYP1A1	[55]
CYP1A1, CYP1A2, and CYP1B1	Polymethoxystilbenes	PDB: 4I8V, PDB: 2HI4, and PDB: 3PM0	Potent and selective inhibitory activity of 2,3', 4'-trimethoxy-$trans$-stilbene	[56]
CYP1A1, CYP1A2, and CYP1B1	30 drugs metabolized by CYPs	PDB: 4I8V, PDB: 2HI4, and PDB: 3PM0	MetaSite	[57]
CYP1A1 and CYP1A2	22 aromatic hydrocarbons and 3 fluorogenic alkoxyaryl compounds	PDB: 4I8V and PDB: 2HI4	CYP1A variants	[58]
CYP1A1, CYP1A2, and CYP1B1	Alkoxyresorufins	Homology modeling	Baikal seal and human CYPs	[59]
CYP1A1, CYP1A2, and CYP1B1	5F-203, 5-aminoflavone, 17-β-estradiol, melatonin, debrisoquine, theophylline, clozapine, and lidocaine	PDB: 4I8V, PDB: 2HI4, and PDB: 3PM0	Differences in substrate specificity among CYPs	[60]
CYP1A1	Naringenin and dihydroxybergamottin	Rat homology model, human PDB: 4I8V, and molecular dynamics		[61]
CYP1A1	Compounds selected by virtual screening of databases	Database screening, Hypo1; metabolite prediction study, MetaSite software; molecular docking studies; and molecular dynamics simulations	Antiproliferative activity on MDA-MB-435 human cells and two lead compounds with antitumor activity against MDA-MB-435 line	[62]
CYP1A1, CYP1A2, and CYP1B1	Polymethoxy- and methylthio-$trans$-stilbene derivatives	PDB: 4I8V, PDB: 2HI4, and PDB: 3PM0		[63]
CYP1A2 and other cytochromes				
CYP1A2	Caffeine and MeIQ	Homology model based on CYP BM3 crystal structure		[64]
CYP1A2 (CYP2D6 and CYP3A4)	Selected substrates	Homology modeling	Substrate selectivity studies	[65]
CYP1A2	7-Methoxyresorufin	Homology model based on the crystal structure of CYP2C5	Hydrogen bonds and π-π stacking with Phe226	[66]

Table 1: Continued.

Cytochrome P450	Ligand	Methods	Notes	References
CYP1A2 (CYP2A6, CYP2C9, CYP3A4, and CYP2E1)	Caffeine, theophylline, acetanilide, phenacetin, 7-methoxycoumarin, 7-ethoxycoumarin, 3-cyano-7'-ethoxycoumarin, naproxen, tacrine, amitriptyline, clozapine, and 7-ethoxyresorufin	PDB: 2HI4	Regioselectivity prediction of CYP1A2-mediated metabolism	[67]
CYP1A2	Methoxyresorufin and ethoxyresorufin	CYP1A2 homology model and crystal structure PDB: 2HI4 and homology structures of CYP1A2 mutants		[68]
CYP1A2	Virtual screening of CYP1A2 ligands	PDB: 2HI4 and automated docking (Gold version 3.2)	Prediction of the site of metabolism	[69]
CYP1A2	Structurally diverse CYP1A2 ligands (substrates and inhibitors)	PDB: 2HI4 and molecular dynamics	Versatility and plasticity of the CYP1A2 active site	[70]
CYP1A2 (CYP2C9)	Chrysin, 7,8-benzoflavone, 7-hydroxyflavone, and warfarin	PDB: 2HI4 and molecular dynamics		[71]
CYP1A2	Phenacetin	PDB: 2HI4	Wild-type and mutant forms of enzyme	[72]
CYP1A2	Virtual screening of 971 herb compounds	Pharmacophore searching and docking procedure to CYP1A2 crystal structure (PDB: 2HI4)	Herb-drug interactions	[73]
CYP1A2 (CYP2A6, CYP2C9, and CYP2D6)		PDB: 2HI4 and molecular dynamics	Flexibility at normal and high-pressure conditions (300 MPa)	[74]
CYP1A2 and CYP1B1	Polymethoxy-trans-stilbenes	PDB: 2HI4 and homology model of CYP1B1	Potent and selective inhibitory activity of 2,4,2', 6'-tetramethoxy-trans-stilbene	[75]
CYP1A2	7,8-Benzoflavone, oroxylin, and wogonin	PDB: 2HI4, binding free energy analysis with the MM-PBSL method, and molecular dynamics		[76]
CYP1A2 and CYP1B1	4'-Methylthio-trans-stilbene derivatives	PDB: 2HI4 and PDB: 3PM0		[77]
CYP1A2	7-Ethoxyresorufin	PDB: 2HI4, ensemble docking, and molecular dynamics	Phe186Leu mutation	[78]
CYP1A2 (CYP2C9, CYP2D6, and CYP3A4)	Kinase inhibitors	PDB: 2HI4	Drug-drug interactions	[79]
CYP1A2 (CYP2A6, CYP2C9, CYP3A4, and CYP2E1)	Acetaminophen	Large-scale 2D umbrella sampling, PDB: 2HI4, and molecular dynamics	Regioselectivity	[80]
CYP1A2		The initial structure of wild-type CYP1A2 (CYP1A2.1) constructed from the CYP1A2 crystal structure PDB: 2HI4, and CYP1A2 mutants constructed from CYP1A2.1 refined after molecular dynamics simulation	Influence of amino acid mutations on the 3D structure and dynamic properties of the enzyme	[81]
CYP1B1				
CYP1B1	17-β-Estradiol, α-naphthoflavone, 7-ethoxycoumarin, 7-ethoxyresorufin, bufuralol, and benzo(a)pyrene-7,8-diol	Homology model based on the structure of CYP2C5	Allelic variant effects on metabolism	[82]
CYP1B1	17-β-Estradiol	Molecular dynamics simulations of homology-modeled structures	PCG-associated mutants	[83]
CYP1B1	7,8-Benzoflavone derivatives	PDB: 3PM0; MOE docking program	Inhibitors that eliminate CYP1B1-mediated drug resistance	[16]

B[a]P: benzo[a]pyrene; TCB: 2,3',4,4'-tetrachlorobiphenyl; TCDD: tetrachlorodibenzo-p-dioxin; PCG: primary congenital glaucoma.

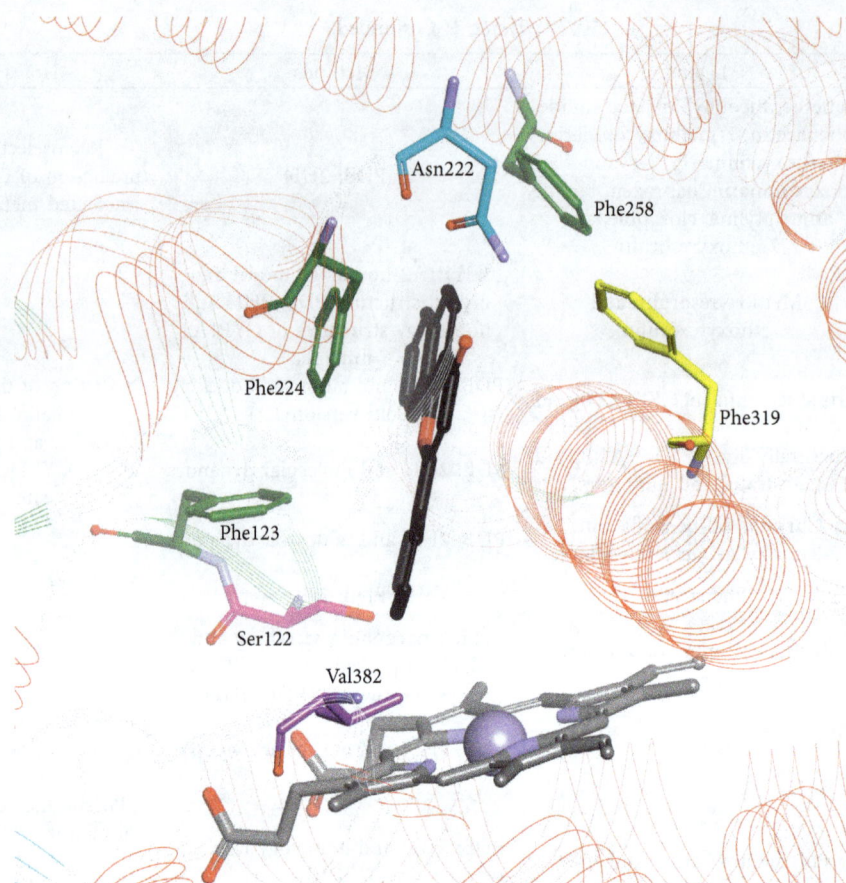

FIGURE 2: ANF bound in the active site of CYP1A1 (PDB ID: 4I8V). ANF: black carbon atoms; conserved phenylalanines 123, 224, and 258: light green; selected nonconserved residues: Ser122, Asn222, Phe319, and Val382.

performed on homology models (Table 1). In the1990s, homology models were based on crystal structures of bacterial enzymes. The first reports involved the CYP1A1 model based on the structure of bacterial CYP102 (CYP BM3) [22]. The disadvantage of this model was a low sequence homology between bacterial and eukaryotic CYPs. A better homology was achieved for the CYP1A1 constructed by Szklarz and Paulsen [28] with the use of mammalian CYP2C5, the first crystal structure of a microsomal CYP 2C5 from a rabbit [84]. Since 2007, the crystal CYP1A2 structure has been available, and the CYP1A1 homology model based on this template was constructed, achieving a better stereochemical quality [34].

The crystal 2.6 Å structure of human CYP1A1 (PDB: 4I8V; Figure 2) in complex with the inhibitor α-naphthoflavone (ANF) was determined by Walsh and coworkers in 2013 as the last of the three members of the CYP1 family [25]. The overall CYP1A1 structure displays the typical cytochrome P450 fold with canonical helices A–L and short F' and G' helices thought to be buried in the membrane and probably involved in enabling the access of hydrophobic ligands to the active site. A characteristic five-residue break in the middle of the F helix and lack of one of the four canonical β sheets that occur in human P450 enzymes were also found in the CYP1A1 crystal structure [25].

Recently, the CYP1A1 structure was used in studies of interactions of the enzyme with a great number of CYP1A1 substrates and inhibitors summarized in Table 1, including phytochemicals, for example, dietary flavonoids [32, 41, 47]; drugs, for example, melatonin, debrisoquine, theophylline, clozapine, and carvedilol [57]; environmental pollutants, for example, aromatic hydrocarbons and their derivatives [22, 28, 35, 37, 45, 55]; and natural and synthetic derivatives of trans-stilbene [25, 46, 63].

The CYP1A1 crystal structure was harnessed to identify new leads exhibiting CYP1A1-mediated anticancer activity [62]. The authors generated and validated a ligand-based pharmacophore model using a series of known anticancer compounds acting via CYP1A1. Selected compounds were subsequently subjected to pharmacokinetic screening, MetaSite screening, a molecular docking study, and a PAINS (pan-assay interference compounds) filter to refine the retrieved hits. Nine compounds capable of generating reactive metabolites and good interactions with CYP1A1 were selected for further studies in vitro. Two compounds showing a potent activity against the MDA-MB-435 cell line with $IC_{50} < 0.1\,\mu M$ and low toxicity to normal cells were selected. These compounds were metabolized by CYP1A1 to the N-hydroxylated products that are potential genotoxic agents and may be responsible for a toxic effect of parent compounds. A molecular dynamics simulation analysis was used to visualize the orientation of hit molecules in a CYP1A1 binding site cavity, which promotes bioactive metabolite formation by N-hydroxylation [62].

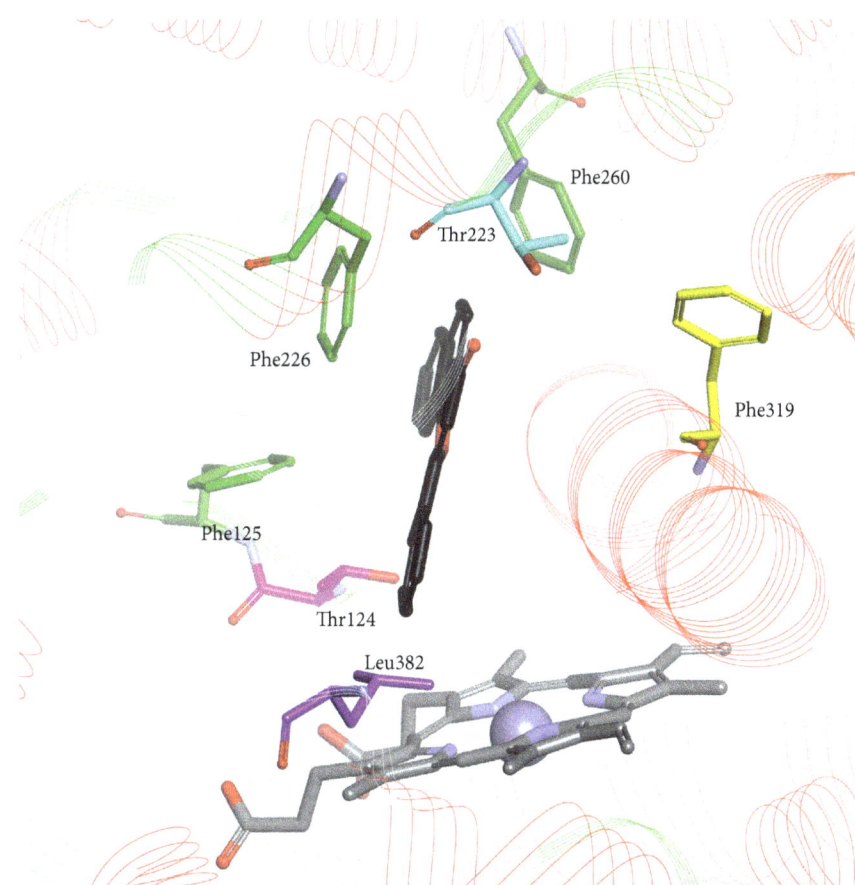

FIGURE 3: ANF bound in the active site of CYP1A2 (PDB ID: 2HI4). ANF: black carbon atoms; conserved phenylalanines 125, 226, and 260: light green; selected nonconserved residues: Thr124, Thr223, Phe319, and Leu382.

2.2. CYP1A2. The crystal structure of CYP1A2 (PDB: 2HI4; Figure 3) was published in 2007, becoming a template for other members of the CYP1 subfamily [26]. This structure had been employed in the homology modeling of CYP1A1 and CYP1B1 until their crystal structures were determined. Site-directed mutagenesis and homology modeling studies of Wang and Zhou [85] revealed a series of residues in the substrate recognition sites (SRSs) of CYP1A2 (Arg108, Thr124, Glu225, Phe226, Lys250, Arg251, Lys253, Asn312, Glu318, Thr319, Asp320, Thr321, Vall322, Leu382, Thr385, and Ile386), which have been shown to play important roles in ligand-enzyme binding.

In the studies of flexibility of human cytochrome P450 enzymes (CYP1A2, CYP2A6, CYP2C9, 2D6, and CYP3A4) with molecular dynamics in combination with UV/Vis and resonance Raman spectroscopy, the active site of CYP1A2 was described as small and rigid in contrast to CYP3A4, which displayed a greater flexibility and the highest substrate promiscuity [74].

The narrow and flat binding pocket of CYP1A2 determines the substrate specificity of the enzyme. Phenacetin, ANF, furafylline, caffeine, and 7-methoxyresorufin serve as standard CYP1A2 inhibitors. Drug metabolism prediction with the use of docking, molecular dynamics, and quantum chemical methods was a good option to screen a library for potential inhibitors and drug-drug interactions [69, 86]. An efficient model for in silico screening was developed to identify CYP1A2 inhibitors in databases of herbal ingredients [73]. A rationale for these studies was herb-drug interactions. First, a pharmacophore model was constructed and validated. Then, the best pharmacophore model was chosen for a virtual screening of 989 herbal compounds. The hits (147 herbal compounds) were investigated through molecular docking and tested *in vitro*. Finally, 5 inhibitors of 18 candidate compounds were found to inhibit CYP1A2 activity. Molecular dynamics simulations provided an insight into the role of molecules of water in the enzyme active site. ANF forms the hydrogen bond with a water molecule, but during the simulations, different water molecules interact with ANF at different points of time [70].

Recently, more than 200 ns MD simulations were performed to investigate the role of water molecules in the active site of CYP1A2 complexed with 7-ethoxyresorufin and ANF [87]. Docking studies followed by MD simulations revealed that water molecules have an effect on hydrogen bond networks formed in the enzyme active site influencing the interactions of the substrate with amino acids in the enzyme active site. It appeared that water molecules were necessary for 7-ethoxyresorufin recognition, while for ligand recognition (ANF), water molecules were not required. The last conclusion is not consistent with the fact that, in the crystal structure of the CYP1A2 binding site (PDB: 2HI4), a water molecule is present. It is likely that the CYP1A2-ANF-WAT complex, with the crystal water molecule, did not

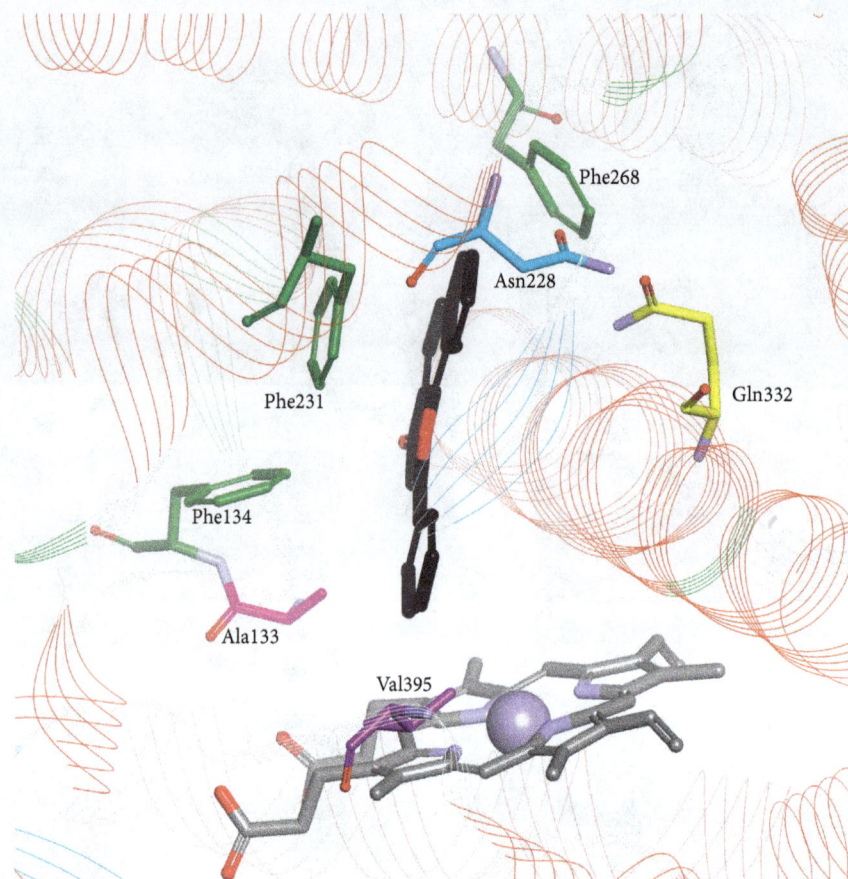

FIGURE 4: ANF bound in the active site of CYP1B1 (PDB ID: 3PM0). ANF: black carbon atoms; conserved phenylalanines 134, 231, and 268: light green; selected nonconserved residues: Ala133, Asn228, Gln332, and Val395.

reach equilibrium even during a longer (400 ns) simulation, which may be indicated by large fluctuations in RMSD for the ANF molecule, and the complex is in an intermediate state, between equilibrium and crystal structure.

2.3. CYP1B1.

The crystallographic structure of CYP1B1 (PDB: 3PM0; Figure 4) was determined by Wang and co-workers [27] with ANF as a ligand bound in the active site cavity. Like the CYP1A subfamily, CYP1B1 has a narrow active site. However, the sequence divergence causes a different orientation of ANF in CYP1B1, CYP1A1, and CYP1A2. Amino acids that line the edges of the cavities modify the substrate and inhibitor binding to CYP1B1 and other CYPs. In the characteristic distortion of the helix F of CYP1B1 and CYP1A2, π-π stacking interactions occur with Phe231 and Phe226, respectively. The amino acid residues Val395 and Ala133 determine the cavity shape in the vicinity of the heme. Val395 plays the role of Val382 in CYP1A1.

2.4. A Comparison of CYP1 Structures.

A comparison of CYP1 family enzyme structures points to similarities and differences among the three active sites that determine their varied substrate specificities. The structures of all members of the CYP1 family were determined as complexes with ANF, making possible the comparison of ANF interactions in the active site cavities. ANF bound in the CYP1 cavities occupies the same plane adjacent to the heme and opposite to the I helix in each enzyme. The orientations of ANF in CYP1A1 and CYP1A2 are similar, whereas in CYP1B1, ANF is flipped by 180° about the long axis of the ligand.

The most meaningful differences in the structure were listed and analyzed by Walsh et al. [25]. The residue at position 382 is known to be important in determining the functional differences between CYP1A1 and CYP1A2; mutations at this position have a significant impact on the catalytic efficiency of enzymes; the smaller 382 residue facilitates ligand placement in the enzyme cavity. In CYP1A1, the only interaction formed by ANF is π-π stacking with Phe224 that is situated on the opposite side of I helix. In the CYP1A2 active site, there is a water molecule that forms hydrogen bonds with carbonyl groups of ANF and Gly316 [25]. A five-residue break in the CYP1A1 F helix has an effect on ligand binding, increasing the flexibility of the active site. The volumes of the active sites of CYP1A1, CYP1A2, and CYP1B1, which are 524, 375, and 398 Å^3, respectively [25–27], and crucial amino acid residues in the enzyme cavity determine the shape of the best-fitted ligands. A triangle with the side length of 9.3, 8.7, and 7.2 Å was proposed as a contour of selective CYP1A2 inhibitors [53]. Moreover, CYP1A1 side chains of amino acid residues lining the active site are smaller in comparison to the corresponding residues in CYP1A2. For example, amino acid residues in the vicinity of the heme, Val382 in CYP1A1 and

Val395 in CYP1B1, are replaced by the branched Leu382 in CYP1A2 narrowing the CYP1A2 cavity, which results in a lesser affinity of polymethoxystilbenes to CYP1A2 in comparison to CYP1A1 [63].

To investigate the active sites of CYP1s, a series of potential inhibitors were synthesized and tested for their inhibitory activity. Synthesizing two series of chemical probes: α-naphthoflavone-like and β-naphthoflavone-like pyranoflavones, Liu et al. created ligand models of CYP1A1 and CYP1A2. The molecular surface images were generated in UCSF Chimera 1.6.2. (UCSF San Francisco, CA) after energy minimization using the conjugate gradient method with the CHARMM force field. The authors concluded that CYP1A1 has a narrow and long cavity, 15.8 Å in length and 4.6 Å in width. The CYP1A2 cavity can accommodate a triangular molecule, showing a planar heart-like structure with a 9.1 Å long side and a 7.0 Å short side [53]. According to that suggestion, a series of 14 flavone and coumarin derivatives exhibiting a triangular planar shape were designed with the use of the computer-assisted alignment assay. Most of the tested compounds (13 out of 14) appeared to be selective CYP1A2 inhibitors. 4-Trifluoromethyl-7,8-pyranocoumarin and 7,8-furanoflavone were found to be the most effective CYP1A2 inhibitors with K_i at a submicromolar level [88].

In conclusion, the detailed topology of the CYP1A1 active site is more similar to the CYP1B1 cavity than to the cavity of CYP1A2 [25]. The closer similarity between the CYP1A1 active site and CYP1B1 compared with the CYP1A2 active site may influence the substrate profiles of these enzymes, which are more similar for CYP1A1 and CYP1B1 than for CYP1A1 and CYP1A2.

3. Mutations

The molecular modeling of enzymes differing in a selected amino acid sequence provides a rationale for substrate specificity. Studies of the relationship between an amino acid sequence and the functionality of CYP1s were possible, thanks to mutagenesis methods. To reduce the number of mutants to be constructed, the residues for replacement should be located in the active site of the enzyme and must be different from the corresponding residues in the enzyme which is being compared. Based on the knowledge of which active site residues are different between CYP1A1 and CYP1A2, the effect of reciprocal mutations on substrate specificity was examined [31]. The residue replacement in the substrate recognition site (SRS) reduced 7-methoxyresorufin (7-MR) and 7-ethoxyresorufin (7-ER) O-dealkylase activities, except for the CYP1A1 S122T mutation which increased both activities. The results confirmed the importance of SRSs for enzyme-substrate interaction, proposed earlier by Gotoh [89].

Functional alterations as a result of genetic polymorphism may influence the therapeutic response of many drugs changing their efficacy and toxicity. CYP1A2 participates in the metabolism of 9% of all medicines [6]. The CYP Allele Nomenclature Committee (http://www.cypalleles.ki.se/cyp1a2.htm) has recognized 40 CYP1A2 variant alleles. A functional characterization of 20 allelic variants of CYP1A2 was performed with two substrates: phenacetin and 7-ethoxyresorufin [90]. Four of the studied alleles, which exhibited the substitutions critical to the enzymatic function (located in: SRS, the heme-binding region, the aromatic region, and the proline-rich region), showed reduced activity toward both substrates. However, the substitution Arg377Gln might cause a change in hydrogen bonds to alternative ones with other amino acids, which resulted in the loss of enzyme activity or decreased holoprotein level. Two variants with the substitutions Thr438Ile and Asp436Asn showed a significantly higher activity toward phenacetin than the wild-type enzyme. Interestingly, the amino acid residues Thr438 and Asp436 are not located in the substrate binding site. They are situated on the surface of CYP1A2 and may influence the interaction of the enzyme with cytochrome b5 [90].

Zhang et al. [78] applied molecular dynamics simulations and structural analyses to elucidate mechanisms of mutation-induced allostery in CYP1A2. They explored the effects of a peripheral mutation, F186L, at ~26 Å away from the enzyme active site on the enzyme catalytic activity. For these mutations, they found a change in protein flexibility and a collective protein motion that caused the main substrate access channel to be mostly closed. Dynamics simulations were used to explain the mechanism of a changed binding of 7-ethoxyresorufin in the catalytic pocket of the F186L mutant enzyme. Ma et al. demonstrated an impact of F186L mutation on the function of CYP1A2 [91]. Despite the fact that the mutation Phe186Leu is located on the surface, a series of changes in the catalytic pocket were observed. Phe186Leu mutation enhanced the binding affinity but lowered the O-deethylation velocity of 7-ethoxyresorufin. It was suggested that channel 2c, which is the main active channel in CYPs [92], is closed in the mutant CYP1A2 enzyme as a result of B'helix/B-C loop stabilization.

Allelic CYP1B1 variants were constructed in studies of the substrate metabolism catalyzed by CYP1B1 [82, 93, 94]. Mutant forms of CYP1B1 have been discovered in the childhood disease primary congenital glaucoma (PCG). Homology-modeled structures of wild-type and disease-associated mutant forms were constructed on the basis of human CYP2C9. In the mutant form of CYP1B1, changes in the geometry of the substrate binding region and the position of the heme were observed. Using molecular dynamics simulations, altered interactions of estradiol with the disease mutant of CYP1B1 in comparison with the wild type of enzyme were demonstrated [83].

More recently, the structures of eight mutants differing only in one residue were generated from the crystal structure of CYP1A2. Mutation of only one amino acid changed the enzyme static structure even in distant regions of the protein and influenced the flexibility of the whole protein and influenced the catalytic activity of the enzyme by changing the conformation of a ligand-enzyme complex [81]. Significant changes in the dynamic properties of CYP1A2 were observed when long-time MD simulations (100 ns or longer) were used.

4. The Molecular Docking of Ligands

Molecular docking is a computational approach which predicts the orientation of a ligand (*pose*) in complex with a protein target and assesses its binding affinity using scoring functions.

Structure-based drug design is performed in order to identify bioactive compounds in the compound pool found in high-throughput virtual screening (HTVS) based on the information from the protein structure. In structure-activity relationship studies, molecular docking helps to elucidate the bioactivity of lead compounds identified at the "hit-to-lead" stage on the basis of a ligand-target interaction analysis.

A computational approach predicts the orientation of ligands in complex with a protein target using scoring functions (specific algorithms). Structure-based virtual screening is a quick and more economical method of lead identification than experimental screening. In research, popular open-source docking software and more advanced commercial packages are used. Nonetheless, in the opinion of many authors, they still need to be improved to obtain a better pose prediction capability. The main factors limiting the accuracy of docking results are protein flexibility and solvation. The affinity of a ligand to a protein target is characterized by scoring functions which represent a relative binding free energy based on protein-ligand interactions. Scoring functions do not consider the contribution of thermodynamic effects on binding free energy like solvation, long-range interactions, and conformational changes. Protein-ligand docking methods are widely used at different stages of the drug design process. They are employed at the beginning for the virtual screening (VS) of large ligands' databases and at the lead optimization stage.

The scoring function is used by the searching algorithm to identify the best pose of a particular ligand, the most energetically favoured orientation inside the active site. It also estimates the binding affinity of a ligand. This allows us to rank ligands in virtual screening, where large and chemically diverse databases should be docked very effectively; so here, the speed of docking is more important than its accuracy [5]. However, in lead optimization, researchers are interested in obtaining docking results that are as accurate as possible for a small set of ligands, which are often structurally related. Besides assessing binding affinities to the macromolecular target for close analogues in lead optimization, docking can also be used for predicting off-target binding to related proteins and to cytochromes as drug-metabolizing enzymes [5].

The role of a docking algorithm is to generate ligand poses inside the binding site. The scoring function should correctly recognize the bioactive orientation and assign a sufficiently high score to it, allowing us to discriminate binders from nonbinders in terms of calculated binding affinity [5]. Scoring functions are classified as force field-based, empirical, and knowledge-based [95]. Force field-based functions account for electrostatic and van der Waals interactions in protein-ligand complexes using force field parameters. In empirical scoring, functions are terms describing specific ligand-protein interactions, for example, hydrogen bonds, ionic interactions, or hydrophobic effects. Another class of scoring functions, knowledge-based, was derived from a statistical analysis of the crystal structures of ligand-protein complexes. It does not use information about

experimental activity but analyzes the distribution of ligand-protein atom pairs giving pairwise potentials [95].

All scoring functions have some limitations. They perform much better in identifying correct poses of individual ligands than in ranking ligands according to their activity for respective targets. Difficulty in differentiating between nano- and micromolar compounds limits the reliability of docking [5]. To overcome this issue, more than one scoring function can be employed in assessing binding affinity. Consensus scoring combines the results of several scoring functions; this approach is in some cases more successful in predicting activity than a single function [95].

Some specific interactions, for example, cation-pi, CH-pi, or weak hydrogen bonds, are not captured by commonly used scoring functions. Also, many simplifications, such as treating solvation effects and contributions of entropy to the binding energy, result in a poor ranking of compounds in VS [95]. Therefore, more advanced and computationally demanding methods for rescoring docked poses are applied. For this purpose, physics-based methods and simulations based on force fields and implicit solvent models are employed. Among them, commonly used approaches are the molecular mechanics-Poisson–Boltzmann surface area (MM-PBSA) and the computationally less demanding molecular mechanics-generalized Born surface area (MM-GBSA).

Many docking algorithms treat the receptor as conformationally rigid, which is a severe approximation influencing the final results. In fact, upon binding to a protein, ligands often induce changes in its conformation [5, 96]. The flexibility of a protein can be included in the macromolecular model in several ways. The simplest one is using "soft" receptors (*soft docking*) with decreased energy penalties for steric clashes between the ligand atom and the receptor. Other docking methods accounting for protein flexibility are a docking using side-chain flexibility, in which side-chain rotations of residues in the binding site are allowed, a docking using an ensemble of receptor structures (experimental or simulated), and on-the-fly docking, where protein conformations are generated "on the fly" during docking by exploring the protein's degrees of freedom [5, 97].

There are many examples of ligand-target interactions via water molecules (e.g., hydrogen bonding), so neglecting water molecules could be an additional source of errors in docking. Usually, before docking, water molecules are removed from the binding site, but there are also other options, such as keeping or displacing water molecules which are placed in the binding site or are important for the binding of ligands [5, 95].

Existing scoring functions are not perfect in ranking compounds in virtual screening and estimating absolute binding affinities in lead optimization. Also, receptor flexibility needs to be taken into account during docking experiments. Therefore, docking methods are still under development regarding aspects such as receptor flexibility, structural water or the solvation, and entropic effect [98].

Docking can predict a plausible orientation and conformation of ligands inside the binding site of the receptor, although this method gives only a static picture of ligand-receptor interactions. A deeper insight into the time-dependent properties of ligand-protein complexes could be obtained with the use of molecular dynamics (MD) simulations. In molecular dynamics simulations, solvent molecules are included explicitly or with the use of implicit solvent models. MD, an invaluable tool in SBDD, has many applications. Before docking, MD simulations could be used to give an ensemble of protein structures, but for postdocking complexes, this method allows for a computational testing of its stability and is often used to rescore docked ligands because of the improvement in the mutual fit and optimization of interactions that occur during the simulation. Calculations of binding free energy (ΔG_{bind}) could be made using different methods, such as thermodynamic integration (TI), free energy perturbation (FEP), linear interaction energy (LIE), and the aforementioned MM-PBSA or MM-GBSA approaches [99]. There are many examples of successful applications of MD in the characterization of ligand-macromolecular target complexes [4, 99].

Protein-ligand binding energy should be determined as a nonadditive effect, which depends on the chemical environment and protein-ligand cooperative dynamic processes. Molecular dynamics simulation improves predictions of binding free energy by considering the time-dependent behaviour of the macromolecular system in response to changes in its molecular environment. However, docking results are not always consistent with MD simulation (different poses observed by docking and MD; a ligand does not form a long-lasting complex). It also happens that docking results are not proved by the biochemical assay in vitro, and vice versa; compounds with high bioactivities are shown to have a poor docking score. Many scientists point to the limitations of docking procedures [2, 69]. Table 1 presents a survey of studies devoted to computer-aided analysis of interactions of CYP1 enzymes with their ligands.

5. Substrates and Inhibitors of CYP1s

A classification of inhibitors and noninhibitors of CYPs is particularly important in relation to drug design and the prediction of drug-drug interactions. CYP1A2 is responsible for the biotransformation of ~5% of currently used drugs. Screening a set of compounds from a database in search of CYP1A2 ligands seems to be more efficient than an experimental determination of catalytic activities of a series of compounds. However, the use of scoring functions did not always give satisfactory results. Better results were achieved with recently developed nonlinear machine learning methods. Seven thousand test compounds from a database were analyzed as CYP1A2 inhibitors. The accuracy of the developed method for the prediction of inhibitory activity was estimated at 73–76% [86], while the decision tree model based on Lipinski's rule of five classified 67% of the test compounds correctly. Binding free energies of structurally diverse CYP1A2 substrates and inhibitors were predicted with the use of the linear interaction energy (LIE) method. For 10 compounds (from the set of 13 test ligands), the difference between the calculated and experimental binding free energies

was smaller than 4.0 kJ/mol [70]. CYP1A2 ligands were identified from a large compound library (16,338 compounds) with the use of two approaches: structure-based and ligand-based virtual screening. As compared to the ligand-based method, the structure-based method identified more inhibitors which were more potent as well [100].

In this review, we present studies on specific interactions of substrates/inhibitors with CYP1 isozymes, which allowed for an analysis of the relationship between the structure of the tested compounds and their inhibitory activities. The studies on ligand-CYP1 enzyme interactions with the use of computational methods are summarized in Table 1. There are some groups of compounds that are particularly interesting, and many reports devoted to their interactions with CYP1s are discussed below. These are endogenic substrates, alkoxyresorufins, polycyclic aromatic hydrocarbons, and compounds that are supposed to play a role in cancer chemoprevention: natural flavonoids and trans-stilbene derivatives. The bioactivity of natural chemopreventive agents inspired researchers to synthesize their derivatives in order to study the structure-activity relationship and to obtain more active and efficient chemopreventive agents.

5.1. Endogenic Substrates. 17-β-Estradiol (E2) is metabolized by CYP1s to 2-hydroxy, 4-hydroxy, or 16-hydroxy derivatives. The order of preference for in vitro 2-hydroxylation by CYP1 isoforms was CYP1A2 > CYP1A1 > CYP1B1; for 4-hydroxylation, it was CYP1B1 > CYP1A2 > CYP1A1; and for 16-hydroxylation, CYP1A2 showed the highest preference followed by CYP1A1 and CYP1B1. In the mammary gland, CYP1A1 catalyzes predominantly 2-hydroxylation. 2-Hydroxyestradiol (2-OHE$_2$) is further methylated by catechol-O-methyltransferase to produce 2-methoxyestradiol, which does not exhibit carcinogenic activity. On the contrary, it inhibits the proliferation of cancer cells. 4-Hydroxyestradiol (4-OHE$_2$) is produced in a reaction catalyzed mainly by CYP1B1 [101, 102]. The product of 4-OHE$_2$ oxidation (estradiol-3,4-quinone) forms quinone-DNA adducts and initiates carcinogenesis [103]. With the use of homology models of CYP1A1 and CYP1B1 based on the crystal structure of CYP1A2, Itoh et al. analyzed the structural causes of different sites of E2 metabolism [44]. The studies revealed one binding mode of E2 (18-methyl group up) to CYP1A1 and CYP1A2 and two binding modes of E2 (18-methyl group up and down) to CYP1B1. Thr124 and Phe260 of CYP1A2 and Ser122 and Phe258 of CYP1A1 were identified as causing steric hindrance with the B-ring of E2. Ala133 and Asn265 of CYP1B1 are critical residues influencing the interaction of E2 with the binding site. Conformations of E2 in enzyme cavities decided on the site of E2 metabolism leading to the hydroxylation preferentially at the position 2 in case of CYP1A1 and CYP1A2, and at the position 4 in CYP1B1 [44].

Fatty acids are an essential class of CYP endogenic substrates whose metabolites are supposed to play a physiological role in the cardiovascular system. With the use of molecular docking, regiospecificity of the metabolism of arachidonic acid (AA) and eicosapentaenoic acid (EPA) catalyzed by human recombinant CYP1A1 in the reconstituted enzymatic system was studied. Interestingly, AA was mainly metabolized to 19-hydroxyarachidonic acid by CYP1A1. With EPA as a substrate, CYP1A1-dependent epoxygenase activity leading to the regiospecific and stereoselective formation of 17(R),18(S)-epoxyeicosatetraenoic acid (68%) and 19-hydroxy-EPA (31%) was demonstrated [30]. The molecular docking of AA and EPA to the CYP1A1 active site revealed that fatty acids interact with the same amino acid residues as alkoxyresorufins and benzo(a)pyrene, although additional residues located in the access channel may interact with AA and EPA owing to their longer molecules. Conformations of fatty acids in the CYP1A1 binding site are stabilized at their carboxy ends by hydrogen bonds, while resorufin and benzo(a)pyrene are mainly stabilized by hydrophobic interactions [30]. The complexes of AA and EPA with CYP1A1 were further examined with MD simulations to obtain productive binding modes. The in silico site scoring of geometric criteria, angles and distances of the substrates to the ferryl oxygen, confirmed that steric factors play a key role in the regiospecificity of CYP1A1-mediated metabolism [33].

5.2. Alkoxyresorufins. Alkoxyresorufins are CYP1 substrates used in activity assays. Lewis and Lake [22] initiated the computational approach in the studies of substrate affinity to CYP1 binding sites. In the 1990s, homology models of CYP1A1 and CYP1A2 binding sites were generated from the bacterial CYP102 crystal structure via residue replacement and energy minimization procedures, and a series of known substrates and inhibitors of the CYP1A subfamily were docked interactively to the active sites [22]. The orientation of 7-MR and 7-ER in binding site cavities was determined with the aim of elucidating their substrate specificity; 7-MR is a specific substrate of CYP1A2, while 7-ER demonstrates a higher affinity to CYP1A1 over CYP1A2. The enzymes share 72% of the amino acid sequence identity; however, the differences in their structures seem to be sufficient to explain their specific affinity to the ligands. Critical changes in the CYP1A1 and CYP1A2 structures were found. In the I helix, the change from aspartate adjacent to Thr268 in CYP1A1 to glutamate in CYP1A2 gives rise to the steric restriction in the CYP1A2; as a result, there is no sufficient space for 7-ER in the binding site of CYP1A2. Moreover, in the F helix, which lies above the heme moiety, there are amino acid residues that are donors of hydrogen for the carbonyl group being located in resorufins on the opposite side of a molecule; in CYP1A1, the carbonyl group of 7-ER can form a hydrogen bond with Thr185, while 7-MR can form a hydrogen bond with Asp184 in CYP1A2 [22].

Szklarz and Paulsen [28] docked 7-MR and 7-ER manually to the CYP1A1 binding site (homology model generated from CYP2C5) in the orientations, leading to the formation of major products. The residues located within 5 Å were identified. Val382 was found as a key residue that stabilized 7-ER in the CYP1A1 binding site through van der Waals interactions [28]. This interaction did not occur in the case of 7-MR. Moreover, a higher activity of CYP1A1 toward 7-ER may be explained by interaction energy, which is significantly higher for 7-MR (lower absolute value). The effects of five key residues—Ser122, Asn221, Gly225, Leu312, and Val382 in CYP1A1, and Thr124, Thr223, Val227, Asn312, and Leu382 in CYP1A2—on the

substrate specificity of enzymes were investigated [31]. Specificity changes were observed, but no single mutation that could confer the activity of one isoform onto another was found. As a continuation of studies, 26 possible multiple mutants of CYP1A2 were constructed and investigated with the molecular dynamics-based scoring method. In 7 mutants, the specificity shift from CYP1A2 to CYP1A1 was predicted. For 5 mutants, the prediction was confirmed by site-directed mutagenesis and biochemical assays [68].

When 7-ER was docked to CYP1A1 generated using CYP1A2 as a template, the amino acid residues found within a 3 Å radius from the substrate were Ser120, Ser12, Phe123, Phe224, Phe258, Tyr259, Asp313, Thr321, Val382, and Ile386. For this CYP1A1 structure, substrate inhibition kinetics was observed, probably due to a nonproductive orientation of 7-ER in the CYP1A1 binding site. Docking studies showed that the symmetrical molecule of 7-ER may be bound in a reverse orientation with the ethoxy group directed in the opposite side of the heme, which is energetically favourable in the CYP1A1 wild type and mutants [34].

5.3. Polycyclic Aromatic Hydrocarbons. Polycyclic aromatic hydrocarbons, present ubiquitously in the environment, are planar aromatic compounds produced mainly in combustion processes. Benzo(a)pyrene is a procarcinogen activated by CYPs to mutagenic products which form adducts with DNA. Its metabolism by CYP1s has been studied with the use of molecular docking since the 1990s [22]. The studies have been continued by Szklarz and collaborators [28, 33], who found a correlation between the numbers of docked orientations within 4 Å of the ferryl oxygen and experimentally determined metabolite ratios. The regiospecificity of B(a)P metabolism was demonstrated with a homology model on the CYP 2C5 crystal structure of CYP1A1 [33] and with the use of multiple models of killifish, scup, rat, and human CYP1A1s [37]. In all the models analyzed, the 8,9-bond was more frequently close to ferryl oxygen than 7,8- or 9,10-positions. However, 8,9-epoxide production has never been observed owing to unfavourable formation energy. The formation of epoxides in the close vicinity of 8,9-position—7,8-epoxide or 9,10-epoxide—is supposed to be a result of a small reposition of a substrate molecule by vibration or rotation within the active site [37].

CYP1B1 inhibition by eleven polycyclic aromatic hydrocarbons (PAHs) and 14 acetylenic PAHs and biphenyls was studied. Five of the potent inhibitors with IC$_{50}$ at the nanomolar level (benzo(a)pyrene, dibenzo[aj]acridine, 1-(1-propynyl)pyrene, 3-(1-propynyl)phenanthrene, and benzo[j]fluoranthene) were docked to the CYP1B1 and CYP1A2 cavities showing different binding modes for selected aromatic hydrocarbons [104].

An alternatively spliced variant of CYP1A1 having a deletion of exon 6 was discovered in human brain tissue [35]. The lack of B(a)P metabolism to genotoxic ultimate carcinogens by the exon 6 del CYP1A1 was elucidated by molecular docking studies. B(a)P docked to the wild CYP1A1 (being modeled with the CYP2C5 crystal structure) was situated in a way that made possible an oxidation reaction in the positions 7, 8, 9, and 10 of the aromatic ring. Two major clusters of orientations were found, out of the 6 observed for the B(a)P molecule docked to the CYP1A1 binding site: the first with 7, 8, 9, and 10 positions near the heme iron (72% of all 50 studied conformations) and the second with position 3 close to the heme (14% of conformations). Among 11 orientations found for B(a)P in the exon 6 del CYP1A1 active site, in two main orientations, C-3 was in a close proximity to the heme [99]. However, the 3-hydroxylated product of B(a)P metabolism is not considered as genotoxic. B(a)P was differentially orientated in the CYP1A2 and CYP1B1 binding sites; positions 7, 8, 9, and 10 of the aromatic scaffold were observed in proximity to the heme iron only in the CYP1A2 binding cavity [104].

A collection of 22 polycyclic aromatic hydrocarbons of increasing size were docked to wild-type and chimeric CYP1A enzymes. The QSAR analysis revealed that the size of the substrate influences its accessibility to the enzyme cavity via access channels. A visualization of CYP1A enzymes with the use of CAVER software showed two regions located close to or within the CYP access channels affecting differentially small and large polycyclic substrates [58].

Polychlorinated dibenzo-p-dioxins (PCDDs) and co-planar polychlorinated biphenyls (PCBs) are a class of aromatic hydrocarbons demonstrating high genotoxicity. The metabolism of dioxins and PCBs shows the species-based differences between humans and rats [55]. Human CYP1s metabolized efficiently low-chlorinated PCDDs, while 2,3,7,8-tetrachlorodibenzo-p-dioxin (TCDD) metabolites were not detected. Rat, but not human, CYP1A1 metabolized 3,3′,4,4′,5-pentachlorobiphenyl, the most toxic PCB, to two hydroxylated derivatives showing lower toxicity than the parent compound. Docking studies with the use of homology models of human and rat CYP1A1 indicated essential amino acids residues (Ala120 and Phe316) for 3,3′,4,4′,5-pentachlorobiphenyl metabolism. The differences in amino acid residues led to changes in the size and shape of the cavities; in the rat CYP1A1 cavity, 3,3′,4,4′,5-pentachlorobiphenyl was close enough to the heme to be metabolized [45].

Species-based differences were studied by the docking of B(a)P, 3,3′,4,4′-tetrachlorobiphenyl (TCB), and TCDD to multiple models of rat, human, killifish, and scup CYP1A1 [37]. Mutating interacting residues of killifish CYP1A1 to corresponding residues of human CYP1A1 led to TCB poses similar to those of human CYP1A1. A slower oxidation of TCDD in comparison to TCB by each species may be explained by structural constraints in the enzyme binding site. A slower metabolism of TCDD by human CYP1A1 than rat CYP1A1 resulted from the lower frequency of productive poses in human CYP1A1.

The molecular docking of 37 polycyclic aromatic hydrocarbons, corresponding diols, and heterocyclic hydrocarbons to homology models of CYP1A1 and CYP1B1 based on the crystal structure of CYP1A2 was performed with LigandFit and CDOCKER algorithms [50]. The analysis of CYP1A1 and CYP1B1 binding sites revealed their hydrophobic character due to hydrophobic residues, mainly the phenylalanines Phe123, 224, and 258 in CYP1A1 and Phe134, 231, and 268 in CYP1B1, which may interact

through π-π stacking with aromatic ligands. However, potential hydrogen bond donor residues, Ser122, Asn221, Leu312, Asp313, Gly316, Ala317, and Asp320, found in the CYP1A1 binding site and corresponding residues in CYP1B1, Ala133, Asn228, Thr325, Asp326, Gly329, Ala330, and Asp333, stabilized the ligand molecules by hydrogen bonds. The amino acid residues which mainly interact with the ligands under study are located in the substrate recognition sites classified by Gotoh [89]. Interestingly, the CDOCKER docking procedure gave the best results for CYP1A1 linear statistical analysis, while LigandFit appeared to be a more suitable procedure for CYP1B1 [50].

5.4. Flavonoids. Flavonoids are a large class of natural bioactive compounds present in fruits and vegetables. Their role in cancer prevention is established in epidemiologic studies [21]. In experimental *in vitro* studies, flavonoids appeared to be potent inhibitors of CYP1s. A correlation was found between the inhibition of CYP1A1 and CYP1A2 activities by flavonoids differing in the position and number of hydroxyl groups and theoretical descriptors obtained from quantum mechanical calculations and molecular dynamics of the ligand-enzyme complex [32]. In this report, quercetin and kaempferol docked to the binding site of CYP1A2 were demonstrated, and amino acid residues responsible for ligand-enzyme interactions that may be useful in site-directed mutagenesis were found. Takemura et al. demonstrated a selective inhibition of CYP1B1 by flavonoids, particularly chrysoeriol and isorhamnetin. To explain their strong effect on CYP1B1, a molecular docking approach was employed [42]. For this purpose, they constructed three-dimensional structures of CYP1A1 and CYP1B1 by homology modeling, using the crystal structure of CYP1A2. The authors concluded that methoxy-flavonoids—chrysoeriol and isorhamnetin—fit well into the active site of CYP1B1, while in active sites of CYP1A1 and CYP1A2, there occurred a steric collision between methoxy substituents and Ser-122 in CYP1A1 and Thr-124 in CYP1A2. The binding specificity of methoxyflavonoids is based on interactions between methoxy groups and specific CYP1 residues. Methoxyflavonoids possessing a 2-3 double bond in the C-ring, as selective inhibitors of CYP1B1, are supposed to be chemopreventive agents against CYP1B1-related carcinogenesis.

Oroxylin and wogonin are biologically active compounds occurring in the extract of roots of *Scutellaria baicalensis*, used in traditional oriental medicines [76]. Oroxylin and wogonin are inhibitors of CYP1A2 with IC_{50} values of 579 and 248 nM. With the use of molecular docking, molecular dynamics simulation, and MM-PBSA, the mechanism of the inhibitory action of flavonoids differing in the position of a hydroxyl group was analyzed. Calculated binding free energies of ANF (−23.5 kcal/mol), wogonin (−21.1 kcal/mol), and oroxylin (−19.8 kcal/mol) are significantly overestimated; however, they are in accordance with the order of experimentally determined inhibitory activities. The difference in the affinity of oroxylin and wogonin to the CYP1A2 active site was explained by molecular dynamics and molecular docking; for ANF and wogonin, noncovalent interactions (van der Waals and hydrophobic interactions) influenced the stability of their complexes with CYP1A2. In the CYP1A2-oroxylin complex, there occurred an energetically unfavourable repulsion between Thr118 and the methoxy group at position 6 in the oroxylin molecule. As a result, conformational changes in the side chain of Thr118 were observed, which caused the formation of a more open and larger binding site cavity of CYP1A2 and a weaker inhibitory activity of oroxylin. Moreover, the O7 atom of oroxylin formed a strong hydrogen bond with Asp313, as the O5 and O6 atoms formed two hydrogen-bonding interactions with a molecule of water. These interactions were not observed in complexes of CYPA2 with wogonin and ANF [76].

Chrysin (5,7-dihydroxyflavone), a natural, biologically active flavonoid extracted from plants and honey, exhibited an inhibitory activity toward CYP1A2 comparable to ANF (IC$_{50}$ values of 54 nm versus 49 nM). With molecular docking and molecular dynamics simulations, the interactions in the enzyme binding site were estimated. The complex of chrysin with CYP1A2 was stabilized with van der Waals interactions, H-bond with Asp313, and stacking interactions with Phe226 [71]. The affinity of chrysin to CYP2C9 was significantly weaker because van der Waals interactions in the larger pocket of CYP2C9 were not as strong as in CYP1A2.

Based on the known inhibitory activity of compounds, more efficient inhibitors can be designed. Flavone derivatives with an acetylene group linked to the flavone backbone showed a comparable ANF inhibitory activity against CYP1A1. Moreover, mechanism-based inactivators of CYP1A1 were found. 4'-Ethynylflavone and 7-ethynylflavone irreversibly inactivated half of the CYP1A1 activity in less than two minutes. The acetylene group is probably responsible for irreversible enzyme inactivation. Docking simulations revealed the orientations of ethynylflavones in the CYP1A1 binding site with the acetylene group toward the heme. Only 2'-ethynylflavone demonstrated another orientation in the CYP1A1 cavity; this compound appeared to be a selective inhibitor of CYP1A2. In all the studied ethynylflavones docked to CYP1A2, acetylene groups were oriented away from the heme [54].

5.5. Stilbenoids.

Since the 1990s, natural stilbenoids—trans-resveratrol (3,4',5-trihydroxy-trans-stilbene; RESV), pterostilbene (3,5-dimethoxy-4'-hydroxy-trans-stilbene), and piceatannol (3,4,4',5-tetrahydroxy-trans-stilbene)—have been extensively studied in relation to chemoprevention [105]. RESV is a natural polyphenol found in grapes, berries, and peanuts, showing well-characterized beneficial bioactivities [106, 107]. It efficiently and selectively inhibits CYP1 activities [108, 109], although its bioavailability in humans was determined as poor [110] due mainly to conjugation reactions with sulphuric acid and glucuronic acid. In the last two decades, natural and synthetic RESV analogues have been studied in the context of their interaction with CYP1s. It has appeared that natural trans-resveratrol analogues—pinostilbene,

rhapontigenin, desoxyrhapontigenin, and pterostilbene, which possess some of the hydroxyl groups substituted by methoxy groups—are more potent CYP1A1 and CYP1A2 inhibitors than trans-resveratrol [111, 112]. Therefore, the substitution of hydroxyl groups with methoxy substituents efficiently influenced the affinity of compounds to active sites of cytochromes and, moreover, improved bioavailability by preventing polyphenol metabolism. Consequently, the interest focused on synthetic derivatives of trans-stilbene appeared to be more promising with regard to their interaction with CYP1 enzymes.

The pattern of substituents linked to the trans-stilbene core exerts a decisive effect on the affinity of stilbenoids to active sites of cytochromes P450 family 1. The positions of some substituents influence the ligand orientation and interactions with amino acid residues in the enzyme active site, affecting the distance to the heme, which determines the course of enzymatic reaction. In the studies of Chun et al. [75, 112–114], 3,5,2',4'-tetramethoxy-trans-stilbene and 2,4,2',6'-tetramethoxy-trans-stilbene were identified as very potent CYP1B1 inhibitors, indicating a distinctive role of methoxy substituents in positions 2 and 4, as well as 2 and 6 in the inhibition of CYP1B1 activity.

The design of the series of polymethoxy-trans-stilbenes with the constant motif of 3,4-dimethoxyphenyl influenced the way the ligands were oriented in the enzyme binding site [56]. The molecular docking of trans-stilbenes to the CYP1A2 active site showed the most favourable orientation with the ring possessing the altering pattern of substituents directed to the heme (orientation A; Figure 5(a)). However, in CYP1B1, 2',3,4-trimethoxystilbene was oriented with 3,4-dimethoxyphenyl directed toward the heme (orientation B; Figure 5(b)). This orientation occurred in 17 out of a total of 20 poses and was energetically favourable in comparison to orientation A; the interaction energy and binding energy for the ligand calculated for orientation B was higher by 12 kcal/mol and 40 kcal/mol, respectively [56]. A very strong affinity of 2',3,4-trimethoxystilbene to the CYP1B1 binding site was expressed by the highest value of binding energy (ΔG) in comparison to other compounds of the series. The analysis of the interaction between ligands and amino acid residues in the CYP1B1 active site demonstrated the occurrence of π-π stacking interactions for both phenyl rings of 2',3,4-trimethoxystilbene with Phe231 (Figure 5), whereas a hydrogen bond was observed only for the opposite ligand orientation (Figure 5(a)), which was less energetically favourable [56]. In this bonding, Gln332 was engaged; the same amino acid residue formed a hydrogen bond with 4'-methylthiostilbenes. Therefore, it may be supposed that the effect of hydrogen bonds on the ligand affinity to the cytochrome P450 active site is not of primary importance. Hydrophobic interactions between methoxy groups and amino acid residues seem to be more essential in determining the inhibitor affinity to cytochrome P450. Moreover, 2',3,4-trimethoxy-trans-stilbene is characterized by a high selectivity of action; it inhibits CYP1B1 90 times more strongly than CYP1A1 and 830 times more strongly than CYP1A2. Thus, the 2',3,4-triMS molecule appeared to be a comparably effective CYP1B1 inhibitor than the molecules designed earlier [106–108], demonstrating an

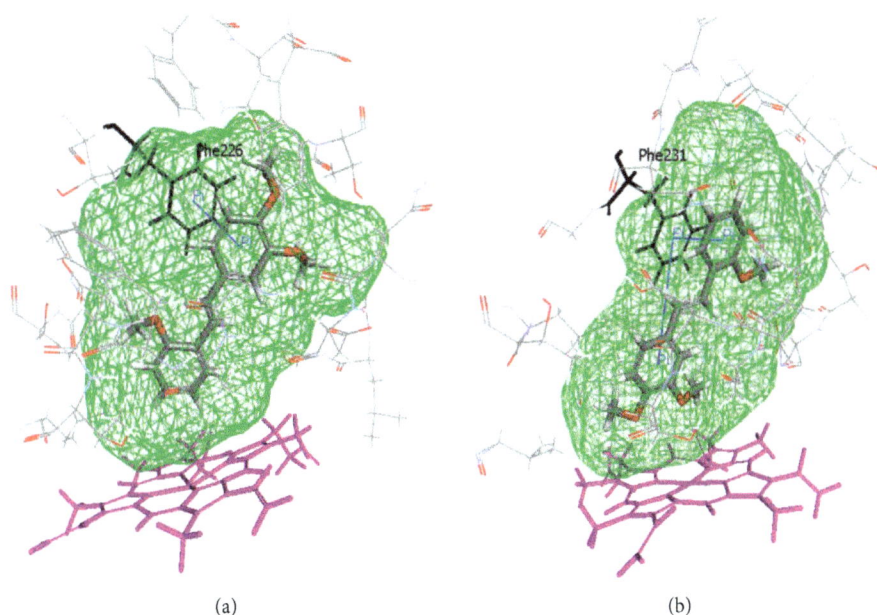

FIGURE 5: $2',3,4$-trimethoxy-*trans*-stilbene docked to the CYP1A2 (a) and CYP1B1 (b) binding sites. Amino acid residues surrounding the active sites are visualized with Phe226 and Phe231 in black colour. The heme is represented as a stick model in pink. The solid blue lines represent π-π stacking interactions.

effective inhibitory action of the compound with a pattern of methoxy groups in positions $2'$, 3, and 4 [56].

In search of novel CYP1 inhibitors, methylthiostilbene derivatives were designed and synthesized [63, 77]. The orientation of a series of polymethoxy-*trans*-stilbene derivatives containing a $4'$-methylthio substituent in the CYP1 active sites was studied, and molecular interactions between ligands and amino acid residues of the enzyme pocket were estimated. The orientation with a $4'$-methylthiophenyl ring toward the heme for the studied derivatives in CYP1A2 and CYP1B1 active sites was the most favoured one. For this series of compounds, Phe226 and Phe260 in CYP1A2 and Phe231 in CYP1B1 were involved in π-π stacking interactions that stabilized the orientation of ligands in the enzyme active sites. Additionally, for some of the examined compounds docked to CYP1B1, an active site hydrogen bond was formed with Gln332. However, it should be mentioned that the occurrence of the hydrogen bond did not correlate with the inhibitory effect on enzyme activity. An important role is assigned to the hydrophobic interactions that may have an effect on the closer contact of docked molecules with the Fe atom of the prosthetic group, resulting in the hydroxylation of ligands. For 3,4,5-trimethoxy-$4'$-MTS and 2,4,5-trimethoxy-$4'$-MTS, the distances of C atoms in $3'$ and $5'$ positions to the Fe atom were shorter than 4.5 and 5.5 Å, respectively.

Stilbene derivatives better fit in the CYP1A1 binding site exhibiting a planar long strip cavity than in the CYP1A2 binding site with a more triangular shape [53], for example, the selective CYP1A1 and CYP1B1 inhibitor, 2,3,4-trimethoxy-$4'$-methylthio-*trans*-stilbene, did not fit the shape of the CYP1A2 binding pocket. The low affinity of 2,3,4,-trimethoxy-$4'$-methylthio-*trans*-stilbene to the CYP1A2 binding site was additionally confirmed by a high strain energy (103.09 kcal/mol). By comparison, in the CYP1B1

binding site, the strain energy for 2,3,4-trimethoxy-$4'$-methylthio-*trans*-stilbene was only 40.70 kcal/mol [77].

Another derivative, 2-methoxy-$4'$-methylthio-*trans*-stilbene, was found to be a selective and potent CYP1A1 inhibitor. Interestingly, its analogue, $2,4'$-dimethoxy-*trans*-stilbene, was not so effective. For this derivative, docked to the CYP1A1 binding site, a high number of nonbonded molecular interactions were observed. However, the binding of 2-methoxy-$4'$-methylthio-*trans*-stilbene was not favourable energetically [63].

6. Other Ligands

Most of the ligands of CYP1s are compounds with established pharmacological activity. They include drugs metabolized by the constitutive liver isozyme CYP1A2. A molecular docking of phenacetin and furafylline to CYP1A1 and CYP1A2 active sites was first performed by Lewis and Lake [22] with the use of homology models based on the CYP102 crystal structure. In 2012, Huang et al. studied the isoform-selective metabolism of phenacetin and acetaminophen with the use of the CYP1A2 crystal structure and homology model of CYP1A1 [49].

More recently, the metabolism of drugs selected from the Drug Bank comprising 1,528 drugs approved by the FDA was analyzed with the use of crystal structures of all isozymes of CYP family 1. The substrates were divided into three groups: substrates having a single site of metabolism (SOM) but showing a different preference to get metabolized by CYP1A1, CYP1A2, and CYP1B (e.g., carvedilol, phenacetin, and bufuralol); substrates that are metabolized by any of the three isoforms (e.g., chloroquine and haloperidol); and substrates that show a different SOM and a different preference to isozymes (17-β-estradiol) [57]. Differences in substrate specificity among CYPs were studied for melatonin, debrisoquine, theophylline, clozapine, and lidocaine [60]. The regioselectivity of CYP1A2-mediated metabolism was investigated for caffeine, theophylline, acetanilide, naproxen, tacrine, amitriptyline, clozapine, and alkoxyresorufins by Jung and coworkers [67]. More recently, regioselective metabolism of acetaminophen catalyzed by CYP1A2 was proved through a molecular dynamics procedure [80].

Although many reports have demonstrated the inhibitory activity of alkaloids against the activities of CYPs [115], only for caffeine, theophylline [80], and rutaecarpine and its derivatives [29, 116], structural modeling has been performed. A good fitting of rutaecarpine with the binding site of the CYP1A2 model based on the rabbit CYP2C5 as a template was found [29]. Two hydrogen bonds can be formed between the keto and N14 groups of rutaecarpine and Thr208 and Thr473 residues of CYP1A2. The planar molecule of rutaecarpine forms π-π stacking interaction between the C-ring and aromatic ring of the Phe205 residue. A possible orientation of coumarin in CYP1A1 and CYP1A2 binding sites for 3,4-epoxidation was demonstrated with enzyme structures based on the CYP2A5 crystallographic template. Key amino acid residues—Ser113, Phe205, Tre298, and Phe352—were identified for coumarin docked to CYP1A1 [116]. In the CYP1A2 binding site, Tre113, Phe205, and Tre298 participate in ligand-enzyme interaction. In both enzymes, Phe205 is responsible for π-π stacking interaction with aromatic rings of the substrates. Both CYP1A1 and CYP1A2 metabolize coumarin in the same molecular positions.

In the context of cancer chemoprevention, naturally occurring isothiocyanates (ITCs) as inhibitors of CYP1s were studied. Sulforaphane, which is one of the most active chemopreventive agents and inhibitors of CYP1A1 activity, was docked to the CYP1A1 active site. Two hydrogen bonds between the nitrogen atom of sulforaphane and the hydrogen of the amino groups of Arg110 were found [51]. Moreover, sulforaphane suppressed the aryl hydrocarbon receptor (AHR) by binding to its ligand binding domain with hydrogen bonds. However, the studies did not explain the lack of potential to reduce the genotoxicity of TCDD.

Emodin is a natural anthraquinone extracted from *Rheum emodi*, a plant used in Chinese medicine. Among the CYPs studied (CYP1A1, CYP1A2, and CYP2B1), this anthraquinone demonstrated the most potent inhibitory activity toward CYP1A2 with the IC$_{50}$ value of 3.73 μM [117]. In the PubChem and ZINC chemical databases, 12 emodin analogues were found for further studies. Two of them (1-amino-4-chloro-2-methylanthracene-9,10-dione (compound 1) and 1-amino-4-hydroxyanthracene-9,10-dione (compound 2)) inhibited CYP1A2 with IC$_{50}$ < 1 μM, but only compound 1 was a mechanism-based inhibitor of both CYP1A1 and CYP1A2. Molecular docking revealed the orientation of molecules in the binding site, which make possible the abstraction of hydrogen from the 2-methyl group present only in compound 1. As a result, a benzylic carbon radical intermediate might be produced, which, after rearrangement, could form an irreversible complex with the enzyme. The radical can react with the iron-bound hydroxyl radical to form a hydroxylated metabolite, which acts as an inactivator of CYPs. A 2-methyl

group of compound 1 docked to the CYP1A1 and CYP1A2 binding sites was found close to the heme moieties [117].

Combining *in vitro* studies with a computational approach enabled us to identify compounds that may interact with other drugs. This strategy appeared to be useful in the investigation of drug-drug interactions which are of great clinical importance in relation to multidrug disease treatment. Inhibitory effects of 91 kinase inhibitors (KIs; 80 KIs are not used clinically and 11 are FDA-approved KIs) on human CYPs—CYP1A2, 2C9, 2D6, and 3A4—were determined. For the majority of the KIs under analysis, a differential inhibitory effect on CYP enzymes was observed; fifteen compounds exhibited a potent inhibitory effect on CYP activities (IC$_{50}$ ≤ 1 μM). Clinically used KIs—nilotinib, sunitinib, and imatinib—appeared to be potent CYP1A2 inhibitors with IC$_{50}$ values of 0.92–1.23 μM [79]. In the docking validation studies, 20 compounds among 22 inhibitors selected in high-throughput *in vitro* studies (90.9%) demonstrated a high docking interaction energy. Three functional residues (Phe226, Phe125, and Asp320) in the active site of CYP1A2 were identified [79].

7. The Site of Metabolism

Identifying the sites of metabolism (SOMs) can play a decisive role in the design of drugs displaying desirable properties. The basic computational methods used for predicting SOMs and the structures of metabolites are QSAR, 3D QSAR, the pharmacophore-based method, molecular docking, molecular dynamics simulation, and a combined approach which is applied in numerous studies [118, 119]. Computational techniques used in studies of xenobiotic metabolism are classified into the ligand-based approach and the structure-based approach. Taking into account the scope and limitations of these techniques, the combination of both ligand-based and structure-based approaches seems to be promising. In order to predict the site of metabolism, a molecular docking of substrates to the binding sites of cytochromes P450 may be performed. Lewis et al. [116] used CYP1A1 and CYP1A2 homology structures based on the CYP2C5 crystallographic template in a study of coumarin metabolism, finding a good correlation for binding energies determined experimentally and with the use of molecular docking.

In studies of stilbene derivatives, the molecular docking of 4'-methylthio-*trans*-stilbene derivatives to the CYP1A2 binding site confirmed the orientation of the 4'-methylthiophenyl ring of 2,4,5-trimethoxy-4'-methylthio-*trans*-stilbene and 3,4,5-trimethoxy-4'-methylthio-*trans*-stilbene in the close vicinity of the heme, allowing the reaction of hydroxylation at C-3' to take place [77].

From all the binding modes obtained as a result of the docking procedure, possible metabolic sites of a substrate are assigned to the atoms located within 5 Å from the Fe atom [120]. Molecular docking takes into account binding affinities and steric effects related to the conformation of an active site. The best results of SOM prediction were obtained with the approach combining molecular docking with semiempirical molecular orbital calculations that provide the activation energy characterizing the reactivity of a substrate [67]. Possible binding modes of CYP1A2 substrates were analyzed using automated docking with the use of the crystal structure of CYP1A2. For caffeine and theophylline, the SOMs found were in accordance with experimental data typing N1-CH$_3$, N7-CH$_3$, and N3-CH$_3$ as sites of the formation of primary and secondary metabolites [67].

Biotransformation studies of drugs can be performed with the use of molecular docking and molecular dynamics. Prediction of the formation of toxic metabolites is particularly important [121]. Two acetylcholinesterase inhibitors, derivatives of *p*-aminophenol and succinic anhydride, were tested in order to determine whether toxic metabolites are generated as in the case of *N*-acetyl-*p*-aminophenol (APAP) which is metabolized by CYP1A1 and CYP2B1 to toxic *N*-acetyl-*N*-hydroxy-*p*-aminophenol. Molecular dynamics confirmed that the amide group of APAP interacted with the heme iron of CYP1A1, and as a result of N-oxidation, a toxic intermediate (*N*-acetyl-*p*-benzoquinone imine) was formed. For both studied inhibitors docked to CYP1A1, this kind of interaction was not found. Instead, an aryl hydroxyl hydrogen interaction with the heme was observed. The results obtained in silico correlated well with the studies *in vitro*, which revealed the formation of only hydroxylated metabolites as a result of the metabolism of the studied inhibitors by rat liver microsomes [121]. The regioselectivity of APAP metabolism was studied with molecular docking and molecular dynamics, followed by 2D USP free energy scanning. CYP1A2 and CYP2E1, the enzymes with compact active sites, were found to be major APAP metabolizers [80]. APAP formed more interactions in CYP1A2 and CYP2E1 binding sites as compared with the more voluminous binding sites of CYP3A4 and CYP2C9, which resulted in stabilized binding states and a longer residence time.

8. Conclusions

Computational docking studies contribute to a better understanding of ligand-enzyme interactions at a molecular level. Studies with the use of computational procedures provide a rationalization of the selectivity of ligands toward CYP1 isozymes. The elucidation of cytochrome P450 family 1-specific activities at a molecular level is of great importance with regard to novel and potent drug design and drug-drug interactions. Although computational methods have been significantly developed, a further improvement of virtual procedures could have an impact on their usefulness in the design of drugs targeting CYP enzymes by predicting the site of metabolism and drug-drug interactions and determining the potential toxicity of substrates and their metabolites.

Molecular docking helped to visualize spatial ligand fitting and molecular interactions occurring in the enzyme active site. The hypothesis is that not a single substituent but a pattern of substituents determines the shape of a molecule and influences a ligand's affinity to a binding site. The pattern of substituents exerts an effect on the ligand orientation in the enzyme active site, which in the case of some ligands is

stabilized by hydrophobic interactions, especially π-π stacking interactions. In the case of ligands that are substrates of enzymatic reactions, the distance between a ligand and the prosthetic group is essential for the course of reaction. Combining experimental studies on enzymatic reactions with a computer analysis of ligand-active site interactions are expected to produce valuable results, useful in the design of molecules with a desired activity.

Summarizing the achievements of the reports reviewed, many authors emphasize the versatility and plasticity of CYPs. In silico methods in the studies of ligand-CYP isozyme interactions provide a predictive model based mainly on van der Waals interactions, whereas electrostatic interactions do not play a considerable role here. The ligand can change its conformation through adaptation to the shape of the enzyme active site. Analyses of the ligand shape revealed the essential role of shape complementarity to the cavity of the enzyme binding site. Amino acid residues and water molecules can form hydrogen bonds that stabilize the ligand-enzyme complex.

Computational structure-based ligand design is a promising technique which enables an efficient analysis of preclinical drug candidates. Docking may be used to provide information about the conformation of a bioactive ligand and its position in the binding site. Knowing the orientation of a ligand helps to predict the site of metabolism.

Acknowledgments

The study was supported by funding from the Poznan University of Medical Sciences (no. 502-01-03313427-08870) and by Nicolaus Copernicus University in Toruń (Collegium Medicum Fund no. DS-UPB-WF-411).

References

[1] L. Chen, J. K. Morrow, H. T. Tran, S. S. Phatak, L. Du-Cuny, and S. Zhang, "From laptop to benchtop to bedside: structure-based drug design on protein targets," *Current Pharmaceutical Design*, vol. 18, no. 9, pp. 1217–1239, 2012.

[2] Y.-C. Chen, "Beware of docking," *Trends in Pharmacological Sciences*, vol. 36, no. 2, pp. 78–95, 2015.

[3] V. Lounnas, T. Ritschel, J. Kelder, R. McGuire, R. P. Bywater, and N. Foloppe, "Current progress in structure-based rational drug design marks a new mind set in drug discovery," *Computational and Structural Biotechnology Journal*, vol. 5, no. 6, article e201302011, 2013.

[4] J. Mortier, C. Rakers, M. Bermudez, M. S. Murgueitio, S. Riniker, and G. Wolber, "The impact of molecular dynamics on drug design: applications for the characterization of ligand-macromolecule complexes," *Drug Discovery Today*, vol. 20, no. 6, pp. 686–702, 2015.

[5] B. Waszkowycz, D. E. Clark, and E. Gancia, "Outstanding challenges in protein-ligand docking and structure-based virtual screening," *WIREs Computational Molecular Science*, vol. 1, no. 2, pp. 229–259, 2011.

[6] U. M. Zanger, K. Klein, M. Thomas et al., "Genetics, epigenetics, and regulation of drug-metabolizing cytochrome P450 enzymes," *Clinical Pharmacology and Therapeutics*, vol. 95, no. 3, pp. 258–261, 2014.

[7] D. W. Nebert, T. P. Dalton, A. B. Okey, and F. J. Gonzalez, "Role of aryl hydrocarbon receptor-mediated induction of the CYP1 enzymes in environmental toxicity and cancer," *Journal of Biological Chemistry*, vol. 279, no. 7, pp. 23847–23850, 2004.

[8] S. Rendic and F. P. Guengerich, "Contributions of human enzymes in carcinogen metabolism," *Chemical Research in Toxicology*, vol. 25, no. 7, pp. 1316–1383, 2012.

[9] K. Gajjar, P. L. Martin-Hirsch, and F. L. Martin, "CYP1B1 and hormone-induced cancer," *Cancer Letters*, vol. 324, no. 1, pp. 13–30, 2012.

[10] J. D. Yager, "Endogenous estrogens are carcinogens through metabolic activation," *JNCI Monographs*, vol. 2000, no. 27, pp. 67–73, 2000.

[11] Y.-J. Kwon, H.-S. Baek, D.-J. Ye, S. Shin, D. Kim, and Y.-J. Chun, "CYP1B1 enhances cell proliferation and metastasis through induction of EMT and activation of Wnt/β-catenin signalling via Sp1 upregulation," *PLoS One*, vol. 11, no. 3, Article ID e0151598, 2016.

[12] D. W. Nebert and D. W. Russell, "Clinical importance of the cytochromes P450," *The Lancet*, vol. 360, no. 9340, pp. 1155–1162, 2002.

[13] R. Santes-Palacios, D. Ornelas-Ayala, N. Cabañas et al., "Regulation of human cytochrome P4501A1 (hCYP1A1): a plausible target for chemoprevention," *BioMed Research International*, vol. 2016, Article ID 5341081, 17 pages, 2016.

[14] H. I. Swanson, V. C. O. Njar, Z. Yu et al., "Targeting drug-metabolizing enzymes for effective chemoprevention and chemotherapy," *Drug Metabolism and Disposition*, vol. 38, no. 4, pp. 539–544, 2010.

[15] R. Dutour and D. Poirier, "Inhibitors of cytochrome P450 (CYP) 1B1," *European Journal of Medicinal Chemistry*, vol. 135, pp. 296–306, 2017.

[16] J. Cui, Q. Meng, X. Zhang, Q. Cui, W. Zhou, and S. Li, "Design and synthesis of new α-naphthoflavones as cytochrome P450 (CYP) 1B1 inhibitors to overcome docetaxel-resistance associated with CYP1B1 overexpression," *Journal of Medicinal Chemistry*, vol. 58, no. 8, pp. 3534–3547, 2015.

[17] I. Bièche, C. Narjoz, T. Asselah et al., "Reverse transcriptase-PCR quantification of mRNA levels from cytochrome (CYP)1, CYP2 and CYP3 families in 22 different human tissues," *Pharmacogenetics and Genomics*, vol. 17, no. 9, pp. 731–742, 2007.

[18] S. Badal and R. Delgod, "Role of the modulation of CYP1A1 expression and activity in chemoprevention," *Journal of Applied Toxicology*, vol. 34, no. 7, pp. 743–753, 2014.

[19] R.-E. Go, K.-A. Hwang, and K.-C. Choi, "Cytochrome P450 1 family and cancers," *Journal of Steroid Biochemistry and Molecular Biology*, vol. 147, pp. 24–30, 2015.

[20] L. W. Wattenberg, "Chemoprophylaxis of carcinogenesis: a review," *Cancer Research*, vol. 26, no. 7, pp. 1520–1526, 1966.

[21] H. Takemura, H. Sakakibara, S. Yamazaki, and K. Shimoi, "Breast cancer and flavonoids–a role in prevention," *Current Pharmaceutical Design*, vol. 19, no. 34, pp. 6125–6132, 2013.

[22] D. F. V. Lewis and B. G. Lake, "Molecular modelling of CYP1A subfamily members based on an alignment with CYP102: rationalization of CYP1A substrate specificity in terms of active site amino acid residues," *Xenobiotica*, vol. 26, no. 7, pp. 723–753, 1996.

[23] D. F. V. Lewis, B. G. Lake, S. G. George et al., "Molecular modelling of CYP1 family enzymes CYP1A1, CYP1A2, CYP1A6 and CYP1B1 based on sequence homology with CYP102," *Toxicology*, vol. 139, no. 1-2, pp. 53–79, 1999.

[24] D. F. V. Lewis, B. G. Lake, and M. Dickins, "Quantitative structure-activity relationships within a homologous series of 7-alkoxyresorufins exhibiting activity towards CYP1A and CYP2B enzymes: molecular modelling studies on key members of the resorufin series with CYP2C5-derived models of human CYP1A1, CYP1A2, CYP2B6 and CYP3A4," *Xenobiotica*, vol. 34, no. 6, pp. 501–513, 2004.

[25] A. A. Walsh, G. Szklarz, and E. E. Scott, "Human cytochrome P450 1A1 structure and utility in understanding drug and xenobiotic metabolism," *Journal of Biological Chemistry*, vol. 288, no. 18, pp. 12932–12943, 2013.

[26] S. Sansen, J. K. Yano, R. L. Reynald et al., "Adaptations for the oxidation of polycyclic aromatic hydrocarbons exhibited by the structure of human P450 1A2," *Journal of Biological Chemistry*, vol. 282, no. 19, pp. 14348–14355, 2007.

[27] A. Wang, U. Savas, C. D. Stout, and E. F. Johnson, "Structural characterization of the complex between α-naphthoflavone and human cytochrome P450 1B1," *Journal of Biological Chemistry*, vol. 286, no. 7, pp. 5736–5743, 2011.

[28] G. D. Szklarz and M. D. Paulsen, "Molecular modeling of cytochrome P450 1A1: enzyme-substrate interactions and substrate binding affinities," *Journal of Biomolecular Structure and Dynamics*, vol. 20, no. 2, pp. 155–162, 2002.

[29] M.-J. Don, D. F. V. Lewis, S.-Y. Wang, M.-V. Tsai, and Y.-F. Ueng, "Effect of structural modification on the inhibitory selectivity of rutaecarpine derivatives on human CYP1A1, CYP1A2 and CYP1B1," *Bioorganic & Medicinal Chemistry Letters*, vol. 13, no. 15, pp. 2535–2538, 2003.

[30] D. Schwarz, P. Kisselev, S. S. Ericksen et al., "Arachidonic and eicosapentaenoic acid metabolism by human CYP1A1: highly stereoselective formation of 17(R),18(S)-epoxyeicosatetraenoic acid," *Biochemical Pharmacology*, vol. 67, no. 8, pp. 1445–1457, 2004.

[31] J. Liu, S. S. Ericksen, M. Sivaneri, D. Besspiata, C. W. Fisher, and G. D. Szklarz, "The effect of reciprocal active site mutations in human cytochromes P450 1A1 and CYP1A2 on alkoxyresorufin metabolism," *Archives of Biochemistry and Biophysics*, vol. 424, no. 1, pp. 33–43, 2004.

[32] F. Iori, R. da Fonseca, M. J. Ramos, and M. C. Menziani, "Theoretical quantitative structure-activity relationships of flavone ligands interacting with cytochrome P450 1A1 and 1A2 isozymes," *Bioorganic & Medicinal Chemistry*, vol. 13, no. 14, pp. 4366–4374, 2005.

[33] S. S. Ericksen and G. D. Szklarz, "Regiospecificity of human cytochrome P450 1A1-mediated oxidations: the role of steric effects," *Journal of Biomolecular Structure and Dynamics*, vol. 23, no. 3, pp. 243–256, 2005.

[34] B. C. Lewis, P. I. Mackenzie, and J. O. Miners, "Comparative homology modeling of human cytochrome P4501A1 (CYP1A1) and conformation of residues involved in 7-ethoxyresorufin O-deethylation by site-directed mutagenesis and enzyme kinetic analysis," *Archives of Biochemistry and Biophysics*, vol. 468, no. 1, pp. 58–69, 2007.

[35] R. P. Kommadi, C. M. Turman, B. Moorthy, L. Wang, H. W. Strobel, and V. Ravindranath, "An alternatively sliced cytochrome P4501A1 in human brain fails to bioactivate polycyclic aromatic hydrocarbons to DNA-reactive metabolites," *Journal of Neurochemistry*, vol. 102, no. 3, pp. 867–877, 2007.

[36] J.-C. Tseng, M.-J. Don, D. F. V. Lewis, S.-Y. Wang, and Y.-F. Ueng, "Inhibition of CYP1 by dehydrorutaecarpine and

its methoxylated derivatives," *Journal of Food and Drug Analysis*, vol. 15, no. 4, pp. 480–487, 2007.

[37] J. C. Prasad, JV Goldstone, CJ Camacho, S. Vajda, and J. J. Stegeman, "Ensemble modeling of substrate binding to cytochromes P450: analysis of catalytic differences between CYP1A orthologs," *Biochemistry*, vol. 46, no. 10, pp. 2640–2654, 2007.

[38] A. T. Sangamwar, L. B. Labhsetwar, and S. V. Kuberkar, "Exploring CYP1A1 as anticancer target: homology modeling and in silico inhibitor design," *Journal of Molecular Modeling*, vol. 14, no. 11, pp. 1101–1109, 2008.

[39] J. Sridhar, P. Jin, J. Liu, M. Foroozesh, and C. L. Klein Stevens, "In silico studies of polyaromatic hydrocarbon inhibitors of cytochrome P450 enzymes 1A1, 1A2, 2A6, and 2B1," *Chemical Research in Toxicology*, vol. 23, no. 3, pp. 600–607, 2010.

[40] J. Pan, G.-Y. Liu, J. Cheng, X.-J. Chen, and X.-L. Ju, "CoMFA and molecular docking studies of benzoxazoles and benzothiazoles as CYP450 1A1 inhibitors," *European Journal of Medicinal Chemistry*, vol. 45, no. 3, pp. 967–972, 2010.

[41] T. Shimada, K. Tanaka, S. Takenaka et al., "Structure-function relationships of inhibition of human cytochromes P450 1A1, 1A2, 1B1, 2C9, and 3A4 by 33 flavonoid derivatives," *Chemical Research in Toxicology*, vol. 23, no. 12, pp. 1921–1935, 2010.

[42] H. Takemura, T. Itoh, K. Yamamoto, H. Sakakibara, and K. Shimoi, "Selective inhibition of methoxyflavonoids on human CYP1B1 activity," *Bioorganic & Medicinal Chemistry*, vol. 18, pp. 6310–6315, 2010.

[43] M. C. Rosales-Hernández, J. E. Mendieta-Wejebe, J. G. Trujillo-Ferrara, and J. Correa-Basurto, "Homology modeling and molecular dynamics of CYP1A1 and CYP2B1 to explore the metabolism of aryl derivatives by docking and experimental assays," *European Journal of Medicinal Chemistry*, vol. 45, no. 11, pp. 4845–4855, 2010.

[44] T. Itoh, H. Takemura, K. Shimoi, and K. Yamamoto, "A 3D model of CYP1B1 explains the dominant 4-hydroxylation of estradiol," *Journal of Chemical Information and Modeling*, vol. 50, no. 6, pp. 1173–1178, 2010.

[45] K. Yamazaki, M. Suzuki, T. Itoh et al., "Structural basis of species differences between human and experimental animal CYP1A1s in metabolism of 3,3′,4,4′,5-pentachlorobiphenyl," *Journal of Biochemistry*, vol. 149, no. 4, pp. 487–494, 2011.

[46] M. K. A. Khan, S. Akhtar, and J. M. Arif, "Homology modelling of CYP1A1, CYP1B1 and its subsequent molecular docking studies with resveratrol and its analogues using AutoDock Tools 4.0," *Biochemical and Cellular Archives*, vol. 11, pp. 49–55, 2011.

[47] V. P. Androutsopoulos, A. Papakyriakou, D. Vourloumis, and D. A. Spandidos, "Comparative CYP1A1 and CYP1B1 substrate and inhibitor profile of dietary flavonoids," *Bioorganic & Medicinal Chemistry*, vol. 19, no. 9, pp. 2842–2849, 2011.

[48] J. Sridhar, J. Ellis, P. Dupart, J. Liu, C. L. Stevens, and M. Foroozesh, "Development of flavone propargyl ethers as potent and selective inhibitors of cytochrome enzymes CYP1A1 and CYP1A2," *Drug Metabolism Letters*, vol. 6, no. 4, pp. 275–284, 2012.

[49] Q. Huang, R. S. Deshmukh, S. S. Ericksen, Y. Tu, and G. D. Szklarz, "Preferred binding orientations of phenacetin in CYP1A1 and CYP1A2 are associated with isoform-selective metabolism," *Drug Metabolism and Disposition*, vol. 40, no. 12, pp. 2324–2331, 2012.

[50] J. Gonzales, N. Marchand-Geneste, J. L. Giraudel, and T. Shimada, "Docking and QSAR comparative studies of polycyclic aromatic hydrocarbons and other procarcinogen interactions with cytochromes P450 1A1 and 1B1," *SAR and QSAR in Environmental Research*, vol. 23, no. 1-2, pp. 87–109, 2012.

[51] F. Yang, S. Zhuang, C. Zhang, H. Dai, and W. Liu, "Sulforaphane inhibits CYP1A1 activity and promotes genotoxicity induced by 2,3,7,8-tetrachlorodibenzo-p-dioxin in vitro," *Toxicology and Applied Pharmacology*, vol. 269, no. 3, pp. 226–232, 2013.

[52] P. P. Nandekar, K. M. Tumbi, N. Bansal et al., "Chembioinformatics and in vitro approaches for candidate optimization: a case study of NSC745689 as a promising antitumor agent," *Medicinal Chemistry Research*, vol. 22, no. 8, pp. 3728–3742, 2013.

[53] J. Liu, S. F. Taylor, P. S. Dupart et al., "Pyranoflavones: a group of small-molecule probes for exploring the active site activities of cytochrome P450 enzymes 1A1, 1A2 and 1B1," *Journal of Medicinal Chemistry*, vol. 56, no. 10, pp. 4082–4092, 2013.

[54] N. Goyal, J. Liu, L. Lovings et al., "Ethynyl flavones, highly potent, and selective inhibitors of CYP1A1," *Chemical Research in Toxicology*, vol. 27, no. 8, pp. 1431–1439, 2014.

[55] H. Inui, T. Itoh, K. Yamamoto, S.-I. Ikushiro, and T. Sakaki, "Mammalian cytochrome P-450-dependent metabolism of polychlorinated dibenzo-p-dioxins and coplanar polychlorinated biphenyls," *International Journal of Molecular Sciences*, vol. 15, no. 8, pp. 14044–14057, 2014.

[56] R. Mikstacka, M. Wierzchowski, Z. Dutkiewicz et al., "3,4,2′-Trimethoxy-trans-stilbene–a potent CYP1B1 inhibitor," *Medicinal Chemistry Communications*, vol. 5, no. 4, pp. 496–501, 2014.

[57] P. Pragyan, S. S. Kesharwani, P. P. Nandekar, V. Rathod, and A. T. Sangamwar, "Predicting drug metabolism by CYP1A1, CYP1A2, and CYP1B1: insights from MetaSite, molecular docking and quantum chemical calculations," *Molecular Diversity*, vol. 18, no. 4, pp. 865–878, 2014.

[58] P. Urban, G. Truan, and D. Pompon, "Access channels to the buried active site control substrate specificity in CYP1 P450 enzymes," *Biochimica et Biophysica Acta*, vol. 1850, no. 4, pp. 696–707, 2015.

[59] H. Iwata, K. Yamaguchi, Y. Takeshita et al., "Enzymatic characterization of in vitro-expressed Baikal seal cytochrome P450 (CYP) 1A1, 1A2, and 1B1: implication of low metabolic potential of CYP1A2 uniquely evolved in aquatic mammals," *Aquatic Toxicology*, vol. 162, pp. 138–151, 2015.

[60] S. S. Kesharwani, P. P. Nandekar, P. Pragyan, V. Rathod, and A. T. Sangamwar, "Characterization of differences in substrate specificity among CYP1A1, CYP1A2 and CYP1B1: an integrated approach employing molecular docking and molecular dynamics simulations," *Journal of Molecular Recognition*, vol. 29, no. 8, pp. 370–390, 2016.

[61] R. Santes-Palacios, A. Romo-Mancillas, R. Camacho-Carranza, and J. J. Espinosa-Aguirre, "Inhibition of human and rat CYP1A1 enzyme by grapefruit juice compounds," *Toxicology Letters*, vol. 258, pp. 267–275, 2016.

[62] P. P. Nandekar, K. Khomane, V. Chaudhary et al., "Identification of leads for antiproliferative activity on MDA-MB-435 human breast cancer cells through pharmacophore and CYP1A1-mediated metabolism," *European Journal of Medicinal Chemistry*, vol. 115, pp. 82–93, 2016.

[63] M. Wierzchowski, Z. Dutkiewicz, A. Gielara-Korzańska et al., "Synthesis, biological evaluation and docking studies of trans-stilbene methylthio derivatives as cytochromes P450 family 1 inhibitors," *Chemical Biology & Drug Design*, vol. 90, no. 6, pp. 1226–1236, 2017.

[64] J. J. Lozano, E. López-de-Briñas, N. B. Centeno, R. Guigó, and F. Sanz, "Three-dimensional modelling of human cytochrome P450 1A2 and its interaction with caffeine and MeIQ," *Journal of Computer-Aided Molecular Design*, vol. 11, no. 4, pp. 395–408, 1997.

[65] F. De Rienzo, F. Fanelli, C. Menziani, and P. G. De Benedetti, "Theoretical investigation of substrate specificity for cytochrome P450 IA2, P450 IID6 and P450 IIIA4," *Journal of Computer-Aided Molecular Design*, vol. 14, no. 1, pp. 93–116, 2000.

[66] D. F. V. Lewis, "Modelling human cytochromes P450 involved in drug metabolism from the CYP2C5 crystallographic template," *Journal of Inorganic Biochemistry*, vol. 91, no. 4, pp. 502–514, 2002.

[67] J. Jung, N. D. Kim, S. Y. Kim et al., "Regioselectivity prediction of CYP1A2-mediated phase I metabolism," *Journal of Chemical Information and Modeling*, vol. 48, no. 5, pp. 1074–1080, 2008.

[68] Y. Tu, R. Deshmukh, M. Sivane, and G. D. Szklarz, "Application of molecular modeling for prediction of substrate specificity in cytochrome P450 1A2," *Drug Metabolism and Disposition*, vol. 36, no. 11, pp. 2371–2380, 2008.

[69] P. Vasanthanathan, J. Hritz, O. Taboureau et al., "Virtual screening and prediction of site of metabolism for cytochrome P450 1A2 ligands," *Journal of Chemical Information and Modeling*, vol. 49, no. 1, pp. 43–52, 2009.

[70] P. Vasanthanathan, L. Olsen, F. S. Jørgensen, N. P. E. Vermeulen, and C. Oostenbrink, "Computational prediction of binding affinity for CYP1A2-ligand complexes using empirical free energy calculation," *Drug Metabolism and Disposition*, vol. 38, no. 8, pp. 1347–1354, 2010.

[71] L. He, F. He, H. Bi et al., "Isoform-selective inhibition of chrysin towards human cytochrome P450 1A2. Kinetics analysis, molecular docking, and molecular dynamics simulations," *Bioorganic & Medicinal Chemistry Letters*, vol. 20, no. 20, pp. 6008–6012, 2010.

[72] M. Huang and G. Szklarz, "Significant increase in phenacetin oxidation on L382V substitution in human cytochrome P450 1A2," *Drug Metabolism and Disposition*, vol. 38, no. 7, pp. 1039–1045, 2010.

[73] R. Zhu, L. Hu, H. Li, J. Su, Z. Cao, and W. Zhang, "Novel natural inhibitors of CYP1A2 identified by *in silico* and *in vitro* screening," *International Journal of Molecular Sciences*, vol. 12, no. 5, pp. 3250–3262, 2011.

[74] T. Hendrychová, E. Anzenbacherová, J. Hudeček et al., "Flexibility of human cytochrome P450 enzymes: molecular dynamics and spectroscopy reveal important function-related variations," *Biochimica et Biophysica Acta*, vol. 1814, no. 1, pp. 58–68, 2011.

[75] Y. J. Chun, C. Lim, S. O. Ohk et al., "*trans*-Stilbenoids: potent and selective inhibitors for human cytochrome P450 1B1," *Medicinal Chemistry Communications*, vol. 2, no. 5, pp. 402–405, 2011.

[76] Y.-X. Shao, P. Zhao, Z. Li et al., "The molecular basis for the inhibition of human cytochrome P450 1A2 by oroxylin and wogonin," *European Biophysics Journal*, vol. 41, no. 17, pp. 297–306, 2012.

[77] R. Mikstacka, A. M. Rimando, Z. Dutkiewicz, T. Stefański, and S. Sobiak, "Design, synthesis and evaluation of the inhibitory selectivity of novel *trans*-resveratrol analogues on human recombinant CYP1A1, CYP1A2 and CYP1B1," *Bioorganic & Medicinal Chemistry*, vol. 20, no. 17, pp. 5117–5126, 2012.

[78] T. Zhang, L. A. Liu, D. F. V. Lewis, and D.-Q. Wei, "Long-range effects of a peripheral mutation on the enzymatic activity of cytochrome P4501A2," *Journal of Chemical Information and Modeling*, vol. 51, no. 6, pp. 1336–1346, 2011.

[79] Z.-X. Wang, J. Sun, C. E. Howell et al., "Prediction of the likelihood of drug interactions with kinase inhibitors based on in vitro and computational studies," *Fundamental and Clinical Pharmacology*, vol. 28, no. 5, pp. 551–582, 2014.

[80] Y. Yang, S. E. Wong, and F. C. Lightstone, "Understanding a substrate's product regioselectivity in a family of enzymes: a case study of acetaminophen binding in cytochrome P450s," *PLoS One*, vol. 9, no. 2, Article ID e87058, 2014.

[81] Y. Watanabe, S. Fukuyoshi, M. Hiratsuka et al., "Prediction of three-dimensional structures and structural flexibilities of wild-type and mutant cytochrome P450 1A2 using molecular dynamics simulations," *Journal of Molecular Graphics & Modelling*, vol. 68, pp. 48–56, 2016.

[82] D. F. V. Lewis, E. M. J. Gillam, S. A. Everett, and T. Shimada, "Molecular modelling of human CYP1B1 substrate interactions and investigation of allelic variant effects on metabolism," *Chemico-Biological Interactions*, vol. 145, no. 3, pp. 281–295, 2003.

[83] M. S. Achary and H. A. Nagarajaram, "Comparative docking studies of CYP1b1 and its PCG-associated mutant forms," *Journal of Biosciences*, vol. 33, no. 5, pp. 699–713, 2008.

[84] P. A. Williams, J. Cosme, V. Sridhar, P. F. Johnson, and D. E. McRee, "Mammalian microsomal cytochrome P450 monooxygenase: structural adaptations for membrane binding and functional diversity," *Molecular Cell*, vol. 5, no. 1, pp. 121–131, 2000.

[85] B. Wang and S.-F. Zhou, "Synthetic and natural compounds that interact with human cytochrome P450 1A2 and implications in drug development," *Current Medicinal Chemistry*, vol. 16, no. 31, pp. 4066–4218, 2009.

[86] P. Vasanthanathan, O. Taboureau, C. Oostenbrink, N. P. E. Vermeulen, L. Olsen, and F. S. Jørgensen, "Classification of cytochrome P450 1A2 inhibitors and noninhibitors by machine learning techniques," *Drug Metabolism and Disposition*, vol. 37, no. 3, pp. 658–664, 2009.

[87] Y. Watanabe, S. Fukuyoshi, K. Kato et al., "Investigation of substrate recognition for cytochrome P450 1A2 mediated by water molecules using docking and molecular dynamics simulations," *Journal of Molecular Graphics & Modelling*, vol. 74, pp. 326–336, 2017.

[88] J. Liu, P. T. Pham, E. V. Skripnikova et al., "A ligand-based drug design. Discovery of 4-trifluoromethyl-7,8-pyranocoumarin as a selective inhibitor of human cytochrome P450 1A2," *Journal of Medicinal Chemistry*, vol. 58, no. 16, pp. 6481–6493, 2015.

[89] O. Gotoh, "Substrate recognition sites in cytochrome P450 family 2 (CYP2) proteins inferred from comparative analyses of amino acid and coding nucleotide sequences," *Journal of Biological Chemistry*, vol. 267, no. 1, pp. 83–90, 1992.

[90] M. Ito, Y. Katono, A. Oda, N. Hirasawa, and M. Hiratsuka, "Functional characterization of 20 allelic variants of CYP1A2," *Drug Metabolism and Pharmacokinetics*, vol. 30, no. 3, pp. 247–252, 2015.

[91] L.-N. Ma, Z.-Z. Du, P. Lian, and D.-Q. Wei, "A theoretical study on the mechanism of a superficial mutation inhibiting the enzymatic activity of CYP1A2," *Interdisciplinary Sciences*, vol. 6, no. 1, pp. 25–31, 2014.

[92] V. Cojocaru, P. J. Winn, and R. C. Wade, "The ins and outs of cytochrome P450s," *Biochimica et Biophysica Acta*, vol. 1770, no. 3, pp. 390–401, 2007.

[93] T. Shimada, J. Watanabe, K. Inoue, F. P. Guengerich, and E. M. J. Gillam, "Specificity of 17β-oestradiol and benzo[*a*]

pyrene oxidation by polymorphic human cytochrome P4501B1 variants substituted at residues 48, 119 and 432," *Xenobiotica*, vol. 31, no. 3, pp. 163–176, 2001.

[94] I. Jansson, I. Stoilov, M. Sarfarazi, and J. B. Schenkman, "Effects of two mutations on cytochrome P4501B1 G61E and R469W, on stability and endogenous steroid metabolism," *Pharmacogenetics*, vol. 11, no. 9, pp. 793–801, 2001.

[95] N. Moitessier, P. Englebienne, D. Lee, J. Lawandi, and C. R. Corbeil, "Towards the development of universal, fast and highly accurate docking/scoring methods: a long way to go," *British Journal of Pharmacology*, vol. 153, no. S1, pp. S7–S26, 2008.

[96] F. Feixas, S. Lindert, W. Sinko, and J. A. McCammon, "Exploring the role of receptor flexibility in structure-based drug discovery," *Biophysical Chemistry*, vol. 186, pp. 31–45, 2014.

[97] D. A. Antunes, D. Devaurs, and L. E. Kavraki, "Understanding the challenges of protein flexibility in drug design," *Expert Opinion on Drug Discovery*, vol. 10, no. 12, pp. 1301–1313, 2015.

[98] E. Yuriev, J. Holien, and P. A. Ramsland, "Improvements, trends, and new ideas in molecular docking: 2012-2013 in review," *Journal of Molecular Recognition*, vol. 28, no. 10, pp. 581–604, 2015.

[99] A. Ganesan, M. L. Coote, and K. Barakat, "Molecular dynamics-driven drug discovery: leaping forward with confidence," *Drug Discovery Today*, vol. 22, no. 2, pp. 249–269, 2017.

[100] P. Vasanthanathan, J. Lastdrager, C. Oostenbrink et al., "Identification of CYP1A2 ligands by structure-based and ligand-based virtual screening," *Medicinal Chemistry Communications*, vol. 2, no. 9, pp. 853–859, 2011.

[101] B. T. Zhu and A. J. Lee, "NADPH-dependent metabolism of 17β-estradiol and estrone to polar and nonpolar metabolites by human tissues and cytochrome P450 isoforms," *Steroids*, vol. 70, no. 4, pp. 225–244, 2005.

[102] Y. Tsuchiya, M. Nakajima, and T. Yokoi, "Cytochrome P450-mediated metabolism of estrogens and its regulation in human," *Cancer Letters*, vol. 227, no. 2, pp. 115–124, 2005.

[103] F. F. Parl, S. Dawling, N. Roodi, and P. S. Crooke, "Estrogen metabolism and breast cancer: a risk model," *Annals of the New York Academy of Sciences*, vol. 1155, no. 1, pp. 68–75, 2009.

[104] T. Shimada, N. Murayama, K. Tanaka et al., "Interaction of polycyclic aromatic hydrocarbons with human cytochrome P450 1B1 in inhibiting catalytic activity," *Chemical Research in Toxicology*, vol. 21, no. 12, pp. 2313–2323, 2008.

[105] J. A. Sirerol, M. L. Rodriguez, S. Mena, M. A. Asensi, J. M. Estrela, and A. L. Ortega, "Role of natural stilbenes in the prevention of cancer," *Oxidative Medicine and Cellular Longevity*, vol. 2016, Article ID 3128951, 15 pages, 2016.

[106] J. K. Kundu and Y.-J. Surh, "Cancer chemopreventive and therapeutic potential of resveratrol: mechanistic perspectives," *Cancer Letters*, vol. 269, no. 2, pp. 243–261, 2008.

[107] M. E. Juan, I. Alfaras, and J. M. Planas, "Colorectal cancer chemoprevention by *trans*-resveratrol," *Pharmacological Research*, vol. 65, no. 6, pp. 584–591, 2012.

[108] T. K. H. Chang, W. B. Lee, and H. H. Ko, "*Trans*-resveratrol modulates the catalytic activity and m-RNA expression of the procarcinogens-activating human cytochrome P450 1B1," *Canadian Journal of Physiology and Pharmacology*, vol. 78, no. 11, pp. 874–881, 2000.

[109] T. K. H. Chang, J. Chen, and W. B. K. Lee, "Differential inhibition and inactivation of human CYP1 enzymes by

trans-resveratrol: evidence for mechanism-based inactivation of CYP1A2," *Journal of Pharmacology and Experimental Therapeutics*, vol. 299, no. 3, pp. 874–882, 2001.

[110] T. Walle, F. Hsieh, M. H. DeLegge, J. E. Oatis Jr., and U. K. Walle, "High absorption but very low bioavailability of oral resveratrol in humans," *Drug Metabolism and Disposition*, vol. 32, no. 12, pp. 1377–1382, 2004.

[111] R. Mikstacka, D. Przybylska, A. M. Rimando, and W. Baer-Dubowska, "Inhibition of human recombinant cytochromes P450 CYP1A1 and CYP1B1 by trans-resveratrol methyl ethers," *Molecular Nutrition & Food Research*, vol. 51, no. 5, pp. 517–524, 2007.

[112] Y.-J. Chun, S. Y. Ryu, T. C. Jeong, and M. Y. Kim, "Mechanism-based inhibition of human cytochrome P450 1A1 by rhapontigenin," *Drug Metabolism and Disposition*, vol. 29, no. 1, pp. 389–393, 2001.

[113] Y.-J. Chun, S. Kim, D. Kim, S. K. Lee, and F. P. Guengerich, "A new selective and potent inhibitor of human cytochrome P450 1B1 and its application to antimutagenesis," *Cancer Research*, vol. 61, no. 22, pp. 8164–8170, 2001.

[114] Y.-J. Chun, Y.-K. Oh, B. J. Kim et al., "Potent inhibition of human cytochrome P450 1B1 by tetramethoxystilbene," *Toxicology Letters*, vol. 189, no. 1, pp. 84–89, 2009.

[115] J. Liu, J. Sridhar, and M. Foroozesh, "Cytochrome P450 family 1 inhibitors and structure-activity relationships," *Molecules*, vol. 18, no. 12, pp. 14470–14495, 2013.

[116] D. F. V. Lewis, Y. Ito, and B. G. Lake, "Metabolism of coumarin by human P450s: a molecular modelling study," *Toxicology in Vitro*, vol. 20, no. 2, pp. 256–264, 2006.

[117] J. Sridhar, J. Liu, M. Foroozesh, and C. L. Klein Stevens, "Inhibition of cytochrome P450 enzymes by quinones and anthraquinones," *Chemical Research in Toxicology*, vol. 25, no. 2, pp. 357–365, 2012.

[118] J. Kirchmair, M. J. Williamson, J. D. Tyzack et al., "Computational prediction of metabolism: sites, products, SAR, P450 enzyme dynamics and mechanisms," *Journal of Chemical Information and Modeling*, vol. 52, no. 3, pp. 617–648, 2012.

[119] L. Olsen, C. Oostenbrink, and F. S. Jorgensen, "Prediction of cytochrome P450 mediated metabolism," *Advanced Drug Delivery Reviews*, vol. 86, pp. 61–71, 2015.

[120] C. de Graaf, C. Oostenbrink, P. H. Keizers, T. van der Wijst, A. Jongejan, and N. P. Vermeulen, "Catalytic site prediction and virtual screening of cytochrome P450 2D6 substrates by consideration of water and rescoring in automated docking," *Journal of Medicinal Chemistry*, vol. 49, no. 8, pp. 2417–2430, 2006.

[121] N. K. Mishra, "Computational modelling of P450s for toxicity prediction," *Expert Opinion on Drug Metabolism & Toxicology*, vol. 7, no. 10, pp. 1211–1231, 2011.

7

Synthesis of Novel VO(II)-Perimidine Complexes: Spectral, Computational, and Antitumor Studies

Gamil A. Al-Hazmi,[1,2] Khlood S. Abou-Melha,[1] Nashwa M. El-Metwaly ⓘ,[3,4] and Kamel A. Saleh[5]

[1]Chemistry Department, Faculty of Science, King Khalid University, P.O. Box 9004, Abha, Saudi Arabia
[2]Chemistry Department, Faculty of Applied Sciences, Taiz University, P.O. Box 82, Taiz, Yemen
[3]Chemistry Department, College of Applied Sciences, Umm Al-Qura University, Makkah, Saudi Arabia
[4]Chemistry Department, Faculty of Science, Mansoura University, Mansoura, Egypt
[5]Biology Department, Faculty of Science, King Khalid University, P.O. Box 9004, Abha, Saudi Arabia

Correspondence should be addressed to Nashwa M. El-Metwaly; n_elmetwaly00@yahoo.com

Academic Editor: Spyros P. Perlepes

A series of perimidine derivatives (L^{1-5}) were prepared and characterized by IR, ^{1}H·NMR, mass spectroscopy, UV-Vis, XRD, thermal, and SEM analysis. Five VO(II) complexes were synthesized and investigated by most previous tools besides the theoretical usage. A neutral tetradentate mode of bonding is the general approach for all binding ligands towards bi-vanadyl atoms. A square-pyramidal is the configuration proposed for all complexes. XRD analysis introduces the nanocrystalline nature of the ligand while the amorphous appearance of its metal ion complexes. The rocky shape is the observable surface morphology from SEM images. Thermal analysis verifies the presence of water of crystallization with all coordination spheres. The optimization process was accomplished using the Gaussian 09 software by different methods. The most stable configurations were extracted and displayed. Essential parameters were computed based on frontier energy gaps with all compounds. QSAR parameters were also obtained to give another side of view about the biological approach with the priority of the L^3 ligand. Applying AutoDockTools 4.2 program over all perimidine derivatives introduces efficiency against 4c3p protein of breast cancer. Antitumor activity was screened for all compounds by a comparative view over breast, colon, and liver carcinoma cell lines. IC_{50} values represent promising efficiency of the L^4-VO(II) complex against breast, colon, and liver carcinoma cell lines. The binding efficiency of ligands towards CT-DNA was tested. Binding constant (K_b) values are in agreement with the electron-drawing character of the p-substituent which offers high K_b values. Also, variable Hammett's relations were drawn.

1. Introduction

Vanadium was widely used as a therapeutic agent in the late eighteenth century, treating a variety of ailments including anemia, tuberculosis, rheumatism, and diabetes [1, 2]. Vanadium compounds exhibit various biological and physiological effects in the human body. Vanadium compounds have been extensively studied for their diverse biological activities such as antitumor, antibacterial, and insulin-enhancing effects and potential capabilities as DNA structural probes [3, 4]. The coordination chemistry of oxovanadium is highly ligand dependent and more important in biological systems [5] as well as catalytic systems [6, 7]. Due to the d1 configuration, vanadium(IV) ionic species are easily identified by EPR spectroscopy. Due to less toxicity [8, 9], the Schiff base complexes of the vanadyl ion are topic of many research reports [10, 11]. In Europe, vanadium is often used as a natural treatment for diabetes. Vanadium has been found in human studies to imitate the effects of insulin in our bodies. This ability may be useful for some of those with diabetes, a natural method to help lower blood sugar, take less insulin, or in some instances stop taking insulin altogether [12, 13]. It is noticeable that complexation of vanadium with organic ligands minimizes unfavorable effects of its inorganic salts such as vanadyl sulfate while even maintains its potential benefits [14]. Furthermore, mimicking the biological activities in natural systems can be achieved by vanadium complexes which contain oxygen and nitrogen donor ligands; so identification of the structure of these complexes is regarded important [15–17]. Bioinorganic chemistry is a fast developing field of modern chemistry that uses Schiff bases and their transition metal complexes for a variety of applications, e.g., in biological, medical, and environmental sciences. This work is interested in preparation of a series of perimidine derivatives by various substituents. New vanadyl complexes will be prepared and well characterized by using different techniques. CT-DNA binding will be tested along the organic series. Theoretical implementation will be accomplished over all prepared compounds by different standard programs. Antitumor activity will be scanned over all new prepared compounds for comparison.

2. Experimental Work

2.1. Chemicals Used. Chemicals essential for preparation of perimidine derivatives such as 1,8-diaminonaphthalene, ethylbenzoyl acetate, 4-methoxyaniline, aniline, 4-chloroaniline, 3-chloroaniline, 4-nitroaniline, $NaNO_2$, NaOH, HCl, and dioxane were purchased from Fluka and used without previous treatments. Also, $VOSO_4 \cdot xH_2O$ salt used for the complexation process was commercially available from Sigma-Aldrich. All handled solvents were from Merck and used without previous purification.

2.2. Synthesis

2.2.1. Synthesis of Compounds 3a–e (Perimidine Derivatives). Ligands **3a–e** were synthesized as reported in the literature [18] from coupling reaction of compound 1 (2.5 mmol) in ethanol (20 mL) with the appropriate arenediazonium chloride 2 in the presence of sodium hydroxide (2.5 mmol) in the ice bath at 0–5°C. The whole mixture was then left in a refrigerator overnight. The precipitated solid was filtered off, washed with water, and finally crystallized from dioxane/EtOH to give the respective hydrazones **3a–e** (Scheme 1). ^{1}H·NMR and mass spectra are displayed in Figures 1, 2, S1, and S2. The analysis is matching completely with that reported in the literature [16]. The structural forms of new perimidine compounds are displayed in Figures 3(a) and 3(b).

(1) *2-[N-phenyl-2-oxo-2-phenylethanehydrazonoyl]-1H-perimidine (3a) (L^1).* IR v: 3402, 3194 (2NH), and 1616 (CO) cm^{-1}. ^{1}H·NMR (DMSO-d$_6$) d: 6.72–7.80 (m, 16H, Ar–H) and 13.62 (s, 2H, 2NH). MS *m/z* (%): 390 (M, 22), 285 (2), 166 (5), 140 (7), 127 (1), 105 (100), 93 (8), and 77 (35). Anal. calcd. for $C_{25}H_{18}N_4O$ (390.42).

(2) *2-[N-(4-methoxyphenyl)-2-oxo-2-phenylethanehydrazonoyl]-1H-perimidine (3b) (L^2).* IR v: 3333, 3167

(2NH), and 1680 (CO) cm^{-1}. ^{1}H·NMR (DMSO-d$_6$) δ: 3.58 (s, 3H, OCH$_3$), 7.59–7.97 (m, 15H, ArH), and 12.20 (br s, 2H, 2NH). MS *m/z* (%): 420 (M, 5), 419 (9), 193 (12), 166 (12), 126 (14), 107(17), 105 (100), 92 (31), and 77 (75). Anal. calcd. for $C_{26}H_{20}N_4O_2$ (420.47).

(3) *2-[N-(4-chlorophenyl)-2-oxo-2-phenylethanehydrazonoyl]-1H-perimidine (3c) (L^3).* IR v: 3422, 3206 (2NH), and 1612 (CO) cm^{-1}. ^{1}H·NMR (DMSO-d$_6$) d: 6.73–7.77 (m, 15H, Ar–H) and 13.27 (s, 2H, 2NH). MS *m/z* (%): 426 (M 2, 5), 425 (M 1, 6), 424 (M, 14), 140 (4), 127 (5), 111 (2), 105 (100), and 77 (37). Anal. calcd. for $C_{25}H_{17}ClN_4O$ (435.44).

(4) *2-[N-(4-nitrophenyl)-2-oxo-2-phenylethanehydrazonoyl]-1H-perimidine (3d) (L^4).* IR v: 3356, 3198 (2NH), and 1670 (CO) cm^{-1}. ^{1}H·NMR (DMSO-d$_6$) d: 6.64–8.21 (m, 15H, Ar–H) and 12.60 (s, 2H, 2NH). MS *m/z* (%): 435 (M, 8), 238 (34), 138 (13), 167 (11), 106 (45), 105 (58), 93 (100), 77 (44), and 66 (76). Anal. calcd. for $C_{25}H_{17}N_5O_3$ (424.88).

(5) *2-[N-(3-chlorophenyl)-2-oxo-2-phenylethanehydrazonoyl]-1H-perimidine (3e) (L^5).* IR v: 3229, 3167 (2NH), and 1622 (CO) cm^{-1}. ^{1}H·NMR (DMSO-d$_6$) d: 6.70–7.91 (m, 15H, Ar–H) and 13.02 (s, 2H, 2NH). MS *m/z* (%): 426 (M2, 7), 425 (M1, 9), 424 (M, 16), 194 (7), 166 (7), 140 (5), 127 (4), 105 (100), 111 (4), and 77 (36). Anal. calcd. for $C_{25}H_{17}ClN_4O$ (424.88).

2.2.2. Synthesis of VO(II) Complexes. New VO(II) complex series was synthesized by using variable derivatives from perimidine ligands. Equimolar (3 mmol) values were used from the perimidine ligand and dissolved fully in dioxane; after that, it was mixed with $VOSO_4 \cdot xH_2O$ which dissolves in the dioxane/H_2O mixture. The weighted molar ratio value from vanadyl salt was calculated attributing to its anhydrous weight. After ≈5 h reflux, 0.5 g sodium acetate was added after dissolving in a little amount of bi-distilled water to precipitate the complexes. Each precipitate was separated out on hot, filtered off, washed several times with ethanol and diethyl ether, and finally dried in a vacuum desiccator.

2.3. DNA Binding Study. The binding attitudes of perimidine derivatives towards calf thymus DNA (CT-DNA) will be studied by using the spectroscopy method. CT-DNA (50 mg) was dissolved by stirring overnight in double deionized water (pH = 7.0) and must be kept at 4°C. Bi-distilled water was used to prepare the buffer (5.0 mM tris (hydroxymethyl)-aminomethane and 50 mM NaCl, pH = 7.2). Tris-HCl buffer was prepared in deionized water. DNA buffering solution gave absorbance ratio at 260/280 nm by 1.8–1.9, and this indicates the absence of protein from DNA [19, 20]. Applying the UV-Vis technique, the DNA concentration was determined (5.10×10^{-4} M) using its known molar absorptivity coefficient value (6600 M$^{-1} \cdot$cm^{-1} at 260 nm). At room temperature, 200–900 nm is the

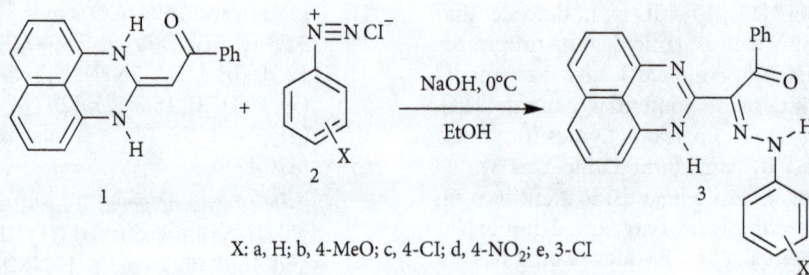

SCHEME 1: Synthesis of perimidine compounds 3a–e.

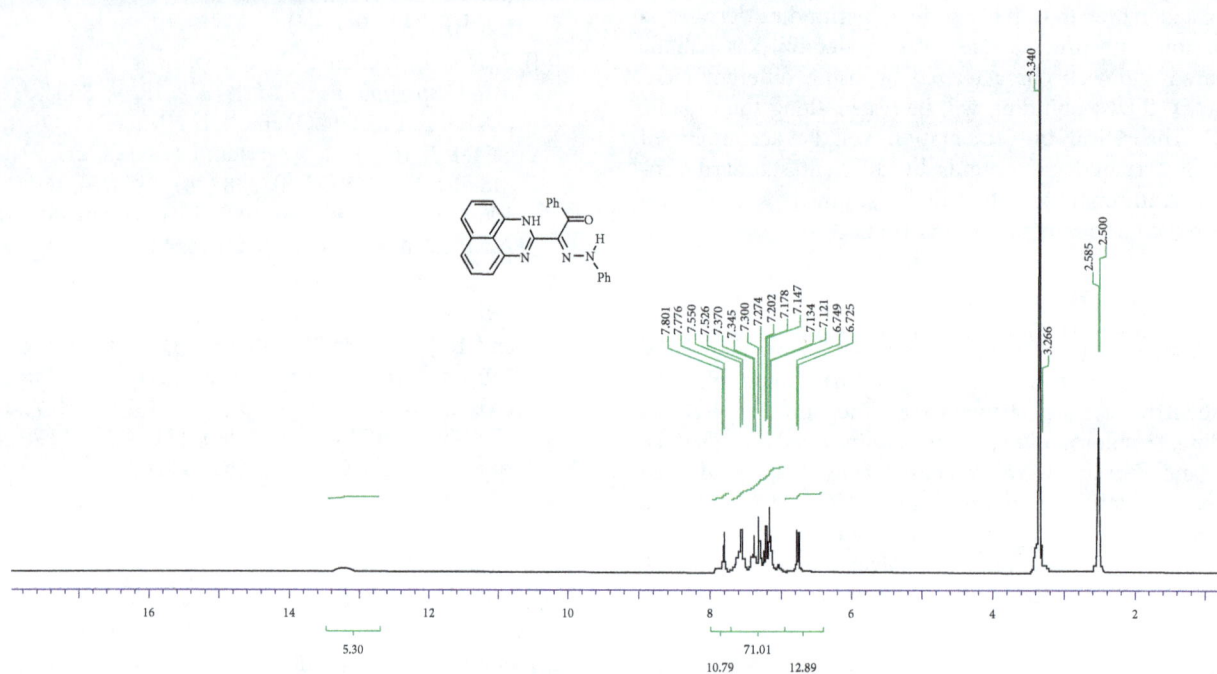

FIGURE 1: ^{1}H·NMR of L^1 ligand (as example).

wavelength range used, and in 1 cm quartz cuvette, a fixed concentration (2.0×10^{-5} M in dioxane) from each ligand was utilized. A scanning process was done after adding CT-DNA by a gradual way from 0.00 to $\approx 2.18 \times 0^{-4}$ mol·L^{-1}. The same DNA amount added to the ligand solution was added also to the reference cell to delete the absorbance of free DNA. A significant binding constant (K_b) for interaction between ligands towards CT-DNA was determined by using the following equation: $[DNA]/(\epsilon_a - \epsilon_f) = [DNA]/(\epsilon_b - \epsilon_f) + 1/K_b (\epsilon_a - \epsilon_f)$ [21], where [DNA] is the concentration of CT-DNA in base pairs, ϵ_a is the extinction coefficient observed for A/[compound] at the used DNA concentration, and ϵ_f is the extinction coefficient for each free compound (HL$^{1-5}$) in the solution. Moreover, ϵ_b is the extinction coefficient of the compound when fully bond to DNA. In plots of $[DNA]/(\epsilon_a - \epsilon_f)$ vs. [DNA], K_b is given by the following ratio: slope/intercept.

2.4. Antitumor Influence. The evaluation of cytotoxicity of candidate anticancer drugs will be performed using the most effective, available SRB method. All molecules and

their derivatives will be tested for their toxicity on different cancer cell lines. In an attempt to evaluate the impact, the samples were prepared with different concentrations: 0.01, 0.1, 1, 10, and 100 μg/ml, respectively. The cells were cultured in the mixture of samples and media (RPMI-FBS + samples) for 72 h; after that, cytotoxicity impact was evaluated compared to the response of doxorubicin as a positive control.

The cytotoxic effect of the composites and ligands will be tested against different cancer cell lines (HepG2, MCF-7, and HCT116) as donor cancer cell lines by means of the SRB cytotoxicity test. To avoid the contamination, the RPMI media of the cells were supplemented with 100 μg/ml streptomycin and 100 units/ml penicillin with 10% FBS and incubated at a 5% CO$_2$ incubator. Growing cells were collected using the trypsin enzyme and then counted using the cell counter in order to distribute equally the number of cells to each well of 69-well plates. The cells will incubate under sterile conditions with different concentrations of both ligands and composites for 72 hours, and subsequently, treated cells and untreated cells and the positive control were fixed with 10% TCA (trichloroacetic acid) and kept at 4°C for

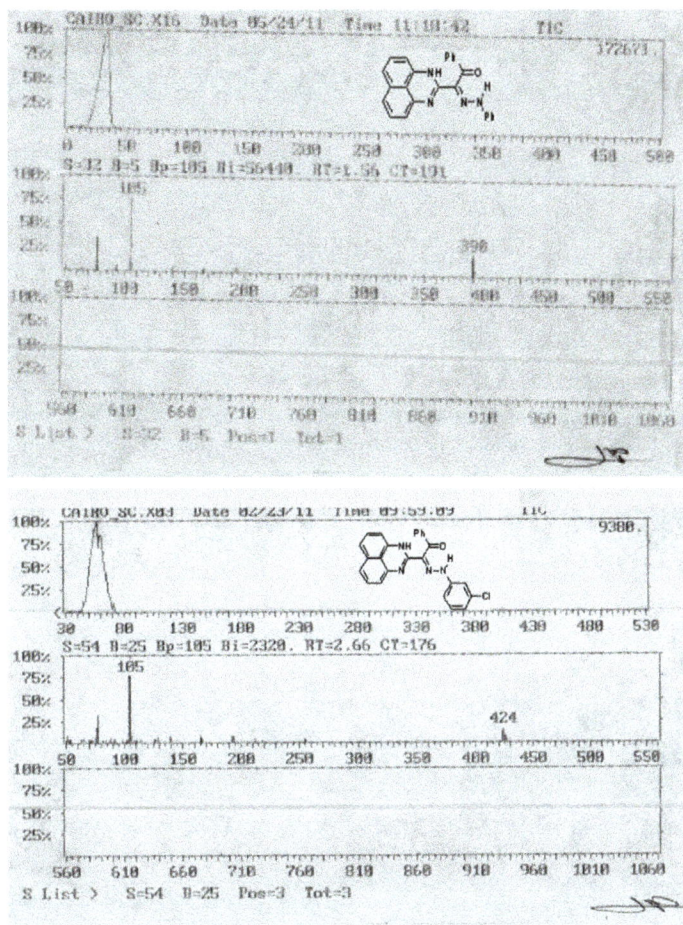

FIGURE 2: Mass spectra of L^1 and L^4 ligands.

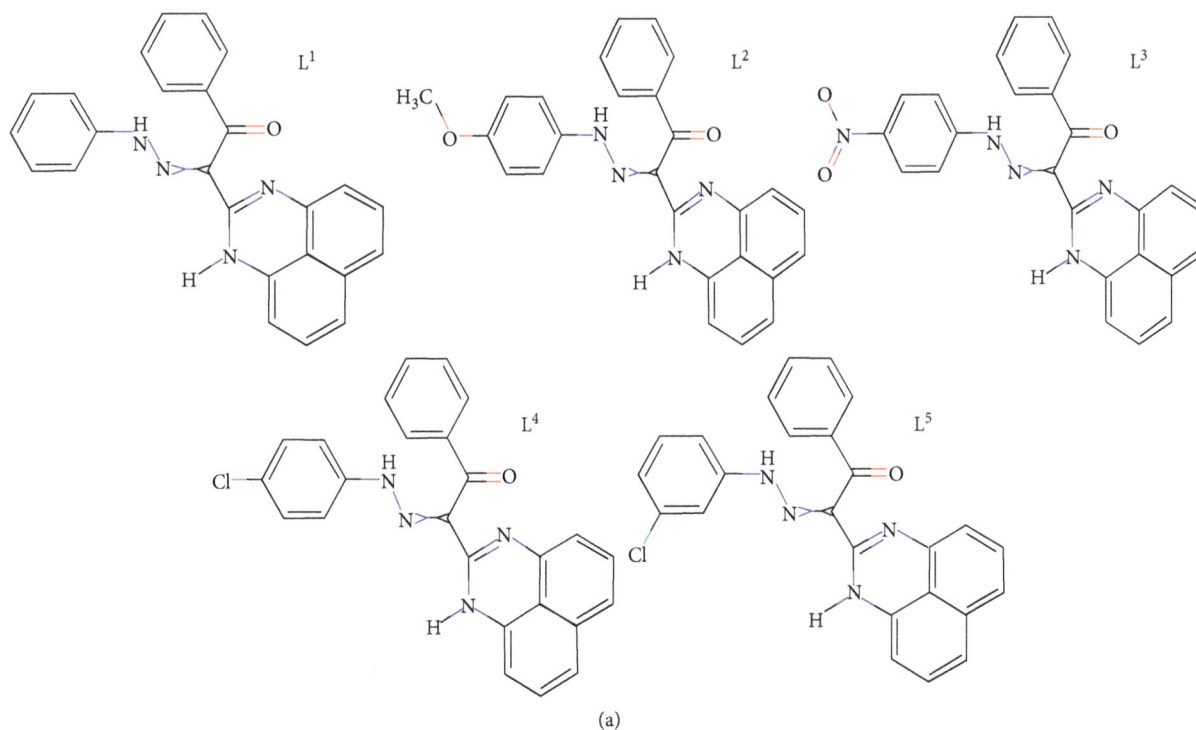

(a)

FIGURE 3: Continued.

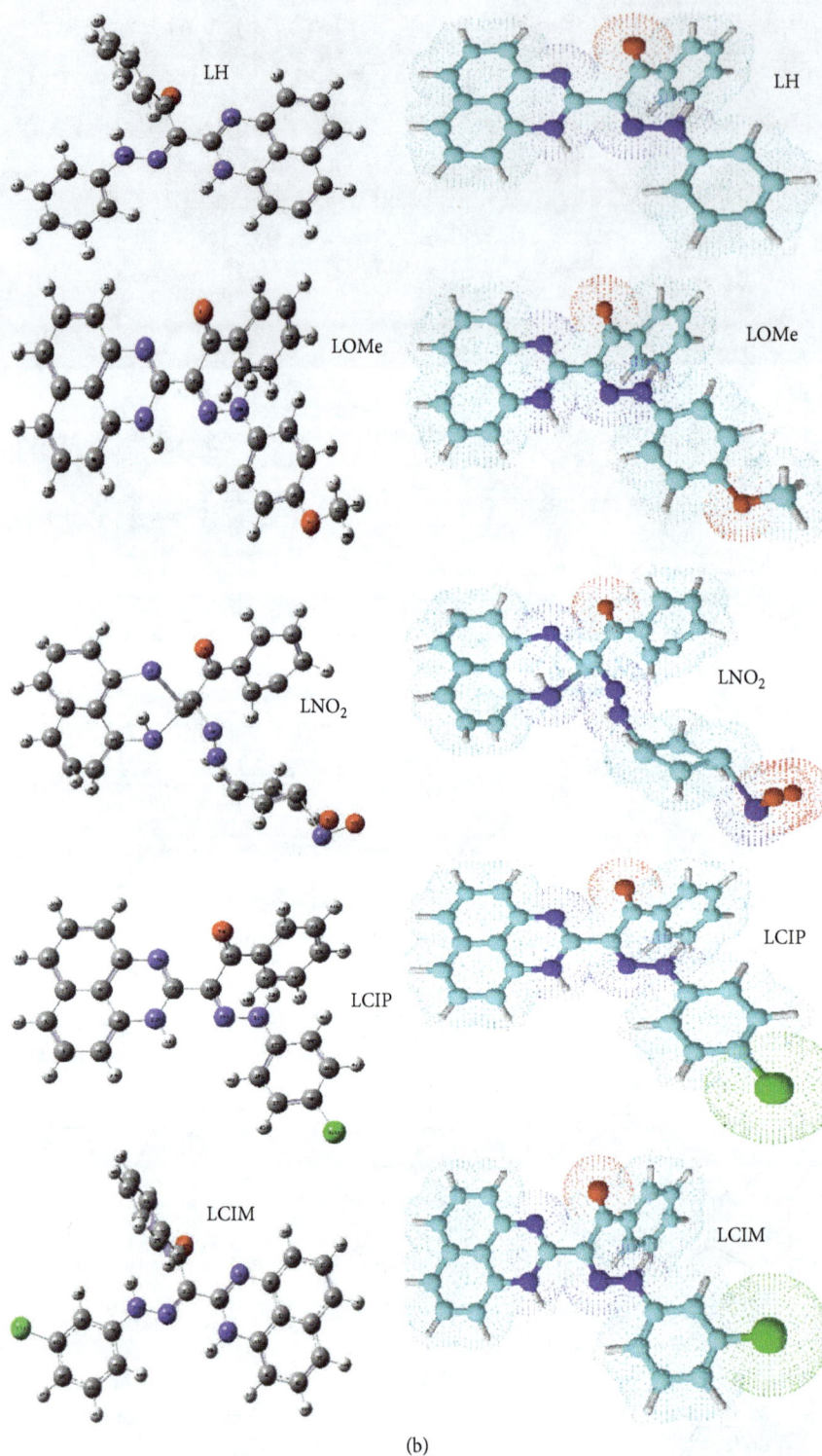

(b)

FIGURE 3: (a) Structures of perimidine ligands (L^{1-5}). (b) Optimized structures of five perimidine ligands.

1 h. After washing few times, fixed and washed cells were stained with 0.4% SRB stain solution for ten minutes, and subsequently, the cells were washed with 1% glacial acetic acid. To dissolve SRB-stained cells, Tris-HCl was used. To detect the density of remaining colors, a plate reader will be used at 540 nm wavelength. In order to determine the IC_{50} value, statistical analysis was accomplished through SigmaPlot version 14.0. The advantage of prepared compounds as potential drugs against different cancer cells was investigated.

2.5. Physical Techniques

2.5.1. Elemental Analysis.
The element contents (carbon, hydrogen, and nitrogen) were determined at the Micro-Analytical Unit of Cairo University. Vanadium, sulfate, and chloride contents were evaluated by known standard methods [22] through complexometric and precipitation methods.

2.5.2. Conductivity Measurements.
Applying the Jenway 4010 conductivity meter, the molar conductivity of freshly prepared 1.0×10^{-3} mol/cm^3 in DMSO solutions was estimated.

2.5.3. X-Ray Diffraction and SEM.
X-ray diffraction manners were recorded on the Rigaku diffractometer using Cu/Kα radiation. Scanning electron microscopy (SEM) images were obtained by using Joel JSM-6390 equipment.

2.5.4. IR, ^{1}H·NMR, and ^{13}CNMR Spectra.
IR spectra were obtained using the JASCO FT/IR-4100 spectrophotometer from 400 to 4000 cm^{-1} in the KBr disc, while ^{1}H·NMR spectra were recorded in deuterated dimethyl sulfoxide using the Varian Gemini 300 NMR spectrometer.

2.5.5. Mass spectra.
Mass spectra were recorded on GCMS-QP1000 EX (Shimadzu) and GCMS 5988-A.

2.5.6. ESR Analysis.
ESR spectra of VO(II)-powdered complexes were obtained on the Bruker EMX spectrometer working in the X-band (9.60 GHz) with 100 kHz modulation frequency. The microwave power was set at 1 mW, and modulation amplitude was set at 4 Gauss. The low field signal was obtained after 4 scans with a 10-fold increase in the receiver gain. A powder spectrum was obtained in a 2 mm quartz capillary at ordinary temperature.

2.5.7. UV-Vis Spectra and Magnetic Measurements.
Electronic spectra for all compounds were recorded using the UV$_2$ Unicam UV/Vis spectrophotometer in the DMSO solvent. Magnetic susceptibility values for VO(II) complexes were conducted by the Johnson Matthey magnetic susceptibility balance at room temperature.

2.5.8. Thermal Analysis.
The Shimadzu thermogravimetric analyzer (20–900°C) at 10°C·min^{-1} heating rate under nitrogen was used for thermal analysis. Theoretical treatments (modeling and docking) were accomplished by known standard programs.

2.5.9. Antitumor Activity.
Antitumor activity was conducted at the Regional Center for Mycology and Biotechnology.

2.6. Computational

2.6.1. DFT/Hartree–Fock Study.
Implementing the Gaussian 09 software [23], the structural optimization process was accomplished over pyrimidine ligands and their VO(II) complexes in the gas phase. Two known methods were found as the most suitable one for the optimization process. The output files were visualized by the GaussView program [24]. According to the numbering scheme, DFT parameters were extracted using frontier energy gaps (E_{HOMO} and E_{LUMO}) for all investigated compounds. Moreover, other significant computations were taken from log files as oscillator strength, excitation energy, charges assigned for coordinating atoms, and some bond lengths.

2.6.2. QSAR Computation.
New perimidine compounds were treated for the optimization process to give the best structural forms. HyperChem (v8.1) software is the tool used for such a purpose. The preoptimization process was executed by molecular mechanics force field (MM$^+$) accompanied by semiempirical AM1 for the soft adjustment procedure. This process was accomplished without fixing any parameter till the equilibrium state for geometric structures. A system for minimizing energy was employed the Polak–Ribiere conjugated gradient algorithm. The QSAR process leads to computing essential parameters including the partition coefficient (log P). Log P value is considered the essential indicator used to predict the biological activity for optimized compounds [25].

2.6.3. Docking Computation.
Applying AutoDockTools 4.2 by using Gasteiger partial charges which added over the elements of pyrimidine ligands, the simulation procedure was executed to give a view on the biological behavior of compounds. Rotatable bonds were cleared, and nonpolar hydrogen atoms were conjoined. Interaction occurred between inhibitors (ligands) and protein receptors (4c3p, 3bch, and 4zdr) for breast, colon, and liver cancer proteins. The docking process was accomplished after addition of fundamental hydrogen atoms, Kollman united atom-type charges, and salvation parameters [26]. Affinity (grid) maps of ×× Å grid points and 0.375 Å spacing were generated applying the AutoGrid program [27]. Van der Waals forces and electrostatic terms were obtained. This is done by applying autodock parameter set-dependent and distance-dependent dielectric functions, respectively. The docking process was executed using the Solis and Wets local search method and Lamarckian genetic algorithm (LGA) [28]. Initial position, orientation, and torsions of the inhibitor molecule were set indiscriminately. All rotatable torsions were expelled during the docking process. Each experiment is the mean value of 10 different runs that are set close after the maximum of 250000 energy assessments. 150 is the used population size. During the process, the translational step of 0.2 Å, quaternion, and torsion steps of 5 were applied.

3. Results and Discussion

3.1. Physical Properties. Essential analytical and physical data for ligands and their VO(II) complexes are summarized in Table 1. All investigated complexes are nonhygroscopic in nature, having high melting point (>300°C). The elemental analysis proposes 1 : 2 (HL : M) molar ratio as the general formula for all complexes. All complexes are completely soluble in DMSO or DMF solvents. The conductivity measured is $5.66–14.22 \, \Omega^{-1} \cdot cm^2 \cdot mol^{-1}$. Such conductivity values are attributed to the nonconducting character of tested complexes [29]. This coincides with sulfate anion which favors covalent attachment with metal ions inside the coordination sphere.

3.2. Comparative IR Study. The assignments of all characteristic bands for five perimidine ligands and their VO(II) complexes are summarized in Table 2. ν(NH), δ(NH), ν(C=N), and ν(C=O) are the significant functional bands for coordinating groups which appear in narrow regions observed in all derivatives. This may refer to the far effect of the aromatic substituent on bond movement inside such groups. A comparative study of ligands and their VO(II) complexes reveals the following observations: (1) lower-shifted appearance of former bands is considered a strong evidence for contribution of C=O, NH, and C=N groups in coordination towards two central atoms. (2) New bands appeared at 1368–1434 and 1140–1179 cm^{-1} assigned for $\nu_{as}(SO_4)$ and $\nu_s(SO_4)$, respectively, through bidentate attachment [30]. (3) Other bands appearing at ≈760 and $\approx690 \, cm^{-1}$ are attributed to $\delta_r(H_2O)$ and $\delta_w(H_2O)$, respectively, for crystal water molecules. (4) ν(M-L) bands appeared at the low wavenumber region belonging to M-O and M-N bonds. These spectral observations suggest a tetradentate mod of coordination towards two vanadyl atoms. Also, the band observed at 966–1074 cm^{-1} range assigns for ν(V = O), significantly pointing to the square-pyramidal configuration [31].

3.3. Electronic Spectra and Magnetic Measurements. Electronic transition bands and magnetic moment values are aggregated in Table 3. UV-Vis spectra were recorded qualitatively in the DMSO solvent to gain smoothly absorption curves. Intraligand transition bands appearing at 31,250–38,168, 25,974–30,769, and 17,544–19,157 cm^{-1} are attributed to $n \longrightarrow \sigma^*$, $\pi \longrightarrow \pi^*$, and $n \longrightarrow \pi^*$ transitions, respectively, inside variable groups [32]. A structural condensed conjugation of chromophores leads to appearance of deep colors for all perimidine ligands. This is accompanied with the appearance of the $n \longrightarrow \pi^*$ band in the middle of the visible region. VO(II) complex spectra display intraligand transitions suffer shift due to coordination. The appearance of charge transfer bands attributes to O $\longrightarrow$ V and N $\longrightarrow$ V transitions. Also, new significant d-d transition bands were observed at $\approx15,300$ and 12,800 cm^{-1} assigned for $^2B_2g \longrightarrow {}^2B_1g$ (E_2, v_2) and $^2B_1g \longrightarrow {}^2Eg$ (E_1, v_1), respectively. These bands are attributed to transition inside the square-pyramidal configuration (Figure 4). Reduced magnetic

moment values ($\mu_{eff} = 1.65–1.68$ BM) recorded for all complexes support the proposal of binuclear complexes [33].

3.4. ESR Spectra. ESR spectra (Figure 5, for example) of VO(II) solid complexes were obtained and investigated to verify the structural forms of them. All spectra demonstrated an eight-line pattern, which attributes to the analogous and vertical ingredients g-tensors and hyperfine (hf) A-tensors. Spin Hamiltonian parameters and molecular orbital values were calculated and are represented in Table 4. The analogous and vertical ingredients are well resolved. Nitrogen super-hyperfine splitting is not observed, which points to the presence of single electron in the d_{xy} orbital. The pattern suggests that g and A are axially symmetric in nature. The factors A and g appear to be in covenant with the values commonly known for vanadyl complex in the square-pyramidal geometry. G factor, which is expressed by $G = (g_{//} - 2.0023)/(g_\perp - 2.0023) = 4$, measures the exchange interaction between metal centers. In agreement with Hathaway [34, 35], $G > 4$ shows negligible exchange interaction, while $G < 4$ is the vice versa. An observable reduction of the values calculated (1.71–2.79) proposes strong interaction inside binuclear complexes [36, 37]. This interaction affects the magnetic moment value of complexes which suffers observable reduction. The tendency of A_{11} to decrease with increasing g_{11} is an index for tetrahedral distortion ($f = g_{//}/A_{//}$) [38–40]. The molecular orbital coefficients α^2 and β^2 are calculated by

$$\beta^2 = \frac{7}{6}\left(-\frac{A_{11}}{P} + \frac{A_\perp}{P} + g_{11} - \frac{5}{14}g_\perp - \frac{9}{14}g_e\right),$$

$$\alpha^2 = \frac{2.0023 - \Delta g}{8\beta^2\lambda}, \quad \text{where } \Delta g = \left(g_\perp - g_{||}\right) \times 10^{-3}. \tag{1}$$

The hyperfine conjunction disciplinarians were calculated by taking $A_{//}$ and $A_\perp$ as negative, which gave positive values of β^2 and α^2. The calculated α^2 and β^2 values introduce the highly ionic character of in-plane σ- and π-bonding. The electronic transition spectra display two significant bands at $\approx15,300$ and 12,800 cm^{-1} assigned for $^2B_2g \longrightarrow {}^2B_1g$ (E_2, v_2) and $^2B_1g \longrightarrow {}^2Eg$ (E_1, v_1), respectively. Assume pure d-orbitals by using first- and second-order perturbation theories. The parameters attributing to transition energy are called the spin Hamiltonian parameters and calculated by the following expression: $g_\perp = g_e - (2\lambda/E_2)$, where g_e is the free-electron g value (2.0023). Using E_2 value, the spin-orbital coupling constant (λ) was evaluated (138.18). A value for λ of 250 cm^{-1} is reported [41] for free V^{+4} ion. The high reduction in the magnitude of λ for the double-bonded oxovanadium complex $(V=O)^{+2}$ is attributed to substantial π-bonding. However, the value falls inside the logical borders announced. The orbital reduction factors, namely, $K_{//}$ and $K_\perp$, are also calculated using $^2K_{//} = (g_{//} - 2.00277) E/8 \lambda$ and $^2K_\perp = (g_\perp - 2.00277) E/2 \lambda$. For pure σ-bonding, $K_{//} \approx K_\perp \approx -0.77$, while $^2K_{//} > {}^2K_\perp$ signifies in-plane σ-bonding, with $^2K_\perp < {}^2K_{//}$ accounting for out-of-plane π-bonding [42].

TABLE 1: Significant analytical and physical data of perimidine compounds and their VO(II) complexes.

Compounds (formula weight) (calcd./found)	Color	Elemental analysis (%) calcd. (found)				
		C	H	N	SO$_4$/Cl	V
(1) (C$_{25}$H$_{18}$N$_4$O) (L^1) (390.44/390.42)	Dark brown	76.91 (76.90)	4.65 (4.66)	14.35 (14.35)	—	—
(2) [(VO)$_2$ (SO$_4$)$_2$ (L^1)]H$_2$O (734.46)	Dark brown	40.88 (40.88)	2.74 (2.73)	7.63 (7.65)	26.16 (26.16)	13.87 (13.88)
(3) (C$_{26}$H$_{20}$N$_4$O$_2$) (L^2) (420.47/420.44)	Dark brown	74.27 (74.27)	4.79 (4.79)	13.32 (13.31)	—	—
(4) [(VO)$_2$ (SO$_4$)$_2$ (L^2)]2H$_2$O (782.51)	Dark green	39.91 (39.90)	3.09 (3.10)	7.16 (7.17)	24.55 (24.54)	13.02 (13.03)
(5) (C$_{25}$H$_{17}$N$_5$O$_3$) (L^3) (435.44/435.42)	Dark brown	68.96 (68.95)	3.93 (3.93)	16.08 (16.09)	—	—
(6) [(VO)$_2$ (SO$_4$)$_2$ (L^3)]H$_2$O (779.46)	Dark green	38.52 (38.52)	2.46 (2.46)	8.99 (8.97)	24.65 (24.66)	13.07 (13.05)
(7) (C$_{25}$H$_{17}$N$_4$OCl) (L^4) (424.88/424.86)	Dark brown	70.67 (70.66)	4.03 (4.02)	13.19 (13.18)	8.34 (8.35)	—
(8) [(VO)$_2$ (SO$_4$)$_2$ (L^4)]H$_2$O (768.90)	Dark green	39.05 (39.05)	2.49 (2.48)	7.29 (7.28)	24.99 (25.02)/4.61 (4.63)	13.25 (13.26)
(9) (C$_{25}$H$_{17}$N$_4$OCl) (L^5) (424.88/424.86)	Dark brown	70.67 (70.68)	4.03 (4.05)	13.19 (13.18)	8.34 (8.35)	—
(10) [(VO)$_2$ (SO$_4$)$_2$ (L^5)]3H$_2$O (804.93)	Dark brown	37.30 (37.31)	2.88 (2.88)	6.96 (6.95)	23.87 (23.88)/4.40 (4.41)	12.66 (12.67)

TABLE 2: Significant IR spectral bands (cm^{-1}) of perimidine compounds and their VO(II) complexes.

Compounds	ν_{NH}, ν_{OH}	δ_{NH}	$\nu_{C=O}$	$\nu_{C=N}$	$\nu_{as(SO4)}$	$\nu_{s(SO4)}$	$\delta r(H_2O)$, $\delta w(H_2O)$	$\nu_{V=O}$	ν_{M-O}	ν_{M-N}
(1) (C$_{25}$H$_{18}$N$_4$O) (L^1)	3155	1473	1618	1518	—	—	—	—	—	—
(2) [(VO)$_2$ (SO$_4$)$_2$ (L^1)]H$_2$O	3110, 3350	1470	1596	1514	1420	1142	765, 670	966	588	476
(3) (C$_{26}$H$_{20}$N$_4$O$_2$) (L^2)	3177	1470	1597	1508	—-	—	—	—	—	—
(4) [(VO)$_2$ (SO$_4$)$_2$ (L^2)]2H$_2$O	3100, 3372	1447	1592	1502	1411	1146	765, 697	1074	572	515
(5) (C$_{25}$H$_{17}$N$_5$O$_3$) (L^3)	3150	1473	1620	1538	—	—	—	—	—	—
(6) [(VO)$_2$ (SO$_4$)$_2$ (L^3)]H$_2$O	3105, 3382	1447	1616	1517	1411	1179	743, 697	966	600	508
(7) (C$_{25}$H$_{17}$N$_4$OCl) (L^4)	3160	1474	1614	1518	—	—	—	—	—	—
(8) [(VO)$_2$ (SO$_4$)$_2$ (L^4)]H$_2$O	3100, 3420	1471	1624	1510	1434	1150	755, 637	985	610	550
(9) (C$_{25}$H$_{17}$N$_4$OCl) (L^5)	3150	1473	1620	1518	—	—	—	—	—	—
(10) [(VO)$_2$ (SO$_4$)$_2$ (L^5)]3H$_2$O	3054, 3384	1446	1616	1512	1368	1140	754, 689	1074	589	508

TABLE 3: Electronic transitions of perimidine compounds and their VO(II) complexes.

Compounds	μ_{eff} (BM)	d-d transition bands (cm^{-1})	Intraligand and charge transfer (cm^{-1})
(1) (C$_{25}$H$_{18}$N$_4$O) (L^1)	—	—	31,746; 26,316; 23,923; 18,868
(2) [(VO)$_2$ (SO$_4$)$_2$ (L^1)]H$_2$O	1.66	15,290; 12800	35,714; 29,412; 25,641; 24,272; 17,857
(3) (C$_{26}$H$_{20}$N$_4$O$_2$) (L^2)	—	—	38,168; 28,249; 23,810; 19,048
(4) [(VO)$_2$ (SO$_4$)$_2$ (L^2)]2H$_2$O	1.68	15,393; 12750	37,037; 30,769; 26,316; 23,256; 17,544
(5) (C$_{25}$H$_{17}$N$_5$O$_3$) (L^3)	—	—	36,364; 30,303; 26,316; 23,810; 19,157
(6) [(VO)$_2$ (SO$_4$)$_2$ (L^3)]H$_2$O	1.66	15,873; 12830	37,037; 28,571; 26,667; 18,182
(7) (C$_{25}$H$_{17}$N$_4$OCl) (L^4)	—	—	31,746; 25,974; 18,587
(8) [(VO)$_2$ (SO$_4$)$_2$ (L^4)]H$_2$O	1.67	15,385; 12,800	35,714; 30,303; 25,974; 24,390; 17,857
(9) (C$_{25}$H$_{17}$N$_4$OCl) (L^5)	—	—	31,250; 26,316; 24,390; 18,182
(10) [(VO)$_2$ (SO$_4$)$_2$ (L^5)]3H$_2$O	1.65	15,873; 12780	35,714; 30,769; 25,000; 23,256; 18,868

3.4.1. Calculation of Dipole Term (p). Dipolar term values can be determined by

$$p = \frac{7(A_{11} - A_\perp)}{6 + (3/2)(\lambda/E_1)}. \tag{2}$$

If A_{11} is taken to be negative and $A_\perp$ positive, the value of p will be more than 270 G, which is far from the expected value. Thus, the signs of both A_{11} and $A_\perp$ are used as negative and are indicated in the form of the isotropic hf constant (A_o). McGarvey theoretically accomplished the p value as

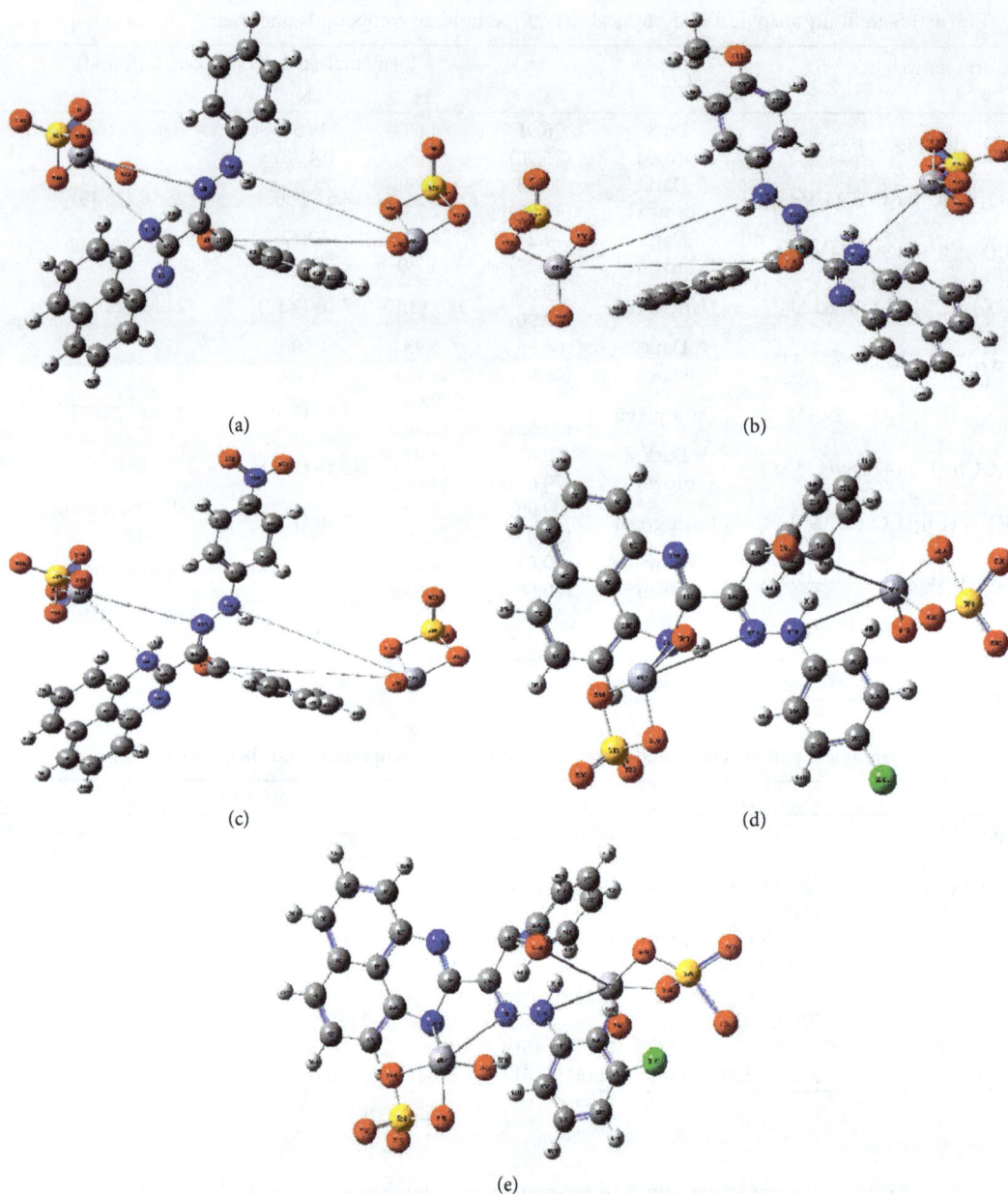

(a)

(b)

(c)

(d)

(e)

FIGURE 4: Geometry optimization of VO(II)-perimidine complexes (a–e, respectively).

+136 G for vanadyl complexes which does not deviate much from the expected value.

3.4.2. Calculation of MO Coefficients and Bonding Parameters. The g values observed are different from the electronic value (2.0023). This assigns to spin-orbit interaction of the d_{xy} ground state level. The isotropic and anisotropic g and A parameters were calculated using the following equations: $A_o = (A_{//} + 2A_{\perp})/3$ and $g_o = (g_{//} + 2g_{\perp})/3$. Taking A_{11} and $A_{\perp}$ to be negative values, the K expression is $K = -(A_o/p) - (g_e - g_o)$.

Thus, K (Fermi contact term) can be determined. The Fermi contact term, k, is a sense of polarization exerted by

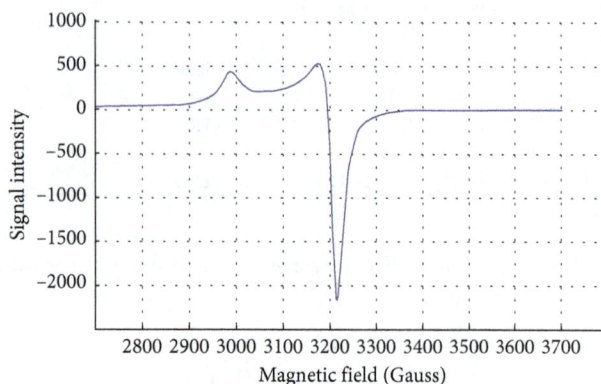

FIGURE 5: ESR spectrum of L^1 + VO(II) complex.

TABLE 4: Spin Hamiltonian parameters of all VO(II) complexes (A and p x10^{-4}).

Complex	$g_{//}$	$g_\perp$	g_o	A_{11}	F	$A_\perp$	A_o	G	p	k	$^2K_{//}$	$^2K_\perp$	α^2	β^2
(1)	1.93	1.96	1.95	167	115.57	66	99.67	1.71	117.52	0.796	−0.843	−1.981	1.959	0.9357
(2)	1.94	1.97	1.96	170	114.12	71	104.00	1.93	115.19	0.861	−0.724	−1.512	1.490	0.9435
(3)	1.92	1.96	1.95	168	114.28	69	105.67	1.95	115.19	0.865	−0.961	−1.985	1.964	0.9243
(4)	1.94	1.98	1.97	171	113.45	73	105.67	2.79	114.00	0.895	−0.727	−1.055	1.033	0.9395
(5)	1.93	1.97	1.96	171	112.86	72	105.00	2.24	115.19	0.869	−0.841	−1.515	1.494	0.9318

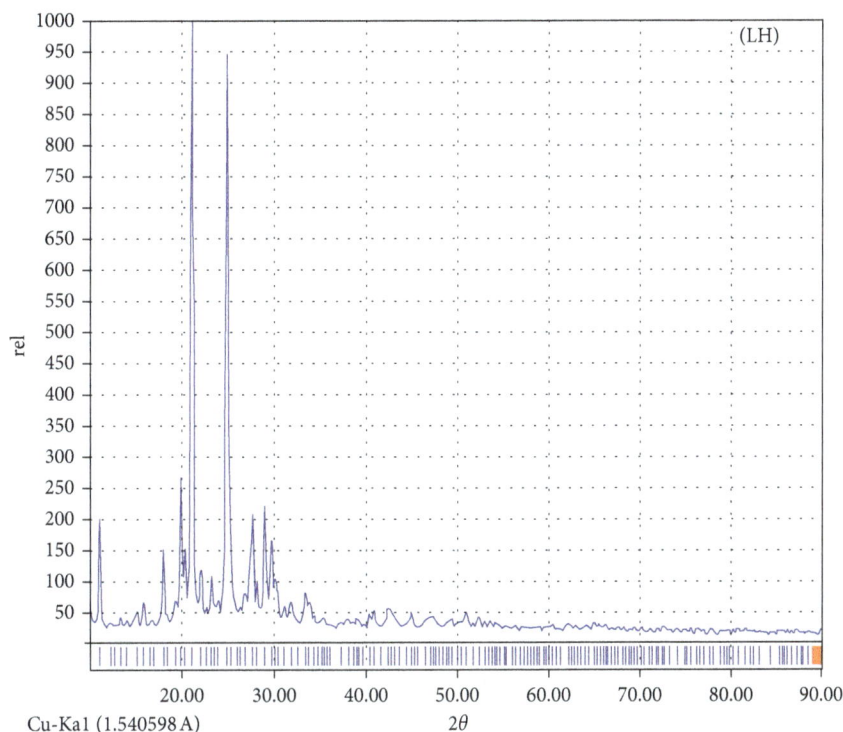

FIGURE 6: X-ray diffraction pattern of high crystalline ligand.

the uneven apportionment of d-electron density on the inner core s-electron.

3.5. X-Ray Diffraction. X-ray diffraction patterns were executed over $10° < 2\theta < 90°$ range (Figures 6 and S3) for ligands under study. This technique gives a considerable view about dynamics of the crystal lattice in solid compounds. Using standard methods, a comparative study of patterns with reactants reflects the purity of isolated compounds [43]. Also, significant parameters related to crystalline compounds can be calculated using the high-intense peak (full width at half maximum (FWHM)). The crystallinity appearing with the LH ligand reflects the isolation of a strictly known irregular crystallite, while the amorphous appearance of others reflects the indiscriminate orientation of atoms inside the 3D space. 2θ (21.18), d spacing (4.1910), FWHM (0.2454), relative intensity (%) (857), and particle size (6.003 E) were calculated for LH compounds. The crystallite size was computed by utilizing the Debye–Scherrer equation: $\beta = 0.94 \ \lambda/(S \cos \theta)$, where S is the crystallite size, θ is the diffraction angle, β is FWHM, and Cu/Kα (λ) = 1.5406 Å. The d-spacing between inner crystal planes was extracted from the Bragg equation: $n\lambda = 2d\sin(\theta)$ at $n = 1$. The size calculated falls in the nanometer range (nm) which expects a widespread application especially for the biological field. Also, crystal strain (ε, 5.027) was calculated by $\beta = (\lambda/S \cos \theta) - \varepsilon \tan \theta$, while dislocation density (δ, 0.0277) was computed by $\delta = 1/S^2$ [44]. The dislocation density and strain are the aspects for network dislocation in compounds. The lower values of them indicate high quality of compounds. The SEM tool is used to give a clear view about the habit and surface morphology of all studied compounds (Figure S4). The images of paramagnetic compounds are not strictly resolved because an insufficient electron beam can meet the surface to provide well resolution. Subsequently, the determination of particle size in an accurate way is strongly absent. It was known about this study that the crystals were grown up from just a single one to several accumulated distributables with particle sizes starting with few nanometers to many hundreds. The formation of extended crystals over a rocky shape may happen by two nucleation processes: by distribution and by piling up of layers which are grown. It was pointed to that if the rate of growth along the C-axis is fast and a great number of grown nuclei are active across the axis in comparison with vertical to the C-axis, the crystals will be extended over patches [45]. The attitude displayed on different crystals may be due to the growth along the strongest bond through anisotropy included in crystal structures. When the amount exceeds to a certain limit, the result is evolution of plates and rock shapes. It is credible to assume that the environmental conditions change the nature and shape of the morphology. Moreover, the rock and plates shaping compounds may have excellent activity towards different applications due to their broad surface area [46]. The homogeneous morphology observed indicates the obtained strict-defined crystals are free from metal ions on the external surface.

3.6. Thermal Study.

The degradation behaviors of all perimidine compounds and their VO(II) complexes were tested. The proposed degradation insights corresponding to all decomposition stages are tabulated (Table 5). The treated perimidine compounds start their successive decomposition at low temperature ($\approx 60°C$) in three stages. A sequenced complete degradation of the organic compound was recorded with or without carbon atoms residue. VO(II)-perimidine complexes display an observable thermal stability for the organic compounds coordinated. The degradation stages varied in between three and four stages. The first degradation process starts at 40–80°C temperature range which starts with the removal of water molecules and is followed by decomposition of the coordinating ligand. Variable residue was proposed with the complexes degradation process but all agree with the presence of biatomic metals. An acceptable conformity between calculated and found weight losses percentage may reflect the exact determination of stage borders.

3.7. DNA Binding.

Appling the spectrophotometric titration method, the binding mod of perimidine derivatives towards CT-DNA was investigated. Electronic absorption of freshly prepared solutions was obtained at 25°C over 200–800 nm range, with a reference solution for each concentration. Scanned solutions include fixed ligand concentration (2×10^{-5} M) with a regular increase of DNA added. The effective binding constant for the interaction of the organic derivatives with DNA was obtained based on observable changes in absorption at 418, 420, 420, 385, and 410 nm for LH, LOMe, LNO$_2$, L^4, and L^5, respectively. A regular increase of DNA amount added to the ligand solution leads to the bathochromic effect for the significant ligand band assigned for transition inside interacting groups. This band is minimized gradually as appeared clearly with the aggregated spectra for each derivative. This minimization is followed by appearance of the slightly shifted peak (1-2 nm) from the free ligand peak, which assigns for the binding complex and suffers a gradual increase in absorbance. This is considered as a sufficient indicator of coupled DNA helix stabilization, after the interaction process. Such an investigation suggests the coupling for binding sites through electrostatic attraction or occluded in major and minor grooves inside DNA. Also, the bathochromic effect can be investigated and explained based on two bases: broad surface area of perimidine molecules and the presence of planar aromatic chromophore, which facilitate well binding towards CT-DNA. This groove binding leads to structural reorganization of CT-DNA. This requires a partial disassembling or deterioration of double helix at the exterior phosphate, which leads to formation of cavity suitable for entering compounds [47]. The bathochromic feature observed is directly proportional to electron withdrawing character for substituents and their position. The binding constants (K_b) for the five derivatives were calculated by known spectral relationships [19] for L^1, L^2, L^3, L^4, and L^5 as 6.10×10^4, 6.07×10^4, 6.75×10^4, 7.99×10^4, and 8.80×10^4 M^{-1}. According to Hammett's constants (σ_R), essential correlation against intrinsic constants will be conducted (Figure 7), and the relation verifies the direct relation in between [48].

3.8. Computational

3.8.1. DFT/Hartree–Fock Study.

Applying the Gaussian 09 software, the optimization process was executed over all new compounds till reaching the best configuration. A known standard method was used for this purpose. Essential parameters will be extracted from the energy levels of frontiers (HOMO and LUMO). The energy gap between E_{HOMO} and E_{LUMO} will give an excellent view about the character of the tested compound. The biological behavior and the ligational mode are most significant features concluded. The frontier images of perimidine ligands and their VO(II) complexes are shown in Figures 8(a) and 8(b), respectively. HOMO-level images display the concentration over the perimidine ring which includes two donor centers, while the LUMO-level images display the concentration over the other side in molecules including the other two coordination sites. This view introduces a good electron relocation between donor atoms which smoothens the donation of coordinating centers. On the other side, the two levels in VO(II) complexes represent the concentration around the two central atoms mainly. This may offer the good role of VO atoms in the application feature interested in this research. This may happen through the smooth charge transfer process that includes the complexes. Electronegativity (χ), chemical potential (μ), global hardness (η), global softness (S), global electrophilicity index (ω), and absolute softness (σ) were calculated by using known standard equations [49, 50]. Toxicity and reactivity of compounds can be clarified by using the electrophilicity index (ω) value. This index gives a clear insight about the expected biological attitude of tested perimidine compounds in comparison with their VO(II) complexes and, also, measures the firmness of the compound which takes an extra negative charge from the environment. Also, the firmness and reactivity of compounds can be tested from two opposite indexes (η and σ) [42, 51].

TABLE 5: Estimated TG data of perimidine compounds and all VO(II) complexes.

Compound	Steps	Temp. range (°C)	Decomposed	Weight loss; calcd. (found %)
L^1	1st	45.1–120.5	-[C$_6$H$_6$ + N$_2$]	27.18 (27.16)
	2nd	122.2–410.1	-[C$_6$H$_5$ + CO]	26.92 (26.95)
	3rd	410.3–670.2	-[C$_8$H$_7$N$_2$]	33.59 (33.54)
	Residue		4C	12.31 (12.35)
[(VO)$_2$ (SO$_4$)$_2$ (L^1)]H$_2$O	1st	80.3–120.3	-[H$_2$O + SO$_4$]	15.53 (15.55)
	2nd	120.6–391.7	-[SO$_4$ + C$_6$H$_6$ + N$_2$]	27.53 (27.55)
	3rd	391.9–798.8	-[C$_{19}$H$_{12}$N$_2$]	36.53 (36.50)
	Residue		V$_2$O$_3$	20.41 (20.40)
L^2	1st	65.6–156.1	-[C$_6$H$_5$OCH$_3$]	25.72 (25.71)
	2nd	156.6–299.9	-[C$_6$H$_5$ + CO + N$_2$]	31.66 (31.69)
	3rd	301.0–663.2	-[C$_9$H$_7$N$_2$]	34.05 (33.89)
	Residue		3C	8.57 (8.71)
[(VO)$_2$ (SO$_4$)$_2$ (L^2)]2H$_2$O	1st	42.1–135.1	-[2H$_2$O + 2SO$_4$]	29.16 (29.16)
	2nd	136.1–270.1	-[C$_6$H$_5$OCH$_3$ + N$_2$]	17.40 (17.29)
	3rd	271.0–485.4	-[C$_6$H$_5$]	9.85 (9.94)
	4th	485.6–797.9	-[C$_{13}$H$_7$N$_2$]	24.43 (24.45)
	Residue		V$_2$O$_3$	19.15 (19.16)
L^3	1st	63.66–230.51	-[C$_6$H$_6$ + NO$_2$]	28.50 (27.90)
	2nd	231.21–410.11	-[C$_7$H$_5$ + CO + N$_2$]	33.33 (33.31)
	3rd	410.52–650.64	-[C$_{11}$H$_6$N$_2$]	38.16 (38.79)
[(VO)$_2$ (SO$_4$)$_2$ (L^3)]H$_2$O	1st	79.2–140.6	-[H$_2$O + SO$_4$]	14.63 (14.62)
	2nd	141.9–278.9	-[SO$_4$ + C$_6$H$_5$ + CO]	25.81 (25.78)
	3rd	279.1–479.5	-[NO$_2$ + C$_6$H$_6$ + N$_2$]	19.52 (19.53)
	4th	480.11–798.8	-[C$_{12}$H$_6$N$_2$]	22.86 (22.90)
	Residue		V$_2$O$_2$	17.18 (17.17)
L^4	1st	64.65–145.46	-[C$_6$H$_5$Cl + CO + N$_2$]	39.68 (39.68)
	2nd	145.68–326.78	-[C$_6$H$_5$ + N$_2$]	24.74 (24.75)
	3rd	330.12–680.23	-[C$_{12}$H$_7$]	35.58 (35.57)
[(VO)$_2$ (SO$_4$)$_2$ (L^4)]H$_2$O	1st	69.1–256.1	-[H$_2$O + C$_6$H$_5$Cl + SO$_4$]	29.48 (29.65)
	2nd	256.9–484.1	-[SO$_4$ + CON$_2$ + C$_6$H$_5$]	29.80 (29.81)
	3rd	484.6–789.4	-[C$_{10}$H$_7$N$_2$]	20.18 (19.98)
	Residue		V$_2$O$_2$ + 2C	20.54 (20.56)
L^5	1st	62.3–169.6	-[C$_6$H$_5$ + N$_2$]	24.74 (24.71)
	2nd	160.1–371.9	-[C$_6$H$_5$Cl + CO + N$_2$]	39.68 (39.59)
	3rd	372.6–666.8	-[C$_9$H$_7$]	27.10 (27.19)
	Residue		3C	8.48 (8.51)
[(VO)$_2$ (SO$_4$)$_2$ (L^5)]3H$_2$O	1st	42.1–266.3	-[3H$_2$O + C$_6$H$_5$Cl]	20.70 (20.71)
	2nd	266.38–482.5	-[2SO$_4$ + CON$_2$ + C$_6$H$_5$]	40.41 (40.61)
	3rd	482.9–793.5	-[C$_8$H$_7$N$_2$]	16.29 (16.36)
	Residue		V$_2$O$_2$ + 4C	22.60 (22.32)

(1) Some Quantum Parameters. Some important quantum parameters are calculated for all treated compounds attributing to frontier energy gaps and are displayed in Table 6. The computed results of ligands introduce the following notices: (i) the degree of converged softness recorded for perimidines offers their compatible flexibility towards the coordination. (ii) Electrophilicity index (χ) and electronic chemical potential index (μ) have two different signs. This is evidence for the ability of compounds to acquire electrons from the surrounding by the following order: L^4 > L^3 > L^5 > L^1 > L^2 ligands. This arrangement agrees by an excellent way with the priority of electron withdrawing substituents (Cl and NO$_2$) in para position which facilitates the compound electron affinity.

Whenever, the extracted data assigning for VO(II) complexes introduce the following observations: (i) frontier energy gaps are completely minimized from original

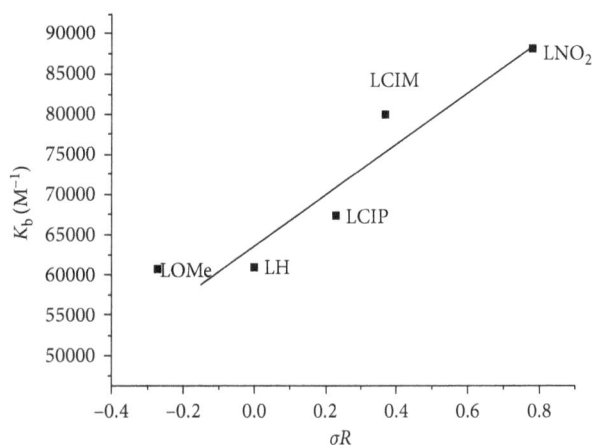

FIGURE 7: Hammett's relation between the effect of p-substituent (σR) and intrinsic binding constants (K_b) of the ligands.

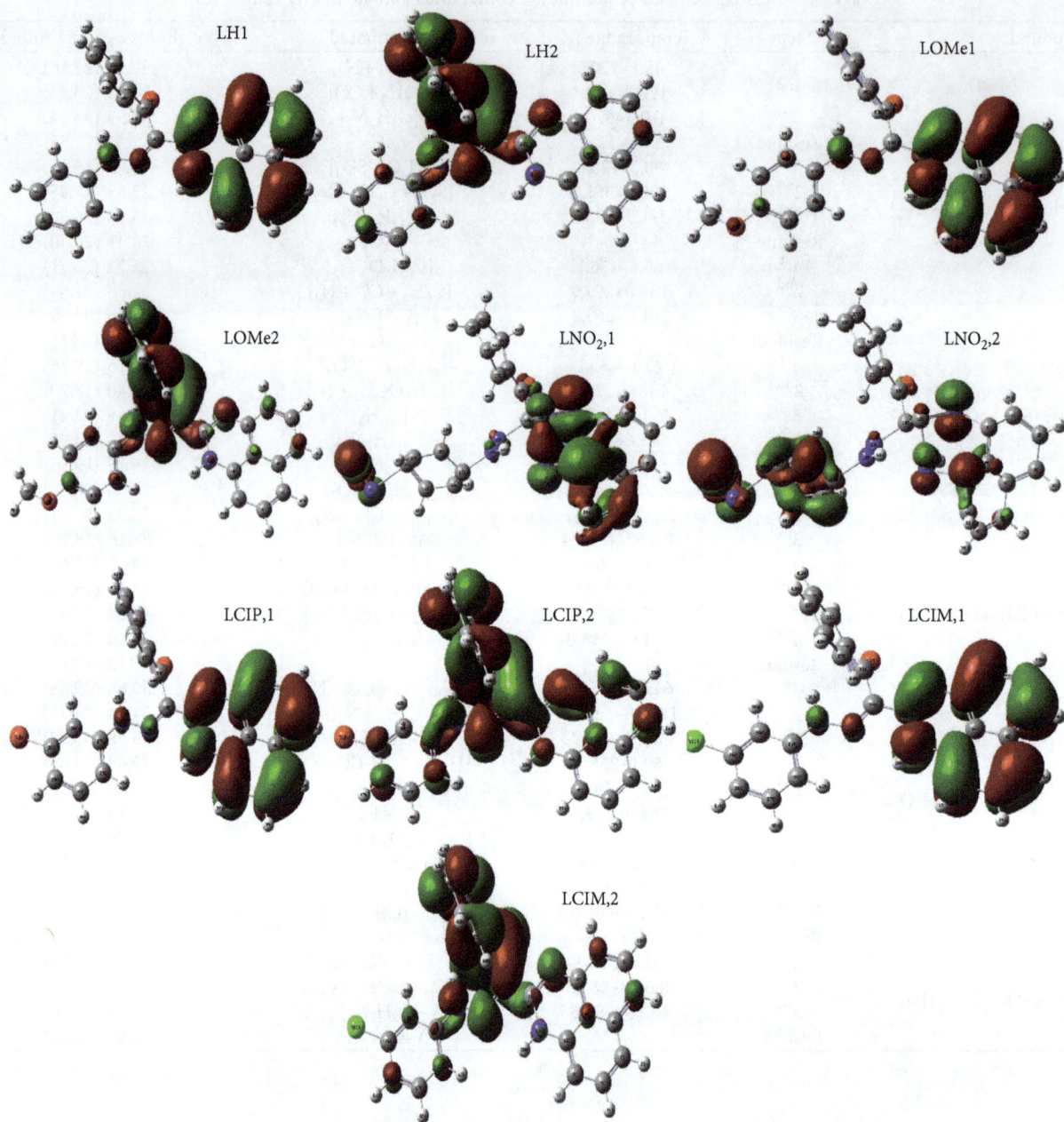

LH1

LH2

LOMe1

LOMe2

LNO$_2$,1

LNO$_2$,2

LCIP,1

LCIP,2

LCIM,1

LCIM,2

(a)

FIGURE 8: Continued.

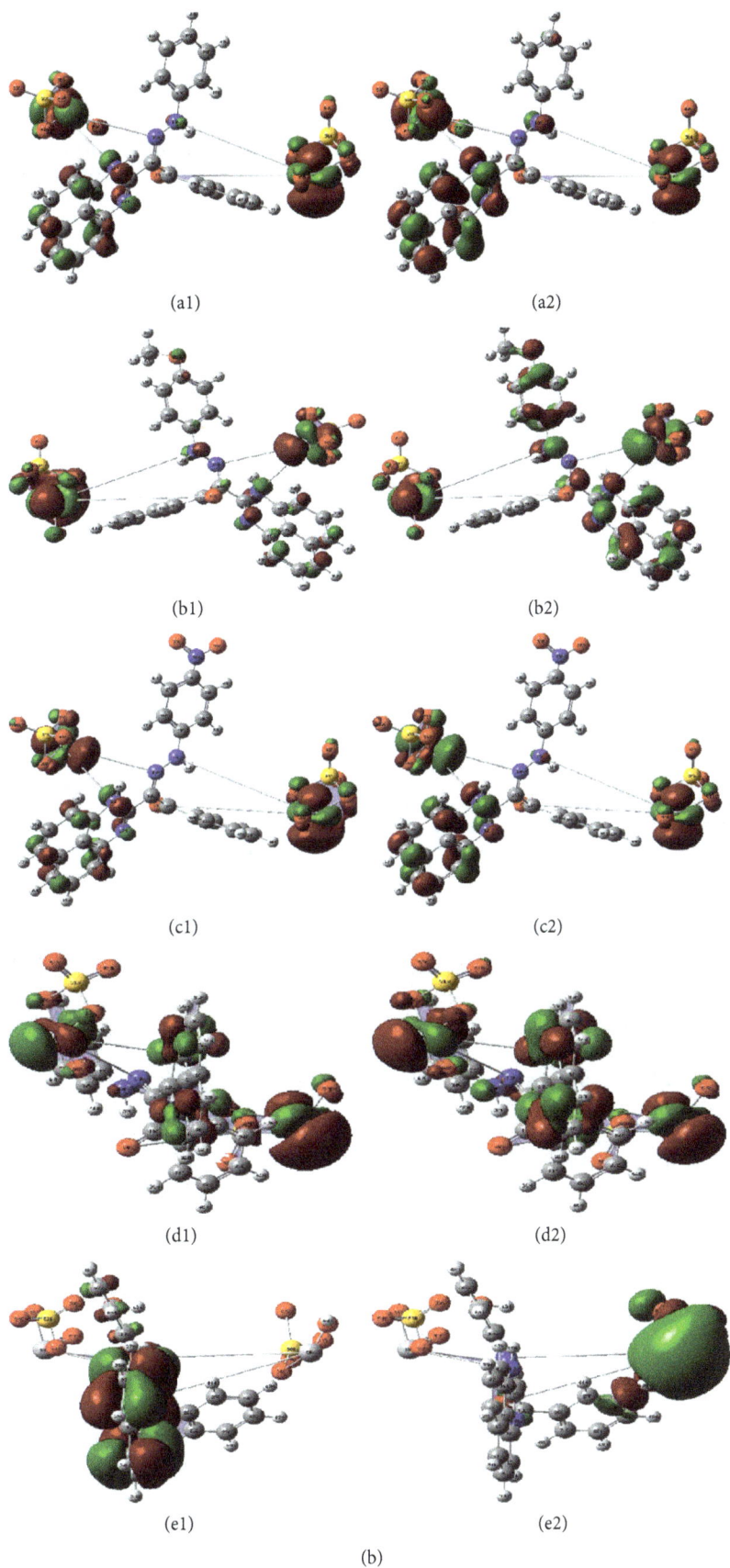

(a1)   (a2)

(b1)   (b2)

(c1)   (c2)

(d1)   (d2)

(e1)   (e2)

(b)

FIGURE 8: (a) Frontier molecular orbitals of HOMO(1) and LUMO(2) pictures of perimidine ligands. (b) Frontier molecular orbitals of HOMO(1) and LUMO(2) pictures of VO(II)-perimidine complexes (A–E, respectively).

TABLE 6: Energy parameters (eV) using the DFT/B3LYP method of optimized structures.

Compound	E_H	E_L	$(E_H - E_L)$	$E_L - E_H$	x	μ	η	S (eV^{-1})	ω	σ
L^1	−0.17417	−0.07574	−0.0984	0.09843	0.124955	−0.12496	0.049215	0.024608	0.158628	20.31900843
L^1 + VO(II)	−0.20433	−0.19545	−0.0089	0.00888	0.19989	−0.19989	0.00444	0.00222	4.499551	225.2252252
L^2	−0.17142	−0.07426	−0.0972	0.09716	0.12284	−0.12284	0.04858	0.02429	0.155307	20.58460272
L^2 + VO(II)	−0.20163	−0.19237	−0.0093	0.00926	0.197	−0.197	0.00463	0.002315	4.191037	215.9827214
L^3	−0.21252	−0.05654	−0.156	0.15598	0.13453	−0.13453	0.07799	0.038995	0.11603	12.82215669
L^3 + VO(II)	−0.21881	−0.21008	−0.0087	0.00873	0.214445	−0.21445	0.004365	0.002183	5.267658	229.0950745
L^4	−0.25291	−0.05654	−0.1964	0.19637	0.154725	−0.15473	0.098185	0.049093	0.121912	10.18485512
L^4 + VO(II)	−0.20433	−0.19545	−0.0089	0.00888	0.19989	−0.19989	0.00444	0.00222	4.499551	225.2252252
L^5	−0.17808	−0.0842	−0.0939	0.09388	0.13114	−0.13114	0.04694	0.02347	0.183188	21.30379207
L^5 + VO(II)	−0.19719	−0.16607	−0.0311	0.03112	0.18163	−0.18163	0.01556	0.00778	1.060073	64.26735219

perimidines leading to red shift inside electronic transitions. Such a behavior may clarify the effect of metal atoms (vanadyl) in stabilizing the compounds. This reduction is preferable in biological attitude of compounds [52]. (ii) The absolute softness values in complexes were enhanced than the ligand values which predicated their high biological activity. From calculated energy gaps, Hammett's relation displays a significant effect of the p-substituent on δE values of ligands or their complexes by two reverse features (Figure 9).

(2) Some Log File Parameters. Essential log file data are summarized and presented in Table 7. The allowed data are varied in between the free ligands and their complexes, due to the difference in methods used for the treatment. A suitable method used for the organic ligand appeared unsuitable for its VO(II) complex. A comparative investigation introduced the following notices: (i) a general reduction in the charges computed for coordinating atoms (O^{19}, N^{11}, N^{15}, and N^{16}). This is due to their participation in coordination with VO(II) atoms. (ii) The computed bond lengths appearing with four perimidine derivatives are comparable with each other except for the L^3 ligand. This displays the inductive effect of the p-substituent (nitro group) on the elongation of bond lengths attributing to the affected function groups. (iii) Oscillator strength values (range 0–1) of complexes are commonly minimized than those of their corresponding ligands. This may indicate the effect of the metal atom (vanadyl) in facilitating the absorption or reemission of electromagnetic radiations inside complex molecules [53]. The values are close to zero and not 1; this may suggest low excitation energy values needed for electronic transitions. (iv) Also, an increase in dipole moment values of complexes over ligands indicates high polarity of covalent bonds surrounding two central atoms except for the L^2 complex. This may refer to the significant difference between all substituents from the methoxy group which has electron-donating feature in opposite with the others [54].

3.8.2. QSAR Calculations. Using the HyperChem (v8.1) program, essential QSAR parameters are calculated and tabulated (Table 8). This computation gives a clear view about some statistics belonging to coordinating agents. Log P value is an indication for the biological feature of the

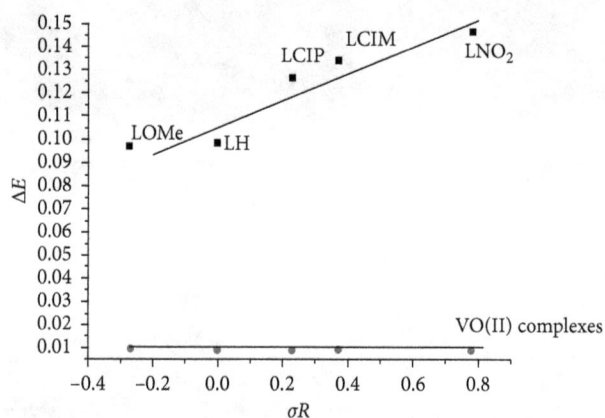

FIGURE 9: Hammett's relation between the effect of p-substituent (σR) and energy gaps (δE) of ligands and their VO(II) complexes.

tested compound by a reverse relation [55]. The values are arranged by the following order: L^1 (2.53) > L^4 = L^5 (2.31) > L^2 (1.53) > L^3 (−1.64). Partition coefficient (log P) values introduce a distinguish biological activity may appear with the L^3 ligand.

3.8.3. Docking Computation. Simulation technique is a new revolution process served in different applications. Drug design is a complicated process that needs significant facilities to establish a view about the expected efficiency of proposed drugs. In last decades, the docking computation process between the proposed drug (inhibitor) and the infected cell proteins is the concern in drug industrial research. AutoDockTools 4.2 software was used for this approach. 4c3p, 3bch, and 4zdr are the BDP files for breast, colon, and liver cancer cell proteins which are used for the docking process with five perimidine derivatives. The extracted energies over PDB files (a format using the Gaussian 09 software) are presented in Table 9. Scanning for the energy values introduces the following observations: (i) there is no interaction observed with the five inhibitors towards 3bch colon cancer protein. (ii) The degree of interaction towards breast colon protein (4c3p) is arranged as L^5 > L^1 > L^2 > L^4, while the arrangement towards 4zdr (liver cancer protein) is as L^5 > L^1 > L^4 > L^3 > L^2. This result displays the priority of L^5 and L^1 ligands in the inhibition

TABLE 7: Considerable bond lengths, charges, dipole moment (D), oscillator strength (f), and excitation energies (E).

Compound	O^{19}	N^{11}	N^{15}	N^{16}	C^{18}–O^{19}	C^{12}–N^{11}	C^{14}–N^{15}	N^{15}–N^{16}	V^1	V^2	D (Debye)	E (nm)	f
L^1	−0.415473	−0.555221	−0.349927	−0.275635	1.224561	1.384229	1.287795	1.366391	—	—	5.1769	567.81	0.0316
L^1 + VO(II)	−0.410181	−0.282766	−0.316313	−0.248429	—	—	—	—	0.933201	0.929331	11.5667	16513.7	0.003
L^2	−0.410237	−0.555363	−0.348134	−0.275746	—	—	—	—	—	—	6.4963	576.26	0.0403
L^2 + VO(II)	−0.420038	−0.397120	−0.322648	−0.250261	—	—	—	—	0.927552	0.915069	3.5504	31387.7	0.0008
L^3	−0.416751	−0.044378	−0.050930	0.113237	1.317259	2.076019	1.772715	1.281712	—	—	5.3595	7613.39	0.002
L^3 + VO(II)	−0.414856	−0.387699	−0.320321	−0.254947	1.223609	1.404653	1.286454	1.367672	0.955124	0.952544	16.6899	17266.3	0.0032
L^4	−0.313357	−0.358945	−0.031267	−0.294059	1.223609	1.384450	1.286411	1.367122	—	—	4.4684	315.92	0.4055
L^4 + VO(II)	−0.410181	−0.282766	−0.316313	−0.248429	—	—	—	—	0.933201	0.929331	11.5667	16513.7	0.003
L^5	−0.354631	−0.480428	−0.209492	−0.312115	1.224058	—	—	—	—	—	3.9661	600.17	0.043
L^5 + VO(II)	−0.405024	−0.408894	−0.304735	−0.210573	—	—	—	—	0.905245	0.730601	8.9706	20988.6	0.002

TABLE 8: QSAR computation for optimized structures of perimidine compounds.

Function	L^1	L^2	L^3	L^4	L^5
Surface area (approx.) (Å^2)	425.73	488.36	496.79	464.04	465.53
Surface area (grid) (Å^2)	623.02	661.15	661.91	644.29	642.82
Volume (Å^3)	1060.57	1138.87	1134.04	1105.99	1105.72
Hydration energy (kcal/mol)	−8.29	−9.97	−17.83	−8.00	−8.02
Log P	2.53	1.53	−1.64	2.31	2.31
Reactivity (Å^3)	132.87	139.25	138.92	137.59	137.59
Polarizability (Å^3)	45.32	47.80	47.62	47.25	47.25
Mass (amu)	390.44	420.47	436.45	424.89	424.89

TABLE 9: Docking energy values (kcal/mol) of perimidine compounds (HL) and protein receptors complexes.

Ligands	pKa	Receptor	Est. free energy of binding	Est. inhibition constant (K_i) (μM)	vdW + bond + desolving energy	Electrostatic energy	Total intercooled energy	Frequency	Interacting surface
L^1	10.96	4c3p	−7.92	1.57	−9.29	−0.06	−9.34	30%	859.778
		3bch	+355.37	—	+349.42	+0.06	+349.48	10%	665.36
		4zdr	−4.72	345.45	−5.97	−0.05	−6.02	20%	723.695
L^2	10.96	4c3p	−7.75	2.09	−9.13	−0.13	−9.26	10%	983.377
		3bch	+490.76	—	+473.39	+0.00	+473.39	10%	662.71
		4zdr	−4.32	686.85	−5.97	−0.02	−5.99	20%	758.018
L^3	10.95	4c3p	+647.56	—	+644.74	+0.01	+644.75	10%	718.318
		3bch	+709.10	—	+699.61	−0.05	+699.56	10%	661.43
		4zdr	−4.66	385.21	−6.46	+0.07	−6.39	10%	621.389
L^4	10.96	4c3p	−7.28	4.62	−8.57	−0.03	−8.60	20%	929.747
		3bch	+552.25	—	+549.59	−0.03	+549.56	30%	710.605
		4zdr	−4.67	376.04	−6.13	−0.19	−6.32	20%	595.541
L^5	10.95	4c3p	−8.41	683.74	−9.81	−0.00	−9.81	20%	925.161
		3bch	+663.87	—	+661.94	+0.01	+661.95	10%	703.598
		4zdr	−4.84	284.49	−6.39	+0.04	−6.35	10%	690.838

process towards breast and liver carcinoma cell lines through a strong interaction (Figures 10 and S5) [55]. Dissociation constant (pK_a) calculated is considered the biopharmaceutical measure of drug-likeness compounds. This constant helps in understanding the ionic form of the drug along the pH range. High pK_a values (>10) reflect their ionization which facilitates their diffusion across the cell membrane to give a well inhibition process. Also, highly reduced energies were recorded for 4c3p and 4zdr receptors. Positive sign of electrostatic energy recorded clarifies high stability of interacting complexes. HP plots (Figures 11 and S6) as well as 2D plots (Figure S7) display prolonged H-bonding appearing with L^5 and LH ligands. This verifies the degree of interaction proposed on extracted energies. Also, high surface area recorded with breast or liver cancer protein complexes introduces a good degree of H-interaction. And hp, 2D, and surface area data verify the absence of interaction recorded with colon cancer protein.

3.9. Antitumor Efficiency.

The results obtained by screening all prepared compounds for comparison confirmed that the complexes exhibit more cytotoxicity against HepG2, MCF-7, and HCT116 *in vitro*. The IC_{50} values are displayed in Table 10 and in Figures 12 and S8. Cells were treated with various concentrations of compounds and incubated for 48 h [56]. Cytotoxicity is considered as a good anticancer parameter if the influence induced apoptotic pathways inside the cell. Apoptotic may be detected by many parameters like the activation of caspase family, DNA fragmentation, or morphology of the cell. The sample L^4 + VO(II) was the best impacted complex on liver, breast, and colon cancer cell lines with IC_{50} values of 1.66, 3.42, and 1.27, respectively, while its relative ligand (L^4) impact was moderate at all cancer types. Hence, these results showed that the present study's effort to improve and enhance the effect of new complexes has achieved a clear, acceptable, and respectable success because the effect was enhanced for 15, 7, and 22 times, respectively, compared to the ligand with clear signs that it is going to be very close to the positive control. On the contrary, unfortunately, the same results were not detected in other complexes; they went in the contrary way: instead, they increased the impact they decreased it dramatically. What our results tell us clearly is that neither VO(II) nor

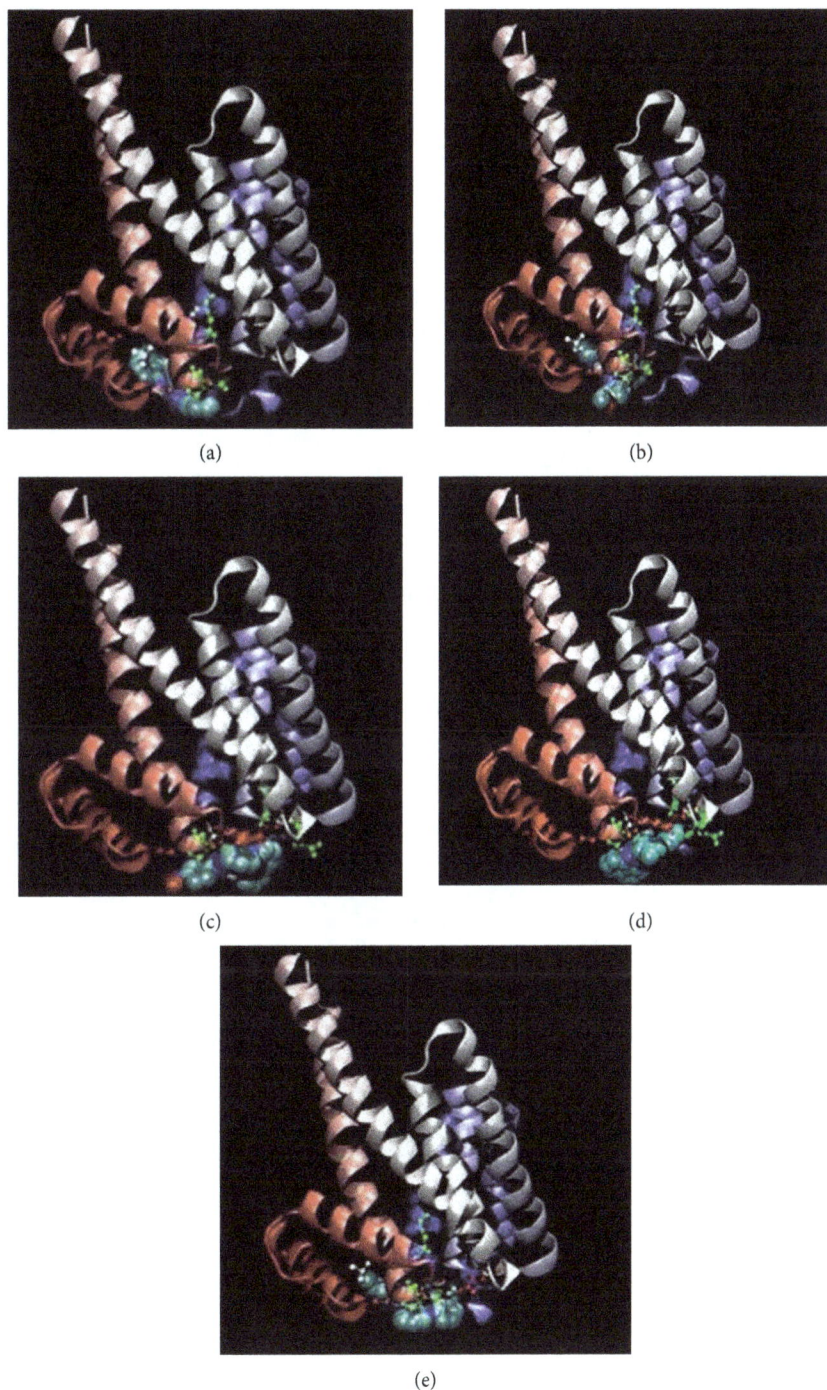

FIGURE 10: Interacting protein-inhibitor complexes (a) L^1, (b) L^2, (c) L^3, (d) L^4, and (e) L^5 with 4zdr receptor (a–e, respectively).

ligands alone can act as an anticancer candidate drug, while only one complex can present that effect. So, the mechanism of action is not related to the ligand or to VO(II) itself, as far as related to the complex itself.

4. Conclusion

This paper presents new VO(II) complexes derived from a series of perimidine ligands. This study focuses on the effect of substituents on the chemistry and applicability of complexes. All the new compounds were well characterized by all possible tools. The complexes were found in a nano-scale comfortably. The different theoretical implementations gave a view about the biological feature of the investigated compounds in a comparative way. The docking process displays the high interaction of organic derivatives against breast cancer, while the experimental investigation displays the priority of the L^4-VO(II) complex against all carcinomas tested. The binding efficiency of ligands towards CT-DNA was tested. Binding constant (K_b) values are in agreement

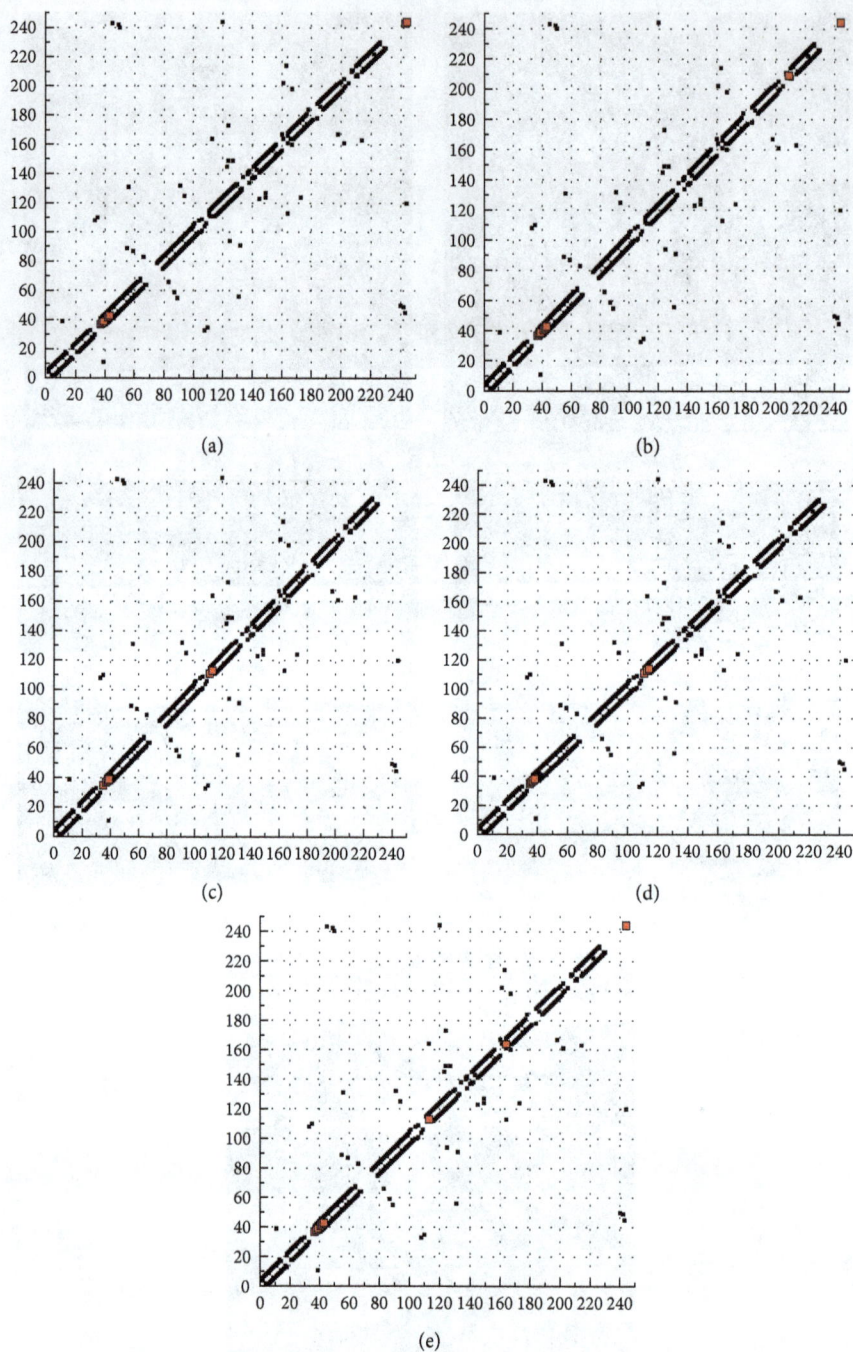

(a)

(b)

(c)

(d)

(e)

FIGURE 11: Interacting complexes hp plot for (a) L^1, (b) L^2, (c) L^3, (d) L^4, and (e) L^5 with 4zdr receptor (a–e, respectively).

TABLE 10: IC$_{50}$ of some tested compounds against liver (HepG2), breast (MCF-7), and colon (HCT116) cancer cell lines.

Cell type	IC$_{50}$ (μg/ml)										
	L^1	VO(II)-L^1	L^2	VO(II)-L^2	L^3	VO(II)-L^3	L^4	VO(II)-L^4	L^5	VO(II)-L^5	Doxorubicin
MCF-7	19.68	23.06	25.92	93.92	15.50	>100	24.96	3.42	11.44	>100	0.60
HepG2	19.79	19.94	27.23	55.67	11.01	>100	28.25	1.27	9.91	>100	0.34
HCT116	19.15	22.93	13.27	95.17	15.53	>100	26.24	1.66	23.30	>100	0.39

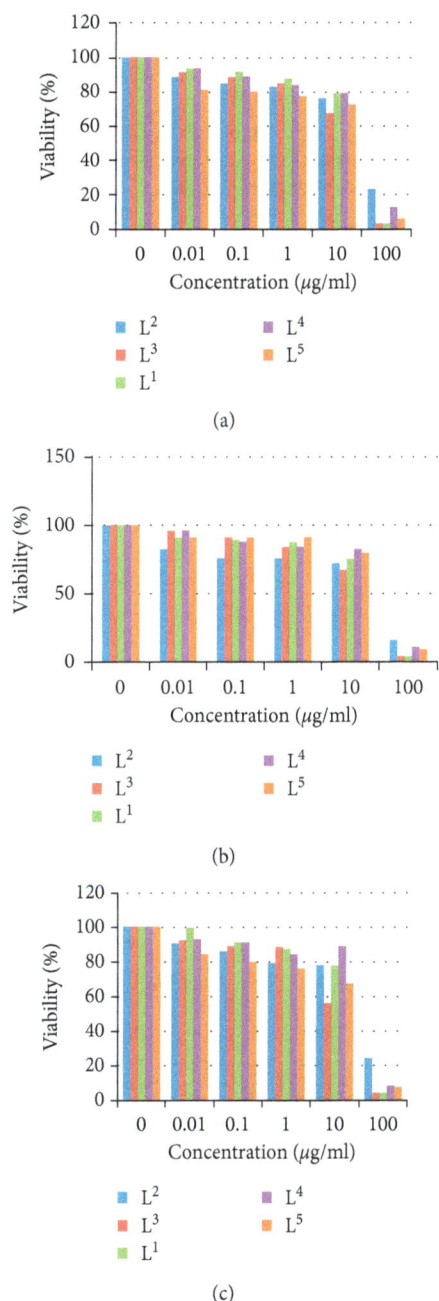

FIGURE 12: Dose response curves of perimidine ligands against MCF-7 (a), HCT116 (b), and HepG2 (c) cancer cells.

with the electron-drawing character of the p-substituent which displayed high K_b values.

Acknowledgments

The authors extend their appreciation to the Deanship of Scientific Research at King Khalid University for funding this work through General Research project under grant number (G.R.P-124-38).

Supplementary Materials

Figure S1: ^{1}H·NMR spectra of L^4 and L^5 perimidine ligands. Figure S2: mass spectra of L^3 and L^4 ligands. Figure S3: X-ray patterns of four perimidine ligands. Figure S4: SEM images of perimidine ligands and their VO(II) complexes. Figure S5: interacting protein-inhibitor complexes for L^1, L^2, L^3, L^4, and L^5 with 4c3p (1) and 3bch (2) receptors (A–E, respectively). Figure S6: interacting hp plot: LH, LOMe, LNO$_2$, LClP, and LClM with 4c3p(1) and 3bch(2) receptors (A–E, respectively). Figure S7: 2D plot forms: L^1, L^2, L^3, L^4, and L^5 with 4c3p (1), 3bch (2), and 4zdr (3) receptors (A–E, respectively). Figure S8: dose response curves of perimidine-VO(II) complexes against MCF-7, HCT116, and HepG2 cancer cells. (*Supplementary Materials*)

References

[1] M. S. Refat, "Synthesis, characterization, thermal and antimicrobial studies of diabetic drug models: complexes of vanadyl(II) sulfate with ascorbic acid (vitamin C), riboflavin (vitamin B2) and nicotinamide (vitamin B3)," *Journal of Molecular Structure*, vol. 969, no. 1–3, pp. 163–171, 2010.

[2] G. R. illsky, "Synthesis, characterization, thermal and antimicrobial studies of diabetic drug models: complexes of vanadyl(II) sulfate with ascorbic acid (vitamin C), riboflavin (vitamin B2) and nicotinamide (vitamin B3)," in *Vanadium in Biological Systems Physiology and Biochemistry*, N. D. Chasteen, Ed., Kluwer Academic Publishers, Dordrect, Netherlands, 1990.

[3] Z. Ye, G. Yan, J. Tang, Q. Fu, J. Lu, and N. Yang, "Synthesis and biological evaluation of two oxidovanadium (IV) complexes as DNA-binding and apoptosis-inducing agents," *American Journal of Biological Chemistry*, vol. 4, no. 1, pp. 6–13, 2016.

[4] J. Farzanfar, K. Ghasemi, A. R. Rezvani et al., "Synthesis, characterization, X-ray crystal structure, DFT calculation and antibacterial activities of new vanadium(IV, V) complexes containing chelidamic acid and novel thiourea derivatives," *Journal of Inorganic Biochemistry*, vol. 147, pp. 54–64, 2015.

[5] K. Kanmani Raja, L. Lekha, R. Hariharan, D. Easwaramoorthy, and G. Rajagopal, "Synthesis, structural, spectral, electrochemical and catalytic properties of VO (IV) complexes containing N, O donors," *Journal of Molecular Structure*, vol. 1075, pp. 227–233, 2014.

[6] D. M. Boghaei and S. Mohebi, "Synthesis, characterization and study of vanadyl tetradentate Schiff base complexes as catalyst in aerobic selective oxidation of olefins," *Journal of Molecular Catalysis A: Chemical*, vol. 179, no. 1-2, pp. 41–51, 2002.

[7] J.-Q. Wu and Y.-S. Li, "Well-defined vanadium complexes as the catalysts for olefin polymerization," *Coordination Chemistry Reviews*, vol. 255, no. 19-20, pp. 2303–2314, 2011.

[8] G. Micera and D. Sanna, *Vanadium in the Environment, Part One: Chemistry and Biochemistry*, J. O. Nriagu, Ed., John Wiley & Sons, New York, NY, USA, 1998.

[9] F. T. G. Hudson, *Vanadium Toxicology and Biological Significance*, Elsevier, NewYork, NY, USA, 1996.

[10] R. C. Maurya and S. Rajput, "Oxovanadium(IV) complexes of bioinorganic and medicinal relevance: synthesis, characterization and 3D molecular modeling and analysis of some oxovanadium(IV) complexes involving the O, N-donor environment of pyrazolone-based sulfa drug Schiff bases," *Journal of Molecular Structure*, vol. 794, no. 1-3, pp. 24-34, 2006.

[11] D. M. Boghaei, A. Bezaatpour, and M. Behzad, "Synthesis, characterization and catalytic activity of novel monomeric and polymeric vanadyl Schiff base complexes," *Journal of Molecular Catalysis A: Chemical*, vol. 245, no. 1-2, pp. 12-16, 2006.

[12] M. S. Refat, M. A. A. Moussa, and S. F. Mohamed, "Synthesis, spectroscopic characterization, thermal analysis and electrical conductivity studies of Mg(II), Ca(II), Sr(II) and Ba(II) vitamin B2 complexes," *Journal of Molecular Structure*, vol. 994, no. 1-3, pp. 194-201, 2011.

[13] M. S. Refat and S. A. El-Shazly, "Identification of a new antidiabetic agent by combining VOSO4 and vitamin E in a single molecule: studies on its spectral, thermal and pharmacological properties," *European Journal of Medicinal Chemistry*, vol. 45, no. 7, pp. 3070-3079, 2010.

[14] R. Takjoo, A. Akbari, S. Yousef Ebrahimipour, M. Kubicki, M. Mohamadi, and N. Mollania, "Synthesis, spectral characterization, DFT calculations, antimicrobial activity and molecular docking of 4-bromo-2-((2-hydroxy-5-methylphenylimino)methyl)phenol and its V(V) complex," *Inorganica Chimica Acta*, vol. 455, pp. 173-182, 2017.

[15] V. Kraehmer and D. Rehder, "Modelling the site of bromide binding in vanadate-dependent bromoperoxidases," *Dalton Transactions*, vol. 41, no. 17, pp. 5225-5234, 2012.

[16] P. Sathyadevi, P. Krishnamoorthy, R. R. Butorac, A. H. Cowley, and N. Dharmaraj, "Synthesis of novel heterobimetallic copper(I) hydrazone Schiff base complexes: a comparative study on the effect of heterocyclic hydrazides towards interaction with DNA/protein, free radical scavenging and cytotoxicity," *Metallomics*, vol. 4, no. 5, pp. 498-511, 2012.

[17] R. Pradhan, M. Banik, D. B. Cordes, A. M. Z. Slawin, and N. C. Saha, "Synthesis, characterization, X-ray crystallography and DNA binding activities of Co(III) and Cu(II) complexes with a pyrimidine-based Schiff base ligand," *Inorganica Chimica Acta*, vol. 442, pp. 70-80, 2016.

[18] T. A. Farghaly and H. K. Mahmoud, "Synthesis, tautomeric structures, and antitumor activity of new perimidines," *Archiv der Pharmazie*, vol. 346, no. 5, pp. 392-402, 2013.

[19] M. E. Reichmann, S. A. Rice, C. A. Thomos, and P. Doty, "A further examination of the molecular weight and size of desoxypentose nucleic acid," *Journal of the American Chemical Society*, vol. 76, no. 11, pp. 3047-3053, 1954.

[20] J. Marmur, "A procedure for the isolation of deoxyribonucleic acid from micro-organisms," *Journal of Molecular Biology*, vol. 3, no. 2, pp. 208-218, 1961.

[21] A. Wolfe, G. H. Shimer, and T. Meehan, "Polycyclic aromatic hydrocarbons physically intercalate into duplex regions of denatured DNA," *Biochemistry*, vol. 26, no. 20, pp. 6392-6396, 1987.

[22] A. I. Vogel, *Text Book of Quantitative Inorganic Analysis*, Longman, London, UK, 1986.

[23] M. J. Frisch, G. W. Trucks, H. B. Schlegel et al., *Gaussian 09, Revision D*, Gaussian, Inc., Wallingford, CT, USA, 2010.

[24] R. Dennington II, T. Keith, and J. Millam, *Gauss View, Version 4.1.2*, Semichem, Inc., Shawnee, KS, USA, 2007.

[25] M. M. Al-Iede, J. Karpelowsky, and D. A. Fitzgerald, "Recurrent diaphragmatic hernia: modifiable and non-modifiable risk factors," *Pediatric Pulmonology*, vol. 51, no. 4, pp. 394-401, 2015.

[26] T. A. Halgren, "Merck molecular force field. I. Basis, form, scope, parametrization, and performance of MMFF94," *Journal of Computational Chemistry*, vol. 17, no. 5-6, pp. 490-519, 1998.

[27] G. M. Morris, D. S. Goodsell, R. S. Halliday et al., "Automated docking using a Lamarckian genetic algorithm and an empirical binding free energy function," *Journal of Computational Chemistry*, vol. 19, no. 14, pp. 1639-1662, 1998.

[28] F. J. Solis and R. J. B. Wets, "Minimization by random search techniques," *Mathematics of Operations Research*, vol. 6, no. 1, pp. 19-30, 1981.

[29] W. J. Geary, "The use of conductivity measurements in organic solvents for the characterisation of coordination compounds," *Coordination Chemistry Reviews*, vol. 7, no. 1, pp. 81-122, 1971.

[30] N. M. El-Metwaly, R. M. El-shazly, I. M. Gabr, and A. A. El-Asmy, "Physical and spectroscopic studies on novel vanadyl complexes of some substituted thiosemicarbazides," *Spectrochimica Acta Part A: Molecular and Biomolecular Spectroscopy*, vol. 61, no. 16, pp. 1113-1119, 2005.

[31] K. S. Abu-Melha and N. M. El-Metwally, "Spectral and thermal studies for some transition metal complexes of bis (benzylthiocarbohydrazone) focusing on EPR study for Cu(II) and VO²⁺," *Spectrochimica Acta Part A: Molecular and Biomolecular Spectroscopy*, vol. 70, no. 2, pp. 277-283, 2008.

[32] A. A. Abou-Hussen, N. M. El-Metwaly, E. M. Saad, and A. A. El-Asmy, "Spectral, magnetic, thermal and electrochemical studies on phthaloyl bis(thiosemicarbazide) complexes," *Journal of Coordination Chemistry*, vol. 58, no. 18, pp. 1735-1749, 2005.

[33] A. B. P. Lever, *Inorganic Electronic Spectroscopy*, Elsevier, Amsterdam, Netherlands, 1986.

[34] B. J. Hathaway and D. E. Billing, "The electronic properties and stereochemistry of mono-nuclear complexes of the copper(II) ion," *Coordination Chemistry Reviews*, vol. 5, no. 2, pp. 143-207, 1970.

[35] B. J. Hathaway, "A new look at the stereochemistry and electronic properties of complexes of the copper(II) ion," *Structure and Bonding*, vol. 57, pp. 55-118, 1984.

[36] H. Montgomery and E. C. Lingefetter, "The crystal structure of Tutton's salts. IV. Cadmium ammonium sulfate hexahydrate," *Acta Crystallographica*, vol. 20, no. 6, pp. 728-730, 1966.

[37] D. Kivelson and R. Neiman, "ESR studies on the bonding in copper complexes," *Journal of Chemical Physics*, vol. 35, no. 1, pp. 149-155, 1961.

[38] J. A. Wellman, F. B. Hulsbergen, J. Verbiest, and J. Reedijk, "Influence of alkyl chain length in N-alkyl imidazoles upon the complex formation with transition-metal salts," *Journal of Inorganic and Nuclear Chemistry*, vol. 40, pp. 143-147, 1978.

[39] U. Sagakuchi and A. W. Addison, "Spectroscopic and redox studies of some copper(II) complexes with biomimetic donor atoms: implications for protein copper centres," *Journal of the Chemical Society, Dalton Transactions*, p. 600, 1979.

[40] H. Yokoi and A. W. Addison, "Spectroscopic and redox properties of pseudotetrahedralcopper(II) complexes. Their relation to copper proteins," *Inorganic Chemistry*, vol. 16, no. 6, pp. 1341-1349, 1977.

[41] B. R. Mc Garvey, "The isotropic hyperfine interaction," *Journal of Physical Chemistry*, vol. 71, no. 1, pp. 51-66, 1967.

[42] R. C. Chikate and S. B. padhye, "Transition metal quinone–thiosemicarbazone complexes 2: magnetism, ESR and redox behavior of iron (II), iron (III), cobalt (II) and copper (II) complexes of 2-thiosemicarbazido-1, 4-naphthoquinone," *Polyhedron*, vol. 24, no. 13, pp. 1689–1700, 2005.

[43] B. D. Cullity, *Elements of X-Ray Diffraction*, Addison–Wesley Inc., Boston, MA, USA, 2nd edition, 1993.

[44] S. Velumani, X. Mathew, P. J. Sebastian, S. K. Narayandass, and D. Mangalaraj, "Structural and optical properties of hot wall deposited CdSe thin films," *Solar Energy Materials and Solar Cells*, vol. 76, no. 3, pp. 347–358, 2003.

[45] F. A. Saad, N. M. El-Metwaly, T. A. Farghaly et al., "Illustration for series of new metal ion complexes extracted from pyrazolone derivative, spectral, thermal, QSAR, DFT/B3LYP, docking and antitumor investigations," *Journal of Molecular Liquids*, vol. 229, pp. 614–627, 2017.

[46] J. H. Al-Fahemi, F. A. Saad, N. M. El-Metwaly et al., "Synthesis of Co(II), Cu(II), Hg(II), UO2(II) and Pb(II) binuclear nanometric complexes from multi-donor ligand: spectral, modeling, quantitative structure–activity relationship, docking and antitumor studies," *Applied Organometallic Chemistry*, vol. 31, no. 11, p. e3787, 2017.

[47] J. M. Chen, W. Wei, W. X. L. Feng, and T. B. Lu, "CO_2 fixation and transformation by a dinuclear copper cryptate under acidic conditions," *Chemistry–An Asian Journal*, vol. 2, no. 6, pp. 710–714, 2007.

[48] A. Z. El-Sonbati, M. A. Diab, A. A. El-Bindary, M. M. Ghoneim, M. T. Mohesien, and M. K. Abd El-Kader, "Polymeric complexes—LXI. Supramolecular structure, thermal properties, SS-DNA binding activity and antimicrobial activities of polymeric complexes of rhodanine hydrazone compounds," *Journal of Molecular Liquids*, vol. 215, pp. 711–739, 2016.

[49] U. El-Ayaan, N. M. El-Metwally, M. M. Youssef, and S. A. A. El Bialy, "Perchlorate mixed–ligand copper(II) complexes of β-diketone and ethylene diamine derivatives: thermal, spectroscopic and biochemical studies," *SpectrochimicaActa Part A*, vol. 68, no. 5, pp. 1278–1286, 2007.

[50] R. K. Ray and G. R. Kauffman, "EPR spectra and covalency of bis(amidinourea/o-alkyl-1-amidinourea)copper (II) complexes. Part II. Properties of the CuN 42-chromophore," *Inorganica Chimica Acta*, vol. 173, no. 12, pp. 207–214, 1990.

[51] S. Sagdinc, B. Köksoy, F. Kandemirli, and S. H. Bayari, "Theoretical and spectroscopic studies of 5-fluoro-isatin-3-(*N*-benzylthiosemicarbazone) and its zinc(II) complex," *Journal of Molecular Structure*, vol. 917, no. 2-3, pp. 63–70, 2009.

[52] I. Fleming, *Frontier Orbital's and Organic Chemical Reactions*, Wiley, London, UK, 1976.

[53] S. K. Tripathi, R. Muttineni, and S. K. Singh, "Extra precision docking, free energy calculation and molecular dynamics simulation studies of CDK2 inhibitors," *Journal of Theoretical Biology*, vol. 334, pp. 87–100, 2013.

[54] M. M. Al-Iede, J. Karpelowsky, and D. A. Fitzgerald, "Recurrent diaphragmatic hernia: modifiable and non-modifiable risk factors," *Pediatric Pulmonology*, vol. 51, no. 4, pp. 394–401, 2015.

[55] N. Terakado, S. Shintani, and Y. Nakahara, "Expression of Cu, Zn-SOD, Mn-SOD and GST-pi in oral cancer treated with preoperative radiation therapy," *Oncology Reports*, vol. 7, no. 5, pp. 1113–1120, 2000.

[56] C. Fosset, B. A. McGaw, and M. D. Reid, "A non-radioactive method for measuring Cu uptake in HepG2 cells," *Journal of Inorganic Biochemistry*, vol. 99, no. 5, pp. 1018–1022, 2005.

8

Metal Complexes of a Novel Schiff Base based on Penicillin: Characterization, Molecular Modeling, and Antibacterial Activity Study

Narendra Kumar Chaudhary and Parashuram Mishra

Bio-Inorganic and Materials Chemistry Research Laboratory, Tribhuvan University, M. M. A. M. Campus, Biratnagar, Nepal

Correspondence should be addressed to Parashuram Mishra; prmmishra@rediffmail.com

Academic Editor: Spyros P. Perlepes

A novel Schiff base ligand of type HL was prepared by the condensation of amoxicillin trihydrate and nicotinaldehyde. The metal complexes of Co^{+2}, Ni^{+2}, Cu^{+2}, and Zn^{+2} were characterized and investigated by physical and spectral techniques, namely, elemental analysis, melting point, conductivity, ^1H NMR, IR, UV-Vis spectra, ESR, SEM, and mass spectrometry measurements. They were further analyzed by thermal technique (TGA/DTA) to gain better insight about the thermal stability and kinetic properties of the complexes. Thermal data revealed high thermal stability and nonspontaneous nature of the decomposition steps. The Coats-Redfern method was applied to extract thermodynamic parameters to explain the kinetic behavior. The molar conductance values were relatively low, showing their nonelectrolytic nature. The powder XRD pattern revealed amorphous nature except copper complex (1c) that crystallized in the triclinic crystal system. The EPR study strongly recommends the tetrahedral geometry of 1c. The structure optimization by MM force field calculation through ArgusLab 4.0.1 software program supports the concerned geometry of the complexes. The in vitro antibacterial activity of all the compounds, at their two different concentrations, was screened against four bacterial pathogens, namely, *E. coli, P. vulgaris, K. pneumoniae, and S. aureus,* and showed better activity compared to parent drug and control drug.

1. Introduction

Schiff bases containing penicillin and heterocyclic structural units with N, N donor atoms are considered the most prominent research area in the field of coordination chemistry [1–3]. The various donor atoms in them offer special ability for binding metals. The incorporated metals in the lattice of donor atoms of Schiff base change the physiological, morphological, and pharmacological activities of the compounds. The penicillin based Schiff base is of promising research interest owing to the widespread antibacterial resistance of the medical science. Moreover, the revival of research is essential to generate new Schiff base metal complexes with a diverse range of applications. Schiff base complexes have been used as drugs and have valuable antibacterial [4, 5], antifungal [6–8], antiviral [9, 10], anti-inflammatory [11], and antitumor activities [12]. Besides these, they also bear strong

catalytic activity in various chemical reactions in chemistry [13] and surfactant activities [14] and as memory storage devices in electronics [15–17]. One of the compounds used to prepare ligand is amoxicillin, a β-lactam antibiotic. It is a broad spectrum, semisynthetic penicillin type antibiotic that has potent bactericidal activity against many gram positive and gram negative bacterial pathogens [18]. It takes action against bacteria by preventing them from forming the cell wall and stopping them from growing. In medical science, it has important application for the treatment of bronchitis, ear infection, pneumonia, throat infections, tonsillitis, typhoid, and urinary tract infections. In combination with other antibiotics, it bears potential applications for the successful treatment of many pathogenic infections. However, synthetic modification in amoxicillin by coordination with metal ions of various types has been found to bear enhanced credibility, as documented in several research papers. Cisplatin is the

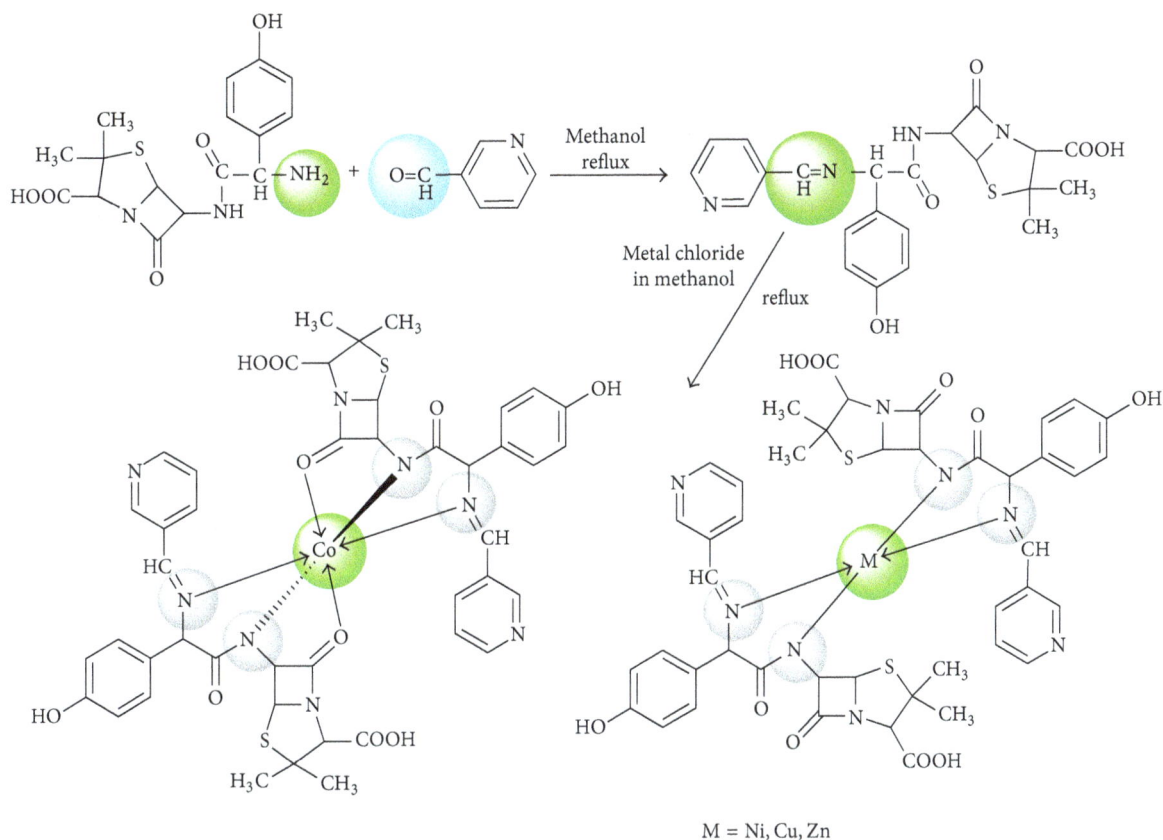

SCHEME 1: Synthetic scheme for the ligand (HL) and its metal complexes.

first metal based drug that emerged in the 20th century and enlightened the world as a promising anticancer drug [19]. Since then several research findings culminated the ideas of inclusion of metals in medicine. Many biological molecules containing pyridine moiety as a part of their structural unit bear enzymatic functions as well as the compounds of diverse biological interest. The pyridine derivatives are reported to have herbicidal, fungicidal, and insecticidal activities and also constitute the major core part of biological enzymes, important vitamins, and toxic alkaloids. Its wide applications in agroindustry and as pharmaceutical ingredients in drug discovery are the key points for this research investigation. Nicotinaldehyde (also called pyridine-3-carboxaldehyde) is a class of heterocyclic compound that has pyridine ring and an aldehyde group at meta-position [20]. Among the other pyridine aldehydes, nicotinaldehyde is suitably preferred for the prevention and treatment of *Acne vulgaris*, a kind of skin disease [21].

In the present paper, we have focused on the synthesis of novel Schiff base ligand, by the condensation of amoxicillin trihydrate and nicotinaldehyde and its four metal complexes with cobalt(II), nickel(II), copper(II), and zinc(II) salts (Scheme 1). The coordination behavior of the ligand towards transition metal ions was fully investigated by various spectral and thermal techniques. The geometry of the complexes was confirmed by energy optimization through MM2 calculation supported in ChemOffice and ArgusLab software program.

In continuation of our antibiotic research, we have also evaluated the antibacterial efficacy of ligand and its metal complexes against *S. aureus*, *E. coli*, *K. pneumoniae*, and *P. vulgaris* bacteria.

2. Experimental Section

2.1. Materials. All the chemicals and solvents used were of analytical reagent grade. The title compounds amoxicillin trihydrate and nicotinaldehyde in extra pure form were procured from Duchefa Biochemie, Netherlands, and Spectrochem, Mumbai, India, and used without further purification. Distilled methanol (Qualigen) was used as solvent for the synthesis. The metal salts (Co^{+2}, Ni^{+2}, Cu^{+2}, and Zn^{+2} chlorides) (Merck) were used for the synthesis of metal complexes.

2.2. Physical Measurements. Elemental microanalysis of the compounds was performed on EURO VECTOR EA 3000 micro analyzer. Melting points of the ligand and its complexes were recorded on an OMEGA melting point apparatus. The pH measurement was done on the Elico-16 pH meter. The infrared (FTIR) spectra of the prepared ligand and metal complexes were recorded on Perkin-Elmer Spectrum version 10.03.06 FT-IR spectrometer that was run as KBr discs in the range 4000–400 cm^{-1}. The ^{1}H NMR spectra were recorded on Bruker Avance III, 400 MHz spectrometer, using

DMSO-d_6 as solvent. The electronic absorption spectra of the complexes were recorded on single beam microprocessor Labtronics UV-Vis spectrophotometer (LT-290 model) in the range 200–1000 nm in DMSO solvent. EPR-JEOL spectra of the complexes were recorded on JES-FA200 ESR spectrometer with X-band at room temperature. ESI-MS spectra were recorded in positive mode on Agilent Q-TOF mass spectrometer equipped with an electron spray ionization source in the mass range of 200 to 1100. X-ray powder diffraction determinations were accomplished using Bruker AXS D8 Advance X-ray diffractometer with monochromatized Cu-Kα line at wavelength 1.5406 Å as the radiation source and the measurements were taken over the range of 2θ (10 to 70°). The thermal events of the compounds (TGA/DTA) were recorded on a Perkin-Elmer thermal analyzer with a linear heating rate of 20°C min^{-1} in the range of 40–730°C. The surface morphology of the synthesized ligand and metal complexes was analyzed by scanning electron microscopy technique. JEOL JSM-6390 LV scanning electron microscope was used for this investigation.

2.3. Synthesis of Ligand (HL).

Amoxicillin trihydrate (2.097 g, 5 mmol) in distilled methanol (30 ml) was stirred under hot conditions for 3 hs. Solubility in methanol was marked at the temperature elevation state. Its pH was adjusted to neutral by adding 0.1 N NaOH solution. Nicotinaldehyde (0.5378 g, 5 mmol) was added slowly to the well stirred amoxicillin trihydrate solution and refluxed under stirring condition for 4 hs. A clear bright yellow solution was left undisturbed for crystallization by slow solvent evaporation process for three days. The resulting solid product was separated, recrystallized with methanol, and dried in desiccator over anhydrous CaCl$_2$. The ligand was stored in the airtight vial in the refrigerator till its further use. M. pt. 140°C. Anal. C$_{22}$H$_{22}$N$_4$O$_5$S (454.13): Calcd. C 58.14, H 4.88, N 12.33, O 17.60, S 7.06; Found C 58.21, H 4.81, N 12.25, O 17.57, S 6.95. IR (KBr pellet, selected bands): $\bar{v}_{max}$ = 3303 (b, N-H and O-H str.), 1640 (s, C=N), 1510, 1443 (s, COOH). ^{1}H NMR (400 MHz, [D$_6$] DMSO): δ = 10.122 (s, 1 H, COOH), 9.425 (b, 1H, Ar-OH), 9.094 (s, 1H imine), 8.535–8.864 (m, 4H pyridine ring), 8.241–8.271 (s, 1H NH-amide), 6.718–7.625 (d, C-H aromatic), 1.118–1.562 (C-H methyl) ppm. UV/Vis: λ_{max} = 206, 262, 356 nm. ESI-MS, positive: m/z = 455 [M + H]$^+$. Conductivity: Λ_M = 10.8 μS/cm.

2.4. Synthesis of Metal Complexes

2.4.1. Co(II) Complex (1a).

A solution of ligand (HL) (0.454 g, 1 mmol) in 10 ml methanol was stirred for 1 h under warm condition and a solution of CoCl$_2$·6H$_2$O (0.119 g, 0.5 mmol) in 5 ml methanol was added dropwise with continuous stirring condition. Then after the mixture solution was refluxed for 1 h over water bath with stirring, till blue colored precipitate resulted. The precipitate was filtered from the supernatant liquid, washed with methanol, and dried over anhydrous calcium chloride, yield (65%). M. pt. 285°C. Anal. C$_{44}$H$_{46}$CoN$_8$O$_{12}$S$_2$ (1001.2): Calcd. C 52.74, H 4.63, N 11.18, O 19.16, S 6.40; Found C 52.69, H 4.69, N 11.26, O 19.20, S 6.64. IR (KBr pellet, selected bands): $\bar{v}_{max}$ = 3417 (b, O-H str.), 1633

(s, C=N), 1510, 1443 (s, COOH), 606 (ρ_w H$_2$O), 526 (Co-O), 425 (Co-N). UV/Vis: λ_{max} = 263, 346–371, 457–488, 549 nm. ESI-MS, positive: m/z = 1001.2 [M + H]$^+$. Conductivity: Λ_M = 21.8 μS/cm.

2.4.2. Ni(II) Complex (1b).

The nickel complex (1b) was prepared according to the procedure adopted for the preparation of 1a. A solution of NiCl$_2$·6H$_2$O (0.1188 g, 0.5 mmol) in 5 ml methanol was used for this purpose. The mixed solution of ligand (HL) and Ni^{+2} salt was refluxed for 1 and 1/2 h over water bath which resulted in green colored complex, yield (62%). M. pt. 270°C. Anal. C$_{44}$H$_{42}$N$_8$NiO$_{10}$S$_2$ (964.18): Calcd. C 54.73, H 4.38, N 11.60, O 16.57, S 6.64; Found C 54.55, H 4.59, N 11.59, O 16.45, S 6.44. IR (KBr pellet, selected bands): $\bar{v}_{max}$ = 3337 (b, O-H str.), 1625 (s, C=N), 1513, 1435 (s, COOH), 687 (ρ_w H$_2$O), 429 (Ni-N). UV/Vis: λ_{max} = 261, 346, 460, 549 nm. ESI-MS, positive: m/z = 964.18 [M + H]$^+$. Conductivity: Λ_M = 19.9 μS/cm.

2.4.3. Cu(II) Complex (1c).

The copper complex (1c) was also prepared according to the procedure adopted for the preparation of 1a and 1b. A solution of CuCl$_2$·2H$_2$O (0.085 g, 0.5 mmol) in 5 ml methanol was used for this purpose. The mixed solution of ligand (HL) and Cu^{+2} salt was refluxed for 1 and 1/2 h over water bath which resulted in green colored complex, yield (65%). M. pt. 260°C. Anal. C$_{44}$H$_{42}$CuN$_8$O$_{10}$S$_2$ (969.18): Calcd. C 54.45, H 4.36, N 11.55, O 16.49, S 6.61; Found C 54.52, H 4.49, N 11.63, O 16.55, S 6.73. IR (KBr pellet, selected bands): $\bar{v}_{max}$ = 3379 (b, O-H str.), 1636 (s, C=N), 1512, 1436 (s, COOH), 686 (ρ_w H$_2$O), 444 (Ni-N). UV/Vis: λ_{max} = 227, 259, 337, 344, 485 nm. ESI-MS, positive: m/z = 969 [M + H]$^+$. Conductivity: Λ_M = 35.2 μS/cm.

2.4.4. Zn(II) Complex (1d).

The zinc complex (1d) was prepared according to the above procedure and by using Zn^{+2} salt (0.07 g, 0.5 mmol). The mixed solution was refluxed for 2 h over water bath which resulted in light yellow colored complex, yield (57%). M. pt. 250°C. Anal. C$_{44}$H$_{42}$N$_8$O$_{10}$S$_2$Zn (970.18): Calcd. C 54.35, H 4.35, N 11.52, O 16.45, S 6.60; Found C 54.41, H 4.43, N 11.57, O 16.49, S 6.57. IR (KBr pellet, selected bands): $\bar{v}_{max}$ = 3340 (b, O-H str.), 1629 (s, C=N), 1512, 1437 (s, COOH), 657 (ρ_w H$_2$O), 415 (Zn-N). ^{1}H NMR (400 MHz, [D$_6$] DMSO): δ = 10.123 (s, 1 H, COOH), 9.425 (b, 1H, Ar-OH), 9.295 (s, 1H imine), 8.534–8.865 (m, 4H pyridine ring), 6.720–7.66 (d, C-H aromatic), 1.118–1.571 (C-H methyl) ppm. UV/Vis: λ_{max} = 263, 346 nm. ESI-MS, positive: m/z = 970 [M + H]$^+$. Conductivity: Λ_M = 5.6 μS/cm.

2.5. Antibacterial Susceptibility Test.

The antimicrobial potency of the synthesized compounds was done by assaying antibacterial activity study. The experimental portion of the study was accomplished in the laboratory of the Department of Microbiology at Mahendra Morang Adarsh Multiple Campus, Biratnagar. The compounds (HL and 1a–1d) were tested in vitro by standard Kirby-Bauer paper disc diffusion method against some gram positive and gram negative human pathogenic bacteria [12, 22, 23]. The recommended NCCLS guideline was followed for the study [24]. Well sterilized filter paper discs of 5 mm size (Whatman-model) were used as antibiotic assay discs for testing of compounds. The discs were loaded with test compounds at two different concentrations (100 and 50 mcg/mcl in DMSO) under UV laminar flow to reduce bacterial contamination [25]. The loaded discs were dried in the laminar flow chamber by blowing hot air through hair drier. Sterilized nutrient agar media were carefully poured in the Petri plate and kept in rest for few hours in the sterilized zone for solidification. Fresh bacterial culture, revived before injection, was swabbed on the media and the loaded discs were stuck over it. One disc soaked with DMSO was used as the solvent control and amikacin (30 mcg/disc) was used as positive control. Inoculated plates were incubated at 37°C for 24 hs, and the diameter of the zone of inhibition was measured by antibiogram zone measuring scale [26].

3. Results and Discussion

3.1. Physical Characterization. The physical properties and the microanalytical data of the ligand (HL) and metal complexes (1a–1d) are summarized in the experimental section. The analytical results show (1 : 2) metal ligand ratio, that is, ML_2 type. The color change from ligand to metal complexes is in support of metal ligand interaction which is further reinforced by conductivity and pH change. The ligand (HL) was soluble in methanol. The complexes were soluble in DMSO and DMF. The nickel complex (1b) was found to be hygroscopic. The suggested molecular formulae of the ligand (HL) and metal complexes (1a–1d) have been achieved by microanalytical results in combination with various spectral techniques. The experimental molar conductivity data of HL and metal complexes was found in the range of 5.6–35.2 μS/cm and suggests their nonelectrolytic nature. The pH of ligand and complexes was almost in the neutral range.

3.2. Spectral Characterization. The formation of HL was confirmed by ESI mass spectrometry, which showed peaks at m/z = 455, attributable to $[M + H]^{+}$. The FTIR spectrum is also in line with the proposed structure of HL, with characteristic stretching vibrations at 1640 cm^{-1} assignable to azomethine group [27]. A broadband with absorption maximum of 3303 cm^{-1} is possibly due to collapse of N-H and O-H stretching peaks. Other significant strong bands at 1510 and 1433 cm^{-1} for HL are attributed to v(COOH) asymmetric and symmetric stretch. The ^{1}H NMR spectrum of HL executes a sharp singlet at 9.09 ppm corresponding to azomethine proton. On complexation, v(C=N) stretching band for HL has shifted to lower absorption frequency of 1633 cm^{-1} (1a), 1625 cm^{-1} (1b), 1636 cm^{-1} (1c), and 1629 cm^{-1} (1d), indicating the coordination of azomethine nitrogen atom to the metal ion [28]. This lower frequency shift of azomethine group in the complexes is due to the decrease in electron density and force constant of the metal with the azomethine nitrogen lone pair. In all the complexes, FTIR absorption bands corresponding to v(O-H) execute in the range 3337–3417 cm^{-1} relative to 3303 cm^{-1} for HL. The complexes exhibit v(COOH) stretching vibrations at the equivalent positions of the ligand, suggesting their noncoordination

with the metal centers. The formation of cobalt complex (1a) was confirmed by ESI-MS peak at m/z = 1002, attributable to $[M + H]^{+}$. The well resolved IR band at 3417 cm^{-1}, for complex (1a), corresponds to the v(O-H) stretching vibration (Figure 1) [29]. The evidence of bonding in 1a is also shown by the observation of new bands in the lower frequency regions at 425 and 526 cm^{-1} characteristic to v(Co-N) and v(Co-O) stretching vibrations that are not observed in the IR spectrum of ligand. The less intense IR band at 606 cm^{-1} is assignable to bending vibration of two lattice water molecules of the outer sphere region. The observed molecular mass of nickel complex (1b) was evidenced by ESI mass spectrum peak value at m/z = 964, assignable to molecular ion peak. The formation of this complex was verified by FTIR spectroscopy, where specific bands are observed at 1625 cm^{-1} v(CH=N), 3337 cm^{-1} v(O-H), 429 cm^{-1} v(Ni-N), and 687 cm^{-1} for outer sphere lattice water molecules. The copper complex (1c) executes a strong azomethine band at 1636 cm^{-1} which has undergone a negative shift by 4 cm^{-1} relative to that of the free ligand. The other significant FTIR bands are observed at 3379 cm^{-1} v(O-H), 444 cm^{-1} v(Cu-N), and 686 cm^{-1} for outer sphere lattice water molecules. The formation of the complex 1c is further evidenced by the ESI-MS peak at m/z = 970, attributed to $[M + H]^{+}$. The positive ion ESI mass spectrum showed peaks at m/z = 971 for zinc complex (1d), attributed to $[M + H]^{+}$. Its formation was strongly evidenced by FTIR and ^{1}H NMR spectral data. The strong azomethine band at 1629 cm^{-1} v(CH=N) for this complex has shifted by 11 cm^{-1} towards a lower wave number relative to that of the free ligand, indicating metal coordination with azomethine nitrogen. The metal nitrogen coordination is further evidenced by a sharp peak at 415 cm^{-1} in the FTIR spectrum of 1d. The ^{1}H NMR spectrum is also consistent with the suggested structure. The downfield shift of ^{1}H NMR signal for azomethine proton from δ 9.094 ppm for ligand to δ 9.295 ppm for zinc complex (1d) also supports coordination of the azomethine nitrogen to the zinc(II) ion. Two doublets observed at δ 6.718–7.625 for HL and δ 6.72–7.66 ppm for zinc complex 1d are attributed to aromatic ring protons. The methyl protons of amoxicillin moiety in both HL and 1d appear as a singlet peak in the region of δ 1.118–1.562 ppm. Amide NH proton for HL executes signal at 8.241–8.271 ppm, which is absent in the spectrum of 1d, and this confirms the coordination of amide N-atom with metal center via deprotonation [30]. In the spectrum of HL, the signal due to carboxylic proton appears at δ 10.122 ppm, which is still present in the ^{1}H NMR spectrum of zinc complex (1d).

3.3. Electronic Absorption Spectra and Magnetic Moment Measurement. The electronic absorption spectrum of ligand (HL) displays high energy bands in the ultraviolet region at 206 and 262 nm, corresponds to $\pi \rightarrow \pi^{*}$ transitions of the aromatic and pyridinium ring, and, at 356 nm, corresponds to $n \rightarrow \pi^{*}$ intraligand charge transfer band with the involvement of C=N group [31]. However, the additional bands in the higher wavelength region are observed in the complexes

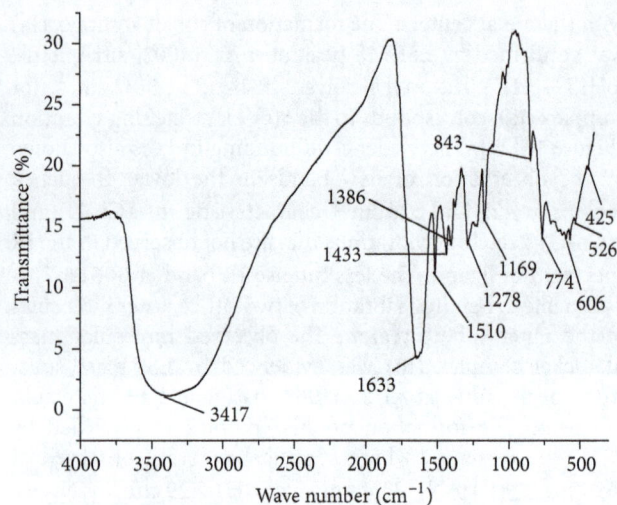

FIGURE 1: IR spectrum of cobalt complex (1a).

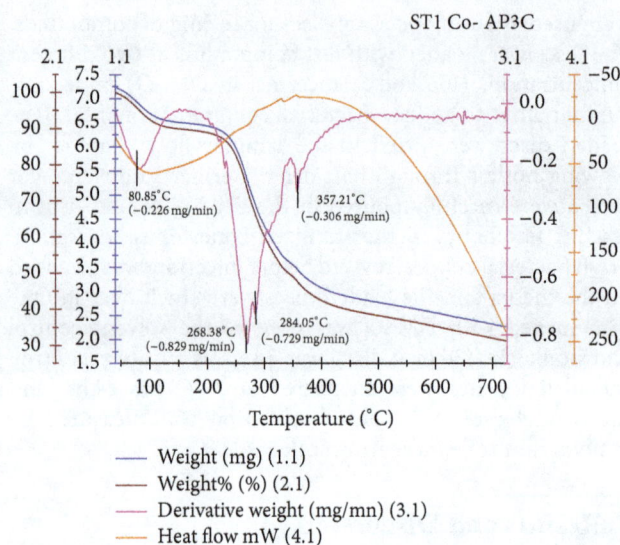

FIGURE 2: Thermogram of cobalt complex (1a).

which signify metal ligand coordination. The cobalt complex (1a) exhibits two distinct bands in high wavelength region of the spectrum at 457–488 nm and 549 nm. The former band is assignable to $^4T_{1g}(F) \rightarrow {}^4T_{1g}(P)$ and latter band indicates $^4T_{1g} \rightarrow {}^4A_{2g}$ transition, confirming its octahedral geometry [32]. The magnetic moment value (4.75 BM) further supports this geometry. The high energy bands for this complex are observed at 263 and 346–371 nm, assignable to $\pi \rightarrow \pi^*$ and $n \rightarrow \pi^*$ LMCT transitions, respectively. The electronic absorption spectrum of nickel complex (1b) displays d-d transition band at 460 nm assignable to $^1A_{1g} \rightarrow {}^1B_{1g}$ transition along with the bands in the low wavelength region at 261 and 346 nm [33, 34]. The diamagnetic nature of this complex is suggestive of the complete distortion of octahedral geometry and confirms its square planar geometry. The magnetic moment value (1.82 BM) and electronic absorption spectrum of paramagnetic copper complex (1c) exhibit absorption band in the high wavelength region at 485 nm, attributed to $^2T_{2g} \rightarrow {}^2E_g$ transition which is suggestive of tetrahedral geometry [35]. Other high energy bands for this complex are observed at 227 and 259 nm for $\pi \rightarrow \pi^*$ transition and 337 and 344 nm for $n \rightarrow \pi^*$ LMCT transition. The zinc complex (1d) displays an absorption band at 346 nm assignable to the LMCT transition, compatible with tetrahedral geometry, and this is further supported by its diamagnetic nature and absence of d-d band, due to its complete d^{10} electronic configuration.

3.4. TGA/DTA Studies. The TGA/DTA curves for the complexes were carried out within the temperature range from room temperature to 700°C with the linear heating rate of 20°C/min in the nitrogen atmosphere. Correlation of the thermal events at elevated temperatures with kinetic parameters provides useful physicochemical information of the compounds. Thermo gravimetric analysis is one such important instrumental technique to observe thermal changes with respect to increase in temperatures [36]. The

computed thermal decomposition data in Table 1 are in good agreement with the suggested microanalytical data. The following findings have been achieved in our research analysis.

The thermogram of cobalt complex (1a) (Figure 2) exhibited four decomposition steps in the temperature range of 50–380°C. The first decomposition step in the temperature range of 50–110°C with % mass loss of 4.826% (0.226 mg) is assignable to the loss of two lattice water molecules from the outer sphere [37, 38]. The second and third decomposition steps with % mass loss of 25.923% (0.829 mg) and 33.837% (0.729 mg) in the temperature range of 241–273°C and 279–300°C have considered the loss of organic ligand moiety. The last decomposition step with % mass loss of 52.247% (0.306 mg) represents a complete loss of ligand from the complex in the temperature range of 338–380°C, leaving cobalt oxide as stable residue. The nickel complex (1b) exhibited thermal decomposition in two distinct steps. The first step with % mass loss of 5.6% (0.153 mg) is assignable to the loss of outer sphere lattice water molecules in the temperature range of 44–113°C. The exothermic peak with 30.86% (0.484 mg) mass loss in the temperature range of 232–403°C is attributed to loss of ligand moiety. The thermograms of other two complexes 1c and 1d are complement to the analyzed data of 1a and 1b. The first step decomposition in 1c and 1d occurred in the temperature range around 45–107°C with T_{DTG} 76.79 and 75.46°C and this again suggests the loss of two lattice water molecules. The final thermal decomposition step in all the metal complexes is noticed above 400°C, which is indicated by the formation of the horizontal TG curve. This step interprets the formation of stable metal oxide residue.

Kinetic Parameters. The thermal dehydration and decomposition of the complexes were studied by using an integral method applying a very popular Coats-Redfern method [39, 40]. The thermodynamic activation parameters of decomposition processes are essential to describe the thermal stability

TABLE 1: Thermal decomposition data of metal complexes.

Comp.	Step	TG range (°C)					DTA	
		$\Delta_{m\%}$ found	T_i	T_f	T_{DTG}	Mass loss	T_{dta}	Peak
1a	1	4.826	50	110.2	80.85	−0.226	103.68	Endo
	2	25.923	241.63	276.73	268.38	−0.829	—	—
	3	33.837	279	300	284.05	−0.729	—	—
	4	52.247	338	380	357.21	−0.306	334.9	Endo
1b	1	5.601	44	113	85.91	−0.153	105.33	Endo
	2	30.861	232	403	278.16	−0.484	382.91	Exo
1c	1	2.699	49	105	76.79	−0.100	111.23	Endo
	2	8.814	148	163	159.86	−0.395	—	—
	3	21.635	187	246	237.13	−0.310	—	—
	4	42.395	293	390	326.01	−0.244	385.92	Exo
1d	1	3.047	46	107	75.46	−0.121	115.75	Endo
	2	18.233	230	287	260.99	−0.491	339.24	Endo
	3	36.441	310	410	330.41	−0.238	365.88	Exo

as well as the nature and rates of thermal decomposition of the complexes. These parameters are evaluated graphically by plotting of data based on Coats-Redfern relation in the following form:

$$\ln\left[-\frac{\ln(1-\alpha)}{T^2}\right] = \ln\left[\frac{AR}{\beta E^*}\right] - \frac{E^*}{RT}, \tag{1}$$

where α represents the decomposition fraction at temperature T K and β denotes linear heating rate (dT/dt). E^* and A denote the activation energy and Arrhenius preexponential factor, respectively. R represents gas constant. Molding the equation for the straight line $(y = mx + c)$, a linear plot of left side versus $1/T$ of Coats-Redfern equation gives a straight line, whose slope E^*/R furnishes activation energy parameter and the preexponential factor (A) can be determined from the intercept. The other thermodynamic parameters such as entropy of activation (ΔS^*), enthalpy of activation (ΔH^*), and free energy of activation (ΔG^*) have been calculated by using the following relation:

$$\Delta S^* = R\ln\left[\frac{Ah}{k_B T}\right]$$

$$\Delta H^* = E^* - RT \tag{2}$$

$$\Delta G^* = \Delta H^* - T\Delta S^*.$$

The computed data of thermodynamic activation parameters of various decomposition steps of the metal complexes are listed in Table 2. In the present work, the plot of left hand side of Coats-Redfern equation versus 1000/T in all the decomposition steps of all complexes shows a best fit for first-order reaction kinetics [41]. The high and increasing values of activation energy in the subsequent steps of all the

complexes reflect high thermal stability, which may be due to covalent bond character. The entropy of activation value of first decomposition step in all the complexes is negative, which indicates nonspontaneous dehydration reaction process. Most of this value of other steps is positive and infers the dissociation character of decomposition [42]. This also attributes more ordered activated state than the reactants. The positive ΔG^* values of all the complexes justify the nonspontaneous nature of decomposition steps. The enthalpy of activation values (ΔH^*) in most of the decomposition steps is positive which reveal endothermic processes. However, this nature also depends upon the value of other thermodynamic activation parameters. The computed data of correlation coefficient (r) obtained from the graphical plot reflect a good fit of the data with linear function [43, 44].

3.5. XRPD Study. Single crystal growth of the synthesized compounds was unsuccessful, so their crystallinity was established by X-ray powder diffraction study. The ligand (HL) and complexes (1a, 1b, and 1d) were found amorphous. The crystal structure of copper complex (1c) was worked out by its well resolved crystalline peaks (Figure 3), which crystallized in a triclinic crystal system with P1 space group. The diffractogram of this complex registered 22 reflection peaks in the range of (2θ) 0 to 50° with maxima at 15.974° with corresponding d spacing value of 5.584 Å. The cell dimensions a (6.2282 Å), b (109390 Å), c (20.3388 Å), α (63.1585°), β (113.5723°), and γ (64.269°) are in good agreement with the refined triclinic crystal system. The unit cell volume of this compound was calculated to be 747.4131 Å^3 with FOM 31. The details of crystallographic data are summarized in Table 3. The particle size was calculated from Scherer's formula $\alpha = 0.9\lambda/\beta\cos\theta$, where λ is the wavelength, β is the full-width half maximum of the characteristic peak, and θ is the

TABLE 2: Kinetic and thermodynamic parameters of metal complexes.

Comp.	Step	r	A (s^{-1})	T_{max} (K)	E^* (kJ/mol)	ΔS^* (J/K·mol)	ΔH^* (kJ/mol)	ΔG^* (kJ/mol)
1a	1	−0.98495	5.62×10^6	353.85	58.36	−117.92	55.4719	97.19789
	2	−0.99209	5.06×10^{19}	541.38	229.22	126.559	224.718	156.202
	3	−0.99052	1.19×10^{36}	557.05	403.95	18.3647	399.319	389.089
	4	−0.99343	1.14×10^{22}	630.21	288.1474	170.339	282.907	175.525
1b	1	−0.99453	7.054×10^6	358.91	59.62	−40.236	56.634	71.075
	2	−0.9887	1.35×10^{13}	551.16	160.7139	35.4267	160.288	140.762
1c	1	−0.99401	4.064×10^7	349.79	63.4725	−110.3698	60.564	64.191
	2	−0.99698	10.06×10^{51}	432.86	447.68	765.803	444.080	112.594
	3	−0.99643	5.98×10^{11}	510.13	129.453	−24.716	125.2115	137.819
	4	−0.99604	6.55×10^5	599.01	93.1345	−140.1638	92.6438	176.603
1d	1	−0.99448	5.89×10^6	348.46	57.8981	−117.3976	55.00	95.908
	2	−0.99491	2.92×10^{15}	533.99	176.7582	45.523	172.318	148.00
	3	−0.99118	7.64×10^5	603.41	96.5845	−103.320	91.567	153.911

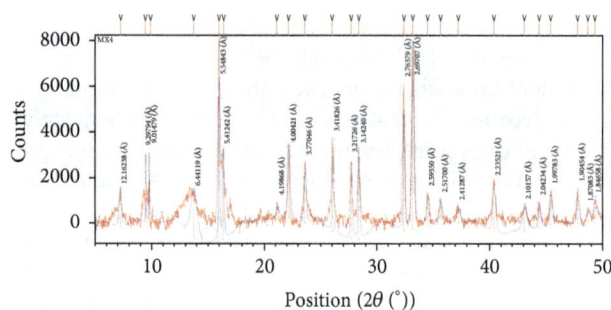

FIGURE 3: X-ray diffractogram of copper complex (1c).

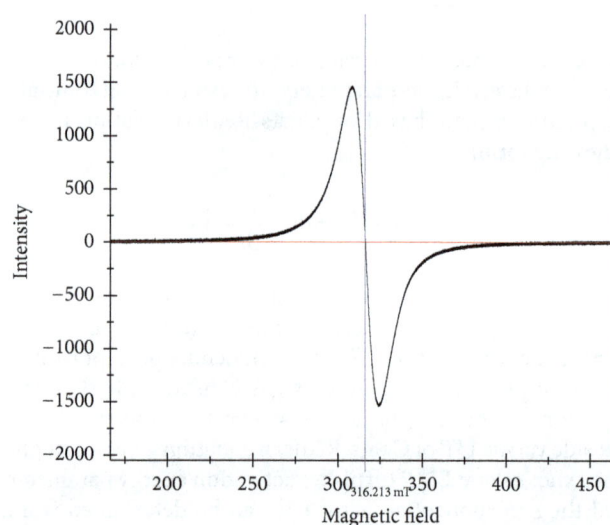

FIGURE 4: EPR spectrum of copper complex (1c) at room temperature.

diffraction angle of the *hkl* plane [45, 46]. The average particle size 69.34 nm suggests its nanocrystalline nature.

3.6. EPR Analysis. The solid state X-band EPR spectrum of the copper complex (1c) was recorded at room temperature under the frequency 9447.606 MHz with no marker lines used and center line at 316.213 mT. The standard lines that are used in EPR model are of Mn, which has been omitted in the graph. The EPR spectrum of complex provides useful information about the metal ion environment within the complex. The highly symmetrical EPR spectrum of copper complex (1c) (Figure 4) delivered a single isotropic signal with $g_\parallel$ value of 2.18 and $g_\perp$ value of 2.08 [47]. The absence of poorly resolved hyperfine signal may be attributed to the considerable exchange coupling interaction of the Cu^{+2} ions in the complex. The order of splitting factors $g_\parallel > g_\perp > 2.0023$ clearly indicates the localized unpaired electron in the d orbital of Cu^{+2} ion and is of axial symmetry [6, 48]. The calculated g_{av} value is 2.11 whose deviation from the free electron (2.003) is due to covalent character of the metal ligand bond. This fact is further supported

by $g_\parallel$ value less than 2.3. The value of exchange coupling interaction parameter "G" = 2.25 is less than 4 and suggests considerable exchange interaction in the complex [49]. All these parameters are in support of tetrahedral geometry of copper complex (1c).

3.7. SEM Analysis. The metal coordination to ligand significantly changes the surface morphology of the complexes and this was investigated by SEM analysis. The SEM micrograph of ligand (HL) and metal complexes are shown in Figure 5 and the differences are seen in surface morphology of the metal complexes due to changes in the metal ions. The SEM micrograph of ligand demonstrates nonuniform platelet-like structure with variable lateral dimensions [50]. Moreover,

TABLE 3: X-ray powder diffraction data of copper complex (1c).

Peak number	2θ	θ	$\text{Sin}\,\theta$	$\text{Sin}^2\theta$	$h^2 + k^2 + l^2$	hkl	d	FWHM	% int.	a in nm
1	7.2685	3.63425	0.06338	0.004017	1	0 0 1	12.16238	0.2244	16.00	61.22
2	13.7483	6.87415	0.11968	0.014323	1	0 1 0	6.44119	0.6731	21.90	20.37
3	15.9738	7.9869	0.13894	0.019304	2	1 1 0	5.54843	0.0935	100.00	146.64
4	16.3780	8.189	0.14243	0.020286	10	1 0 −3	5.41242	0.1496	42.23	91.65
5	21.1607	10.58035	0.18361	0.033712	9	1 2 2	4.19868	0.2244	6.47	61.102
6	22.2011	11.10055	0.19253	0.03706	11	1 −1 −3	4.00421	0.2244	33. 02	61.102
7	23.5966	11.7983	0.20446	0.041803	1	1 0 0	3.77046	0.1870	28.72	73.323
8	26.0686	13.0343	0.22553	0.050863	6	1 1 2	3.41826	0.1309	44.73	104.747
9	27.7288	13.8644	0.23962	0.057417	4	0 2 0	3.21726	0.1496	38.35	91.654
10	28.4032	14.2016	0.24533	0.060186	10	1 3 0	3.14240	0.1496	42.24	91.654
11	32.3941	16.19705	0.27894	0.077807	20	2 0 −4	2.76379	0.1122	70.54	122.205
12	33.2184	16.6092	0.28584	0.081704	40	2 0 −6	2.69707	0.2057	91.73	66.657
13	34.5584	17.2792	0.29702	0.088220	13	2 0 −3	2.59550	0.2244	13.44	61.103
14	35.6718	17.8359	0.30629	0.093813	2	1 −1 0	2.51700	0.2244	8.91	61.103
15	37.2666	18.6333	0.31951	0.102086	20	2 4 0	2.41287	0.3739	7.05	36.671
16	40.3520	20.176	0.34490	0.11895	80	0 4 8	2.23521	0.2617	23.89	52.394
17	43.0415	21.52075	0.36683	0.13456	9	1 −2 −2	2.10157	0.4487	11.08	30.5585
18	44.3550	22.1775	0.37747	0.14248	30	2 5 1	2.04234	0.2244	11.90	61.103
19	45.3976	22.6988	0.38588	0.14890	69	2 1 −8	1.99783	0.2244	15.90	61.103
20	47.7559	23.87795	0.40478	0.16384	25	3 4 0	1.90454	0.1870	16.97	73.324
21	48.6714	24.3357	0.41211	0.16983	58	3 0 −7	1.87083	0.2991	5.79	45.843
22	49.3094	24.6547	0.41714	0.174	18	3 3 0	1.84658	0.2736	7.79	50.116
								Average particle size		69.3474

inhomogeneous matrix with broken ice-like structure has been observed in the SEM micrograph of nickel complex (1b). The SEM micrograph of copper complex (1c) displays agglomerated morphology with small sized grains scattered in homogenous matrix and gives the appearance of coral-like structure. In the SEM micrograph of zinc complex (1d), small sized particles crumbled together to give rock-like structure with somewhat cotton-like appearance.

3.8. Molecular Modeling. The computational study of the compounds furnishes a clear idea about the three-dimensional arrangement of different atoms in the molecules. The energy optimization of the ligand (HL) and metal complexes (1a–1d) was done by Universal Force Field (UFF) technique with minimum RMS gradient 0.100, supported in ArgusLab 4.0.1 version software [51, 52]. The details of the bonding and energy parameters optimized by molecular modeling calculations of the metal complexes are depicted in Table 4. For ligand, single point energy calculation with Hamiltonian AM1 revealed final SCF energy and heat of formation, −132288.8349 and 45.0637 kcal/mol, respectively [53]. After the geometry optimization by molecular mechanics (UFF) technique, the final geometrical energy of HL has been reported to 139.2725 kcal/mols. On ESP mapped electron density surface of HL (Figure 6), red color indicates the highest electron density region which is around O-atom.

The second highest electron density region is around an azomethine N-atom which is shown by mixed green and yellow colors. This is the region for stability of coordinated metal ions and supports its linkage with azomethine N-atom. It seems clear that the coordination with O-atoms is restricted due to greater electronic repulsion and field obstruction. In the nickel complex (1b), the high electron density (Figure 7) around the coordinated azomethine N-atom, indicated by red color, is in favor of its proposed geometry. Similar study and computational data of the complexes (1a to 1d) are in good support of their proposed structures.

3.9. Antibacterial Activity Study. The antibacterial efficacy of the ligand (HL) and metal complexes (1a–1d) was tested against *S. aureus, E. coli, K. pneumoniae,* and *P. vulgaris* bacteria. The antibacterial results are presented in the bar graph (Figure 8). Two different concentrations (100 and 50 mcg/mcl) of the compounds have been selected for antibacterial assay. The results suggest enhanced antibacterial activity of the ligand (HL) and metal complexes (1a–1d). The compound (1c) showed little activity against all the bacterial pathogens, compared to ligand and other metal complexes. The ligand bears activity, even greater than parent drug amoxicillin and control drug amikacin at higher concentration. This higher activity of ligand is possibly due to interference in the normal cell process of organism caused by the formation

FIGURE 5: SEM micrographs of HL, 1b, 1c, and 1d.

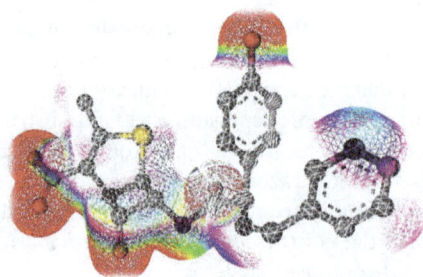

FIGURE 6: Electrostatic potential mapped electron density surface of HL.

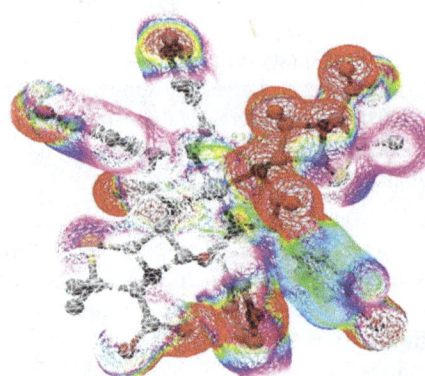

FIGURE 7: Electrostatic potential mapped electron density surface of nickel complex (1b).

of hydrogen bond through the azomethine group with the active center of cell constituents [54]. Further, the uncoordinated heteroatom of pyridine moiety also contributes to microbial growth inhibition. Moreover, the complexes deliver better antibacterial activity at their higher concentration. Precise observation reveals that the compounds are less active against *S. aureus* and more active against *E. coli* and *P. vulgaris* bacteria. This enhanced activity of the complexes may be attributed to chelation of Schiff base with metal ions that provide stability and more susceptibility against the bacterial

pathogens [55, 56]. It has been suggested that the structural components possessing additional (C=N) bond with nitrogen and oxygen donor systems inhibit enzyme activity due to their deactivation by metal coordination. This permits their efficient permeation through the lipid layer of organisms and destroys their activity [6].

TABLE 4: Selected bond lengths and bond angles of metal complexes.

Complex	Atoms	Bond length (Å)	Bond energy (Kcal/mol)	Atoms	Bond angle	Bond angle energy	Final geom. energy
1a	N(3)-Co(34)	1.957	273.796	N(3)-Co(34)-N(20)	90.00	300.46	349.2538 (Kcal/mol) (0.556 au)
	N(20)-Co(34)	1.972	267.453	N(3)-Co(34)-N(35)	90.00	304.025	
	Co(34)-N(35)	1.957	273.796	N(3)-Co(34)-N(52)	90.00	300.46	
	Co(34)-N(52)	1.972	267.453	N(3)-Co(34)-O(104)	90.00	273.401	
	Co(34)-O(104)	1.964	244.913	N(3)-Co(34)-O(113)	90.00	273.401	
	Co(34)-O(113)	1.964	244.913	N(35)-Co(34)-N(20)	90.00	300.46	
				N(52)-Co(34)-N(20)	90.00	296.982	
				O(104)-Co(34)-N(20)	90.00	270.214	
				O(113)-Co(34)-N(20)	90.00	270.214	
				N(35)-Co(34)-N(52)	90.00	300.46	
				N(35)-Co(34)-N(104)	90.00	273.401	
				N(35)-Co(34)-N(113)	90.00	273.401	
				N(52)-Co(34)-O(104)	90.00	270.214	
				N(52)-Co(34)-N(113)	90.00	270.214	
1b	N(2)-Ni(33)	1.870	313.824	N(2)-Ni(33)-N(19)	90.00	344.228	324.5763 (Kcal/mol) (0.517 au)
	N(19)-Ni(33)	1.885	306.275	N(2)-Ni(33)-N(34)	90.00	348.473	
	N(34)- Ni(33)	1.870	313.824	N(2)-Ni(33)-N(51)	90.00	344.228	
	N(51)- Ni(33)	1.885	306.275	N(34)-Ni(33)-N(19)	90.00	344.228	
				N(51)-Ni(33)-N(19)	90.00	340.090	
				N(34)-Ni(33)-N(51)	90.00	344.228	
1c	N(2)-Cu(33)	2.016	181.007	N(2)-Cu(33)-N(19)	109.47	158.764	373.488 (Kcal/mol) (0.595 au)
	N(19)-Cu(33)	2.031	176.938	N(2)-Cu(33)-N(34)	109.47	160.587	
	N(34)-Cu(33)	2.016	181.007	N(2)-Cu(33)-N(51)	109.47	158.764	
	N(51)-Cu(33)	2.031	176.938	N(34)-Cu(33)-N(19)	109.47	158.764	
				N(51)-Cu(33)-N(19)	109.47	156.977	
				N(34)-Cu(33)-N(51)	109.47	158.764	
1d	N(2)-Zn(33)	1.888	164.142	N(2)-Zn(33)-N(19)	109.47	193.167	352.3697 (Kcal/mol) (0.561 au)
	N(19)- Zn(33)	1.903	160.260	N(2)-Zn(33)-N(34)	109.47	195.503	
	N(34)- Zn(33)	1.888	164.142	N(2)-Zn(33)-N(51)	109.47	193.167	
	N(51)- Zn(33)	1.903	160.260	N(34)-Zn(33)-N(19)	109.47	193.167	
				N(51)-Zn(33)-N(19)	109.47	190.879	
				N(34)-Zn(33)-N(51)	109.47	193.167	

4. Conclusion

The novel ligand (HL) and the metal complexes (1a–1d) were successfully synthesized. The ligand can complex the metal ion via N donor atoms. The electronic absorption spectral analysis in combination with ESR data revealed octahedral geometry for cobalt complex (1a), square planar geometry of nickel complex (1b), and tetrahedral geometry for both copper complex (1c) and zinc complex (1d). Several spectral data nicely support the above concerned geometry of the complexes. Furthermore, the metal complexes were screened in vitro for antibacterial assay. Based on the results of this study of synthesized compounds, it has been concluded that the ligand bears greater potency than amoxicillin and control drug amikacin. The complexes 1a, 1b, and 1d were even highly active against all the bacterial pathogens at their higher concentration; however the copper complex (1c) was less active than others. This greater activity might be due to azomethine linkage and heteroatoms present in these compounds.

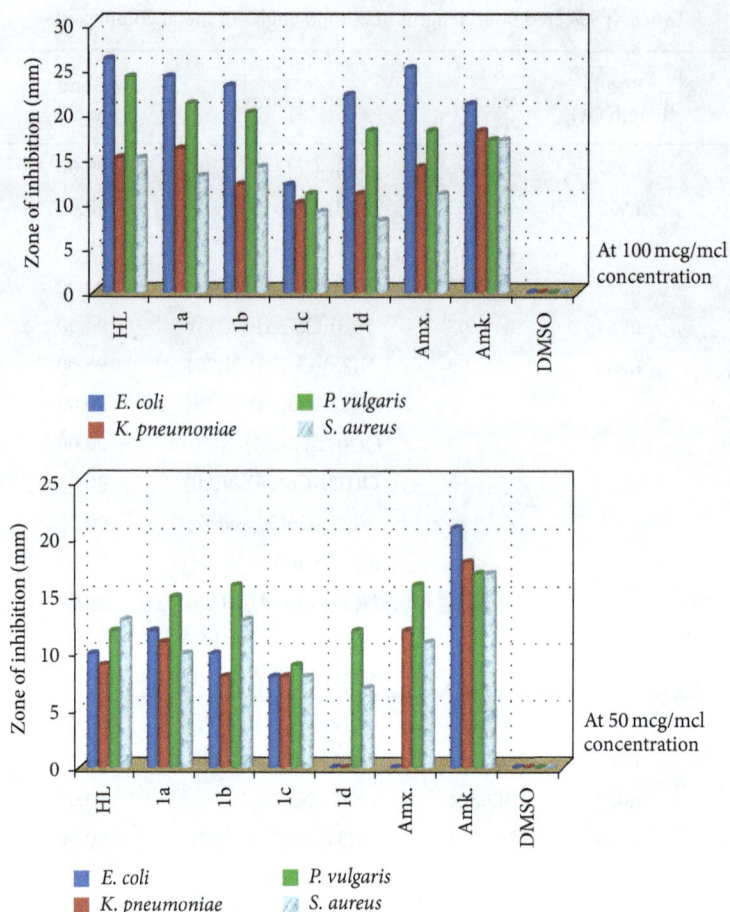

FIGURE 8: Bar graph of antibacterial evaluation study.

Acknowledgments

This research work was financially supported by Nepal Academy of Science and Technology (NAST) by providing fellowship, so one of the authors is highly grateful to this organization. The authors express great honor to the entire team of Solid State and Structural Unit, Indian Institute of Science, Bangalore, India, for their cooperation in recording the spectra of the compounds.

References

[1] M. Gulcan, S. Özdemir, A. Dündar, E. Ispir, and M. Kurtoğlu, "Mononuclear complexes based on pyrimidine ring azo schiff-base ligand: Synthesis, characterization, antioxidant, antibacterial, and thermal investigations," *Zeitschrift fur Anorganische und Allgemeine Chemie*, vol. 640, no. 8-9, pp. 1754–1762, 2014.

[2] M. S. Nair, D. Arish, and R. S. Joseyphus, "Synthesis, characterization, antifungal, antibacterial and DNA cleavage studies of some heterocyclic Schiff base metal complexes," *Journal of Saudi Chemical Society*, vol. 16, no. 1, pp. 83–88, 2012.

[3] W. Al Zoubi, A. A. S. Al-Hamdani, and M. Kaseem, "Synthesis and antioxidant activities of Schiff bases and their complexes: a review," *Applied Organometallic Chemistry*, vol. 30, no. 10, pp. 810–817, 2016.

[4] J. R. Anacona, N. Noriega, and J. Camus, "Synthesis, characterization and antibacterial activity of a tridentate Schiff base derived from cephalothin and sulfadiazine, and its transition metal complexes," *Spectrochimica Acta - Part A: Molecular and Biomolecular Spectroscopy*, vol. 137, pp. 16–22, 2015.

[5] R. Nair, A. Shah, S. Baluja, and S. Chanda, "Synthesis and antibacterial activity of some Schiff base complexes," *Journal of the Serbian Chemical Society*, vol. 71, no. 7, pp. 733–744, 2006.

[6] G. B. Bagihalli, P. G. Avaji, S. A. Patil, and P. S. Badami, "Synthesis, spectral characterization, in vitro antibacterial, antifungal and cytotoxic activities of Co(II), Ni(II) and Cu(II) complexes with 1,2,4-triazole Schiff bases," *European Journal of Medicinal Chemistry*, vol. 43, no. 12, pp. 2639–2649, 2008.

[7] N. Raman, A. Sakthivel, and K. Rajasekaran, "Synthesis and Spectral Characterization of Antifungal Sensitive Schiff Base Tran- sition Metal Complexes," *Mycobiology*, vol. 35, no. 3, pp. 150–153, 2007.

[8] B. S. Creaven, E. Czeglédi, M. Devereux et al., "Biological activity and coordination modes of copper(II) complexes of Schiff base-derived coumarin ligands," *Dalton Transactions*, vol. 39, pp. 10854–10865, 2010.

[9] A. Jarrahpour, D. Khalili, E. De Clercq, C. Salmi, and J. M. Brunel, "Synthesis, antibacterial, antifungal and antiviral activity evaluation of some new bis-Schiff bases of isatin and their derivatives," *Molecules*, vol. 12, pp. 1720–1730, 2007.

[10] S. K. Bharti, S. K. Patel, G. Nath, R. Tilak, and S. K. Singh, "Synthesis, characterization, DNA cleavage and in vitro antimicrobial activities of copper(II) complexes of Schiff bases containing a 2,4-disubstituted thiazole," *Transition Metal Chemistry*, 2010.

[11] M. Manjunatha, V. H. Naik, A. D. Kulkarni, and S. A. Patil, "DNA cleavage, antimicrobial, antiinflammatory anthelmintic activities, and spectroscopic studies of Co(II), Ni(II), and Cu(II) complexes of biologically potential coumarin Schiff bases," *Journal of Coordination Chemistry*, vol. 64, no. 24, pp. 4264–4275, 2011.

[12] S. Amer, N. El-Wakiel, and H. El-Ghamry, "Synthesis, spectral, antitumor and antimicrobial studies on Cu(II) complexes of purine and triazole Schiff base derivatives," *Journal of Molecular Structure*, vol. 1049, pp. 326–335, 2013.

[13] P. G. Cozzi, "Metal-Salen Schiff base complexes in catalysis: practicalaspects," *Chemical Society Reviews*, vol. 33, pp. 410–421, 2004.

[14] R. K. Singh, A. Kukrety, R. C. Saxena, G. D. Thakre, N. Atray, and S. S. Ray, "Novel Triazine Schiff Base-Based Cationic Gemini Surfactants: Synthesis and Their Evaluation as Antiwear, Antifriction, and Anticorrosive Additives in Polyol," *Industrial and Engineering Chemistry Research*, vol. 55, no. 9, pp. 2520–2526, 2016.

[15] N. Sathya, G. Raja, N. Padma Priya, and C. Jayabalakrishnan, "Ruthenium(II) complexes incorporating tridentate schiff base ligands: Synthesis, spectroscopic, redox, catalytic and biological properties," *Applied Organometallic Chemistry*, vol. 24, no. 5, pp. 366–373, 2010.

[16] V. Arun, N. Sridevi, P. P. Robinson, S. Manju, and K. K. M. Yusuff, "Ni(II) and Ru(II) Schiff base complexes as catalysts for the reduction of benzene," *Journal of Molecular Catalysis A: Chemical*, vol. 304, no. 1-2, pp. 191–198, 2009.

[17] J. L. Segura, M. J. Mancheño, and F. Zamora, "Covalent organic frameworks based on Schiff-base chemistry: synthesis, properties and potential applications," *Chemical Society Review*, 2016.

[18] J. R. Anacona, J. Calvo, and O. A. Almanza, "Synthesis, Spectroscopic, and Magnetic Studies of Mono- and Polynuclear Schiff Base Metal Complexes Containing Salicylidene-Cefotaxime Ligand," *International Journal of Inorganic Chemistry*, pp. 1–7, 2013, http://dx.doi.org/10.1155/2013/108740.

[19] I. Rama and R. Selvameena, "Synthesis, structure analysis, antibacterial and in vitro anti-cancer activity of new Schiff base and its copper complex derived from sulfamethoxazole," *Journal of Chemical Sciences*, vol. 127, no. 4, pp. 671–678, 2015.

[20] S. Jyothi, K. Sreedhar, D. Nagaraju, and S. J. Swamy, "Synthesis and spectral investigation of Co(II), Ni(II), Cu(II) and Zn(II) complexes with novel N4 Ligands," *Canadian Chemical Transactions*, vol. 3, no. 4, pp. 368–380, 2015.

[21] S. P. Jose and S. Mohan, "FT-IR and FT-RAMAN investigations of nicotinaldehyde," *Spectrochimica Acta - Part A: Molecular and Biomolecular Spectroscopy*, vol. 64, no. 1, pp. 205–209, 2006.

[22] A. K. Ghosh, M. Mitra, A. Fathima et al., "Antibacterial and catecholase activities of Co(III) and Ni(II) Schiff base complexes," *Polyhedron*, vol. 107, pp. 1–8, 2016.

[23] H. F. Abd El-Halim, M. M. Omar, G. G. Mohamed, and M. A. El-Ela Sayed, "Spectroscopic and biological activity studies on tridentate Schiff base ligands and their transition metal complexes," *European Journal of Chemistry*, vol. 2, no. 2, pp. 178–188, 2011.

[24] M. A. Pfaller, S. A. Messer, L. Boyken et al., "Evaluation of the NCCLS M44-P Disk Diffusion Method for Determining Susceptibilities of 276 Clinical Isolates of Cryptococcus neoformans to Fluconazole," *Journal of Clinical Microbiology*, vol. 42, no. 1, pp. 380–383, 2004.

[25] M. Balouiri, M. Sadiki, and S. K. Ibnsouda, "Methods for in vitro evaluating antimicrobial activity: A review," *Journal of Pharmaceutical Analysis*, vol. 6, pp. 71–79, 2016.

[26] S. M. Bell, J. N. Pham, and G. T. Fisher, "Antibiotic susceptibility testing by the CDS method," in *The prince of Wales Hospital, South Eastern Area Laboratory Services, Randwick NSW 2031*, A manual for Medical and Veterinary Laboratories, Fifth edition, 2009.

[27] S. Hazra, A. Karmakar, M. D. F. C. Guedes da Silva, L. Dlháň, R. Boča, and A. J. L. Pombeiro, "Sulfonated Schiff base dinuclear and polymeric copper(II) complexes: crystal structures, magnetic properties and catalytic application in Henry reaction," *New Journal of Chemistry*, vol. 39, pp. 3424–3434, 2015.

[28] A. Soroceanu, M. Cazacu, S. Shova et al., "Copper(II) complexes with Schiff bases containing a disiloxane unit: Synthesis, structure, bonding features and catalytic activity for aerobic oxidation of benzyl alcohol," *European Journal of Inorganic Chemistry*, no. 9, pp. 1458–1474, 2013.

[29] E. Pahonţu, D.-C. Ilieş, S. Shova et al., "Synthesis, characterization, crystal structure and antimicrobial activity of copper(II) complexes with the Schiff base derived from 2-hydroxy-4-methoxybenzaldehyde," *Molecules*, vol. 20, no. 4, pp. 5771–5792, 2015.

[30] A. M. Naglah, H. M. Awad, M. A. Bhat, M. A. Al-omar, and A. E. E. Amr, "Microwave-Assisted Synthesis and Antimicrobial Activity of Some Novel Isatin Schiff Bases Linked to Nicotinic Acid via Certain Amino Acid Bridge," *Journal of chemistry*, vol. 2015, 2015, http://dx.doi.org/10.1155/2015/364841.

[31] D. Dey, G. Kaur, A. Ranjani et al., "A trinuclear zinc-Schiff base complex: Biocatalytic activity and cytotoxicity," *European Journal of Inorganic Chemistry*, no. 21, pp. 3350–3358, 2014.

[32] O. E. Sherif and N. S. Abdel-Kader, "DFT calculations, spectroscopic studies, thermal analysis and biological activity of supramolecular Schiff base complexes," *Arabian Journal of Chemistry*, 2015.

[33] H. Amiri Rudbari, M. R. Iravani, V. Moazam et al., "Synthesis, characterization, X-ray crystal structures and antibacterial activities of Schiff base ligands derived from allylamine and their vanadium(IV), cobalt(III), nickel(II), copper(II), zinc(II) and palladium(II) complexes," *Journal of Molecular Structure*, vol. 1125, pp. 113–120, 2016.

[34] S. Kundu, A. K. Pramanik, A. S. Mondal, and T. K. Mondal, "Ni(II) and Pd(II) complexes with new N,O donor thiophene appended Schiff base ligand: Synthesis, electrochemistry, X-ray structure and DFT calculation," *Journal of Molecular Structure*, vol. 1116, pp. 1–8, 2016.

[35] S. Jana, P. Bhowmik, M. Das, P. P. Jana, K. Harms, and S. Chattopadhyay, "Synthesis and characterisation of two double EE azido and thiocyanato bridged dimeric Cu(II) complexes with tridentate Schiff bases as blocking ligands," *Polyhedron*, vol. 37, no. 1, pp. 21–26, 2012.

[36] Z. Asadi, M. Asadi, and M. R. Shorkaei, "Synthesis, characterization and DFT study of new water-soluble aluminum(III), gallium(III) and indium(III) Schiff base complexes: Effect of

metal on the binding propensity with bovine serum albumin in water," *Journal of the Iranian Chemical Society*, vol. 13, no. 3, pp. 429–442, 2016.

[37] A. S. Gaballa, S. M. Teleb, M. S. Asker, E. Yalçin, and Z. Seferoğlu, "Synthesis, spectroscopic properties, and antimicrobial activity of some new 5-phenylazo-6-aminouracil-vanadyl complexes," *Journal of Coordination Chemistry*, vol. 64, no. 24, pp. 4225–4243, 2011.

[38] H. Abu Ali, H. Fares, M. Darawsheh, E. Rappocciolo, M. Akkawi, and S. Jaber, "Synthesis, characterization and biological activity of new mixed ligand complexes of Zn(II) naproxen with nitrogen based ligands," *European Journal of Medicinal Chemistry*, vol. 89, pp. 67–76, 2014.

[39] G. G. Mohamed, M. M. Omar, and A. M. Hindy, "Metal Complexes of Schiff Bases: Preparation, Characterization, and Biological Activity," *Turkey Journal of Chemistry*, vol. 30, pp. 361–382, 2006.

[40] S. I. Al-Resayes, "Kinetics analysis for non-isothermal decomposition γ-irradiated indium acetate," *Arabian Journal of Chemistry*, vol. 3, no. 3, pp. 191–194, 2010.

[41] H. P. Ebrahimi, J. S. Hadi, Z. A. Abdulnabi, and Z. Bolandnazar, "Spectroscopic, thermal analysis and DFT computational studies of salen-type Schiff base complexes," *Spectrochim Acta A*, vol. 117, pp. 485–492, 2014.

[42] M. Montazerozohori, S. Zahedi, A. Naghiha, and M. M. Zohour, "Synthesis, characterization and thermal behavior of antibacterial and antifungal active zinc complexes of bis (3(4-dimethylaminophenyl)-allylidene-1, 2-diaminoethane," *Materials Science and Engineering C*, vol. 35, no. 1, pp. 195–204, 2014.

[43] W. H. Mahmoud, G. G. Mohamed, and M. M. I. El-dessouky, "Synthesis , Characterization and in vitro Biological Activity of Mixed Transition Metal Complexes of Lornoxicam with 1,10-phenanthroline," *International Journal of Electrochemical Science*, vol. 9, pp. 1415–1438, 2014.

[44] A. H. Kianfar, H. Farrokhpour, P. Dehghani, and H. R. Khavasi, "Experimental and theoretical spectroscopic study and structural determination of nickel(II) tridentate Schiff base complexes," *Spectrochimica Acta - Part A: Molecular and Biomolecular Spectroscopy*, vol. 150, pp. 220–229, 2015.

[45] G. Grivani, V. Tahmasebi, A. D. Khalaji, V. Eigner, and M. Dušek, "Synthesis, characterization, crystal structure, catalytic activity in oxidative bromination, and thermal study of a new oxidovanadium Schiff base complex containing O, N-bidentate Schiff base ligand," *Journal of Coordination Chemistry*, vol. 67, no. 22, pp. 3664–3677, 2014.

[46] B. B. Mahapatra, R. Mishra, and A. K. Sarangi, "Synthesis, characterisation, XRD, molecular modelling and potential antibacterial studies of Co(II), Ni(II), Cu(II), Zn(II), Cd(II) and Hg(II) complexes with bidentate azodye ligand," *Journal of Saudi Chemical Society*, 2013.

[47] O. A. El-Gammal, G. M. A. El-Reash, T. A. Yousef, and M. Mefreh, "Synthesis, spectral characterization, computational calculations and biological activity of complexes designed from NNO donor Schiff-base ligand," *Spectrochimica Acta - Part A: Molecular and Biomolecular Spectroscopy*, vol. 146, pp. 163–176, 2015.

[48] S. A. Patil, S. N. Unki, A. D. Kulkarni, V. H. Naik, U. Kamble, and P. S. Badami, "Spectroscopic, invitro antibacterial, and antifungal studies of Co(II), Ni(II), and Cu(II) complexes with 4-chloro-3-coumarinaldehyde Schiff bases," *Journal of Coordination Chemistry*, vol. 64, no. 2, pp. 323–336, 2011.

[49] A. A. Abdel Aziz, A. N. M. Salem, M. A. Sayed, and M. M. Aboaly, "Synthesis, structural characterization, thermal studies, catalytic efficiency and antimicrobial activity of some M(II) complexes with ONO tridentate Schiff base N-salicylidene-o-aminophenol (saphH 2)," *Journal of Molecular Structure*, vol. 1010, pp. 130–138, 2012.

[50] M. Shakir, A. Abbasi, M. Faraz, and A. Sherwani, "Synthesis, characterization and cytotoxicity of rare earth metal ion complexes of N,N*l*-bis-(2-thiophenecarboxaldimine)-3,3*l*-diaminobenzidene, Schiff base ligand," *Journal of Molecular Structure*, vol. 1102, pp. 108–118, 2015.

[51] V. Sharma, E. K. Arora, and S. Cardoza, "4-Hydroxy-benzoic acid (4-diethylamino-2-hydroxy-benzylidene)hydrazide: DFT, antioxidant, spectroscopic and molecular docking studies with BSA," *Luminescence*, vol. 31, no. 3, pp. 738–745, 2016.

[52] N. Kavitha and P. Anantha Lakshmi, "Synthesis, characterization and thermogravimetric analysis of Co(II), Ni(II), Cu(II) and Zn(II) complexes supported by ONNO tetradentate Schiff base ligand derived from hydrazino benzoxazine," *Journal of Saudi Chemical Society*, 2015.

[53] N. K. Chaudhary and P. Mishra, "Spectral Investigation and in Vitro Antibacterial Evaluation of NiII and CuII Complexes of Schiff Base Derived from Amoxicillin and α-Formylthiophene (α ft)," *Journal of Chemistry*, vol. 2015, Article ID 136285, 2015.

[54] R. S. Joseyphus and M. S. Nair, "Antibacterial and Antifungal Studies on Some Schiff Base Complexes of Zinc (II)," *Mycobiology*, vol. 36, no. 2, pp. 93–98, 2008.

[55] A. A. Al-Amiery, Y. K. Al-Majedy, H. H. Ibrahim, and A. A. Al-Tamimi, "Antioxidant, antimicrobial, and theoretical studies of the thiosemicarbazone derivative Schiff base 2-(2-imino-1-methylimidazolidin-4-ylidene)hydrazinecarbothioamide (IMHC)," *Organic and Medicinal Chemistry Letters*, vol. 2, no. 4, 2012.

[56] M. Jesmin, M. M. Ali, M. S. Salahuddin, M. R. Habib, and J. A. Khanam, "Antimicrobial Activity of Some Schiff Bases Derived from Benzoin, Salicylaldehyde, Aminophenol and 2,4 Dinitrophenyl Hydrazine," *Mycobiology*, vol. 36, pp. 70–73, 2008.

Luminescent Iridium Complex-Peptide Hybrids (IPHs) for Therapeutics of Cancer: Design and Synthesis of IPHs for Detection of Cancer Cells and Induction of their Necrosis-Type Cell Death

Abdullah-Al Masum,[1] Yosuke Hisamatsu,[1] Kenta Yokoi,[1] and Shin Aoki ⓘ[1,2]

[1]*Faculty of Pharmaceutical Sciences, Tokyo University of Science, 2641 Yamazaki, Noda, Chiba 278-8510, Japan*
[2]*Imaging Frontier Center, Tokyo University of Science, 2641 Yamazaki, Noda, Chiba 278-8510, Japan*

Correspondence should be addressed to Shin Aoki; shinaoki@rs.noda.tus.ac.jp

Academic Editor: Viktor Brabec

Death receptors (DR4 and DR5) offer attractive targets for cancer treatment because cancer cell death can be induced by apoptotic signal upon binding of death ligands such as tumor necrosis factor-related apoptosis-inducing ligand (TRAIL) with death receptors. Cyclometalated iridium(III) complexes such as fac-Ir(tpy)$_3$ (tpy = 2-(4-tolyl)pyridine) possess a C_3-symmetric structure like TRAIL and exhibit excellent luminescence properties. Therefore, cyclometalated Ir complexes functionalized with DR-binding peptide motifs would be potent TRAIL mimics to detect cancer cells and induce their cell death. In this study, we report on the design and synthesis of C_3-symmetric and luminescent Ir complex-peptide hybrids (IPHs), which possess cyclic peptide that had been reported to bind DR5. The results of 27 MHz quartz-crystal microbalance (QCM) measurements of DR5 with IPHs and costaining experiments of IPHs and anti-DR5 antibody, suggest that IPHs bind with DR5 and undergo internalization into cytoplasm, possibly via endocytosis. It was also found that IPHs induce slow cell death of these cancer cells in a parallel manner to the DR5 expression level. These results indicate that IPHs may offer a promising tool as artificial luminescent mimics of death ligands to develop a new category of anticancer agents that detect and kill cancer cells.

1. Introduction

Death receptors (DRs) are often overexpressed on the cell membrane of cancer cells and bind with death ligands such as tumor necrosis factor-related apoptosis-inducing ligand (TRAIL) to send the cell extrinsic apoptotic signal [1]. TRAIL receptors comprise five categories, plasma membrane-expressed TRAIL-R1 (DR4), TRAIL-R2 (DR5), TRAIL-R3 (DcR1), TRAIL-R4 (DcR2), and a soluble receptor, osteoprotegerin (OPG). DR4 and DR5 contain death domain (DD) to transduce apoptotic signals and hence named as death receptors, whereas DcR1 and DcR2 are unable to induce cell death and regarded as decoy receptors [1–7]. TRAIL is a C_3-symmetric protein, consists of three monomeric units, and binds with three DRs [8, 9]. Upon binding of DR with TRAIL,

TRAIL-DR cluster sets up death-inducing signaling complex (DISC) at their cytoplasmic death domain (DD) and recruits adaptor protein (FADD) via death-effector domain (DED). The signaling activates procaspase-8 to caspase-8 and then caspase-3 in order to cleave multiple substrates to execute cell death [10]. The nonsignaling DcR1 does not contain cytoplasmic DD, and DcR2 is very similar to death receptors but contains a truncated form of DD. Therefore, both DcR1 and DcR2 are unable to assemble DISC [4–6].

Since death receptors are overexpressed in various types of cancer cells, TRAIL is capable of selectively inducing apoptosis of cancer cells with low cytotoxicity in normal cells [11–19]. Therefore, death receptors have been conceived as promising targets for the treatment and imaging of cancer cells. To date, only limited examples of artificial death

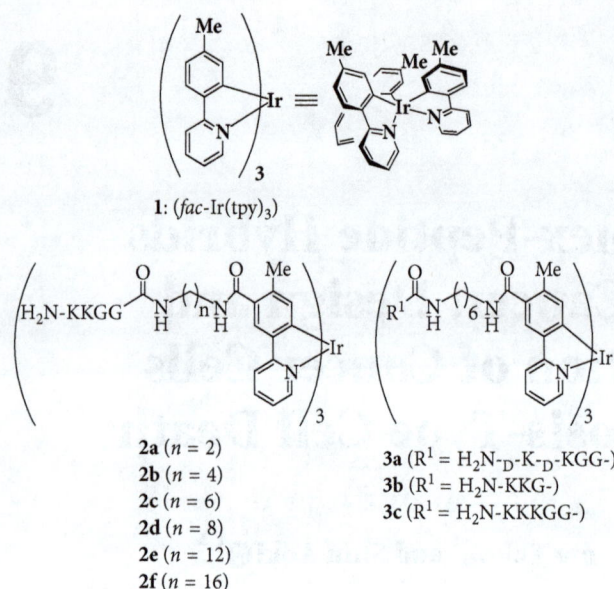

FIGURE 1: Ir complexes having cationic peptides.

receptor binders having TRAIL-like functionalities are reported. Representative examples include DR5 binding peptides [20–26], zinc-binding site peptide of TRAIL (RNSCWSKD that was screened out from TRAIL (227–234)) [27], and small molecular TRAIL mimics (bioymifi) [28].

Meanwhile, cyclometalated iridium (Ir(III)) complexes such as fac-Ir(tpy)$_3$ 1 (tpy = 2-(4-tolyl)pyridine) (Figure 1) draw increasing attention as one of the imaging tools to study extra- and intracellular events, in addition to organic light-emitting diodes (OLEDs) such as phosphorescent emitters [29–33] because of their significant stability and excellent photophysical properties under physiological conditions [31, 33–39]. Such types of Ir complex analogues have been widely applied to oxygen sensors [40, 41], chemosensors [42–47], and luminescent probes for biological systems [48–66]. These advantages originate from their high-luminescence quantum yields, the long luminescence lifetimes ($\tau \sim \mu s$) that can eliminate the short-lived autofluorescence ($\tau \sim ns$) from biological samples in cellular imaging, and significant stokes shift that minimizes self-quenching process [29–39]. We previously reported some examples of Ir complexes that can be functionalized as blue~green~red and white color emitters [67–71], pH sensors [68–70], photosensitizers [68–70], and cell death inducer of cancer cells [68–70]. Recently, we have reported on Ir complexes (2a-f and 3a-c) having cationic peptides (typically, H$_2$N-KKGG-) (Figure 1) as inducers and detectors of cell death of Jurkat cells (a human T-lymphoma cell line) [72–74]. These results suggest that Ir complexes are potential agents for the diagnosis and treatment of cancer and related diseases and even for mechanistic study of cell death processes.

Because cyclometalated iridium (Ir(III)) complexes such as fac-Ir(tpy)$_3$ 1 possess a C_3-symmetric structure like TRAIL (the top of Figure 2), it is hypothesized that these complexes could be good scaffolds to mimic TRAIL. In this manuscript, we report on the design and synthesis of some new C_3-symmetric tris-cyclometalated Ir complexes having cyclic peptides (Figure 2) [20–22], which had been reported

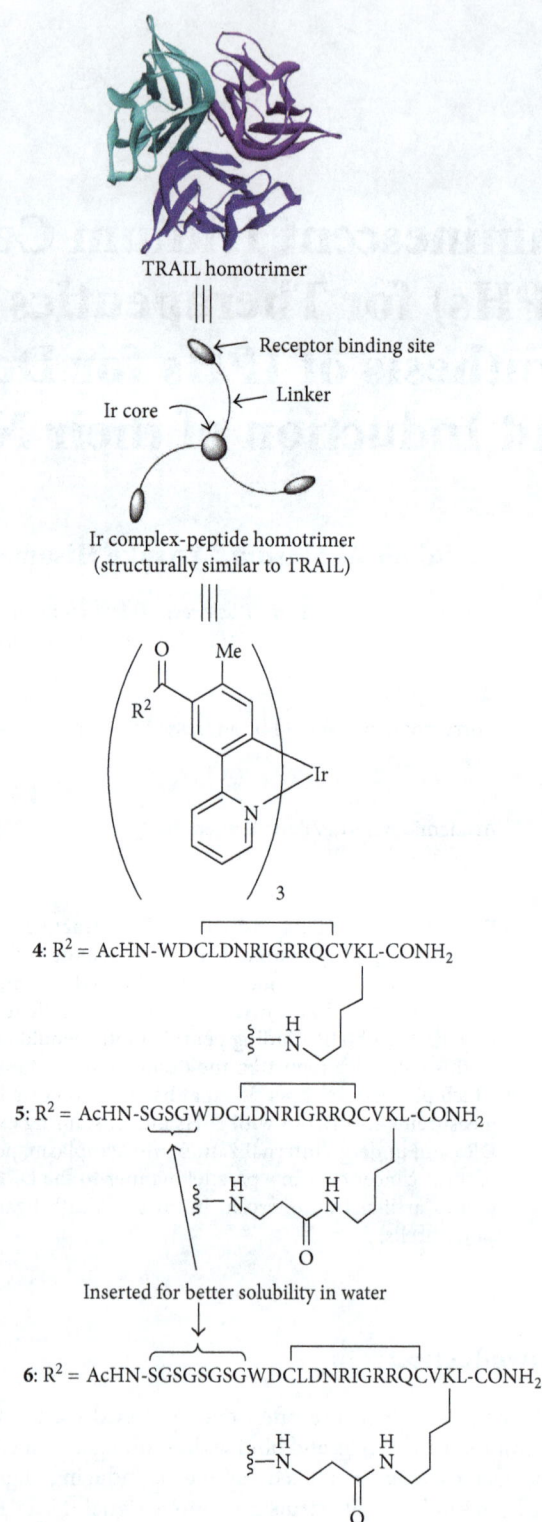

FIGURE 2: C_3-symmetric tris-cyclometalated Ir complex-peptide hybrids (IPHs).

to be able to bind DR5, for selective staining and induction of cell death of cancer cells. The Ir complexes 4–6 were synthesized by regioselective substitution reactions reported by us [75] and the successive coupling reactions with cyclic peptides [20–22]. Due to low solubility of 4 in water, Ser-Gly-Ser-Gly

(SGSG) was inserted at the *N*-terminus of the peptide parts of **5-6** to improve their solubility. The results of 27 MHz quartz-crystal microbalance (QCM) measurements of DR5 with **5** and **6** and costaining experiments of Jurkat cells with **5** and anti-DR5 antibody that **5** and **6** bind to DR5 at the different sites from the recognition sites of anti-DR5 antibody. Some cancer cell lines such as Jurkat cells, K562 cells, and Molt-4 cells were stained with **5** for luminescence microscopic observation, showing that Ir complexes exhibit green emitting spots localized inside the cell. In addition, it was turned out that **5** induces cell death of Jurkat cells more slowly than those by **2c-d** (**5** requires almost 24 h to induce considerable cell death, while **2c**, **2d**, **3a**, and **3c** induce cell death in a couple of hours). Binding of **5** to cancer cells and its cytotoxicity against cancer cells are dependent on the DR5 expression level of cancer cells. Further mechanistic studies suggest that cell death induced by **5** is necrotic type. Interestingly, the treatment of cancer cells with **5** and then with anti-DR5 antibody lowers the staining level on cell membrane of Jurkat cells by anti-DR5 antibody, indicating that DR5 undergoes endocytosis upon binding with **5**. Furthermore, it was found that DR5 moved from cytoplasm to the cell membrane or reproduced on the cell membrane after additional incubation for 6 h. Finally, detection of Jurkat cells spiked in bovine blood is demonstrated. To the best of our knowledge, **5** is the first example of artificial luminescent death ligand that detects cancer cells and induces their necrosis-type cell death.

2. Experimental Section

2.1. General Information. All reagents and solvents were purchased from commercial suppliers and were used without further purification, unless otherwise noted. MTT (3-(4,5-dimethyl-2-thiazolyl)-2,5-diphenyl-2*H*-tetrazolium bromide) was purchased from Dojindo. Z-VAD-FMK (Z-Val-Ala-Asp (OMe) fluoromethylketone) was purchased from the Peptide Institute. Necrostatin-1 and IM-54 were purchased from Enzo Life Sciences. Anti-DR5 antibody [DR5-01-1] (phycoery-thrin) (ab55863) was purchased from Abcam. TRAIL/Apo2L (human recombinant), chloroquine diphosphate, verapamil hydrochloride, and nicardipine hydrochloride were purchased from Wako Pure Chemical Industries. The oligomycin complex was purchased from Cayman Chemical Co., quinidine and 4-aminopyridine were purchased from TCI. CCCP (carbonyl cyanide 3-chlorophenylhydrazone), and bafilomycin A1 and amiloride hydrochloride were purchased from Sigma-Aldrich. Propidium iodide and NaN_3 were purchased from Nacalai Tesque. Annexin V-Cy3 was purchased from BioVision, Inc. All aqueous solutions were prepared using deionized and distilled water. UV spectra were recorded on a JASCO V-550 spectrophotometer, equipped with a temperature controller unit at $25 \pm 0.1°C$. Emission spectra were recorded on a JASCO FP-6200 and FP-6500 spectrometers. IR spectra were recorded on a Perkin-Elmer FT-IR spectrophotometer (Spectrum100). ^1H NMR (300 MHz) spectra were recorded on a JEOL Always 300 spectrometer. Tetramethylsilane (TMS) was used as an internal reference for ^1H measurements in $CDCl_3$ and CD_3OD, and 3-(Trimethylsilyl)-propionic-2,2,3,3-d_4 acid (TSP) sodium salt was used as an

external reference for ^1H NMR measurement in D_2O. Mass spectral measurements were performed on a JEOL JMS-SX102A and Varian TQ-FT. Luminescence imaging studies were performed using fluorescent microscope (Bio-revo, BZ-9000, Keyence). Thin-layer chromatographies (TLC) and silica gel column chromatographies were performed using Merck Art. 5554 (silica gel) TLC plate and Fuji Silysia Chemical FL-100D, respectively. HPLC experiments were carried out using a system consisting of two PU-980 intelligent HPLC pumps (JASCO, Japan), a UV-970 intelligent UV-visible detector (JASCO), a Rheodine injector (Model no. 7125), and a Chromatopak C-R6A (Shimadzu, Japan). For analytical HPLC, the Senshu Pak Pegasil ODS column (Senshu Scientific Co., Ltd.) ($4.6\varphi \times 250$ mm, No. 07051001) was used. For preparative HPLC, the Senshu Pak Pegasil ODS SP100 column (Senshu Scientific Co., Ltd.) ($20\varphi \times 250$ mm, No. 1302014G) was used. Lyophilization was performed with the freeze-dryer FD-5N (EYELA).

2.2. Synthesis of Ir Complexes and Peptide Units. Ir Complexes **1** and **7**: These complexes were synthesized according to our previously reported procedure [75].

Ir Complex **8**: A solution of Ir complex **7** (50 mg, 0.06 mmol), DIEA (190 μL, 1.08 mmol), and PyBOP (283 mg, 0.54 mmol) in distilled DMF (3 mL) was stirred for 10 min at room temperature, to which β-alanine ethyl ester (84 mg, 0.54 mmol) was added. The whole solution was stirred for 18 h and concentrated under reduced pressure. The remaining residue was extracted with $CHCl_3/H_2O$, dried over Na_2SO_4, concentrated under reduced pressure, and purified by silica gel column chromatography. (hexanes: $AcOEt = 2:3 \rightarrow 1:2 \rightarrow 1:4 \rightarrow 1:6$) to obtain Ir complex **8** as a yellow solid. This Ir complex (30 mg, 0.026 mmol) and 5 M LiOH (25 mg, 1.06 mmol) in H_2O/THF (1/1, 9 mL) was stirred at 60°C for 17 h, and then 2 N HCl (pH = 1) was added to form precipitate. The precipitate was filtrated and washed with H_2O to give Ir complex **8** as a yellow solid (26 mg, 41% from **7**). IR (ATR): $\nu = 3279, 2924, 2579, 1975, 1712, 1586, 1527, 1471, 1259, 1186, 1068, 1022, 892, 781,$ and 750 cm^{-1}. ^1H NMR (CD_3OD, 300 MHz): $\delta = 8.04$ (d, 3H, $J = 8.1$ Hz), 7.74 (m, 6H), 7.47 (d, 3H, $J = 5.4$ Hz), 6.98 (t, 3H, $J = 7.5$ Hz), 6.69 (s, 3H), 3.59 (t, 6H, $J = 5.1$ Hz), 2.63 (t, 6H, $J = 6.9$ Hz), and 2.12 (s, 9H) ppm. ESI-MS (m/z): calcd. for $C_{48}H_{45}IrN_6O_9 [M]^+$: 1042.28773 and found: 1042.28784.

Ir Complex **9** [72]: DIEA (63 mg, 0.5 mmol), PyBOP (126.6 mg, 0.24 mmol), and mono-Boc-protected hexamethylenediamine (103 mg, 0.5 mmol) were added to a solution of **7** (33 mg, 0.039 mmol) in distilled DMF (1 mL). The reaction mixture was stirred at room temperature for 36 h and then concentrated under reduced pressure. The remaining residue was purified by silica gel column chromatography ($CHCl_3/MeOH$, 1/0 to 50/1 to 0/1), gel permeation chromatography ($CHCl_3$), and recrystallization from hexanes/$CHCl_3$ to afford the Boc-protected **9** as a yellow solid. A mixture of TMSCl (21 mg, 0.46 mmol) and NaI (70 mg, 0.46 mmol) in CH_3CN (1 mL) was added to a suspension of Boc-protected **9** (21.6 mg, 15 μmol) in CH_3CN (3.3 mL). The mixture was stirred at room temperature for 10 min and sonicated for 2 min. The insoluble compound was centrifuged and washed with CH_3CN to give **9** as the HI salt. The product was purified by preparative HPLC (H_2O (0.1% TFA)/CH_3CN (0.1% TFA) = 80/20 to 50/50 (30 min), t_r = 21 min, 6.0 mL/min), followed by lyophilization to give **9** as a yellow solid (41 mg, 62% as 3 TFA salt). IR (ATR): ν = 2934, 2862, 2035, 1674, 1600, 1531, 1472, 1425, 1262, 1199, 1069, 892, 781, and 721 cm^{-1}. ^{1}H NMR (D_2O, 300 MHz): δ = 8.04 (d, 3H, J = 8.41 Hz), 7.82–7.76 (m, 6H), 7.69 (d, 3H J = 5.4 Hz), 7.09–7.05 (t, 3H J = 6.0 Hz), 6.58 (s, 3H), 3.38–3.33 (m, 6H), 3.01–2.96 (t, 6H, J = 6.9 Hz), 2.11 (s, 9H), 1.68–1.62 (m, 12H), and 1.43 (m, 12H) ppm. ESI-MS (m/z): calcd for $C_{57}H_{73}IrN_9O_3$ [M + H]$^+$: 1124.54656 and found: 1124.54487.

Fmoc-Leu-NH-SAL-Trt(2-Cl)-resin: NH_2-SAL-Trt (2-Cl)-resin (500 mg, 0.54 mmol/g) was suspended in DMF (2.5 mL), to which a mixture of Fmoc-Leu-OH (3 eq.), DIC (4 eq.), and HOBt (4 eq.) was added. After stirring for 2 h, the residue was filtrated, washed with DMF, CH_2Cl_2, and ether, and dried in vacuo. The amount of Fmoc-Leu-OH loaded on the resin was determined by UV absorption of the Fmoc derivative at 301 nm (ε_{301nm} = 7800 M^{-1}·cm^{-1}) after treatment with 20% (v/v) piperidine in DMF. Yield: 593 mg (loading: 0.43 mmol/g resin).

Cyclic peptide **CP1**: Fmoc-protecting group of Fmoc-Leu-NH-SAL-Trt(2-Cl)-resin (500 mg, 0.43 mmol) was deprotected by treatment with 20% piperidine in DMF. Each Fmoc-Xaa-OH (4 eq.) was coupled at 45°C for 1 h to the Fmoc-deprotected resin in the presence of DIC (4 eq.) and HOBt (8 eq.) in DMF (2.5 mL). After the last deprotection, the *N*-terminus was acetylated using Ac_2O (4 eq.) and DIEA (4 eq.) in DMF (2.5 mL). The peptide was cleaved from the resin and also deprotected using a mixture of TFA/H_2O/TIPS/thioanisole (85/2.5/10/2.5). After 4 h of stirring, the resin was filtered and washed with TFA. After evaporation of TFA, the precipitation was obtained by adding cooled Et_2O and collected by centrifugation. The crude product was purified by preparative HPLC (H_2O (0.1% TFA)/CH_3CN (0.1% TFA) = 80/20→50/50 (30 min) t_r = 11 min, 1 mL/min), lyophilized to give **CP1** as a linear form. The linear form of **CP1** was subjected to cyclization using 0.1 mM aq. NH_4HCO_3 (1 mg/1 mL). After cyclization reaction for 12–24 h, the solution was concentrated and the remaining residue was purified again by preparative HPLC (H_2O (0.1% TFA)/CH_3CN (0.1% TFA) = 80/20→50/50 (30 min), t_r = 7.5 min, 1 mL/min), lyophilized to give **CP1** as white powder (217 mg, 26%). IR (ATR): ν = 3217, 2992, 2291, 2167, 2097, 2031, 1980, 1637, 1508, 1434, 1173, 1118, 839, 800, and 723 cm^{-1}. ^{1}H NMR (D_2O, 300 MHz): δ = 7.62 (m, 1H), 7.46 (m, 1H), 7.23 (m, 2H), 7.22 (m, 1H), 5.59 (s, 1H), 4.29 (m, 17H), 3.91 (s, 1H), 3.34 (s, 1H), 2.80 (s, 8H), 2.55 (m, 3H), 2.49 (m, 3H), 2.34 (m, 6H), 1.95 (m, 5H), 1.59 (m, 7H), 1.41 (m, 31H), 1.29 (m, 2H), and 0.98 (m, 50H) ppm. ESI-MS (m/z): calcd for $C_{85}H_{142}N_{30}O_{23}S_6$ [M + 2H]$^{2+}$: 1008.01809 Found: 1008.01709.

Cyclic peptides **CP2** and **CP3** were prepared according to the same procedure described for cyclic peptide **CP1**.

Cyclic peptide **CP2**: white powder (371 mg, 29% over two steps). HPLC: (H_2O (0.1% TFA)/CH_3CN (0.1%

TFA) = 90/10→60/40 (30 min), t_r = 16 min, 1 mL/min). IR (ATR): ν = 3283, 2939, 2167, 2027, 1980, 1651, 1533, 1431, 1177, 1119, 1044, 891, 840, 798, 721, and 655 cm^{-1}. ^{1}H NMR (D_2O, 300 MHz): δ = 7.49 (m, 1H), 7.38 (m, 1H), 7.20 (m, 2H), 7.09 (m, 1H), 4.97 (m, 24H), 3.86 (m, 12H), 3.17 (m, 24H), 2.09 (m, 24H), 1.62 (m, 29H), and 0.90 (m, 38H) ppm. ESI-MS (m/z): calcd. for $C_{95}H_{159}N_{34}O_{29}S_2$ [M + 3H]$^{3+}$: 768.71559 and found: 768.71568.

Cyclic peptide **CP3**: white powder (18 mg, 23% over two steps). HPLC: (H_2O (0.1% TFA)/CH_3CN (0.1% TFA) = 90/10→60/40 (30 min), t_r = 11 min, 1 mL/min). IR (ATR): ν = 3287, 3071, 2965, 2547, 2045, 1778, 1651, 1532, 1440, 1290, 1155, 1042, 845, 700, and 578 cm^{-1}. ^{1}H NMR (D_2O, 300 MHz): δ = 7.51 (m, 1H), 7.38 (m, 1H), 7.20 (m, 2H), 7.10 (m, 1H), 4.97 (m, 31H), 3.86 (m, 18H), 3.17 (m, 27H), 2.09 (m, 24H), 1.62 (m, 29H), and 0.90 (m, 38H) ppm. ESI-MS (m/z): calcd. for $C_{105}H_{175}N_{38}O_{35}S_2$ [M + 3H]$^{3+}$: 864.41581 and found: 864.41735.

Ir Complex 4: EDC (832 mg, 4.34 mmol) and NHS (500 mg, 4.34 mmol) was added to a solution of **7** (120 mg, 0.14 mmol) in DMF (12 mL). The reaction mixture was stirred for 24 h at room temperature and concentrated under reduced pressure. After addition of $CHCl_3$, organic layer was washed with sat. NH_4Cl. The combined organic layer was washed with H_2O, dried over Na_2SO_4, filtered, and concentrated under reduced pressure to give NHS ester of **7** as a yellow solid (110 mg, 67%). IR (ATR): ν = 2941, 2162, 1705, 1581, 1519, 1380, 1293, 1206, 1066, 976, 886, and 648 cm^{-1}. ^{1}H-NMR ($CDCl_3$, 300 MHz): δ = 8.45 (s, 3H), 8.00 (d, 3H, J = 7.8), 7.73 (t, 3H, J = 7.8), 7.42 (d, 3H, J = 5.7), 6.97 (t, 3H, J = 6.0), 6.82 (s, 3H), 2.89 (s, 12H), and 2.42 (s, 9H) ppm. ESI-MS (m/z): calcd for $C_{51}H_{39}IrN_6O_{12}$ [M]$^+$: 1120.22552. Found: 1120.22641. NHS ester of Ir complex **7** (1 mg, 0.8 μmol) was added to a solution of **CP1** (5.39 mg, 2.6 μmol) and DIEA (4.7 μL, 0.026 mmol) in DMF (100 μL) and stirred for 24 h at room temperature in the dark. After that, 0.1% TFA H_2O was added to the reaction mixture and the crude product was purified by preparative HPLC (H_2O (0.1% TFA)/CH_3CN (0.1% TFA) = 80/20→50/50 (30 min), t_r = 21 min, (1 mL/min)), lyophilized to give **4** as a yellow powder (3.8 mg, 43% from **7**). IR (ATR): ν = 32879, 2320, 1981, 1638, 1535, 1426, 1264, 1201, 1134, 923, 835, 800, and 722 cm^{-1}. ^{1}H NMR (DMSO, 300 MHz): δ = 8.26 (s, 3H), 8.22 (s, 7H), 8.15 (m, 10H), 7.83 (m, 8H), 7.40 (m, 8H), 7.30 (m, 12H), 7.13 (m, 21H), 6.96 (s, 13H), 6.55 (s, 3H), 4.23 (m, 6H), 4.12 (m, 14H), 3.11 (s, 3H), 2.51 (m, 29H), 2.49 (m, 10H), 2.48 (m, 24H), 2.22 (m, 12H), 1.76 (s, 7H), 1.47 (m, 36H), 1.3 (s, 3H), and 0.80 (m, 41) ppm. ESI-MS (m/z); calcd for $C_{294}H_{450}IrN_{93}O_{72}S_6$ [M + 6H]$^{6+}$: 1137.2104 and found: 1137.20847.

Ir Complex 5: EDC (137 mg, 0.71 mmol) and NHS (83 mg, 0.71 mmol) was added to a solution of **8** (25 mg, 0.023 mmol) in DMF (2.5 mL). The resulting solution was stirred for 24 h at room temperature. The reaction mixture was concentrated under reduced pressure. After addition of $CHCl_3$, the organic layer was washed with sat. NH_4Cl. The combined organic layer was washed with H_2O, and then dried over Na_2SO_4, filtered, and concentrated under reduced pressure to give NHS ester of **8** as a yellow solid (20 mg,

62%). IR (ATR): $\nu = 3743$, 3318, 2925, 1815, 1780, 1706, 1471, 1375, 1260, 1204, 1131, 1067, 994, 781, and 648 cm^{-1}. ^{1}H-NMR (CDCL$_3$, 300 MHz): δ 7.94 (d, 3H, $J = 8.1$), 7.73 (s, 3H), 7.58 (t, 3H, $J = 7.8$), 7.40 (d, 3H, $J = 5.1$), 6.84 (t, 3H, $J = 6.3$), 6.67 (s, 3H), 6.50 (t, 3H, $J = 6.6$), 3.81 (d, 6H, $J = 5.1$), 2.95 (t, 6H, $J = 6.3$), 2.91 (s, 12H), and 2.23 (s, 9H). ESI-MS (m/z): calcd for C$_{60}$H$_{54}$IrN$_9$O$_{15}$ [M]$^+$: 1333.33686 and found: 1333.33747. NHS ester of Ir complex **8** (6 mg, 0.0044 mmol) was added to a solution of **CP2** (31.06 mg, 0.013 mmol) and DIEA (23 μL, 0.134 mmol) in DMF (600 μL) and stirred for 24 h at room temperature in the dark. The reaction mixture was diluted with 0.1% TFA H$_2$O and purified by preparative HPLC (H$_2$O (0.1% TFA)/CH$_3$CN (0.1% TFA) = 80/20→50/50 (30 min), t_r = 10 min, 1 mL/min), lyophilized to give **5** as a yellow powder (15.45 mg, 27% from **8**). IR (ATR): $\nu = 3282$, 3074, 2964, 2054, 1980, 1639, 1531, 1472, 1425, 1261, 1181, 915, 799, and 720 cm^{-1}. ^{1}H NMR. (D$_2$O, 300 MHz): $\delta = 7.68$ (s, 3H), 7.46 (s, 3H), 7.08 (m, 6H), 6.89 (m, 3H), 6.68 (s, 3H), 3.79 (m, 18H), 3.73 (m, 7H), 3.71 (m, 11H), 3.25 (m, 18H), 3.23 (m, 12H), 3.18 (m, 13H), 2.73 (m, 5H), 2.24 (m, 193H), 2.23 (m, 20H), 2.00 (m, 11H), 1.63 (m, 45), 1.35 (m, 50H) 1.15 (m, 12H), and 0.89 (m, 74H) ppm. ESI-MS (m/z): calcd. for C$_{333}$H$_{513}$IrN$_{108}$O$_{93}$S$_6$ [M + 6H]$^{6+}$: 1316.94104. Found: 1316.94569.

Ir complex **6** was prepared according to the same procedure described for **5**.

Ir Complex **6**: yellow powder (8.3 mg, 21% from **8**). HPLC: (H$_2$O (0.1% TFA)/CH$_3$CN (0.1% TFA) = 90/10→60/40 (30 min), t_r = 12 min, 1 mL/min). IR (ATR): $\nu = 3383$, 2963, 2014, 1984, 1638, 1535, 1475, 1262, 1200, 1057, 836, 799, and 720 cm^{-1}. ^{1}H NMR (D$_2$O, 300 MHz): $\delta = 7.72$ (s, 3H), 7.42 (s, 3H), 7.17 (m, 6H), 6.95 (m, 3H), 6.78 (s, 3H), 3.86 (m, 23H), 3.71 (m, 38H), 3.23 (m, 42H), 2.73 (m, 31H), 2.07 (m, 12H), 1.92 (m, 70H), 1.62 (m, 69H), 1.34 (m, 132H), and 0.88 (m, 120H) ppm. ESI-MS (m/z): calcd for C$_{363}$H$_{563}$IrN$_{120}$O$_{111}$S$_6$ [M + 8H]$^{8+}$: 1096.00145 and found: 1096.00136.

2.3. UV/Vis Absorption and Luminescence Spectra Measurements. UV/Vis spectra were recorded on a JASCO V-550 UV/Vis spectrophotometer equipped with a temperature controller, and emission spectra were recorded on a JASCO FP-6200 spectrofluorometer at 25°C. Before the luminescence measurements, sample aqueous solutions were degassed by Ar bubbling for 10 min in quartz cuvettes equipped with Teflon septum screw caps. Concentrations of all the Ir complexes in stock solutions (DMSO) were determined based on a molar extinction coefficient of 380 nm ($\varepsilon_{380nm} = 1.08 \pm 0.07 \times 10^4$ M^{-1}·cm^{-1}). Quantum yields for luminescence (Φ) were determined by comparing with the integrated corrected emission spectrum of a quinine sulfate standard, whose emission quantum yield in 0.1 M H$_2$SO$_4$ was assumed to be 0.55 (excitation at 366 nm). Equation (1) was used to calculate the emission quantum yields, in which Φ_s and Φ_r denote the quantum yields of the sample and reference compounds, η_s and η_r are the refractive indexes of the solvents used for the measurements of the sample and reference, A_s and A_r are the absorbance of the sample and the reference, and I_s and I_r stand for the integrated areas under

the emission spectra of the sample and reference, respectively (all of the Ir compounds were excited at 366 nm for luminescence measurements in this study):

$$\Phi_s = \frac{\Phi_r \left(\eta_s^2 A_r I_s \right)}{\left(\eta_r^2 A_s I_r \right)}. \tag{1}$$

The luminescence lifetimes of sample solutions were measured on a TSP1000-M-PL (Unisoku, Osaka, Japan) instrument by using THG (355 nm) of Nd:YAG laser, Minilite I (Continuum, CA, USA), at 25°C in degassed aqueous solutions. The R2949 photomultiplier were used to monitor the signals. Data were analyzed using the nonlinear least-squares procedure.

2.4. 27 MHz Quartz Crystal Microbalance (QCM) Analysis. QCM analysis was performed on an Affinix-Q4 apparatus (Initium Inc., Japan). The clean Au (4.9 mm^2) electrode equipped on the quartz crystal was incubated with an aqueous solution of 3,3'-dithiodipropionic acid (3 mM, 4 μL) at room temperature for 60 min. After washing with distilled water, the surface was activated by a mixture of EDC·HCl (0.52 M) and N-hydroxy succinimide (0.87 M) for 30 min, washed with distilled water, and then treated with DR5 (100 μg/mL, 4 μL) at room temperature for 60 min. After washing with distilled water, an aqueous solution of 1 M ethanolamine (5 μL) was added as a blocking reagent. After washing with distilled water, cell was filled with phosphate-buffered saline (PBS) (500 μL). The apparent binding constants (K_{app}) for IPHs with DR5 in PBS were calculated from the decrease in frequency. The nonspecific response was subtracted from frequency decrease curve to obtain apparent complexation constants K_{app} and dissociation constants K_d (=1/K_{app}).

2.5. Cell Culture. All cell lines (Jurkat, Molt-4, and K562 cells) were cultured in RPMI 1640 medium supplemented with 10% heat-inactivated fetal calf serum (FCS), L-glutamine, HEPES (2-[4-(2-hydroxyethyl)-1-piperazinyl]ethanesulfonic acid, pK_a = 7.5), 2-mercaptoethanol, and penicillin/streptomycin and monothioglycerol (MTG) in a humidified 5% CO$_2$ incubator at 37°C.

2.6. MTT Assay. Jurkat cells (1.0×10^5 cells/mL) were incubated in 1% DMSO, 10% FCS RPMI 1640 medium (MTG free) containing solution of Ir complexes **4–6** (0–75 μM) under 5% CO$_2$ at 37°C for 1 to 24 h in 96-well plates (BD Falcon), then 0.5% MTT reagent in PBS buffer (10 μL) was added to the cells. After incubation under 5% CO$_2$ at 37°C for 4 h, formazan lysis solution (10% SDS in 0.01 N HCl) (100 μL) was added and incubated overnight under same conditions, followed by measurement of absorbance at 570 nm by a microplate reader (Bio-Rad). MTT assay of Molt-4 cells and K562 cells with Ir complexes was also performed according to the same procedure described above.

2.7. MTT Assay in the Presence of Caspase Inhibitor (Z-VAD-FMK) and Necroptosis Inhibitor (Necrostatin-1) and Oxidative Stress Induce Necrosis Inhibitor (IM-54). Jurkat cells

$(1.0 \times 10^5$ cells/mL) were incubated in 10% FCS RPMI 1640 medium (MTG-free) containing a solution Z-VAD-fmk $(15 \mu M)$/necrostatin-1 $(30 \mu M)$/IM-54 $(10 \mu M)$/(under 5% CO_2 at 37°C for 1 h in 96-well plates (BD Falcon). Then, Ir complexes $(75 \mu M)$ were added and incubated under 5% CO_2 at 37°C for 24 h, and then 0.5% MTT reagent in PBS buffer $(10 \mu L)$ was added to the cells and incubated under 5% CO_2 at 37°C for 4 h. A formazan lysis solution (10% SDS in 0.01 N HCl) $(100 \mu L)$ was added, and the resulting solution incubated overnight under the same conditions, followed by measurement of absorbance at 570 nm with a microplate reader (Bio-Rad).

2.8. Fluorescent Microscopy Studies of Jurkat Cells, K562 Cells, and Molt-4 Cells with Ir Complexes. Jurkat cells, K562 cells, or Molt-4 cells $(1.0 \times 10^6$ cells/mL) were incubated in the absence or presence of Ir complexes in 10% FCS RPMI 1640 medium (MTG free) for specified time under 5% CO_2 at 37°C. The cells were then washed twice with ice-cold PBS with 0.1% NaN_3 and 0.5% FCS and taken on a Greiner CELLview™ petri dish $(35 \times 10$ mm) and mounted on fluorescent microscope for observation (Biorevo, BZ-9000, Keyence) (excitation: 377 ± 25 nm; emission: 520 ± 35 nm; FF01 filter).

2.9. Fluorescent Microscopy Studies of Jurkat Cells with Ir Complexes in the Presence of Inhibitors. Jurkat cells $(1.0 \times 10^6$ cells/mL) were incubated with the given inhibitors in RPMI 1640 medium (MTG free) with 10% FCS under 5% CO_2 at 37°C for 1 h, and then Ir complexes were added and then again incubated at 37°C for 24 h. The cells were then washed twice with ice-cold PBS containing 0.5% FCS and 0.1% NaN_3 and incubated with PI in PBS at room temperature for 10–15 min. The cells were then again washed with PBS buffer and observed by fluorescent microscope (Biorevo, BZ-9000, Keyence) (excitation: 377 ± 25 nm; emission: 520 ± 35 nm; FF01 filter).

2.10. Propidium Iodide (PI) Staining. The cells were incubated with Ir complexes for specified time, washed with PBS buffer, and then incubated with PI in PBS buffer at room temperature for 10–15 minutes. The cells were then again washed with PBS buffer and observed by fluorescent microscope (Biorevo, BZ-9000, Keyence) (excitation: 540 ± 25 nm; emission: 605 ± 55 nm; TRICT filter) or analyzed by flow cytometer (Beckman Coulter Gallios Flow Cytometer, detector: FL2, excitation: 488 nm, emission: 575 ± 20 nm).

2.11. Anti-DR5 Antibody Staining. Given cells were incubated with anti-DR5 antibody at 4°C for 15 min on ice. The cells were then washed twice with ice-cold PBS containing 0.5% FCS and 0.1 % NaN_3 and were mounted on fluorescent microscope (Biorevo, BZ-9000, Keyence) (excitation: 540 ± 25 nm; emission: 605 ± 55 nm; TRICT filter) or analyzed by flow cytometer (Beckman Coulter Gallios Flow Cytometer, detector: FL2, excitation: 488 nm; emission: 575 ± 20 nm).

2.12. Annexin V-Cy3 Staining. Jurkat cells were incubated with Ir complexes for 2 h, washed with PBS buffer, and then suspended in 1X binding buffer. Annexin V-Cy3 was added to the cell suspension and then incubated at room temperature in the dark for 5–10 minutes. The cells were then washed with PBS buffer and observed by fluorescent microscope (Biorevo, BZ-9000, Keyence) (excitation 540 ± 25 nm, emission 605 ± 55 nm, TRICT filter).

2.13. Flow Cytometry Analysis of Staining and Cell Death Induction Assay. Jurkat cells, K562 cells, or Molt-4 cells $(3.0 \times 10^5$ cells) were incubated in the absence or the presence of Ir complexes in 10% FCS RPMI 1640 medium (MTG free) for the specified time under 5% CO_2 at 37°C. After that, the cells were washed twice with ice-cold FACS buffer and then suspended in 450 μl FACS buffer. The cells were analyzed by flow cytometer (Beckman Coulter Gallios Flow Cytometer, detector: FL2, excitation: 488 nm, emission: 575 ± 20 nm) to detect PI staining or anti-DR5 antibody staining (detector: FL10, excitation: 405 nm, emission: 550 ± 40 nm) to detect Ir complexes staining).

3. Results and Discussion

3.1. Design and Synthesis of Ir Complex-Peptide Hybrids (IPHs). Synthesis of the Ir complex-peptide hybrids (IPHs) **4–6** is shown in Figure 3. The Vilsmeier reaction of **1** (*fac*-Ir (tpy)$_3$) and the following Pinnick oxidation [72, 73, 75] gave **7**. Condensation of **7** with β-alanine ethyl ester hydrochloride and the following ester hydrolysis yielded **8**. Both **7** and **8** were converted to the corresponding N-hydroxy succinimide (NHS) esters and then reacted with the peptide units, **CP1**, **CP2**, and **CP3** that had been prepared by Fmoc solid-phase peptide synthesis, to afford **4–6**, respectively. Because **4** is poorly soluble in water, a hydrophilic Ser-Gly-Ser-Gly (H_2N-SGSG-CO) sequence was incorporated to the N-terminus of cyclic peptide **CP1** to afford **CP2-3**. All the Ir complexes **4–6** were purified by reversed-phase HPLC column with a continuous gradient of H_2O (0.1% TFA)/CH_3CN (0.1%TFA) and lyophilized to give yellow powders as the corresponding TFA salts. It should be mentioned that negligible conversion of the facial form of **4–6** to the corresponding meridional form was observed during their synthesis and the following biological assays.

3.2. UV/Vis and Luminescence Spectra of IPHs. UV/Vis and luminescence spectra of Ir complexes **2c**, **4**, **5**, and **6** $(10 \mu M)$ in DMSO at 25°C are shown in Figure 4, and their photophysical data are summarized in Table 1. The concentrations of the Ir complexes in stock solutions (DMSO) were determined by the molar extinction coefficient at 380 nm $(\varepsilon_{380nm} = 1.43 \pm 0.03 \times 10^4 \cdot M^{-1} \cdot cm^{-1})$ of **4**, **5**, and **6**, which are almost identical to those of typical Ir(tpy)$_3$ derivatives having peptides **2a–f** and **3a–c** (Figure 1), as we previously reported [72, 73]. The strong absorption bands at 270–300 nm were assigned to the $^1\pi$-π^* transition of tpy ligands and weak shoulder bands at 320–450 nm were assigned as spin-allowed singlet-to-singlet metal-to-ligand charge

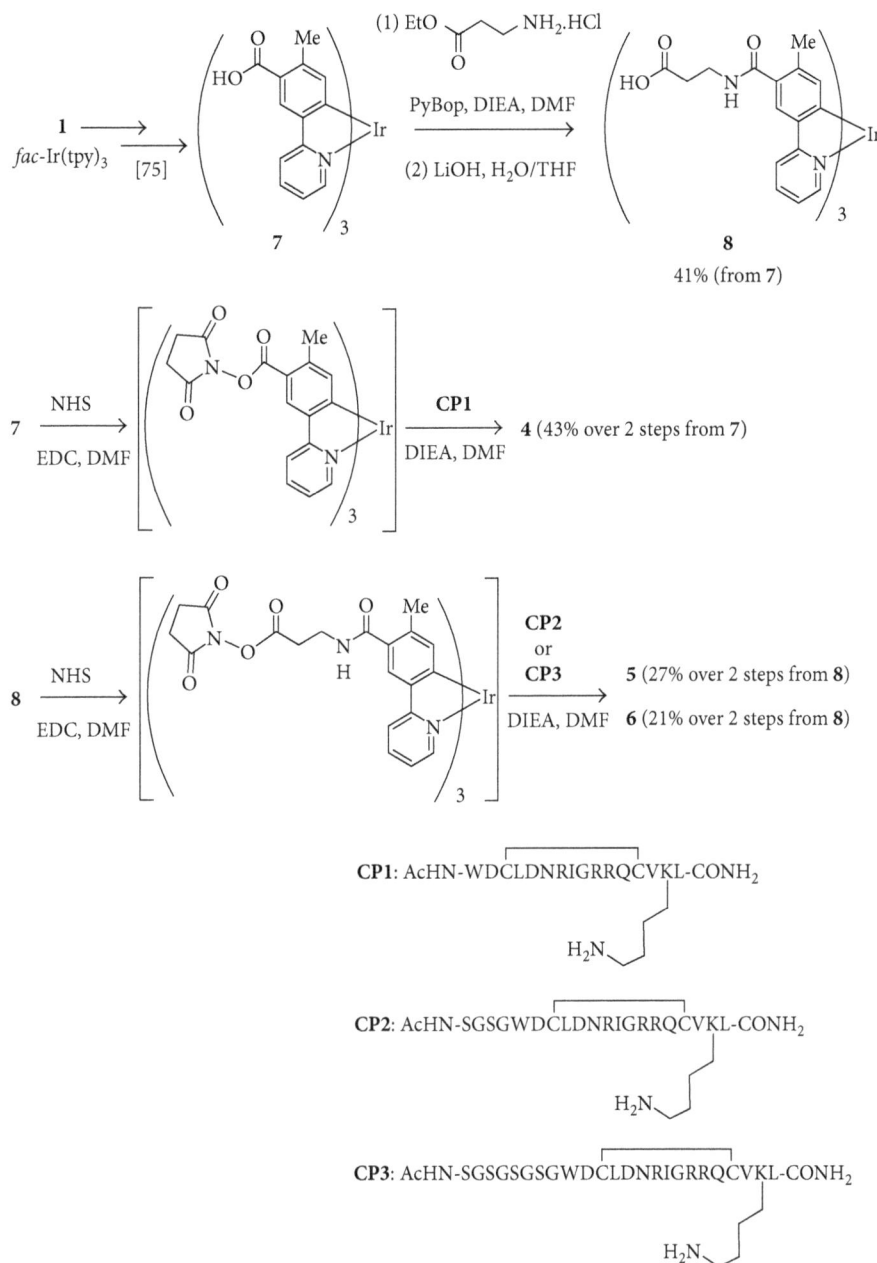

FIGURE 3: Synthesis of the Ir complex-peptide hybrids (IPHs).

transfer (^{1}MLCT) transitions, spin-forbidden singlet-to-triplet (^{3}MLCT) transitions, and $^3\pi$-π^* transitions. Strong green emission of 4, 5, and 6 is observed with emission maxima at ca. 506 nm, which are almost same as that of 2c (Figure 4(b)). The luminescence quantum yields (Φ) of 4, 5, and 6 were determined to be 0.39, 0.33, and 0.36, respectively, and their luminescence lifetimes (τ) are 1.1–1.3 μs, which are almost same as those of 1–3.

3.3. Determination of Complexation Properties of IPHs with DR5. First, complexation of IPHs with DR5 was checked by 27 MHz quartz-crystal microbalance (QCM). DR5 was immobilized on a sensor chip, to which IPHs were added on a certain time interval. Upon addition of IPHs, frequency

change was observed (Figure 5) indicating the interaction of IPHs with DR5. The apparent complexation constants (K_{app}) (and dissociation constant, K_d) for TRAIL, 5, and 6 with DR5 were determined to be $(2.3 \pm 0.05) \times 10^8$ M^{-1} ($K_d = 4.3 \pm 0.1$ nM), $(3.8 \pm 0.1) \times 10^5$ M^{-1} ($K_d = 2.7 \pm 0.1$ μM), and $(4.0 \pm 0.2) \times 10^5$ M^{-1} ($K_d = 2.5 \pm 0.1$ μM), respectively, assuming 1 : 1 complexation (Table 2). Negligible interaction was observed for 9, which lacks the receptor-binding peptide (Figure 6) and 2c that contains a KKGG peptide (Figure 1) [72, 73].

3.4. Cancer Cell Death Induced by IPHs, as Evaluated by MTT Assay and Fluorescence Microscopy. Next, cell death inducing activity of Ir complexes 4–6 against Jurkat cells was

(a)

(b)

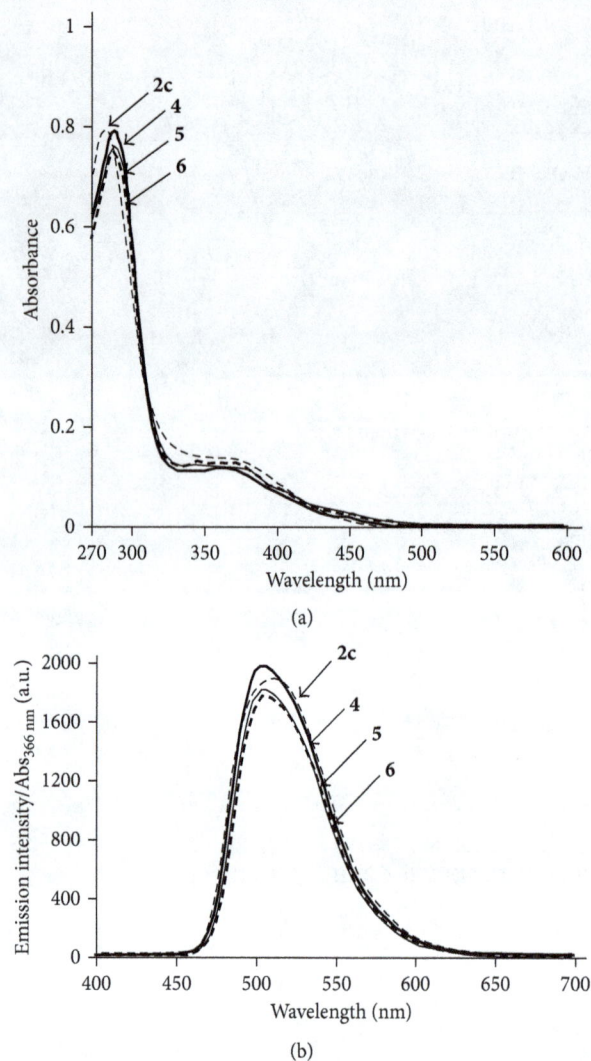

FIGURE 4: UV/Vis spectra of (a) **2c** (dashed curve), **4** (bold curve), **5** (plain curve), and **6** (bold dashed curve). Emission spectra of (b) **2c** (dashed curve), **4** (bold curve), **5** (plain curve), and **6** (bold dashed curve), in degassed DMSO at 25°C ([Ir complex] = 10 μM, excitation at 366 nm) (a.u. is the arbitrary unit).

TABLE 1: Photophysical properties of Ir complexes **2c**, **4**, **5**, and **6** in degassed DMSO at 25°C ([Ir complex] = 10 μM, excitation at 366 nm).

Compound	λ_{max} (absorption) (nm)	λ_{max} (emission) (nm)	Φ^a	τ^b (μs)
2c	280, 362	509	0.55	1.7
4	285, 363	505	0.39	1.1
5	285, 361	506	0.33	1.3
6	286, 360	506	0.36	1.2

[a]Quinine sulfate in 0.1 M H_2SO_4 ($\Phi = 0.55$) was used as a reference. [b]Lifetime of luminescence emission.

evaluated by MTT assay (MTT = 3-(4,5-dimethyl-2-thiazolyl)-2,5-diphenyl-2H-tetrazolium bromide). Jurkat cells were incubated with the given concentrations of **5** or **6** in 10% FCS (fetal calf serum) RPMI 1640 (MTG free) medium at 37°C and treated with MTT reagent. It was observed that the induction of cell death of Jurkat cells by **5**

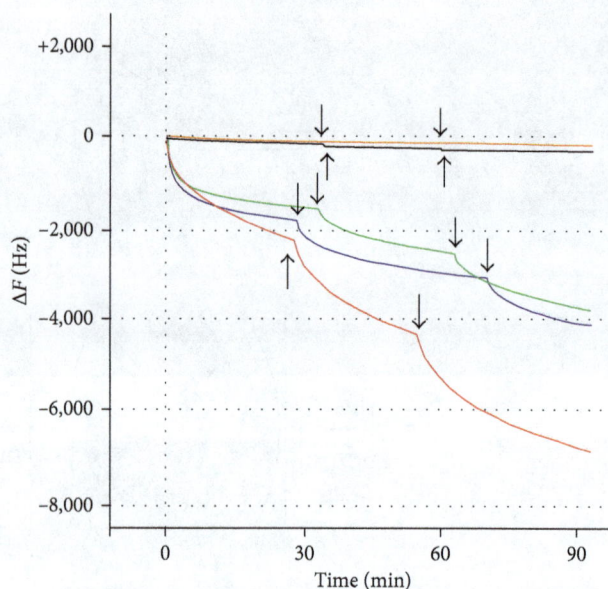

FIGURE 5: Time course of frequency change (ΔF (Hz)) of IPHs-DR5 complexation. Conditions: temperature, 25°C; solvent, phosphate-buffered saline (PBS). An aliquot of solutions of TRAIL (red curve) (200 μg/mL), **5** (green curve), **6** (blue curve), **9** (orange curve), and **2c** (black curve) ([Ir complex] = 10 mM solution in DMSO) was added to DR5 fixed on the sensor chip. Plain arrows indicate the time when solutions of these analytes were added to DR5.

TABLE 2: Complexation constants of IPHs (assuming 1 : 1 complexation).

Analyte	K_{app} (M^{-1})	K_d
TRAIL	$(2.3 \pm 0.05) \times 10^8$	4.3 ± 0.1 nm
5	$(3.8 \pm 0.1) \times 10^5$	2.7 ± 0.1 μM
6	$(4.0 \pm 0.2)10^5$	2.5 ± 0.1 μM
9	$<10^4$	>100 μM
2c	$<10^4$	>100 μM

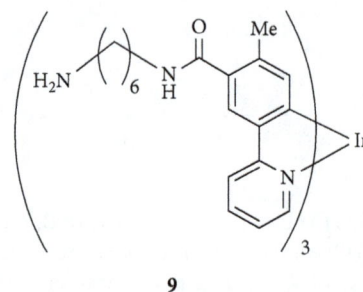

9

FIGURE 6: Ir complex having no peptide.

and **6** is much slower than that by our previous Ir complexes **2c**, **2d**, and **3c**, which induce cell death of Jurkat cells in 1 h. Because both **5** and **6** interact with DR5 to the same extent, as described in Figure 5 and Table 2, the following assays were carried out with **5**. Jurkat cells were observed by luminescence microscopy after incubation with **5** for 1 h, 6 h, 12 h, and 24 h and stained with propidium iodide (PI) to check their cell death. As summarized in Figures 7 and 8, **5** requires ca. 24 h to induce considerable cell death

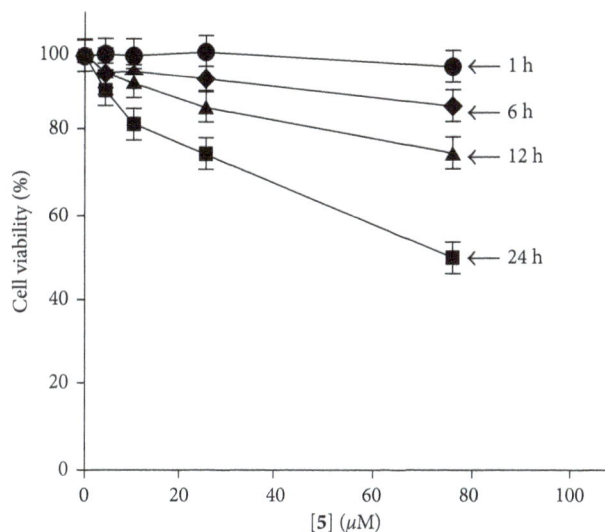

FIGURE 7: The results of MTT assay: cell viability of Jurkat cells after incubation in the presence of **5** (5–75 μM) for 1 h (filled circles), 6 h (filled diamonds), 12 h (filled triangles), and 24 h (filled squares) at 37°C.

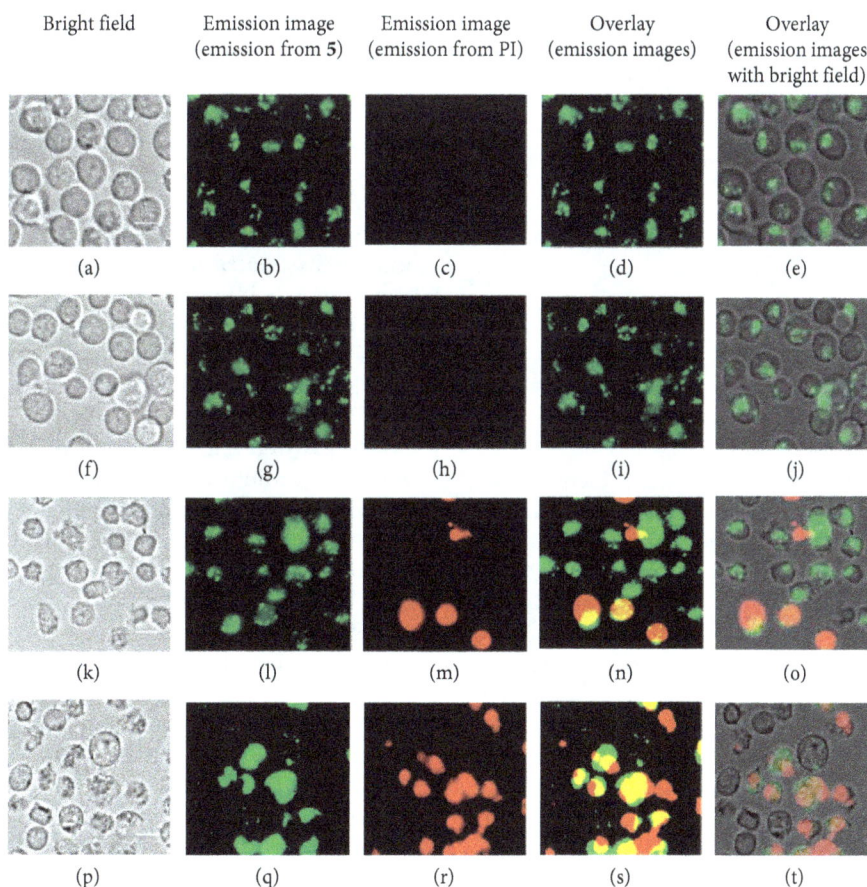

FIGURE 8: Time lapse luminescence microscopy images (Biorevo, BZ-9000, Keyence) of Jurkat cells (×40) treated with **5** (75 μM) at 37°C. Cell deaths were confirmed by staining with propidium iodide (PI), (a–e) after incubation for 1 h, (f–j) after incubation for 6 h, (k–o) after incubation for 12 h, and (p–t) after incubation for 24 h. Scale bar (white) = 10 μm.

(ca. 50% cell death at [**5**] = 72 μM). For comparison, negligible **4**-induced cell death, even after incubation for 16 h, was possibly due to its low solubility in water (Figure S1 in Supplementary Materials).

3.5. Luminescence Staining of Jurkat Cells with IPHs. Slow induction of cell death by IPHs allowed us to conduct luminescence staining experiments of Jurkat cells. Figures 9(a)–9(c) and 9(d)–9(f) indicate that

Bright field	Emission image	Overlay

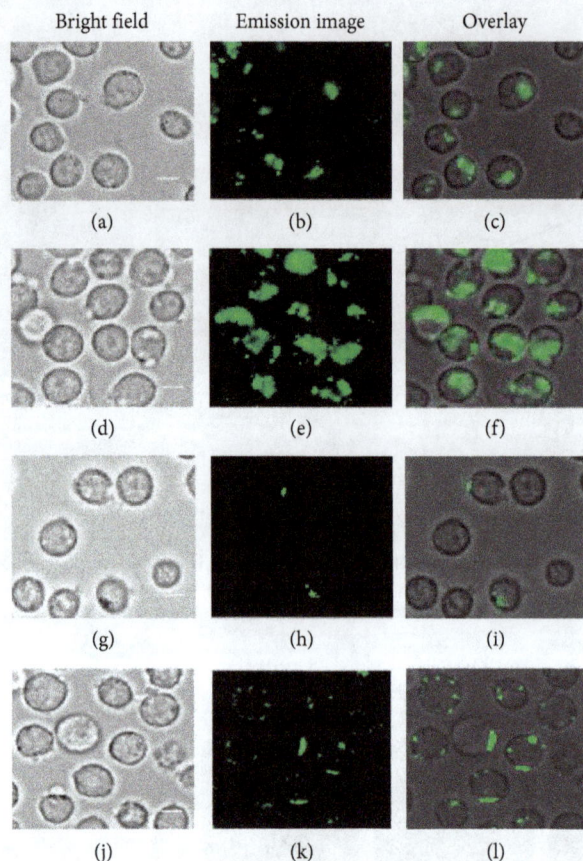

FIGURE 9: Luminescence microscopy images (Biorevo, BZ-9000, Keyence) of Jurkat cells (×40) stained with **5**. (a–c) Jurkat cells after incubation with **5** (5 μM) at 37°C for 1 h; (d–f) Jurkat cells after incubation with **5** (10 μM) at 37°C for 1 h; (g–i) Jurkat cells after incubation with **5** (20 μM) at 4°C for 1 h; (j–l) Jurkat cells after incubation with NaN$_3$ (5 mM) at 4°C for 15 min and then with **5** (5 μM) at 37°C for 1 h. Scale bar (white) = 10 μm.

5 (5 and 10 μM) binds to Jurkat cells after incubation at 37°C for 1 h. Passive uptake of **5** through the cell membrane is unlikely, due to the facts that negligible emission was observed after incubation at 4°C for 1 h ([**5**] = 20 μM) (Figures 9(g)–9(i)) and that the pretreatment with NaN$_3$ (a metabolic inhibitor) at 4°C for 15 min ([NaN$_3$] = 5 mM, a concentration required to inhibit metabolic activity) exhibits the localization of **5** on the cell membrane (Figures 9(j)–9(l)).

Competitive staining of Jurkat cells with **5** and DR5 binding peptide [20–22] (**CP1**, Figure 3) was conducted. Jurkat cells were incubated with **CP1** for 1 h at 37°C to which **5** was added and incubated again for 1 h at 37°C. Green emission from the cells was considerably reduced by the pretreatment with **CP1** (100 μM) than those treated with only **5** (Figure S2 in Supplementary Materials). Similar results were obtained by flow cytometry (Figure S3 in Supplementary Materials), implying that **5** competes with **CP1** for DR5.

3.6. Costaining of Jurkat Cells by IPHs and Anti-DR5 Antibody and Observation of the Movement of DR5 between Cell Surface

and Cytoplasm. In Figures 11(aa)–11(ee), red emission was observed on the cell membrane of Jurkat cells that were treated with anti-DR5 antibody (conjugated with a red color fluorochrome) at 4°C for 15 min (Method a in Figure 10). Next, we observed luminescence images of Jurkat cells that were treated with anti-DR5 antibody at 4°C for 15 min and then incubated with **5** at 37°C for 1 h (Method b in Figure 10). In Figures 11(bb) and 11(bc), green emission from **5** and red emission from anti-DR5 antibody were observed, respectively. The overlays of Figures 11(bb) and 11(bc) show yellow spots (Figures 11(bd) and 11(be)), suggesting that anti-DR5 antibody and **5** stain the same or similar area on the cells. Besides, it was found that there is negligible difference between emission intensity (from anti-DR5 antibody) of Jurkat cells treated with only anti-DR5 antibody and then **5** (Figure S4 in Supplementary Materials), suggesting that both of **5** and anti-DR5 antibody bind with DR5, possibly at the different sites.

Next, the order of the treatment of Jurkat cells with anti-DR5 antibody and **5** was reversed. In this experiment, Jurkat cells were treated with **5** (5 μM) and then with anti-DR5 antibody (Method c in Figure 10). In this case, very weak red emission was observed from Jurkat cells (Figures 11(ca)–11(ce)), suggesting that the expression level of DR5 on cell membrane was decreased by the treatment with **5**. The increase in the concentration of **5** to 10 μM resulted in the considerable decrease in the red emission from anti-DR5 antibody (Figures 11(da)–11(de)). Similar results were observed in flow cytometry assay, in which Jurkat cells treated with only anti-DR5 antibody exhibit high emission intensity (red curve in Figure S5 in Supplementary Materials).

In the previous experiments, it was hypothesized that DR5 is internalized from the cell membrane to the cell cytoplasm after binding with IPH. To check this hypothesis, Jurkat cells were incubated with **5** at 37°C for 1 h, washed with fresh medium, and incubated again in fresh medium at 37°C for 1 or 6 h (Method d in Figure 10) (Figures 11(ea)–11(ee) for 1 h incubation and Figures 11(fa)–11(fe) for 6 h incubation). Interestingly, the red emission from anti-DR5 antibody was restored after an additional incubation for 6 h, as displayed in Figures 11(fa)–11(fe), showing a good contrast to Figures 11(da)–11(de) obtained without incubation in fresh medium for 6 h. In flow cytometry assay, red emission intensity from anti-DR5 antibody in cells was reduced after 1 h incubation with **5** and then anti-DR5 antibody in comparison to Jurkat cells treated with only anti-DR5 antibody. After incubation of Jurkat cells with **5** and then again in fresh medium for 6 h, emission from anti-DR5 antibody was restored (Figure S6 in Supplementary Materials).

Our assumption based on the results of these experiments is summarized in Figure 12. (i) Upon incubation of Jurkat cells with IPH at 37°C for 1 h, IPH-DR5 complex undergoes internalization into the cytoplasm from the cell surface (Figure 12(a)). (ii) As a result, the binding of anti-DR5 antibody to the cell membrane becomes weak (Figure 12(b)). (iii) After incubation of Jurkat cells (in which IPH-DR5 complexes is internalized in the cytoplasm)

Method (a)

Jurkat cells $\xrightarrow[\substack{\text{At 4°C} \\ \text{for 15 min}}]{\substack{\text{Anti-DR5} \\ \text{antibody}}}$ Microscopic observation (Figures 11(aa)–11(ae))

Method (b)

Jurkat cells $\xrightarrow[\substack{\text{At 4°C} \\ \text{for 15 min}}]{\substack{\text{Anti-DR5} \\ \text{antibody}}}$ $\xrightarrow[\substack{\text{At 37°C} \\ \text{for 1h}}]{\textbf{5}}$ Microscopic observation (Figures 11(ba)–11(be))

Method (c)

Jurkat cells $\xrightarrow[\substack{\text{At 37°C} \\ \text{for 1h}}]{\textbf{5} \text{ (5 and 10 } \mu\text{M)}}$ $\xrightarrow[\substack{\text{At 4°C} \\ \text{for 15 min}}]{\substack{\text{Anti-DR5} \\ \text{antibody}}}$ Microscopic observation (Figures 11(ca)–11(ce) and 11(da)–11(de))

Method (d)

Jurkat cells $\xrightarrow[\substack{\text{At 37°C} \\ \text{for 1 h}}]{\textbf{5}}$ $\xrightarrow[\substack{\text{At 37°C} \\ \text{for 1 or 6 h}}]{\substack{\text{Fresh medium} \\ \text{(RPMI 1640)}}}$ $\xrightarrow[\substack{\text{At 4°C} \\ \text{for 15 min}}]{\substack{\text{Anti-DR5} \\ \text{antibody}}}$ Microscopic observation (Figures 11(ea)–11(ee)) for 1 h incubation and Figures 11(fa)–11(fe) for 6 h incubation)

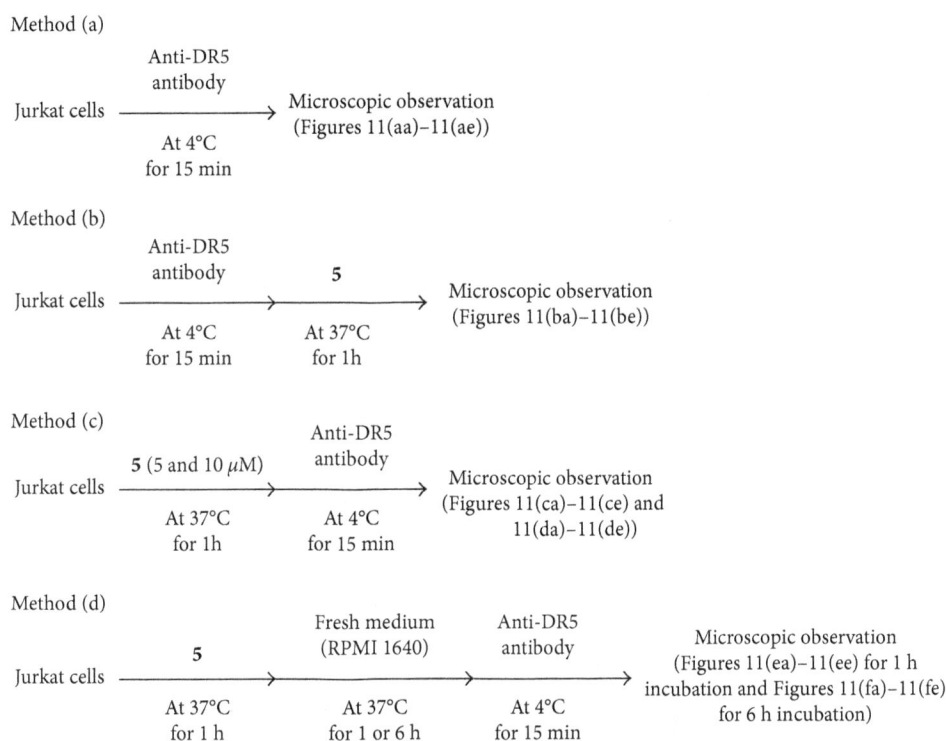

FIGURE 10: Costaining assay protocol.

in fresh medium at 37°C for 1 or 6 h, a considerable amount of DR5 is restored on the cell membrane (Figure 12(c)), and then (iv) anti-DR5 antibody is able to bind DR5 (Figure 12(d)).

To assess the affinity of IPH to DR5 expressed on cancer cell membrane, we carried out cell staining of different types of cancer cell lines that express different levels of DR5. DR5 expression of Jurkat cells, K562 cells, and Molt-4 cells was evaluated by staining with anti-DR5 antibody and flow cytometer analysis. It was found that Jurkat cells express comparatively high level of DR5 than K562 cells and Molt-4 cells, as confirmed by staining with anti-DR5 antibody (the second left in Figure 13). These three cell lines were incubated with 5 (5 μM and 10 μM) at 37°C for 1 h and then analyzed by flow cytometer. Figure 13 shows that, Jurkat cells are highly stained by 5, while K562 cells and Molt-4 cells are weakly stained. Luminescence microscopic observation of different types of cancer cell lines also show similar results (Figures S7(a)–S7(i) in Supplementary Materials) and Ir complex 9 having no peptide [72] (Figure 6) negligibly stains Jurkat cells (Figures S7(j)–S7(l) in Supplementary Materials). Together with the results of the aforementioned QCM experiments (Figure 2), these facts strongly suggest that the binding of 5 is dependent on the DR5 expression level of cancer cells.

Relationship between cell death inducing activity of IPH and DR5 expression of cancer cells was studied. Flow cytometry assay was conducted with Jurkat cells, K562 cells, and Molt-4 cells that express different level of DR5, as mentioned above. These three cell lines were incubated with 5 (25/75 μM) at 37°C for 24 h, to which PI (30 μM) was added to stain the dead cells for analysis by flow cytometer.

As summarized in Figure 14, cell death induction of these three cell lines is parallel to expression level of DR5.

3.7. Mechanistic Studies of Cell Death Induced by IPH. The aforementioned results strongly suggest that 5 is able to detect Jurkat cells and induce their cell death via complexation with DR5. The stability of 5 in RPMI 1640 (MTG Free) medium (Incubation at 37°C for 24 h) was checked, as shown in Figure S8 in the Supplementary Materials. Since it is well established that DR5 initiates apoptosis signal after binding with TRAIL or other reported artificial TRAIL mimics, it was initially presumed that the cell death induced by 5 would be apoptosis. MTT assay of Jurkat cells with 5 was carried out in the presence of Z-VAD-FMK (15 μM), which is a broad caspase inhibitor [76], necrostation-1 (30 μM), which is a necroptosis inhibitor [77], and IM-54 (10 μM), which is an inhibitor of oxidative stress induced necrosis [78]. However, these three drugs negligibly inhibited the cell death induced by 5 (Figures S9 and S10 in Supplementary Materials). On the other hand, TRAIL-induced cell death was inhibited by Z-VAD-FMK almost completely (Figures S9 and S10 in Supplementary Materials). In addition, the staining experiments of the dead Jurkat cells with annexin V-Cy3 [79, 80], which is a well-known reagent to detect apoptosis, disclosed that only few cells were stained by annexin V-Cy3. These facts have allowed us to conclude that it is unlikely that 5 induces apoptosis of Jurkat cells.

Aforementioned studies suggest that 5 internalizes into cells by DR5-mediated endocytosis. Therefore, 5 may interact with cytoplasmic organelles or interfere and/or

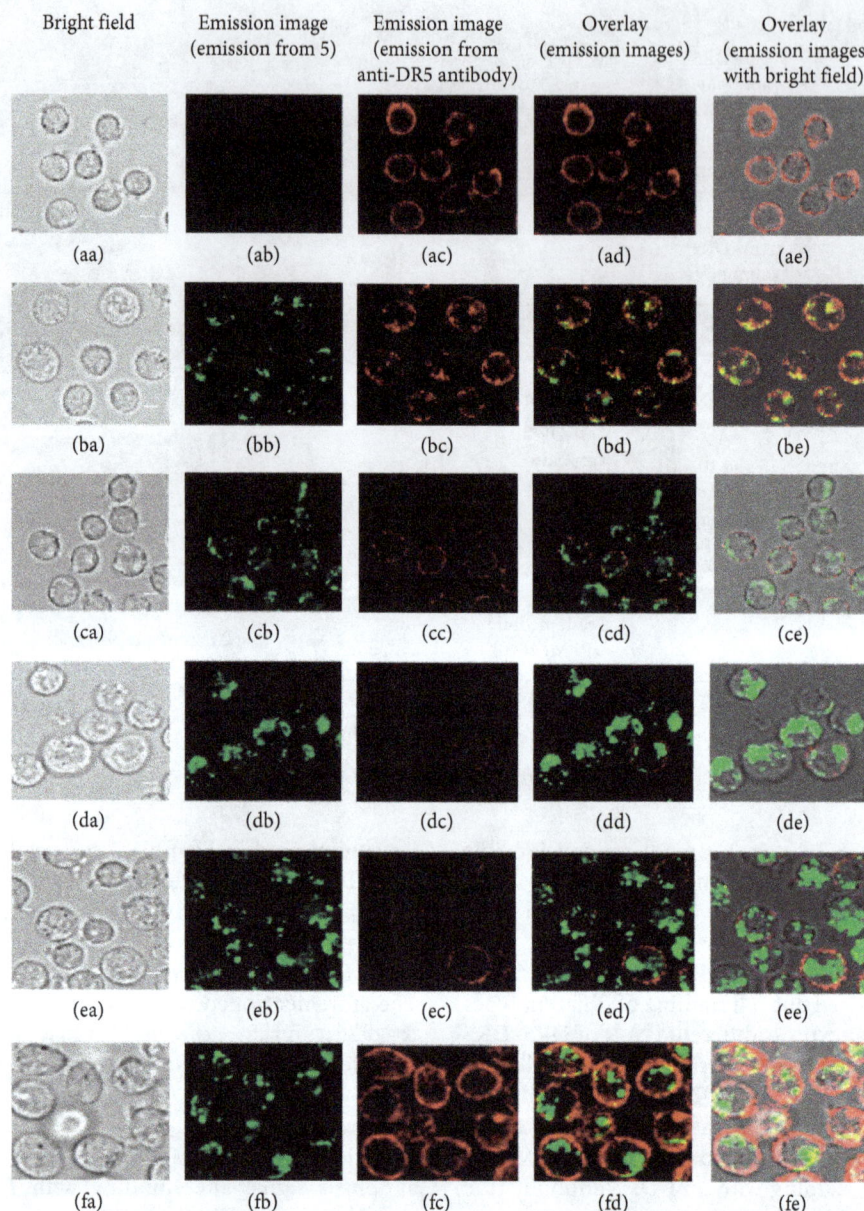

FIGURE 11: Luminescence microscopic images of (Biorevo, BZ-9000, Keyence) of Jurkat cells stained with **5** and anti-DR5 antibody obtained by protocol presented in Figure 10. (aa–ae) Jurkat cells incubated with anti-DR5 antibody (15 μg/mL) at 4°C for 15 min. (ba–be) Jurkat cells incubated with anti-DR5 antibody (15 μg/mL) at 4°C for 15 min and then with **5** (5 μM) at 37°C for 1 h. (ca–ce) Jurkat cells incubated with **5** (5 μM) at 37°C for 1 h and then with anti-DR5 antibody (15 μg/mL) at 4°C for 15 min. (da–de) Jurkat cells incubated with **5** (10 μM) at 37°C for 1 h and then with anti-DR5 antibody (15 μg/mL) at 4°C for 15 min. (ea–ee) Jurkat cells incubated with **5** (10 μM) at 37°C for 1 h and then in fresh medium for 1 h (2 h in total) and then with anti-DR5 antibody (15 μg/mL) at 4°C for 15 min. (fa–fe) Jurkat cells incubated with **5** (10 μM) at 37°C for 1 h and then in fresh medium for 6 h (7 h in total) and then with anti-DR5 antibody (15 μg/mL) at 4°C for 15 min. Scale bar (white) = 10 μm.

activate any cellular events to induce cell death. In order to examine this mechanism, we evaluated the cell death in the presence of the inhibitors of several cellular events or metabolisms, channel blockers, or receptor antagonist, namely, carbonyl cyanide 3-chlorophenylhydrazone (CCCP) (a mitochondrial uncoupling reagent) [81, 82], amiloride (Na+ channel blocker, inhibitors of macropinocytosis) [83], chloroquine (inhibitors of autophagy) [84], bafilomycin A1 (an inhibitor of vacular ATPase (V-ATPase) [85], oligomycin

(an inhibitor of the mitochondrial F1/F0-ATP synthase to cause ATP depletion) [86], 4-aminopyridine (K+ channel blocker), verapamil (L-type voltage-operated Ca2+ channel blocker) [87–89], nicardipine, (L-type voltage-operated Ca2+ channel blocker) [87–89], and quinidine, (Na+ and K+ channel blocker, antagonist of α-adrenergic receptors, and an inhibitor of the mitochondrial uptake of Ca2+) [90–94], according to our previous studies [72, 73]. Jurkat cells were incubated with these inhibitors at their

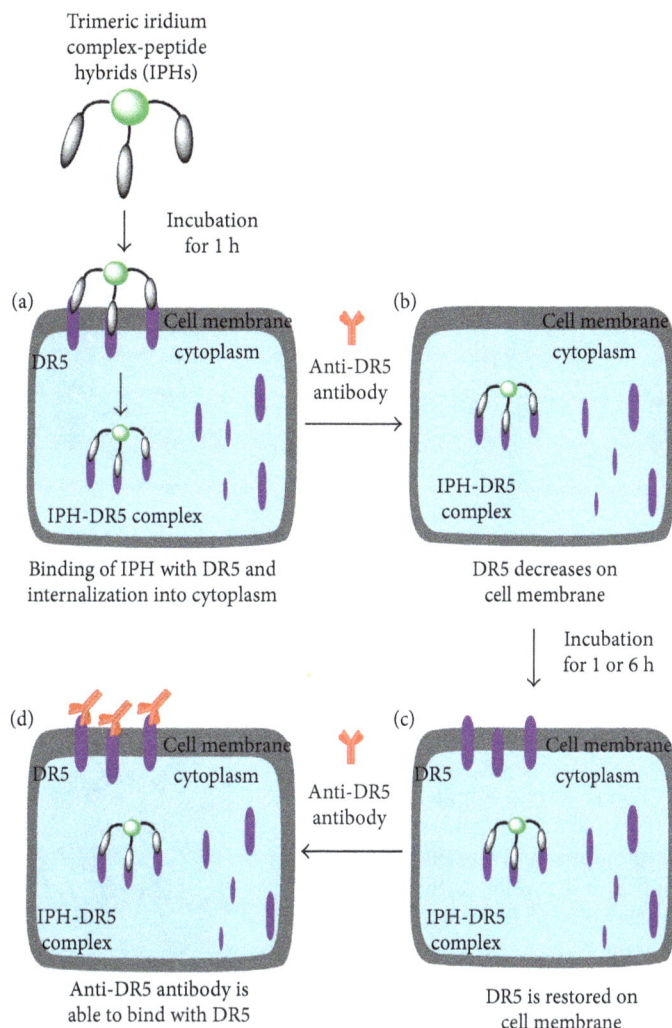

FIGURE 12: Schematic presentation of the behavior of DR5 after complexation with IPH.

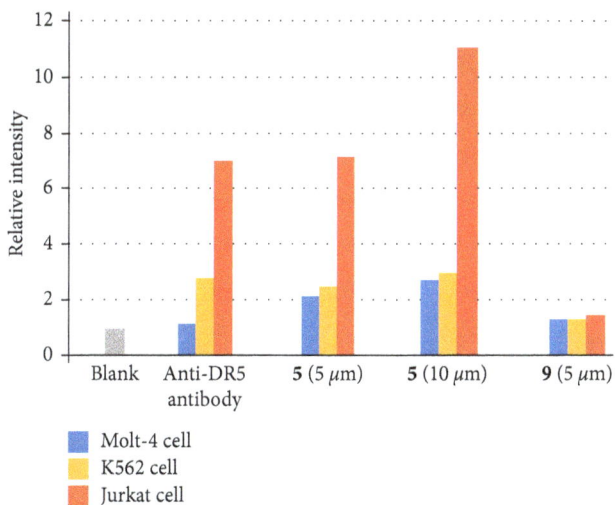

FIGURE 13: Summary of flow cytometry assay of DR5 expression of Molt-4 cells (blue bar), K562 cells (orange bar), and Jurkat cells (red bar). The cells were stained with anti-DR5 antibody (15 μg/mL) at 4°C for 15 min. **5** (5/10 μM) and **9** (5 μM) at 37°C for 1 h. The relative intensity is the ratio of the geometric mean values of luminescence intensity to the blank.

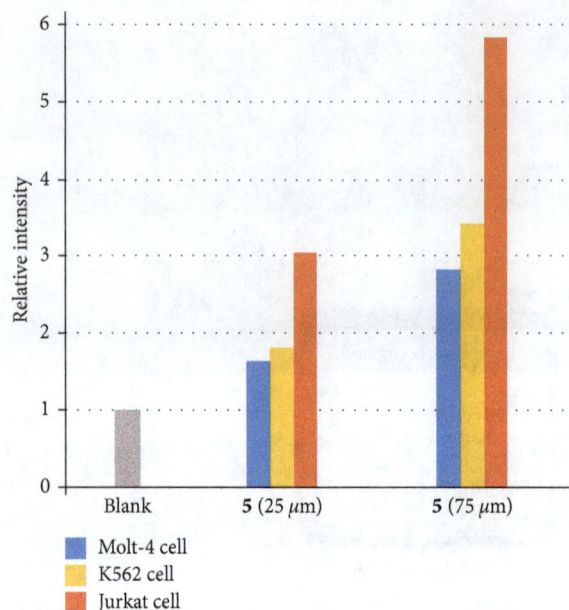

FIGURE 14: Summary of cell death assay (PI staining of dead cells) of Molt-4 cells (blue bar), K562 cells (orange bar), and Jurkat cells (red bar). The relative intensity is the ratio of the geometric mean values of the luminescence intensity to the blank.

recommended concentrations (Figure S11 in Supplementary Materials) at 37°C for 1 h, to which **5** was added and incubated again at the same temperature for 24 h. The cells were then washed with PBS buffer and stained with propidium iodide (PI) for microscopic observation. Interestingly, it was found that only voltage-operated Ca^{2+} channel blocker (nicardipine and verapamil) considerably inhibited cell death induced by **5**. Therefore, we conclude that the cell death induced by **5** can be considered as necrosis-type cell death via Ca^{2+}-mediated intracellular signaling pathway.

3.8. Detection of Jurkat Cells in Bovine Blood. The detection of cancer cells by use of IPHs was carried out. Jurkat cells (10×10^5 cells) were suspended in bovine blood (100 μL) and then incubated with **5** (10 μM) in medium (RPMI 1640) at 37°C for 6 h, resulting in the successful detection of Jurkat cells in bovine blood, as shown in Figure 15.

4. Conclusions

In this study, we report on the design and synthesis of Ir complex-peptide hybrids (**4–6**) consisted of C_3-symmetric tris-cyclometalated Ir complexes equipped with DR5 binding cyclic peptides. Among these complexes, **5** and **6** were found to be cytotoxic against Jurkat cells. Studies demonstrate that **5** binds with DR5 on the cell membrane (by 27 MHz QCM and costaining experiments with anti-DR5 antibody), and their complex is internalized into the cytoplasm by DR5-mediated endocytosis. Our previous Ir complexes **2–3** having basic peptides such as KKGG sequence induce Jurkat cell death in a few hours and exhibit strong emission from dead cells. In contrast, the IPHs

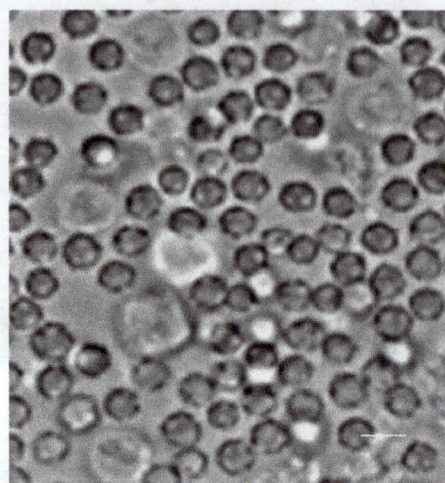

(a)

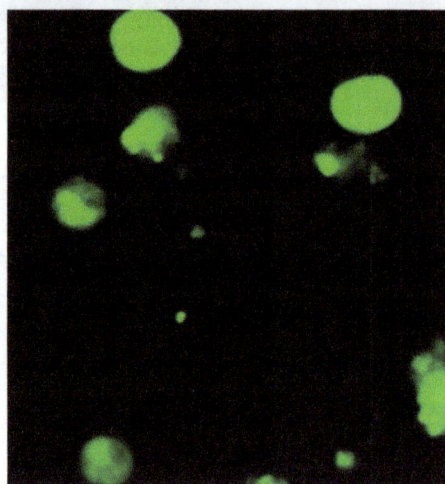

(b)

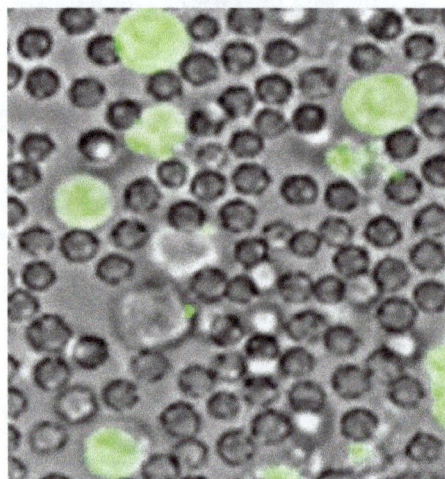

(c)

FIGURE 15: Luminescence microscopy images (×40) (Biorevo, BZ-9000, Keyence) of Jurkat cells spiked in bovine blood in presence of **5**. (a–c) Jurkat cells spiked in bovine blood after incubation with **5** (10 μM) at 37°C for 6 h: (a) bright field image, (b) emission image, and (c) overlay image of (a) and (b). Scale bar (white) = 10 μm.

reported in this study (**5** and **6**) detect Jurkat cells first and then induce their death very slowly (after 16 to 24 h). Moreover, DR5 is transferred and/or reproduced on the cell surface after additional incubation. Jurkat cells can be detected even in bovine blood which may offer an easy and convenient method for cancer diagnosis. To the best of our knowledge, **5** is the first example of the artificial compound that achieve imaging and cell death induction of cancer cells associated with DR5, due to its moderate cytoxicity and slow cell death induction property.

The aforementioned results postulate that IPHs can be good candidates to study death receptor biology, cancer cell imaging, induction of cancer cell death, and understanding of the mechanism of cell death mediated by death receptor. Additionally, IPHs induce slow cell death, which may provide new approaches in the treatment of cancer and related diseases. Improvement of the anticancer activity of IPHs and attempts at controlling cell death types are now in progress.

Abbreviations

DIC: 　*N,N'*-diisopropylcarbodiimide
DIEA: 　*N,N*-diisopropylethylamine
DMF: 　*N,N*-dimethylformamide
EDC: 　1-Ethyl-3-(3-dimethylaminopropyl)carbodiimide
HOBt: 　1-Hydroxybenzotriazole
NHS: 　*N*-hydroxy succinimide
PyBOP: 　Benzotriazol-1-yl-oxytripyrrolidinophosphonium hexafluorophosphate
TFA: 　Trifluoroacetic acid
THF: 　Tetrahydrofuran
TIPS: 　Triisopropylsilane
TMSCl: 　Trimethylsilyl chloride.

Acknowledgments

This work was supported by grants-in-aid from the Ministry of Education, Culture, Sports, Science and Technology (MEXT) of Japan (nos. 22890200, 24890256, 26860016, and 16K18851 to Yosuke Hisamatsu and nos. 19659026, 22390005, 24590025, 24650011, 24640156, and 15K00408 to Shin Aoki), the Uehara Memorial Foundation (Tokyo, Japan) to Yosuke Hisamatsu, the Tokyo Biochemical Research Foundation (Tokyo, Japan) to Shin Aoki, "Academic Frontiers" Project for Private Universities (matching fund subsidy from MEXT), and TUS (Tokyo University of Science) fund for strategic research areas. The authors thank Prof. Dr. Takeshi Inukai (Department of Pediatrics, School of Medicine, University of Yamanashi) for the valuable suggestion. The authors sincerely acknowledge Ms. Fukiko Hasegawa and Ms. Noriko Sawabe (Faculty of Pharmaceutical Sciences, Tokyo University of Science) for measurement of mass spectra and NMR.

Supplementary Materials

Figure S1: the results of MTT assay. Figure S2: microscopic images of competitive staining. Figure S3: flow cytometry assay results of competitive staining. Figures S4–S6: flow cytometry assay results of costaining Figure S7: microscopic images of Jurkat, K562, and Molt-4 cells staining. Figure S8: HPLC charts of **5** after incubation in RPMI 1640 medium. Figures S9 and S10: the results of MTT assay in presence of inhibitors. Figure S11: microscopic images of staining in presence of inhibitors. Figure S12–13: ^{1}H NMR and ESI mass charts of NHS ester of Ir complex **7**. Figures S14–S15: ^{1}H NMR and ESI mass charts of Ir complex **8**. Figures S16 and S17: ^{1}H NMR and ESI mass charts of NHS ester of Ir complex **8**. Figures S18 and S19: ^{1}H NMR and ESI mass charts of Ir complex **9**. Figures S20–S22: HPLC, ^{1}H NMR, and ESI mass charts of **CP1**. Figures S23–S25: HPLC, ^{1}H NMR, and ESI mass charts of **CP2**. Figures S26–S28: HPLC, ^{1}H NMR, and ESI mass charts of **CP3**. Figure S29–S31: HPLC, ^{1}H NMR, and ESI mass charts of Ir complex **4**. Figures S32–S34: HPLC, ^{1}H NMR, and ESI mass charts of Ir complex **5**. Figures S35–S37: HPLC, ^{1}H NMR, and ESI mass charts of Ir complex **6**. (*Supplementary Materials*)

References

[1] A. Ashkenazi and V. M. Dixit, "Death receptors: signaling and modulation," *Science*, vol. 281, no. 5381, pp. 1305–1308, 1998.

[2] G. Pan, K. O'Rourke, A. M. Chinnaiyan et al., "The receptor for the cytotoxic ligand TRAIL," *Science*, vol. 276, no. 5309, pp. 111–113, 1997.

[3] H. Walczak, M. A. Delgi-Esposti, R. S. Johnson et al., "TRAIL-R2: a novel apoptosis-mediating receptor for TRAIL," *EMBO Journal*, vol. 16, no. 17, pp. 5386–5397, 1997.

[4] G. Pan, J. Ni, Y. F. Wei, G. Yu, R. Gentz, and V. M. Dixit, "An antagonist decoy receptor and a death domain-containing receptor for TRAIL," *Science*, vol. 277, no. 5327, pp. 815–818, 1997.

[5] M. A. Delgi-Esposti, P. J. Smolak, H. Walczak et al., "Cloning and characterization of TRAIL-R3, a novel member of the emerging TRAIL receptor family," *Journal of Experimental Medicine*, vol. 186, no. 7, pp. 1165–1170, 1997.

[6] M. A. Delgi-Esposti, W. C. Dougall, P. J. Smolak, J. Y. Waugh, C. A. Smith, and R. G. Goodwin, "The novel receptor TRAIL-R4 induces NF-kB and protects against TRAIL-mediated apoptosis, yet retains an incomplete death domain," *Immunity*, vol. 7, no. 6, pp. 813–820, 1997.

[7] W. S. Simonet, D. L. Lacey, C. R. Dunstan et al., "Osteoprotegerin: a novel secreted protein involved in the regulation of bone density," *Cell*, vol. 89, no. 2, pp. 309–319, 1997.

[8] S. G. Hymowitz, H. W. Christinger, G. Fuh et al., "Triggering cell death: the crystal structure of Apo2L/TRAIL in a complex with death receptor 5," *Molecular Cell*, vol. 4, no. 4, pp. 563–571, 1999.

[9] S. G. Hymowitz, M. O'Connell, M. H. Ultsch et al., "A unique zinc-binding site revealed by a high-resolution X-ray structure of homotrimeric Apo2L/TRAIL," *Biochemistry*, vol. 39, no. 4, pp. 633–640, 2000.

[10] S. W. Fesik, "Insight into programmed cell death through structural biology," *Cell*, vol. 103, no. 2, pp. 273–282, 2000.

[11] A. Ashkenazi, R. C. Pai, S. Fong et al., "Safety and anti-tumor activity of recombinant soluble Apo2 ligand," *Journal of Clinical Investigation*, vol. 104, no. 2, pp. 155–162, 1999.

[12] H. Walczak, R. E. Miller, K. Ariall et al., "Tumoricidal activity of tumor necrosis factor-related apoptosis-inducing ligand in vivo," *Nature Medicine*, vol. 5, no. 2, pp. 157–163, 1999.

[13] S. K. Kelly and A. Ashkenazi, "Targeting death receptors in cancer with Apo2L/TRAIL," *Current Opinion in Pharmacology*, vol. 4, no. 4, pp. 333–339, 2004.

[14] A. Ashkenazi, P. Holland, and S. G. Eckhardt, "Ligand-based targeting of apoptosis in cancer: the potential of recombinant human apoptosis ligand 2/tumor necrosis factor-related apoptosis-inducing ligand (rhApo2L/TRAIL)," *Journal of Clinical Oncology*, vol. 26, no. 21, pp. 3621–3630, 2008.

[15] K. Ichikawa, W. Liu, L. Zhao et al., "Tumoricidal activity of a novel anti-human DR5 monoclonal antibody without hepatocyte cytotoxicity," *Nature Medicine*, vol. 7, no. 8, pp. 954–960, 2001.

[16] H. Jin, R. Yang, S. Fong et al., "Apo2 ligand/tumor necrosis factor-related apoptosis-inducing ligand cooperates with chemotherapy to inhibit orthotopic lung tumor growth and improve survival," *Cancer Research*, vol. 64, no. 14, pp. 4900–4905, 2004.

[17] D. R. Camidge, "Apomab: an agonist monoclonal antibody directed against death receptor 5/TRAIL-receptor 2 for use in the treatment of solid tumors," *Expert Opinion on Biological Therapy*, vol. 8, no. 8, pp. 1167–1176, 2008.

[18] J. Li, D. A. Knee, Y. Wang et al., "LBY135, a novel anti-DR5 agonistic antibody induces tumor cell–specific cytotoxic activity in human colon tumor cell lines and xenografts," *Drug Development Research*, vol. 69, no. 2, pp. 69–82, 2008.

[19] J. Wiezorek, P. Holland, and J. Graves, "Death receptor agonists as a targeted therapy for cancer," *Clinical Cancer Research*, vol. 16, no. 6, pp. 1701–1708, 2011.

[20] Y. M. Angell, A. Bhandari, A. Chakrabarti et al., "Discovery and optimization of a TRAIL R2 agonist for cancer therapy," in *Understanding Biology Using Peptides*, S. E. Blondelle, Ed., pp. 405-406, Springer, New York, NY, USA, 2006.

[21] Y. M. Angell, A. Bhandari, M. N. De Francisco, et al., "Peptides for youth," in *Advances in Experimental Medicine and Biology*, S. Del Valle, E. Escher, and W. D. Lubell, Eds., pp. 101-103, Springer, New York, NY, USA, 2009.

[22] Y. M. Angell, M. Bhandari, M. N. De Francisco et al., "Discovery and optimization of a TRAIL R2 agonist for cancer therapy," *Advances in Experimental Medicine and Biology*, vol. 611, pp. 101-103, 2009.

[23] V. Pavet, J. Beyrath, C. Pardin et al., "Multivalent DR5 peptides activate the TRAIL death pathway and exert tumoricidal activity," *Cancer Research*, vol. 70, no. 3, pp. 1101–1110, 2010.

[24] G. Lamanna, C. R. Smulski, N. Chekkat et al., "Multimerization of an apoptogenic TRAIL-mimicking peptide by using adamantane-based dendrons," *Chemistry-A European Journal*, vol. 19, no. 5, pp. 1762–1768, 2013.

[25] K. Pulka-Ziach, V. Pavet, N. Chekkat et al., "Thioether analogues of disulfide-bridged cyclic peptides targeting death receptor 5: conformational analysis, dimerisation and consequences for receptor activation," *ChemBioChem*, vol. 16, no. 2, pp. 293–301, 2015.

[26] B. Valldorf, H. Fittler, L. Deweid et al., "An apoptosis-inducing peptidic heptad that efficiently clusters death receptor 5," *Angewandte Chemie International Edition*, vol. 55, no. 16, pp. 1–6, 2016.

[27] C. Kaga, M. Okochi, M. Nakanishi, H. Hayashi, R. Kato, and H. Honda, "Screening of a novel octamer peptide, CNSCWSKD, that induces caspase-dependent cell death," *Biochemical and Biophysical Research Communications*, vol. 362, no. 4, pp. 1063–1068, 2007.

[28] G. Wang, X. Wang, H. Yu et al., "Small-molecule activation of the TRAIL receptor DR5 in human cancer cells," *Nature Chemical Biology*, vol. 9, no. 2, pp. 84–89, 2013.

[29] S. Lamansky, P. Djurovich, D. Murphy et al., "Highly phosphorescent bis-cyclometalated iridium complexes: synthesis, photophysical characterization, and use in organic light emitting diodes," *Journal of the American Chemical Society*, vol. 123, no. 18, pp. 4304–4312, 2001.

[30] M. A. Baldo, D. F. O'Brien, Y. You et al., "Highly efficient phosphorescent emission from organic electroluminescent devices," *Nature*, vol. 395, no. 6698, pp. 151–154, 1998.

[31] G. M. Farinola and R. Ragni, "Electroluminescent materials for white organic light emitting diodes," *Chemical Society Reviews*, vol. 40, no. 7, pp. 3467–3482, 2011.

[32] H. Yersin, *Highly Efficient OLEDs with Phosphorescent Materials*, Wiley-VCH, Weinheim, Germany, 2008.

[33] C. Ulbricht, B. Beyer, C. Friebe, A. Winter, and U. S. Schubert, "Recent developments in the application of phosphorescent iridium(iii) complex systems," *Advanced Materials*, vol. 21, no. 44, pp. 4418–4441, 2009.

[34] A. B. Tamayo, B. D. Alleyne, P. I. Djurovich et al., "Synthesis and characterization of facial and meridional tris-cyclometalated iridium(III) complexes," *Journal of the American Chemical Society*, vol. 125, no. 24, pp. 7377–7387, 2003.

[35] K. Dedeian, P. I. Djurovich, F. O. Garces, G. Carlson, and R. J. Watts, "A new synthetic route to the preparation of a series of strong photoreducing agents: *fac*-tris-ortho-metalated complexes of iridium(III) with substituted 2-phenylpyridines," *Inorganic Chemistry*, vol. 30, no. 8, pp. 1685–1687, 1991.

[36] M. S. Lowry and S. Bernhard, "Synthetically tailored excited states: phosphorescent, cyclometalated iridium(III) complexes and their applications," *Chemistry-A European Journal*, vol. 12, no. 31, pp. 7970–7977, 2006.

[37] L. Flamigni, A. Barbieri, C. Sabatini, B. Ventura, and F. Barigelletti, "Photochemistry and photophysics of coordination compounds: iridium," *Topics in Current Chemistry*, vol. 281, pp. 143–203, 2007.

[38] R. C. Evans, P. Douglas, and C. J. Winscom, "Coordination complexes exhibiting room-temperature phosphorescence: evaluation of their suitability as triplet emitters for light-emitting diodes," *Coordination Chemistry Reviews*, vol. 250, no. 15-16, pp. 2093–2126, 2006.

[39] Y. Chi and P. T. Chou, "Transition-metal phosphors with cyclometalating ligands: fundamentals and applications," *Chemical Society Reviews*, vol. 39, no. 2, pp. 638–655, 2010.

[40] G. D. Marco, M. Lanza, A. Mamo et al., "Luminescent mononuclear and dinuclear iridium(III) cyclometalated complexes immobilized in a polymeric matrix as solid-state oxygen sensors," *Analytical Chemistry*, vol. 70, no. 19, pp. 5019–5023, 1998.

[41] M. C. DeRosa, P. J. Mosher, G. P. A. Yap, K. S. Focsaneanu, R. J. Crutchley, and C. E. B. Evanc, "Synthesis, characterization, and evaluation of $[Ir(ppy)_2(vpy)Cl]$ as a polymer-bound oxygen sensor," *Inorganic Chemistry*, vol. 42, no. 16, pp. 4864–4872, 2003.

[42] M. Licini and J. A. G. Williams, "Iridium(iii) bis-terpyridine complexes displaying long-lived pH sensitive luminescence," *Chemical Communications*, no. 19, pp. 1943-1944, 1999.

[43] K. J. Arm, W. Leslie, and J. A. G. Williams, "Synthesis and pH-sensitive luminescence of bis-terpyridyl iridium(III) complexes incorporating pendent pyridyl groups," *Inorganica Chimica Acta*, vol. 359, no. 4, pp. 1222-1232, 2006.

[44] M. L. Ho, F. M. Hwang, P. N. Chen et al., "Design and synthesis of iridium(III) azacrown complex: application as a highly sensitive metal cation phosphorescence sensor," *Organic and Biomolecular Chemistry*, vol. 4, no. 1, pp. 98-103, 2006.

[45] K. Konishi, H. Yamaguchi, and A. Harada, "Synthesis of a water-soluble iridium(iii) complex with pH and metal cation sensitive photoluminescence," *Chemistry Letters*, vol. 35, no. 7, pp. 720-721, 2006.

[46] M. Schmittel and H. Lin, "Luminescent iridium phenanthroline crown ether complex for the detection of silver(I) ions in aqueous media," *Inorganic Chemistry*, vol. 46, no. 22, pp. 9139-9145, 2007.

[47] Q. Zhao, T. Cao, F. Li et al., "A highly selective and multi-signaling optical-electrochemical sensor for Hg^{2+} based on a phosphorescent iridium(III) complex," *Organometallics*, vol. 26, no. 8, pp. 2077-2081, 2007.

[48] K. K. W. Lo, M. W. Louie, and K. Y. Zhang, "Design of luminescent iridium(III) and rhenium(I) polypyridine complexes as in vitro and in vivo ion, molecular and biological probes," *Coordination Chemistry Reviews*, vol. 254, no. 21-22, pp. 2603-2622, 2010.

[49] K. K. W. Lo, S. P. Y. Li, and K. Y. Zhang, "Development of luminescent iridium(III) polypyridine complexes as chemical and biological probes," *New Journal of Chemistry*, vol. 35, no. 2, pp. 265-287, 2011.

[50] S. K. Leung, K. Y. Kwok, K. Y. Zhang, and K. K. W. Lo, "Design of luminescent biotinylation reagents derived from cyclometalated iridium(III) and rhodium(III) bis(pyridylbenzaldehyde) complexes," *Inorganic Chemistry*, vol. 49, no. 11, pp. 4984-4995, 2010.

[51] K. K. W. Lo and S. P. Y. Li, "Utilization of the photophysical and photochemical properties of phosphorescent transition metal complexes in the development of photofunctional cellular sensors, imaging reagents, and cytotoxic agents," *RSC Advances*, vol. 4, no. 21, pp. 10560-10585, 2014.

[52] K. Y. Zhang, S. P. Y. Li, N. Zhu et al., "Structure, photophysical and electrochemical properties, biomolecular interactions, and intracellular uptake of luminescent cyclometalated iridium-(III) dipyridoquinoxaline complexes," *Inorganic Chemistry*, vol. 49, no. 5, pp. 2530-2540, 2010.

[53] P. K. Lee, W. H. T. Law, H. W. Liu, and K. K. W. Lo, "Luminescent cyclometalated iridium(III) polypyridine di-2-picolylamine complexes: synthesis, photophysics, electrochemistry, cation binding, cellular internalization, and cytotoxic activity," *Inorganic Chemistry*, vol. 50, no. 17, pp. 8570-8579, 2011.

[54] K. Y. Zhang, H. W. Liu, T. T. H. Fong, X. G. Chen, and K. K. W. Lo, "Luminescent dendritic cyclometalated iridium (III) polypyridine complexes: synthesis, emission behavior, and biological properties," *Inorganic Chemistry*, vol. 49, no. 12, pp. 5432-5443, 2010.

[55] Q. Zhao, C. Huang, and F. Li, "Phosphorescent heavy-metal complexes for bioimaging," *Chemical Society Reviews*, vol. 40, no. 5, pp. 2508-2524, 2011.

[56] Q. Zhao, M. Yu, L. Shi et al., "Cationic iridium(III) complexes with tunable emission color as phosphorescent dyes for live

cell imaging," *Organometallics*, vol. 29, no. 5, pp. 1085-1091, 2010.

[57] C. Li, M. Yu, Y. Sun, Y. Wu, C. Huang, and F. Li, "Anonemissive iridium(III) complex that specifically lights-up the nuclei of living cells," *Journal of the American Chemical Society*, vol. 133, no. 29, pp. 11231-11239, 2011.

[58] Y. You, S. Lee, T. Kim et al., "Phosphorescent sensor for biological mobile zinc," *Journal of the American Chemical Society*, vol. 133, no. 45, pp. 18328-18342, 2011.

[59] Y. Wu, H. Jing, Z. Dong, Q. Zhao, H. Wu, and F. Li, "Ratiometric phosphorescence imaging of hg(II) in living cells based on a neutral iridium(III) complex," *Inorganic Chemistry*, vol. 50, no. 16, pp. 7412-7420, 2011.

[60] L. Murphy, A. Congreve, L. O. Palsson, and J. A. G. Williams, "The time domain in co-stained cell imaging: time- resolved emission imaging microscopy using a protonatable luminescent iridium complex," *Chemical Communications*, vol. 46, no. 46, pp. 8743-8745, 2010.

[61] E. Baggaley, J. A. Weinstein, and J. A. G. Williams, "Lighting the way to see inside the live cell with luminescent transition metal complexes," *Coordination Chemistry Reviews*, vol. 256, no. 15-16, pp. 1762-1785, 2012.

[62] Y. You, "Phosphorescence bioimaging using cyclometalated ir (III) complexes," *Current Opinion in Chemical Biology*, vol. 17, no. 4, pp. 699-707, 2013.

[63] R. Cao, J. Jia, X. Ma, M. Zhou, and H. Fei, "Membrane localized iridium(III) complex induces endoplasmic reticulum stress and mitochondria-mediated apoptosis in human cancer cells," *Journal of Medicinal Chemistry*, vol. 56, no. 9, pp. 3636-3644, 2013.

[64] Y. Zhou, J. Jia, W. Li, H. Fei, and M. Zhou, "Luminescent biscarbene iridium(III) complexes as living cell imaging reagents," *Chemical Communications*, vol. 49, no. 31, pp. 3230-3232, 2013.

[65] S. Zhang, M. Hosaka, T. Yoshihara et al., "Phosphorescent light–emitting iridium complexes serve as a hypoxia-sensing probe for tumor imaging in living animals," *Cancer Research*, vol. 70, no. 11, pp. 4490-4498, 2010.

[66] S. Tobita and T. Yoshihara, "Intracellular and *in vivo* oxygen sensing using phosphorescent iridium(III) complexes," *Current Opinion in Chemical Biology*, vol. 33, pp. 39-45, 2016.

[67] Y. Hisamatsu and S. Aoki, "Design and synthesis of blue-emitting cyclometalated iridium-(III) complexes based on regioselective functionalization," *European Journal of Inorganic Chemistry*, vol. 2011, no. 35, pp. 5360-5369, 2011.

[68] S. Moromizato, Y. Hisamatsu, T. Suzuki, Y. Matsuo, R. Abe, and S. Aoki, "Design and synthesis of a luminescent cyclometalated iridium(III) complex having N,N-diethyl amino group that stains acidic intracellular organelles and induces cell death by photoirradiation," *Inorganic Chemistry*, vol. 51, no. 23, pp. 12697-12706, 2012.

[69] A. Nakagawa, Y. Hisamatsu, S. Moromizato, M. Kohno, and S. Aoki, "Synthesis and photochemical properties of pH responsive tris-cyclometalated iridium(III) complexes that contain a pyridine ring on the 2-phenylpyridine ligand," *Inorganic Chemistry*, vol. 53, no. 1, pp. 409-422, 2014.

[70] A. Kando, Y. Hisamatsu, H. Ohwada, S. Moromizato, M. Kohno, and S. Aoki, "Photochemical properties of red-emitting tris(cyclometalated) iridium(III) complexes having basic and nitro groups and application to pH sensing and photo induced cell death," *Inorganic Chemistry*, vol. 54, no. 11, pp. 5342-57, 2015.

[71] S. Kumar, Y. Hisamatsu, Y. Tamaki, O. Ishitani, and S. Aoki, "Design and synthesis of heteroleptic cyclometalated iridium

(III) complexes containing quinoline-type ligands that exhibit dual phosphorescence," *Inorganic Chemistry*, vol. 55, no. 8, pp. 3829–3843, 2016.

[72] Y. Hisamatsu, A. Shibuya, N. Suzuki, R. Abe, and S. Aoki, "Design and synthesis of amphiphilic and luminescent tris-cyclometalated iridium(III) complexes containing cationic peptides as inducers and detectors of cell death via a calcium-dependent pathway," *Bioconjugate Chemistry*, vol. 26, no. 5, pp. 857–879, 2015.

[73] Y. Hisamatsu, N. Suzuki, A. Masum et al., "Cationic amphiphilic tris-cyclometalated iridium(iii) complexes induce cancer cell death via interaction with Ca^{2+}-calmodulin complex," *Bioconjugate Chemistry*, vol. 28, no. 2, pp. 507–523, 2017.

[74] K. Yokoi, Y. Hisamatsu, K. Naito, and S. Aoki, "Design, synthesis and anticancer activity of cyclometalated tris(ppy) iridium(III) complexes having cationic peptides at the 4'-position of the ppy ligand (Ppy=2-phenylpyridine)," *European Journal of Inorganic Chemistry*, vol. 2017, no. 44, pp. 5295–5309, 2017.

[75] S. Aoki, Y. Matsuo, S. Ogura et al., "Regioselective aromatic substitution reactions of cyclometalated ir(III) complexes: synthesis and photochemical properties of substituted ir(III) complexes that exhibit blue, green, and red color luminescence emission," *Inorganic Chemistry*, vol. 50, pp. 806–818, 2011.

[76] E. A. Slee, H. Zhu, S. C. Chow, M. MacFarlane, D. W. Nicholson, and G. M. Cohen, "Benzyloxycarbonyl-Val-Ala-Asp(OMe) fluoromethylketone (Z-VAD-FMK) inhibits apoptosis by blocking the processing of CPP32," *Biochemical Journal*, vol. 315, no. 1, pp. 21–24, 1996.

[77] A. Degterev, J. Hitomi, M. Germscheid et al., "Identification of RIP1 kinase as a specific cellular target of necrostatins," *Nature Chemical Biology*, vol. 4, no. 5, pp. 313–321, 2008.

[78] K. Dodo, M. Katoh, T. Shimizu, M. Takahashi, and M. Sodeoka, "Inhibition of hydrogen peroxide-induced necrotic cell death with 3-amino-2-indolylmaleimide derivatives," *Bioorganic and Medicinal Chemistry Letters*, vol. 15, no. 12, pp. 3114–3118, 2005.

[79] G. Koopman, C. P. Reutelingsperger, G. A. Kuijten, R. M. Keehnen, S. T. Pals, and M. H. van Oers, "Annexin V for flow cytometric detection of phosphatidylserine expression on B cells undergoing apoptosis," *Blood*, vol. 84, pp. 1415–1420, 1994.

[80] I. Vermes, C. Haanen, H. Steffens-Nakken, and C. Reutellingsperger, "A novel assay for apoptosis Flow cytometric detection of phosphatidylserine expression on early apoptotic cells using fluorescein labelled Annexin V," *Journal of Immunological Methods*, vol. 184, no. 1, pp. 39–51, 1995.

[81] P.-K. Lee, H.-W. Liu, S.-M. Yiu, M. M.-W. Louie, and K. K.-W. Lo, "Luminescent cyclometallated iridium(III) bis (quinolylbenzaldehyde) diimine complexes–synthesis, photophysics, electrochemistry, protein cross-linking properties, cytotoxicity and cellular uptake," *Dalton Transactions*, vol. 40, no. 10, pp. 2180–2189, 2011.

[82] S. P. Y. Li, T. S. M. Tang, K. S. M. Yiu, and K. K. W. Lo, "Cyclometalated iridium(III)-polyamine complexes with intense and long-lived multicolor phosphorescence: synthesis, crystal structure, photophysical behavior, cellular uptake, and transfection properties," *Chemistry-A European Journal*, vol. 18, no. 42, pp. 13342–13354, 2012.

[83] V. Z. Sun, Z. Li, T. J. Deming, and D. T. Kamei, "Intracellular fates of cell-penetrating block copolypeptide vesicles," *Biomacromolecules*, vol. 12, no. 1, pp. 10–13, 2011.

[84] T. Kimura, Y. Takabatake, T. Takahashi, and Y. Isaka, "Chloroquine in cancer therapy: a double-edged sword of autophagy," *Cancer Research*, vol. 73, no. 1, pp. 3–7, 2013.

[85] S. Dröse and K. Altendorf, "Bafilomycins and concanamycins as inhibitors of V-ATPases and P-ATPases," *Journal of Experimental Biology*, vol. 200, pp. 1–8, 1997.

[86] V. Novohradsky, Z. Liu, M. Vojtiskova, P. J. Sadler, V. Brabec, and J. Kasparkova, "Mechanism of cellular accumulation of an iridium(III) pentamethyl cyclopentadienyl anticancer complex containing a C,N-chelating ligand," *Metallomics*, vol. 6, no. 3, pp. 682–690, 2014.

[87] W. A. Catterall and J. Striessnig, "Receptor sites for Ca^{2+} channel antagonists," *Trends in Pharmacological Sciences*, vol. 13, pp. 256–262, 1992.

[88] G. H. Hockerman, B. Z. Peterson, B. D. Johnson, and W. A. Catterall, "Molecular determinants of drug binding and action on l-type calcium channels," *Annual Review of Pharmacology and Toxicology*, vol. 37, no. 1, pp. 361–396, 1997.

[89] D. J. Triggle, "Calcium channel antagonists: clinical uses: past, present and future," *Biochemical Pharmacology*, vol. 74, no. 1, pp. 1–9, 2007.

[90] H. J. Motulsky, A. S. Maisel, M. D. Snavely, and P. A. Insel, "Quinidine is a competitive antagonist at alpha 1- and alpha 2-adrenergic receptors," *Circulation Research*, vol. 55, no. 3, pp. 376–381, 1984.

[91] K. Shibata, A. Hirasawa, R. Foglar, S. Ogawa, and G. Tsujimoto, "Effects of quinidine and verapamil on human cardiovascular α1-adrenoceptors," *Circulation*, vol. 97, no. 13, pp. 1227–1230, 1998.

[92] G. Schreiber, A. Barak, and M. Sokolovsky, "Disopyramide and quinidine bind with inverse selectivity to muscarinic receptors in cardiac and extra cardiac rat tissues," *Journal of Cardiovascular Pharmacology*, vol. 7, no. 2, pp. 390–393, 1985.

[93] J. A. Harrow and N. S. Dhalla, "Effects of quinidine on calcium transport activities of the rabbit heart mitochondria and sarcotubular vesicles," *Biochemical Pharmacology*, vol. 25, no. 8, pp. 897–902, 1976.

[94] W. Van Driessche, "Physiological role of apical potassium ion channels in frog skin," *Journal of Physiology*, vol. 356, no. 1, pp. 79–95, 1984.

10

The Anticancer Activities of some Nitrogen Donor Ligands Containing bis-Pyrazole, Bipyridine, and Phenanthroline Moiety using Docking Methods

Adebayo A. Adeniyi and Peter A. Ajibade🄳

School of Chemistry and Physics, University of KwaZulu-Natal, Private Bag X01, Scottsville, Pietermaritzburg 3201, South Africa

Correspondence should be addressed to Peter A. Ajibade; ajibadep@ukzn.ac.za

Academic Editor: Viktor Brabec

The anticancer study of nitrogen-chelating ligands can be of tremendous help in choosing ligands for the anticancer metal complexes design especially with ruthenium(II). The inhibitory anticancer activities of some nitrogen-chelating ligands containing bis-pyrazole, bipyridine, and phenanthroline were studied using experimental screening against cancer cell and theoretical docking methods. *In vitro* anticancer activities showed compound **11** as the most promising inhibitor, and the computational docking further indicates its strong inhibitory activities towards some cancer-related receptors. Among the twenty-one modelled ligands, pyrazole-based compounds **7**, **11**, and **15** are the most promising inhibitors against the selected receptors followed by **18** and **21** which are derivatives of pyridine and phenanthroline, respectively. The presence of the carboxylic unit in the top five ligands that displayed stronger inhibitory activities against the selected receptors is an indication that the formation of non-covalent interactions such as hydrogen bonding and a strong electron-withdrawing group in these compounds are very important for their receptor interactions. The thermodynamic properties, the polarizabilities, and the LUMO energy of the compounds are in the same patterns as the observed inhibitory activities.

1. Introduction

Nitrogen-chelating ligands such as bis-pyrazole (pz), bipyridine (bpyr), and phenanthroline (phn) derivatives are being used in the design of several metal complexes for ranges of applications from biological to nanostructured materials, especially the development of anticancer complexes of ruthenium [1–19]. Other studies have shown that these nitrogen-chelating compounds play significant roles in the experimentally observed anticancer activities [20] that might be ascribed to the electronic interactions between the metal centre and the π-electrons in rings [21–23]. However, despite the use of these nitrogen-chelating ligands in the design of anticancer metal complexes, there has been very few or no attention on the anticancer activities of these ligands individually. To this end, we have selected some derivatives of common nitrogen-chelating ligands as shown in Figure 1 to study their individual anticancer activities with particular focus on their interactions with cancer-related receptors using various docking methods. We also report the experimental *in vitro* anticancer activities of some of these ligands.

The selected receptors used for the docking studies are carbonic anhydrase II (CA-II), cathepsins B (Cat B) [24], two different DNAs (DNA-1 [25] and DNA-2), DNA gyrase (Gyrase) [26], histone deacetylase7 (HDAC7) [27], histone protein in the nucleosome core particle (HIS) [28], BRAF kinase (Kinase) [29], recombinant human albumin (rHA) [30], ribonucleotide reductase (RNR) [31], topoisomerase II (Top II) [32], thioredoxin reductase (TrxR) [33], and thymidylate synthase (TS) [34]. These receptors play significant roles in cancer growth and are thus a unique target in cancer therapy. For instance, rHA plays a significant role in the pharmacokinetic availability, bioavailability, and toxicology

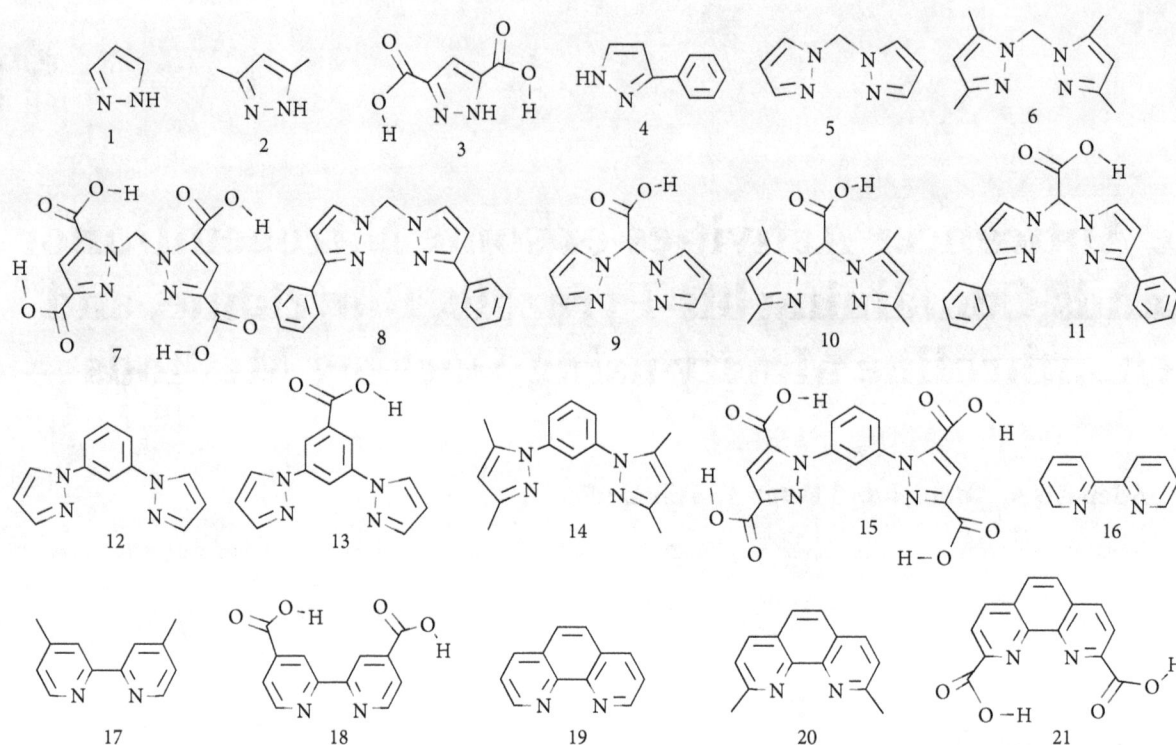

FIGURE 1: The schematic representation of the studied nitrogen-chelating ligands.

[35] and helps either in delivery of metal-based anticancer drugs to their cellular targets or in deactivating them even before reaching the target(s) [36]. The RNR is responsible for the synthesis of DNA from the corresponding building blocks of RNA [37]. Top II plays a key role in relaxing supercoiled DNA for replication and transcription in the absence of inhibitors [38], while the presence of inhibitors forms a stable complex with the enzyme and keeps it from DNA cleavage [39]. Thioredoxin reductase (TrxR) regulates the cellular reduction/oxidation (redox) status [40, 41]. Thymidylate synthase (TS) is a critical enzyme in maintaining a balanced supply of deoxynucleotides required for DNA synthesis and repair [42]. It is a target of chemotherapy to test the vulnerability of cancer cells to the inhibition of TMP synthesis [37]. Gyrase was considered to establish possible dual roles for the ligands as potential antibacterial agents.

2. Synthesis and Structural Elucidation of the Ligands

The synthesis and careful structural elucidation of these ligands have been reported in our previous works [43–45].

3. Experimental Methods

The *in vitro* anticancer activities of the ligands against the cancer cell line HT29 and normal cell line KMST were examined using the MTT colorimetric assay. All the ligands before their docking to the receptors were first optimized with DFT functional PBEPBE [46] and the basis

set 6-31 + G(d,p) for all atoms using the Gaussian 09 package [47]. The docking analyses were carried out using Molegro [48] and Vina and AutoDock [49] packages. The docking of each ligand against the receptors was done five times: twice in Molegro first using the quantum Mulliken atomic charges for all atoms of each ligand (subsequently referred to as Molegro-QC) and second using the predicted atomic charges from the Molegro package (referred to as Molegro). One time in the Vina package using the predicted atomic charges from AutoDock tools and twice in AutoDock using both the QM charges (AutoDock-QC) and package-predicted atomic charges (AutoDock) has done in Molegro.

The default parameters were used in the Vina docking package but with little modifications in Molegro and AutoDock dockings. The scoring function used in Molegro was MolDock because it takes care of the hydrogen bonding, intermolecular protein ligand, and intramolecular ligand interactions and has been successfully applying for molecular docking [50]. The maximum interaction was set to 2500 instead of the default value of 1500, and the population number was increased from the default value of 50 to 100. In using AutoDock, the number of grid points in x-, y-, and z-axes was set to $60 \times 60 \times 60$ with each point separated by 0.375 Å. The Lamarckian genetic algorithm was chosen based on its efficiency and reliability in comparison with others like simulated annealing (SA) and generic genetic algorithm (GA) methods in AutoDock [51, 52]. The maximum number of energy evaluations was set to 2,500,000 for each of the 20 independent runs, a maximum number of 27,000 GA operations were generated on a single population

TABLE 1: The experimental anticancer activities of selected ligands.

Name	Compound	IC_{50}-KMST	IC_{50}-HT29
5	bpzm	>50	6.68
6	bdmpzm	>50	7.12
9	bpza	>50	6.25
11	bphpza	>50	<6.25
12	bpzpy	>50	15.87
14	bdmpzpy	>50	<6.25

of 100 individuals, and step sizes of 2 Å for translation and 50° for rotation were chosen.

All the graphical representations of the docking results are prepared using the package Chimera [53].

4. Results and Discussion

Twenty-one ligands were modelled which comprise fifteen models of bis-pyrazole, three models of bipyridine (bpyr), and three models of phenanthroline (phn) (Figure 1). Six of the modelled compounds were screened to determine their *in vitro* anticancer activities, and their results are presented in Table 1. The *in vitro* activities of these ligands were tested against the cancer cell line HT29 and the normal cell line KMST using the MTT colorimetric assay. The results clearly showed that the six ligands pose no threat to the normal cell line as their inhibitory activities (IC_{50}) in KMST are found to be greater than 50 μm as shown in Table 1. Ligands **11** and **14** are the most potent (<6.25 mm) against the cancer cell line HT29, while ligand **12** is the least active. Also, the same ligand **11** was found among the three best inhibitors of the selected receptors from the docking results (Table 2).

4.1. The Binding Site Predictions of the Compound from the Docking Methods.
The results obtained from the different docking methods and packages: Molegro-QC (grey), Molegro (green), Vina (brown), AutoDock-QC (yellow), and AutoDock (cyan), are shown in Figure 2. The features of the interactions of the ligands with the receptors using different docking methods are similar. In most of the receptors, the same binding sites were located by docking methods, and very similar conformational orientations of the ligands in the binding sites were predicted with some found to overlap each other (Figure 2). Ligands **11** and **15** have similar binding orientation in CA-II, Cat B, and HDAC7 based on the results obtained from the binding site interactions. Compound **11** is of interest because it shows promising anticancer activities *in vitro*, and ligand **15** is predicted as the best inhibitor of many of the receptors according to the results obtained from Molegro-QC and Molegro. The interaction of **11** with CA-II shows that all the docking methods gave similar orientation for the compounds except Molegro-QC (grey) that is slightly different in orientation (Figure 2).

The orientation obtained from Molegro-QC (grey) gives the best interacting energy compared to others (Table 2). Also in the interaction of **15** with CA-II, all of the methods locate the same binding site and give similar orientation of

the ligand except Vina (brown) which locates a different binding site compared to others. The Molegro-QC and Molegro orientations of **15** are superimposed and likewise AutoDock and AutoDock-QC. All the five methods except Vina also locate the same binding site for the interaction of **11** with Cat B, and AutoDock prediction was found to be superimposed with that of Molegro-QC and Molegro. The results obtained from AutoDock predict different binding sites for the interaction of **15** with Cat B, while all other methods predict the same site for its binding. Another receptor of interest is HDAC7 in which strong inhibitory activities are displayed by **11** and **15**. HDAC7 was predicted as one of the most targeted receptors for many of the ligands according to the results obtained from Molegro-QC and Molegro. Vina locates a completely different binding site for the interaction of **11** with HDAC7 when compared to the rest of the methods. The binding site orientation of **11** from all the methods besides Vina is very similar to that of Molegro-QC completely superimposed with that of Auto-Dock. All methods predicted a similar binding site orientation of the ligand **15**, of which the Molegro-QC and Molegro orientations are found to be superimposed and in close orientation with those of Vina and AutoDock-QC, which are different from those of AutoDock. There is a significant difference between the results from AutoDock-QC and AutoDock methods in terms of the features of the ligand interactions with receptors just because of changes in the accuracy of the atomic charges. The correlation of the results of AutoDock-QC and AutoDock ranges from −0.32 to 0.46, while that of Molegro-QC and Molegro ranges from 0.96 to 1.00. It is obvious that the accuracy of atomic charges plays a very strong role in the determination of the ligand interaction with the receptors especially when using AutoDock. The results show that the reason for the inconsistency from Vina, AutoDock-DC, and AutoDock could be ascribed to different binding sites which were predicted for many of the ligand interactions with the receptors, while high consistency and similarity were obtained in Molegro-QC and Molegro because they both predicted the same binding site as cocrystallized inhibitors of the receptors. In addition, the differences in the ligand-receptor inhibitory energies obtained from AutoDock methods for many of the ligands are within the standard error margin of ~2.177 kcal/mol [43, 44] that makes the order obtained from AutoDock unreliable.

4.2. The Inhibitory Activities.
The ligands **7**, **11**, and **15** have the best inhibitory activities towards many of the receptors according to the results obtained from Molegro-QC and Molegro (Table 2 and Figure 3). Besides the first three bis-pyrazole ligands, next promising ligands in interaction with the selected receptors are **18** and **21** which are bpyr and phn compounds. The common feature possessed by all the promising ligands is carboxylic acid moieties which are predicted to enhance the noncovalent interactions such as hydrogen bonding and play the role of electron withdrawing. Since the results from the two methods Molegro-QC and Molegro are highly correlated (0.96 to 1.00), Molegro-QC is

Table 2: The interacting free energy of the ligands with the receptors using the five docking methods approached.

Ligands	CA-II	Cat B	DNA-1	DNA-2	Gyrase	HDAC7	HIS	Kinase	rHA	RNR	Top II	TrxR	TS
Molegro-QC													
1	−46.28	−41.89	−40.28	−46.26	−47.44	−50.56	−40.36	−45.16	−53.17	−50.49	−48.58	−51.94	−49.68
2	−60.37	−60.30	−48.43	−58.25	−59.81	−67.45	−52.16	−56.96	−64.71	−60.84	−64.03	−65.38	−59.07
3	−77.18	−85.66	−79.92	−88.62	−83.09	−94.83	−70.38	−83.19	−89.26	−79.78	−97.24	−86.21	−89.74
4	−94.47	−86.37	−67.14	−82.41	−85.22	−95.87	−64.14	−71.90	−82.84	−82.26	−92.74	−78.40	−86.99
5	−85.49	−86.16	−67.78	−79.76	−82.00	−94.12	−72.68	−81.62	−90.88	−90.19	−89.03	−78.29	−92.50
6	−105.08	−107.91	−85.51	−92.88	−104.83	−123.77	−88.68	−97.99	−124.96	−103.38	−107.50	−101.76	−102.61
7	−156.65	−158.15	−130.71	−127.71	−141.14	−165.05	−110.63	−143.43	−155.81	−134.63	−163.91	−147.53	−143.67
8	−142.47	−144.76	−129.84	−144.30	−146.90	−147.75	−105.85	−130.09	−142.98	−127.19	−137.97	−141.72	−124.60
9	−93.35	−93.62	−78.27	−92.25	−83.92	−104.76	−78.17	−96.30	−87.97	−85.27	−107.07	−96.84	−90.63
10	−113.53	−109.57	−88.24	−104.42	−108.46	−129.40	−86.09	−112.86	−121.09	−108.90	−117.22	−119.46	−109.56
11	−142.68	−138.71	−135.09	−148.33	−148.01	−142.87	−115.42	−133.47	−140.45	−128.89	−153.61	−132.24	−126.99
12	−111.01	−104.12	−92.91	−106.53	−101.00	−121.63	−78.65	−92.82	−105.68	−97.98	−104.24	−105.62	−90.49
13	−127.70	−113.80	−104.98	−117.41	−115.64	−141.14	−87.77	−108.12	−121.31	−114.24	−114.03	−118.08	−100.97
14	−116.59	−116.18	−108.02	−120.02	−119.72	−139.34	−91.47	−113.85	−136.08	−115.29	−118.61	−122.42	−111.92
15	−151.38	−148.61	−154.18	−146.92	−153.19	−183.25	−127.38	−152.66	−164.79	−150.54	−163.67	−157.02	−150.51
16	−67.83	−79.06	−70.27	−90.64	−66.12	−82.86	−52.94	−61.67	−76.02	−77.82	−80.37	−72.42	−81.01
17	−85.51	−85.09	−76.06	−93.14	−82.76	−91.71	−63.40	−76.46	−90.42	−79.56	−99.96	−82.29	−87.81
18	−108.41	−104.64	−104.01	−119.95	−112.01	−110.74	−83.28	−98.06	−105.46	−89.43	−118.24	−104.13	−90.52
19	−94.06	−74.75	−77.07	−93.66	−79.56	−89.73	−54.71	−69.36	−78.56	−79.77	−75.01	−71.60	−78.72
20	−85.01	−77.39	−78.52	−85.22	−82.85	−97.48	−62.38	−76.30	−91.29	−79.35	−75.30	−79.85	−71.57
21	−99.68	−94.52	−107.46	−114.19	−108.07	−116.87	−79.39	−87.55	−106.93	−102.49	−97.32	−97.06	−103.61
Molegro													
1	−45.66	−42.21	−40.15	−46.30	−47.06	−49.96	−40.40	−43.52	−52.79	−53.06	−48.11	−52.08	−49.85
2	−59.33	−61.35	−48.45	−57.94	−59.04	−66.37	−52.56	−56.81	−65.42	−60.80	−61.96	−65.28	−59.28
3	−78.69	−86.29	−80.14	−87.82	−76.80	−91.25	−70.80	−84.05	−87.22	−88.79	−95.38	−87.92	−89.71
4	−73.50	−82.40	−65.33	−81.76	−78.29	−89.64	−63.70	−71.11	−80.07	−80.36	−88.19	−76.57	−85.37
5	−84.53	−86.07	−67.14	−81.79	−81.35	−93.52	−68.36	−82.27	−91.04	−83.31	−88.91	−82.65	−91.05
6	−104.61	−108.91	−86.76	−93.07	−103.26	−122.10	−85.12	−96.98	−116.04	−99.66	−108.78	−101.37	−107.32
7	−153.41	−163.83	−133.00	−120.14	−139.49	−165.52	−118.00	−139.06	−149.63	−140.05	−157.74	−142.84	−162.70
8	−122.26	−141.64	−123.08	−138.05	−137.15	−144.97	−100.60	−131.59	−134.89	−127.66	−139.28	−135.21	−123.90
9	−92.83	−93.73	−88.77	−89.42	−89.03	−106.92	−76.99	−93.14	−107.05	−94.00	−106.00	−96.60	−90.55
10	−113.37	−109.72	−94.83	−102.39	−105.85	−129.22	−90.12	−115.29	−131.03	−116.72	−130.17	−119.51	−110.91
11	−118.48	−139.99	−133.84	−150.77	−145.99	−142.17	−110.66	−134.65	−150.39	−129.23	−159.95	−142.94	−120.18
12	−105.71	−104.13	−92.70	−97.49	−102.97	−123.80	−77.55	−92.01	−107.15	−101.17	−103.01	−106.00	−102.62
13	−120.23	−113.93	−104.08	−111.83	−114.14	−138.90	−90.10	−107.68	−119.68	−119.93	−113.92	−115.29	−109.72
14	−115.97	−116.58	−108.38	−111.39	−118.66	−138.22	−90.15	−112.18	−135.19	−114.99	−119.08	−123.39	−109.29
15	−154.64	−153.59	−143.40	−152.91	−147.30	−185.69	−120.53	−149.49	−162.06	−155.65	−169.29	−155.60	−148.29
16	−65.39	−78.88	−69.92	−90.53	−66.07	−83.24	−52.38	−61.49	−76.15	−76.08	−81.25	−72.75	−80.87
17	−79.36	−85.66	−75.61	−91.56	−81.05	−92.70	−60.22	−75.95	−91.47	−79.43	−98.19	−82.84	−86.38
18	−112.40	−105.11	−108.50	−132.26	−108.90	−113.23	−76.97	−98.29	−110.76	−98.19	−118.44	−99.45	−108.10
19	−72.31	−68.09	−73.83	−87.89	−73.29	−94.22	−53.90	−67.46	−80.10	−83.46	−74.85	−66.48	−61.55
20	−75.31	−75.67	−79.35	−90.25	−84.93	−100.31	−61.62	−74.38	−88.95	−77.73	−75.50	−78.49	−70.52
21	−91.41	−91.74	−105.27	−115.93	−105.56	−120.80	−74.26	−87.73	−106.98	−102.54	−97.56	−98.39	−101.47
Vina													
1	−7.30	−7.00	−7.80	−6.40	−7.60	−7.50	−7.20	−7.00	−7.90	−7.60	−7.40	−8.00	−8.50
2	−7.70	−7.40	−7.80	−6.90	−7.50	−7.50	−7.40	−7.30	−8.50	−7.50	−8.00	−8.60	−9.10
3	−6.00	−5.50	−5.70	−5.60	−5.60	−5.90	−4.90	−5.70	−5.90	−6.40	−6.90	−6.50	−5.00
4	−7.90	−7.60	−8.80	−6.50	−8.00	−7.50	−6.90	−8.10	−8.60	−7.60	−7.60	−8.70	−8.80
5	−7.60	−7.40	−8.40	−6.40	−8.10	−7.80	−7.10	−7.60	−8.00	−7.60	−6.60	−7.80	−8.80
6	−7.60	−7.50	−8.70	−7.20	−8.90	−8.50	−7.90	−8.20	−8.50	−8.50	−6.90	−8.30	−9.60
7	−6.60	−7.20	−6.90	−6.30	−6.70	−6.10	−5.80	−6.00	−7.10	−6.90	−7.80	−7.30	−6.70
8	−7.10	−7.70	−7.40	−6.30	−7.00	−7.50	−6.50	−6.90	−8.70	−7.50	−7.00	−8.50	−9.30
9	−6.90	−7.60	−7.90	−6.20	−7.30	−6.70	−6.70	−6.50	−8.10	−7.50	−7.10	−7.50	−7.70
10	−7.20	−6.90	−7.70	−6.90	−7.30	−7.30	−6.10	−7.10	−8.40	−7.80	−7.00	−7.70	−8.60
11	−5.00	−5.20	−5.70	−6.70	−7.10	−5.70	−4.40	−7.20	−7.90	−6.50	−6.90	−5.80	−6.20
12	−7.30	−7.70	−7.90	−6.30	−7.30	−6.50	−6.50	−6.60	−7.50	−7.50	−8.10	−7.20	−7.70
13	−7.50	−7.60	−7.90	−6.40	−7.80	−7.90	−6.30	−8.00	−8.10	−8.40	−6.80	−7.60	−8.70
14	−7.10	−7.10	−7.90	−7.40	−7.50	−7.60	−6.80	−7.30	−8.30	−7.60	−6.80	−7.90	−8.90

TABLE 2: Continued.

Ligands	CA-II	Cat B	DNA-1	DNA-2	Gyrase	HDAC7	HIS	Kinase	rHA	RNR	Top II	TrxR	TS
15	−7.20	−7.80	−8.60	−7.50	−8.00	−8.50	−6.80	−8.10	−8.20	−7.60	−9.50	−7.90	−7.70
16	−5.50	−6.00	−5.70	−5.70	−6.80	−6.70	−5.60	−6.80	−6.80	−6.20	−6.60	−6.80	−6.60
17	−6.30	−7.20	−7.10	−5.80	−6.80	−7.10	−5.20	−7.10	−8.00	−7.00	−6.30	−7.30	−7.60
18	−5.70	−5.50	−5.10	−5.30	−6.20	−6.70	−5.50	−5.90	−6.40	−6.20	−6.10	−6.30	−6.20
19	−5.80	−5.60	−5.10	−4.90	−6.10	−6.90	−5.50	−5.90	−6.30	−6.20	−6.10	−5.80	−6.20
20	−7.00	−7.30	−7.80	−6.80	−6.90	−7.50	−6.90	−7.60	−7.80	−7.60	−9.10	−7.50	−8.40
21	−6.90	−5.90	−6.30	−6.00	−6.80	−7.30	−6.50	−6.70	−7.40	−6.90	−7.20	−7.40	−6.60
AutoDock-QC													
1	−3.67	−7.64	−7.56	−5.58	−5.13	−7.08	−5.40	−4.64	−4.65	−4.68	−5.00	−4.18	−5.99
2	−3.65	−7.36	−7.43	−5.27	−5.70	−7.42	−5.57	−4.79	−4.65	−4.54	−4.78	−4.77	−6.46
3	−4.12	−2.96	−2.81	−3.02	−4.31	−3.62	−3.55	−3.94	−3.54	−3.65	−6.04	−3.78	−3.63
4	−4.29	−7.72	−7.53	−5.19	−5.70	−7.09	−5.93	−4.70	−4.67	−5.05	−5.38	−5.20	−5.81
5	−3.93	−3.91	−3.74	−3.46	−4.23	−3.70	−3.66	−3.84	−3.61	−4.05	−3.17	−4.78	−4.41
6	−4.14	−4.30	−4.23	−4.18	−5.60	−4.17	−3.84	−4.54	−3.77	−4.85	−2.09	−5.56	−5.44
7	−7.13	−6.24	−5.95	−4.93	−7.01	−6.02	−4.76	−6.20	−5.86	−6.43	−7.22	−6.13	−6.11
8	−3.27	−5.58	−5.76	−4.25	−4.84	−5.36	−5.34	−3.80	−3.45	−4.65	−5.89	−4.98	−5.90
9	−3.34	−6.34	−5.09	−4.24	−5.01	−5.15	−4.99	−3.91	−3.55	−4.78	−4.84	−4.13	−4.88
10	−3.04	−7.12	−6.41	−4.58	−4.82	−5.79	−4.87	−4.69	−4.06	−4.25	−5.27	−5.00	−4.84
11	−5.04	−4.10	−3.15	−3.85	−4.56	−4.16	−3.79	−4.35	−3.30	−4.06	−4.62	−5.14	−4.86
12	−2.93	−6.40	−6.44	−5.13	−5.48	−6.28	−5.47	−4.56	−4.79	−3.47	−4.70	−3.83	−5.76
13	−3.07	−6.79	−6.42	−5.14	−5.74	−5.50	−5.50	−4.41	−4.58	−4.14	−4.34	−4.33	−5.98
14	−3.19	−7.40	−6.67	−5.30	−5.45	−6.01	−5.37	−4.43	−3.67	−4.34	−5.02	−4.59	−6.51
15	−4.36	−4.21	−2.29	−2.91	−4.15	−4.28	−0.94	−1.59	−2.86	−2.58	−4.26	−3.93	−3.11
16	−3.79	−2.98	−2.84	−3.04	−3.75	−3.72	−2.97	−3.54	−2.98	−3.62	−3.92	−3.62	−3.90
17	−3.04	−3.38	−2.92	−2.44	−3.44	−3.29	−2.72	−3.18	−2.64	−3.67	−3.97	−3.97	−4.20
18	−3.98	−4.09	−3.21	−3.35	−3.81	−3.88	−3.40	−4.01	−3.31	−4.63	−4.01	−4.07	−4.67
19	−5.27	−2.87	−2.54	−2.70	−3.51	−2.95	−3.41	−3.30	−2.50	−2.72	−2.62	−2.96	−2.96
20	−2.12	−1.04	−1.26	−1.57	−2.33	−1.53	−2.43	−2.14	−2.67	−1.95	−2.01	−3.10	−2.43
21	−4.24	−1.64	−2.01	−2.32	−1.91	−1.86	−3.17	−3.13	−2.71	−2.32	−2.51	−3.22	−2.76
AutoDock													
1	−2.13	−5.32	−5.94	−4.59	−4.13	−4.99	−4.33	−3.91	−3.66	−2.46	−2.81	−2.64	−4.41
2	−3.99	−8.02	−7.99	−5.90	−6.81	−6.55	−5.06	−5.08	−5.19	−5.13	−5.72	−6.34	−7.17
3	−9.22	−2.68	−3.98	−3.67	−3.65	−3.65	−4.83	−4.70	−5.19	−6.23	−8.47	−5.08	−4.16
4	−2.03	−5.30	−5.41	−4.24	−4.26	−5.18	−4.25	−3.68	−3.66	−2.89	−2.62	−2.89	−4.05
5	−4.45	−4.22	−4.49	−3.85	−5.07	−4.33	−4.37	−4.58	−4.48	−5.04	−3.59	−5.44	−4.98
6	−4.20	−3.58	−3.69	−3.82	−4.55	−3.83	−3.63	−4.27	−3.21	−3.96	−0.70	−4.64	−4.33
7	−7.79	−1.14	−2.48	−2.05	−2.97	−0.53	−5.76	−4.36	−4.99	−7.07	−6.86	−5.61	−3.31
8	−3.99	−6.18	−7.49	−5.30	−6.32	−6.18	−5.65	−4.92	−4.08	−5.22	−6.41	−6.31	−7.84
9	−3.98	−7.81	−7.28	−5.25	−6.77	−5.64	−5.93	−4.11	−5.07	−5.12	−5.36	−5.42	−6.13
10	−2.32	−5.97	−5.77	−4.27	−3.41	−4.39	−3.69	−3.40	−2.79	−2.42	−2.95	−3.24	−4.17
11	−6.67	−9.90	−9.92	−7.88	−8.20	−9.19	−8.43	−7.47	−5.89	−7.95	−7.55	−9.80	−9.35
12	−3.27	−6.35	−6.30	−4.78	−5.76	−5.77	−4.89	−3.76	−5.33	−3.90	−4.85	−4.05	−6.06
13	−4.57	−7.87	−7.24	−6.00	−7.06	−6.02	−5.59	−6.06	−5.92	−6.06	−6.75	−5.89	−7.19
14	−3.20	−6.78	−6.39	−5.23	−5.35	−5.30	−4.51	−4.08	−3.68	−3.74	−4.48	−4.20	−6.16
15	−9.45	−2.73	−2.45	−3.20	−3.14	−2.25	−4.05	−5.84	−3.28	−7.44	−8.14	−5.60	−5.35
16	−5.75	−5.50	−4.59	−4.98	−5.91	−5.99	−4.29	−5.67	−4.88	−6.61	−6.21	−6.42	−6.77
17	−2.92	−3.50	−3.00	−2.97	−3.50	−3.06	−2.95	−2.86	−2.78	−4.04	−3.44	−3.88	−4.36
18	−5.27	−5.44	−4.03	−4.29	−5.01	−5.38	−4.17	−5.08	−4.14	−6.17	−5.48	−5.88	−6.03
19	−5.17	−5.42	−4.31	−4.52	−5.30	−5.55	−4.21	−5.42	−4.66	−5.83	−5.21	−5.46	−6.27
20	−8.96	−3.53	−3.87	−5.28	−4.15	−4.62	−5.39	−5.56	−6.39	−5.48	−6.94	−6.46	−5.56
21	−5.99	−4.99	−4.82	−4.71	−5.47	−6.25	−4.79	−5.88	−5.04	−6.30	−5.69	−6.21	−7.05

then used for the analysis of the receptor interaction of ligands 7, **11**, and **15**. Although in many of the ligand interactions with the receptors, AutoDock and Vina gave similar features in their interaction, the ranking of the ligands according to their inhibitory interaction is vague. It has been pointed out that, to have a good ranking in AutoDock, the error values of ~2.177 kcal/mol [43, 44] have

to be considered, and this makes AutoDock unsuitable for ranking of these ligands because the differences in their interacting energy are lower.

The three best inhibitors among all the modelled compounds are the bis-pyrazole-based compounds 7, **11**, and **15** (Figure 3). The binding site interactions of these three ligands with each of the receptors are shown in Figure 4,

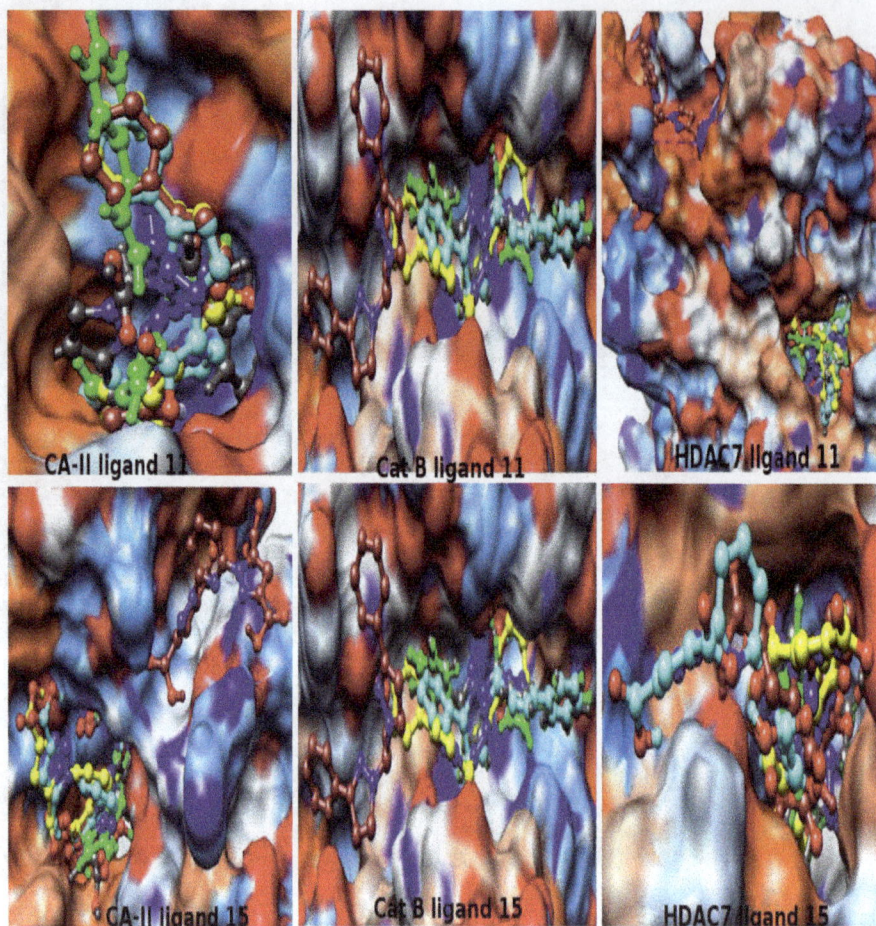

FIGURE 2: The binding site interaction of the ligands **11** and **15** with CA-II, ligand **11** with Cat B, and ligands **11** and **15** with HDAC7 using Molegro-QC (grey), Molegro (green), Vina (brown), AutoDock-QC (yellow), and AutoDock (cyan) methods.

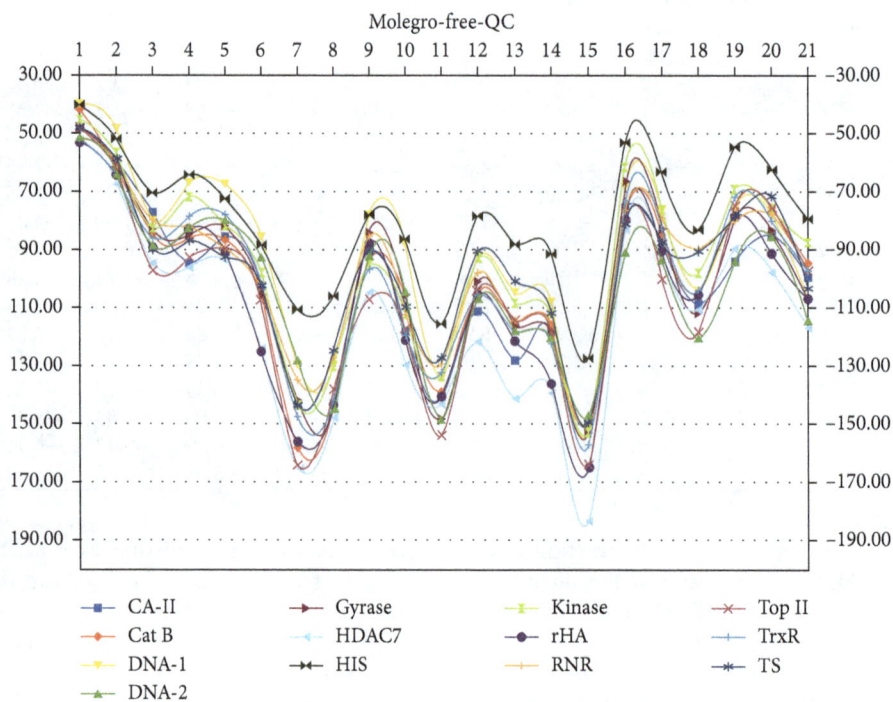

(a)

FIGURE 3: Continued.

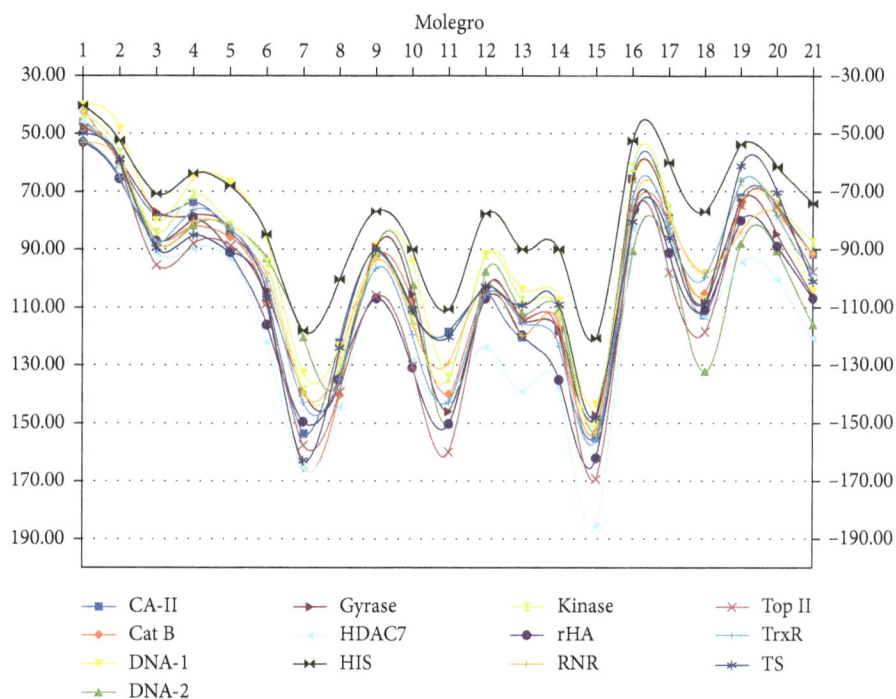

(b)

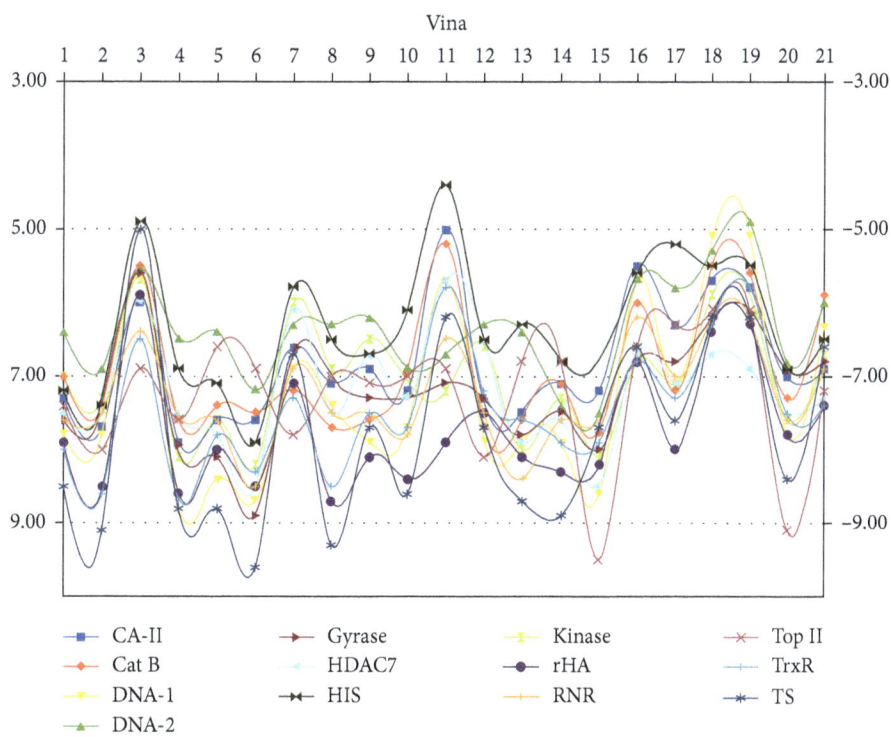

(c)

FIGURE 3: Continued.

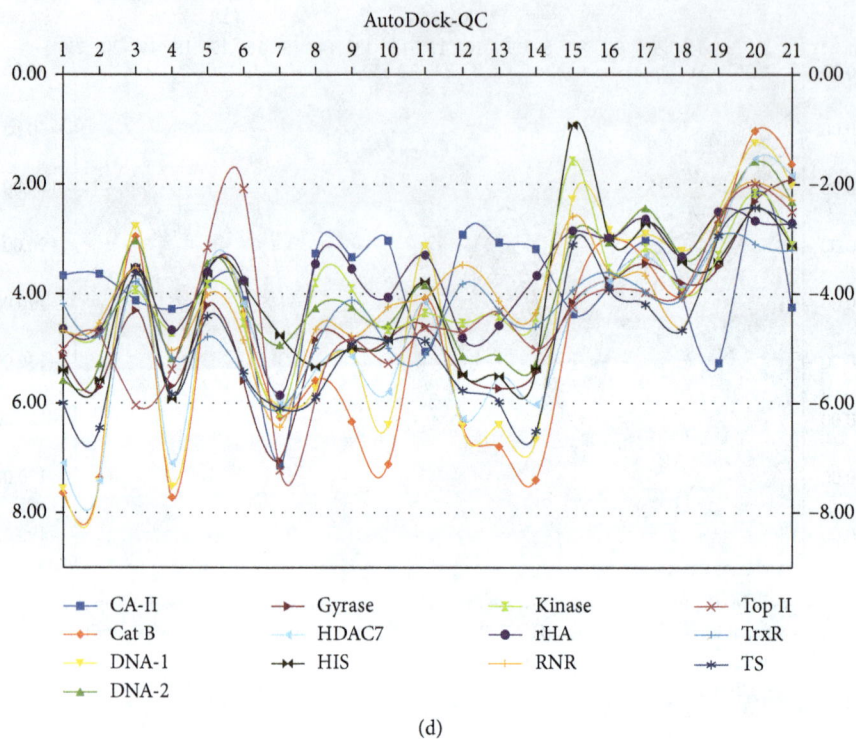

(d)

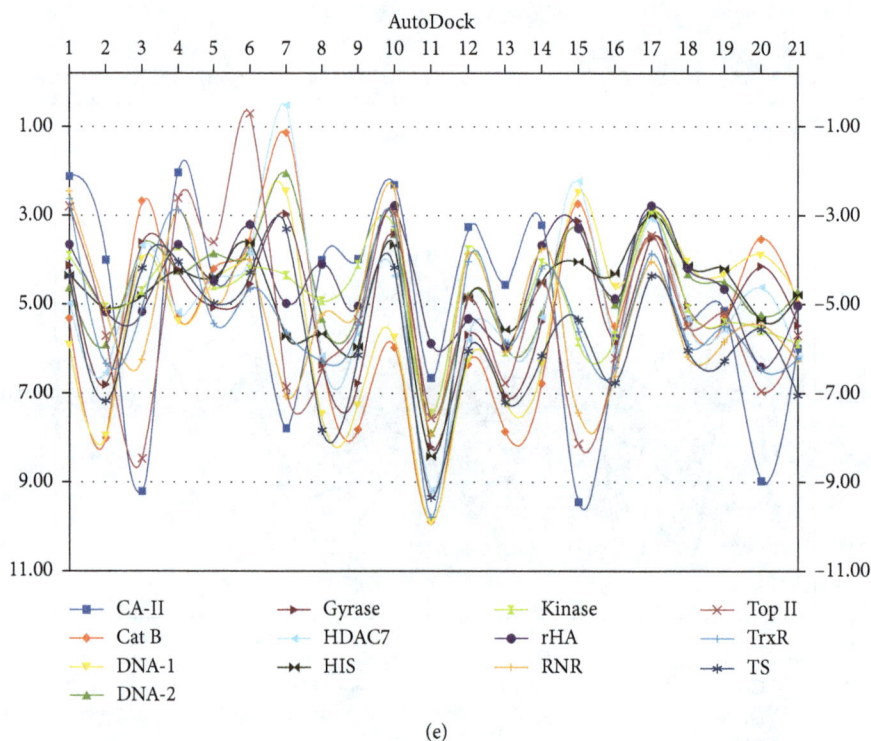

(e)

FIGURE 3: The plots of the energy of interaction of the ligands with each of the receptors using the five methods.

while the interactions of each of the compounds **7**, **11**, and **15** with each of the receptors are shown in supplementary Figure S1. In most of the receptors, the three compounds **7** (grey), **11** (green), and **15** (yellow) are found to be clustering around the same point and sharing common angles of interaction with the receptors CA-II, Cat B, DNA-2, Gyrase,

HDAC7, HIS, Kinase, rHA, RNR, Top II, and TS (Figure 4). The few exceptions to these common interaction features are found in the receptors DNA-1 and TrxR (Figure 4). A different binding site was predicted for the interaction of **11** with DNA-1 where **7** and **15** have inner groove interaction but the interaction of **11** is out of groove. In addition, the

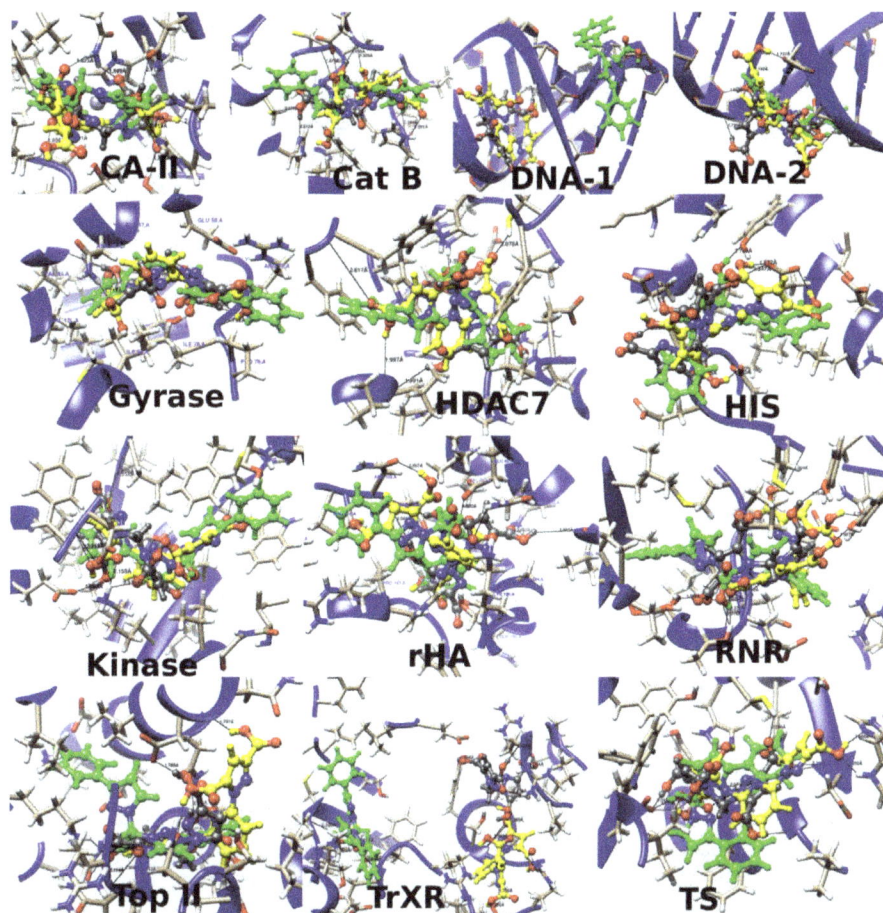

FIGURE 4: The binding site interactions of the three ligands **7** (grey), **11** (green), and **15** (yellow) with each of the receptors.

binding site predicted for **11** is separated from the one predicted for **7** and **15** in the receptor TrxR. In all of the receptor interactions especially for **7** and **15**, the predicted binding sites are the same as the binding site for their cocrystallized inhibitors except in the receptor DNA-2 where the three ligands bind to a groove outside of the binding surface of the cocrystallized *cis*-platin.

The receptor interactions of the three most promising compounds **7**, **11**, and **15** follow a different order based on their receptor preferences. The order of their inhibitory activities in HDAC7, Kinase, rHA, RNR, TrxR, and TS is **15** > **7** > **11**, while they follow the order of **7** > **5** > **11** in their interaction with CA-II, Cat B, and Top. The interaction of **11** was found to be better than that of **7** with the DNA-1, Gyrase, and HIS in the order of **15** > **11** > **7**, and also, the inhibitory activity of **11** against DNA-2 is found to be higher than that of **7** and **15** in the order of **11** > **15** > **7**. The presence of more than one carboxylic unit in **7** and **15** compared to **11** results in a greater number of HB interactions (Table 3), which could also be responsible for their greater inhibitory activities compared to **11** in many of the receptors. The number of the HB interactions of the ligand **11** in most of the receptors ranges from 0 to 2 except in the receptors Top II, TrxR, and TS where their binding site residues support more hydrogen bond interactions with the ligands (Table 3 and Figure S1).

TABLE 3: The number of hydrogen bonds (HBs) in the interactions of the ligands **7**, **11**, and **15** with the receptors.

Receptor	Ligand **7**	Ligand **11**	Ligand **15**
CA-II	6	0	5
Cat B	8	1	3
DNA-1	5	2	7
DNA-2	2	2	5
Gyrase	3	1	6
HDAC7	3	0	3
HIS	4	1	4
Kinase	5	0	4
rHA	4	2	2
RNR	6	1	7
Top II	9	4	7
TrxR	6	4	3
TS	7	4	7

4.3. Molecular Properties of the Compound. The molecular properties of the modelled compounds are shown in Table 4. Compounds **13** and **14** have the highest hyperpolarizability values, while **13** and **21** have the lowest band gap, but **14** and **21** have the highest dipole. The hyperpolarizability of the ligands follows the order of **13** > **14** > **12** > **21**. These four ligands can be useful as building blocks for nonlinear optical materials, but only the ligand **21** appears among the best five inhibitors obtained from the docking results. In order to

TABLE 4: The molecular properties of the modelled ligands.

	Energy	Zero energy	Thermal energy	Enthalpy	Gibbs free energy	CV	Entropy	HOMO	LUMO	Gap	Pol	Pol W	Hyp (esu)	Dipole
1	−226.22	−225.88	−225.88	−225.88	−225.91	13.89	65.46	−0.26	0.00	668.85	18.37	47.89	0.86	2.42
2	−304.87	−304.36	−304.35	−304.35	−304.39	26.38	85.47	−0.24	0.00	446.54	31.27	74.42	2.94	2.61
3	−603.38	−602.65	−602.64	−602.64	−602.68	34.79	98.26	−0.30	−0.09	563.27	42.55	92.52	1.84	3.55
4	−457.29	−456.58	−456.57	−456.57	−456.61	33.90	92.59	−0.22	−0.03	618.28	59.15	124.37	0.34	2.49
5	−490.54	−489.82	−489.81	−489.81	−489.85	33.65	96.73	−0.26	−0.02	622.64	106.83	109.44	1.69	3.58
6	−647.83	−646.77	−646.76	−646.76	−646.82	58.31	129.10	−0.23	−0.01	588.78	157.70	162.68	1.36	4.02
7	−1244.83	−1243.32	−1243.30	−1243.30	−1243.38	76.24	160.36	−0.30	−0.10	519.29	187.11	196.10	1.31	3.42
8	−952.69	−951.21	−951.20	−951.20	−951.27	73.73	150.85	−0.22	−0.04	473.61	265.57	273.88	2.71	3.21
9	−679.11	−678.19	−678.18	−678.18	−678.23	43.92	112.07	−0.26	−0.05	545.07	125.02	129.47	2.40	2.76
10	−836.40	−835.14	−835.12	−835.12	−835.19	68.58	141.09	−0.24	−0.05	503.87	175.52	182.12	2.47	3.59
11	−1141.26	−1139.59	−1139.57	−1139.57	−1139.65	84.04	166.49	−0.23	−0.06	447.56	284.93	296.55	4.19	2.35
12	−698.33	−697.32	−697.31	−697.31	−697.36	48.10	112.13	−0.24	−0.06	485.39	173.62	179.84	7.91	5.16
13	−886.90	−885.70	−885.68	−885.68	−885.74	58.91	129.78	−0.25	−0.10	411.44	196.81	205.34	11.21	3.21
14	−855.62	−854.28	−854.26	−854.26	−854.33	72.66	145.44	−0.23	−0.05	470.74	224.31	233.64	10.05	5.19
15	−1452.64	−1450.85	−1450.82	−1450.82	−1450.91	90.34	178.55	−0.27	−0.10	457.20	253.91	267.46	2.99	4.66
16	−495.41	−494.64	−494.63	−494.63	−494.67	35.89	94.59	−0.25	−0.06	502.88	133.18	136.67	0.48	3.24
17	−574.05	−573.11	−573.10	−573.10	−573.16	48.23	115.27	−0.24	−0.05	500.41	160.72	166.03	0.49	3.99
18	−872.57	−871.41	−871.39	−871.39	−871.45	54.72	120.56	−0.26	−0.10	428.47	177.26	184.39	1.61	2.67
19	−571.64	−570.77	−570.76	−570.76	−570.81	40.19	95.76	−0.24	−0.07	452.54	161.52	165.31	0.35	3.43
20	−650.29	−649.25	−649.24	−649.24	−649.29	52.75	116.19	−0.23	−0.06	500.97	191.88	197.84	3.40	2.30
21	−948.80	−947.54	−947.52	−947.52	−947.58	61.06	128.51	−0.26	−0.10	407.98	210.05	218.27	6.48	6.06

study the possible effects of changes in the molecular properties on the docking interactions with the receptors, the molecular properties of the compound correlated with the docking results obtained from the Molegro-QC method.

All the thermodynamic energy, zero energy, thermal energy, enthalpy, and the Gibbs free energy of the compound have a correlation range of 0.92 to 0.98 with the docking results. In addition, the CV and entropy highly correlate with the inhibitory values of the compounds in the ranges −0.90 to −0.96 for CV and −0.92 to −0.97 for entropy. Besides the thermodynamic properties, other properties that show high correlation with receptors' inhibitory activities are their polarizabilities (range from 0.74 to 0.90) and polarizabilities W (range from 0.78 to 0.93). Other computed properties like LUMO (0.46 to 0.64), gap (0.32 to 0.60), Hyp (−0.20 to −0.40), and dipole (−0.22 to −0.37) have very low correlation (as shown in the parenthesis). The HOMO has the poorest correlation with the binding activities of the ligands which range from 0.00 to 0.18. The better correlation of the LUMO with the receptor interactions further supports the hypothesis that the lower the LUMO energy of inhibitors the easier the overlapping of it with the HOMO of the DNA [54]. However, other factors like the hydrophobicity and electronic effect of the ligands play significant roles in their inhibitory activities [54].

5. Conclusion

The inhibitory potentials of nitrogen donor ligands in their interaction with twelve cancer-related receptors were studied using docking methods. Twenty-one derivatives of nitrogen-chelating ligands consisting of fifteen pyrazole, three bipyridine, and three phenanthroline derivatives were modelled. In addition, the experimental *in vitro* anticancer

activities of the six ligands **5**, **6**, **9**, **11**, **12**, and **14** against the cancer cell line HT29 are discussed. The result of the *in vitro* study shows that the ligands **11** and **14** are the most active ones compared to other ligands. Among the two, only the ligand **11** showed promising inhibitory activities towards the selected receptors. In many of the receptors, different docking methods locate similar binding sites and similar ligand orientations. The order of the ligand interactions with the receptors using Vina, AutoDock-DC, and AutoDock is inconsistent because they locate different binding sites in many cases and their energy difference is also within the range of the standard error of AutoDock (~2.177 kcal/mol), while the order from Molegro-QC and Molegro is highly consistent and similar.

The results from the Molegro-QC and Molegro showed that **7**, **11**, and **15** have better inhibitory activities towards many of the receptors and have similar angles of interaction with CA-II, Cat B, DNA-2, Gyrase, HDAC7, HIS, Kinase, rHA, RNR, Top II, and TS. This is an evidence that they have common residue interactions at the binding sites of each of the receptors. The receptor interactions of the three most promising ligands **7**, **11**, and **15** follow a different order based on their receptor preference. Apart from these three, **18** and **21** with bipyridine and phenanthroline moieties also show promising interaction with many of the receptors. The most common feature of the best five inhibitors is the carboxylic unit, which indicates that the carboxylic units enhance the binding site interaction of the ligands through formation of stronger HB interactions with the receptor residues. The thermodynamic properties have a high correlation with the docking results of the ligands' inhibitory activities. Other molecular properties, which have a high correlation with the ligands' inhibitory activities are polarizabilities. The LUMO of the ligands shows a good correlation with their receptor-binding interaction which further supports the hypothesis that the lower the LUMO energy of inhibitors the easier the overlapping of it with the HOMO of the DNA.

Acknowledgments

The authors gratefully acknowledge the financial support of Sasol and ESKOM TESP, South Africa. The authors also gratefully acknowledge the CHPC of the Republic of South Africa for providing the G09 and computing facilities.

References

[1] T. Bugarcic, A. Habtemariam, J. Stepankova et al., "The contrasting chemistry and cancer cell cytotoxicity of bipyridine and bipyridinediol ruthenium(II) arene complexes," *Inorganic Chemistry*, vol. 47, no. 24, pp. 11470–11486, 2008.

[2] M. Castellano-Castillo, H. Kostrhunova, V. Marini et al., "Binding of mismatch repair protein MutS to mispaired DNA adducts of intercalating ruthenium(II) arene complexes," *Journal of Biological Inorganic Chemistry*, vol. 13, no. 6, pp. 993–999, 2008.

[3] S. J. Dougan, A. Habtemariam, S. E. McHale, S. Parsons, and P. J. Sadler, "Catalytic organometallic anticancer complexes," *Proceedings of the National Academy of Sciences*, vol. 105, no. 33, pp. 11628–11633, 2008.

[4] S. J. Dougan, M. Melchart, A. Habtemariam, S. Parsons, and P. J. Sadler, "Phenylazo-pyridine and phenylazo-pyrazole chlorido ruthenium(II) arene complexes: arene loss, aquation, and cancer cell cytotoxicity," *Inorganic Chemistry*, vol. 45, no. 26, pp. 10882–10894, 2006.

[5] R. Fernández, M. Melchart, A. Habtemariam, S. Parsons, and P. J. Sadler, "Use of chelating ligands to tune the reactive site of half-sandwich ruthenium(II)-arene anticancer complexes," *Chemistry*, vol. 10, no. 20, pp. 5173–5179, 2004.

[6] Y. Fu, A. Habtemariam, A. M. Basri, D. Braddick, G. J. Clarkson, and P. J. Sadler, "Structure-activity relationships for organometallic osmium arene phenylazopyridine complexes with potent anticancer activity," *Dalton Transactions*, vol. 40, no. 40, pp. 10553–10562, 2011.

[7] A. Habtemariam, M. Melchart, R. Fernandez et al., "Structure-activity relationships for cytotoxic ruthenium(II) arene complexes containing N,N−, N,O−, and O,O-chelating ligands," *Journal of Medicinal Chemistry*, vol. 49, no. 23, pp. 6858–6868, 2006.

[8] H. Kostrhunova, J. Florian, O. Novakova, A. F. Peacock, P. J. Sadler, and V. Brabec, "DNA interactions of monofunctional organometallic osmium(II) antitumor complexes in cell-free media," *Journal of Medicinal Chemistry*, vol. 51, no. 12, pp. 3635–3643, 2008.

[9] A. F. Peacock, A. Habtemariam, S. A. Moggach, A. Prescimone, S. Parsons, and P. J. Sadler, "Chloro half-sandwich osmium(II) complexes: influence of chelated N,N-ligands on hydrolysis, guanine binding, and cytotoxicity," *Inorganic Chemistry*, vol. 46, no. 10, pp. 4049–4059, 2007.

[10] L. Xu, N. J. Zhong, H. L. Huang, Z. H. Liang, Z. Z. Li, and Y. J. Liu, "Synthesis, characterization, cellular uptake, apoptosis, cytotoxicity, dna-binding, and antioxidant activity studies of ruthenium(II) complexes," *Nucleosides Nucleotides Nucleic Acids*, vol. 31, no. 8, pp. 575–591, 2012.

[11] H. J. Yu, Y. Chen, L. Yu, Z. F. Hao, and L. H. Zhou, "Synthesis, visible light photocleavage, antiproliferative and cellular uptake properties of ruthenium complex [Ru(phen)2 (mitatp)]$^{2+}$," *European Journal of Medicinal Chemistry*, vol. 55, pp. 146–154, 2012.

[12] H. Niyazi, J. P. Hall, K. O'Sullivan et al., "Crystal structures of Λ-[Ru(phen)₂dppz]$^{2+}$ with oligonucleotides containing TA/TA and AT/AT steps show two intercalation modes," *Nature Chemistry*, vol. 4, no. 8, pp. 621–628, 2012.

[13] X. W. Liu, Z. G. Chen, L. Li, Y. D. Chen, J. L. Lu, and D. S. Zhang, "DNA-binding, photocleavage studies of ruthenium(II) complexes with 2-(2-quinolinyl)imidazo[4,5-f] [1,10]phenanthroline," *Spectrochimica Acta Part A: Molecular and Biomolecular Spectroscopy*, vol. 102, pp. 142–149, 2012.

[14] S. Gu, B. Liu, J. Chen, H. Wu, and W. Chen, "Synthesis, structures, and properties of ruthenium(II) complexes of N-(1,10-phenanthrolin-2-yl)imidazolylidenes," *Dalton Transactions*, vol. 41, no. 3, pp. 962–970, 2012.

[15] H. Henke, W. Kandioller, M. Hanif, B. K. Keppler, and C. G. Hartinger, "Organometallic ruthenium and osmium compounds of pyridin-2- and -4-ones as potential anticancer agents," *Chemistry and Biodiversity*, vol. 9, no. 9, pp. 1718–1727, 2012.

[16] S. David, R. S. Perkins, F. R. Fronczek, S. Kasiri, S. S. Mandal, and R. S. Srivastava, "Synthesis, characterization, and anticancer activity of ruthenium-pyrazole complexes," *Journal of Inorganic Biochemistry*, vol. 111, pp. 33–39, 2012.

[17] R. W. Sun, M. F. Ng, E. L. Wong et al., "Dual anti-angiogenic and cytotoxic properties of ruthenium(III) complexes containing pyrazolato and/or pyrazole ligands," *Dalton Transactions*, vol. 48, pp. 10712–10716, 2009.

[18] X. J. Zhu and B. J. Holliday, "Electropolymerization of a ruthenium(II) bis(pyrazolyl)pyridine complex to form a novel Ru-containing conducting metallopolymer," *Macromolecular Rapid Communications*, vol. 31, no. 9-10, pp. 904–909, 2010.

[19] S. S. Bhat, A. S. Kumbhar, A. A. Kumbhar, and A. Khan, "Efficient DNA condensation induced by ruthenium(II) complexes of a bipyridine-functionalized molecular clip ligand," *Chemistry*, vol. 18, no. 51, pp. 16383–16392, 2012.

[20] P. U. Maheswari, V. Rajendiran, M. Palaniandavar, R. Thomas, and G. U. Kulkarni, "Mixed ligand ruthenium(II) complexes of 5,6-dimethyl-1,10-phenanthroline: the role of ligand hydrophobicity on DNA binding of the complexes," *Inorganica Chimica Acta*, vol. 359, no. 14, pp. 4601–4612, 2006.

[21] I. N. Stepanenko, A. Casini, F. Edafe et al., "Conjugation of organoruthenium(II) 3-(1H-benzimidazol-2-yl)pyrazolo[3,4-b]pyridines and indolo[3,2-d]benzazepines to recombinant human serum albumin: a strategy to enhance cytotoxicity in cancer cells," *Inorganic Chemistry*, vol. 50, no. 24, pp. 12669–12679, 2011.

[22] G. Sava, A. Bergamoa, and P. J. Dyson, "Metal-based antitumour drugs in the post-genomic era: what comes next?," *Dalton Transactions*, vol. 40, no. 36, pp. 9069–9075, 2011.

[23] E. Meggers, "Targeting proteins with metal complexes," *Chemical Communications*, no. 9, pp. 1001–1010, 2009.

[24] D. Musil, D. Zucic, D. Turk et al., "The refined 2.15 A X-ray crystal structure of human liver cathepsin B: the structural basis for its specificity," *EMBO Journal*, vol. 10, pp. 2321–2330, 1991.

[25] Y. Zhao, C. Biertümpfel, M. T. Gregory, Y. J. Hua, F. Hanaoka, and W. Yang, "Structural basis of human DNA polymerase η-mediated chemoresistance to cisplatin," *PNAS*, vol. 109, no. 19, pp. 7269–7274, 2012.

[26] G. A. Holdgate, A. Tunnicliffe, W. H. Ward et al., "The en-

tropic penalty of ordered water accounts for weaker binding of the antibiotic novobiocin to a resistant mutant of DNA gyrase: a thermodynamic and crystallographic study," *Biochemistry*, vol. 36, no. 32, pp. 9663–9673, 1997.

[27] A. Schuetz, J. Min, A. Allali-Hassani et al., "Human HDAC7 harbors a class IIa histone deacetylase-specific zinc binding motif and cryptic deacetylase activity," *Journal of Biological Chemistry*, vol. 283, no. 17, pp. 11355–11363, 2008.

[28] B. Zu, M. S. Ong, M. Groessl et al., "A ruthenium antimetastasis agent forms specific histone protein adducts in the nucleosome core," *Chemistry*, vol. 17, no. 13, pp. 3562–3566, 2011.

[29] P. Xie, C. Streu, J. Qin et al., "The crystal structure of BRAF in complex with an organoruthenium inhibitor reveals a mechanism for inhibition of an active form of BRAF kinase," *Biochemistry*, vol. 48, no. 23, pp. 5187–5198, 2009.

[30] S. Sugio, A. Kashima, S. Mochizuki, M. Noda, and K. Kobayashi, "Crystal structure of human serum albumin at 2.5 A resolution," *Protein Engineering, Design and Selection*, vol. 12, no. 6, pp. 439–446, 1999.

[31] M. Eriksson, U. Uhlin, S. Ramaswamy et al., "Binding of allosteric effectors to ribonucleotide reductase protein R1: reduction of active-site cysteines promotes substrate binding," *Structure*, vol. 5, no. 8, pp. 1077–1092, 1997.

[32] S. Classen, S. Olland, and J. M. Berger, "Structure of the topoisomerase II ATPase region and its mechanism of inhibition by the chemotherapeutic agent ICRF-187," *PNAS*, vol. 100, no. 19, pp. 10629–10634, 2003.

[33] T. Sandalova, L. Zhong, Y. Lindqvist, A. Holmgren, and G. Schneider, "Three-dimensional structure of a mammalian thioredoxin reductase: implications for mechanism and evolution of a selenocysteine-dependent enzyme," *PNAS*, vol. 98, no. 17, pp. 9533–9538, 2001.

[34] Z. Newby, T. T. Lee, R. J. Morse et al., "The role of protein dynamics in thymidylate synthase catalysis: variants of conserved 2′-deoxyuridine 5′-monophosphate (dUMP)-binding Tyr-261," *Biochemistry*, vol. 45, no. 24, pp. 7415–7428, 2006.

[35] W. Hu, Q. Luo, X. Ma et al., "Arene control over thiolate to sulfinate oxidation in albumin by organometallic ruthenium anticancer complexes," *Chemistry–A European Journal*, vol. 15, no. 27, pp. 6586–6594, 2009.

[36] M. Hanif, H. Henke, S. M. Meier et al., "Is the reactivity of M(II)-arene complexes of 3-hydroxy-2(1H)-pyridones to biomolecules the anticancer activity determining parameter?," *Inorganic Chemistry*, vol. 49, no. 17, pp. 7953–7963, 2010.

[37] J. M. Berg, J. L. Tymoczko, and L. Stryer, *Biochemistry*, W H Freeman, New York, NY, USA, 5th edition, 2002.

[38] V. Kettmann, D. Kost'alova, and H. Holtje, "Human topoisomerase I poisoning: docking protoberberines into a structure-based binding site model," *Journal of Computer-Aided Molecular Design*, vol. 18, no. 12, pp. 785–796, 2004.

[39] S. Mokesch, M. S. Novak, A. Roller, M. A. Jakupec, W. Kandioller, and B. K. Keppler, "1,3-dioxoindan-2-carboxamides as bioactive ligand scaffolds for the development of novel organometallic anticancer drugs," *Organometallics*, vol. 34, no. 5, pp. 848–857, 2015.

[40] A. Casini, C. Gabbiani, F. Sorrentino et al., "Emerging protein targets for anticancer metallodrugs: inhibition of thioredoxin reductase and cathepsin B by antitumor ruthenium(II)-arene compounds," *Journal of Medicinal Chemistry*, vol. 51, no. 21, pp. 6773–6781, 2008.

[41] C. J. World, H. Yamawaki, and B. C. Berk, "Thioredoxin in the cardiovascular system," *Journal of Molecular Medicine*, vol. 84, no. 12, pp. 997–1003, 2006.

[42] W. Zhuang, X. Wu, Y. Zhou et al., "Polymorphisms of thymidylate synthase in the 5′- and 3′-untranslated regions and gastric cancer," *Digestive Diseases and Sciences*, vol. 54, no. 7, pp. 1379–1385, 2009.

[43] A. A. Adeniyi and P. A. Ajibade, "The spectroscopic and electronic properties of dimethylpyrazole and its derivatives using the experimental and computational methods," *Spectroscopy and Spectral Analysis*, vol. 34, no. 9, pp. 2305–2319, 2014.

[44] A. A. Adeniyi and P. A. Ajibade, "Experimental and theoretical investigation of the spectroscopic and electronic properties of pyrazolyl ligands," *Spectrochimica Acta Part A: Molecular and Biomolecular Spectroscopy*, vol. 133, pp. 831–845, 2014.

[45] A. A. Adeniyi and P. A. Ajibade, "Synthesis of pyrazole derivatives and their spectroscopic properties using both experimental and theoretical approaches," *New Journal of Chemistry*, vol. 38, no. 9, pp. 4120–4129, 2014.

[46] C. Adamo and V. Barone, "Toward reliable density functional methods without adjustable parameters: the PBE0 model," *Journal of Chemical Physics*, vol. 110, no. 13, pp. 6158–6170, 1999.

[47] M. J. Frisch, G. W. Trucks, H. B. Schlegel et al., *Gaussian 03, Revision C. 02*, Gaussian Inc., Wallingford, CT, USA, 2009.

[48] R. Thomsen and M. H. Christensen, "MolDock: a new technique for high-accuracy molecular docking," *Journal of Medicinal Chemistry*, vol. 49, no. 11, pp. 3315–3321, 2006.

[49] G. M. Morris, R. Huey, W. Lindstrom et al., "AutoDock4 and AutoDockTools4: automated docking with selective receptor flexiblity," *Journal of Computational Chemistry*, vol. 30, no. 16, pp. 2785–2791, 2009.

[50] N. S. Sapre, S. Gupta, and N. Sapre, "Assessing ligand efficiencies using template-based molecular docking and Tabu-clustering on tetrahydroimidazo-[4,5,1-jk][1,4]-benzodiazepin-2(1H)-one and -thione (TIBO) derivatives as HIV-1RT inhibitors," *Journal of Chemical Sciences*, vol. 120, no. 4, pp. 395–404, 2008.

[51] G. M. Morris, D. S. Goodsell, R. S. Halliday et al., "Automated docking using a Lamarckian genetic algorithm and an empirical binding free energy function," *Journal of Computational Chemistry*, vol. 19, no. 14, pp. 1639–1662, 1998.

[52] N. A. Caballero, F. J. Meléndez, A. Niño, and C. Muñoz-Caro, "Molecular docking study of the binding of aminopyridines within the K+ channel," *Journal of Molecular Modeling*, vol. 13, no. 5, pp. 579–586, 2007.

[53] E. F. Pettersen, T. D. Goddard, C. C. Huang et al., "UCSF Chimera–a visualization system for exploratory research and analysis," *Journal of Computational Chemistry*, vol. 25, no. 13, pp. 1605–1612, 2004.

[54] X. Chen, F. Gao, Z. X. Zhou, W. Y. Yang, L. T. Guo, and L. N. Ji, "Effect of ancillary ligands on the topoisomerases II and transcription inhibition activity of polypyridyl ruthenium (II) complexes," *Journal of Inorganic Biochemistry*, vol. 104, no. 5, pp. 576–582, 2010.

11

Impact of Different Serum Potassium Levels on Postresuscitation Heart Function and Hemodynamics in Patients with Nontraumatic Out-of-Hospital Cardiac Arrest

Yan-Ren Lin [ID],[1,2,3] Yuan-Jhen Syue [ID],[4] Tsung-Han Lee,[1] Chu-Chung Chou,[1,3] Chin-Fu Chang,[1] and Chao-Jui Li [ID][5,6]

[1]Department of Emergency Medicine, Changhua Christian Hospital, Changhua, Taiwan
[2]School of Medicine, Kaohsiung Medical University, Kaohsiung, Taiwan
[3]School of Medicine, Chung Shan Medical University, Taichung, Taiwan
[4]Department of Anaesthesiology, Kaohsiung Chang Gung Memorial Hospital,
 Chang Gung University College of Medicine, Kaohsiung, Taiwan
[5]Department of Emergency Medicine, Kaohsiung Chang Gung Memorial Hospital,
 Chang Gung University College of Medicine, Kaohsiung, Taiwan
[6]Department of Public Health, College of Health Science, Kaohsiung Medical University, Kaohsiung, Taiwan

Correspondence should be addressed to Chao-Jui Li; chaojui@cgmh.org.tw

Academic Editor: Spyros P. Perlepes

Background. Sustained return of spontaneous circulation (ROSC) can be initially established in patients with out-of-hospital cardiac arrest (OHCA); however, the early postresuscitation hemodynamics can still be impaired by high levels of serum potassium (hyperkalemia). The impact of different potassium levels on early postresuscitation heart function has remained unclear. We aim to analyze the relationship between different levels of serum potassium and postresuscitation heart function during the early postresuscitation period (the first hour after achieving sustained ROSC). *Methods.* Information on 479 nontraumatic OHCA patients with sustained ROSC was retrospectively obtained. Measures of early postresuscitation heart function (rate, blood pressure, and rhythm), hemodynamics (urine output and blood pH), and the duration of survival were analyzed in the case of different serum potassium levels (low: <3.5; normal: 3.5–5; high: >5 mmol/L). *Results.* Most patients (59.9%, n = 287) had previously presented with high levels of potassium. Bradycardia, nonsinus rhythm, urine output <1 ml/kg/hr, and acidosis (pH < 7.35) were more common in patients with high levels of potassium (all $p < 0.05$). Compared with hyperkalemia, a normal potassium level was more likely to be associated with a normal heart rate (OR: 2.97, 95% CI: 1.74–5.08) and sinus rhythm (OR: 2.28, 95% CI: 1.45–3.58). A low level of potassium was more likely to be associated with tachycardia (OR: 3.54, 95% CI: 1.32–9.51), urine output >1 ml/kg/hr (OR: 5.35, 95% CI: 2.58–11.10), and nonacidosis (blood pH >7.35, OR: 7.74, 95% CI: 3.78–15.58). The duration of survival was shorter in patients with hyperkalemia than that in patients whose potassium levels were low or normal ($p < 0.05$). *Conclusion.* Early postresuscitation heart function and hemodynamics were associated with the serum potassium level. A high potassium level was more likely to be associated with bradycardia, nonsinus rhythm, urine output <1 ml/kg/hr, and acidosis. More importantly, a high potassium level decreased the duration of survival.

1. Introduction

The average survival rate of patients with out-of-hospital cardiac arrest (OHCA) is only 3 to 15% [1–6]. Although a few patients can be successfully resuscitated by cardiopulmonary resuscitation (CPR) in the emergency department (ED), up to only 39% of them ultimately survive to discharge [5, 7, 8]. Some previous studies reported that the potential predictors of outcomes were early prehospital resuscitation, high-quality CPR, early percutaneous coronary intervention, and restoration of electrolyte balance [5, 9–15]. Among these factors, the serum potassium level was recognized as a key factor reflecting certain current body conditions, including abnormal cellular metabolism, the degree of cell death, electrolyte exchange, and insufficient kidney function [16–19]. Clinically, a high level of serum potassium (hyperkalemia) might potentially result in life-threatening complications by changing the action potential of heart rhythm cells and further inducing critical cardiac arrhythmia. Once cardiac arrhythmia has occurred, the impact on hemodynamic status is substantial. Cardiac output and vital organ perfusion can both collapse in a short period of time [16, 20–22].

The heart cells of OHCA patients unavoidably suffer hypoxic injury or ischemia/reperfusion injury during the cardiac arrest period (i.e., a period of no or low perfusion) [23–25]. The resting membrane potential of these damaged heart cells is more easily affected than that of normal heart cells [26–29]. Since a high level of serum potassium changes the resting membrane potential and decreases the ability to reach the action potential, we suspect that the impact of hyperkalemia on the regulation of the heart rhythm would be more obvious during the early postresuscitation period (the first hour after achieving sustained return of spontaneous circulation (ROSC)). We also suspect that the postresuscitation heart function and hemodynamic status (i.e., blood pressure, heart rate, heart rhythm, and the amount of urine output) established by CPR might be more easily destroyed by hyperkalemia [30, 31].

However, the relationship between different levels of serum potassium and the outcomes of OHCA patients had not been well addressed. In particular, the impact on early postresuscitation heart function was still unclear. Therefore, in this study, we aimed to analyze the relationships between different serum potassium levels and measures of postresuscitation heart function during the early postresuscitation period.

2. Methods

2.1. Study Design. Adult patients presenting with non-traumatic OHCA at the EDs of the Changhua Christian Hospital (CCH) medical system (one medical center and eight satellite hospitals) from January 1, 2009, to December 31, 2012, were included in this study. Patient characteristics, demographics, and serum potassium levels (low, normal, and high) that might correlate with postresuscitation heart function and hemodynamics were analyzed.

2.2. Ethics Statement. This paper reports a retrospective study. The protocol of this study was approved by the Institutional Review Board (IRB) of Changhua Christian Hospital (permission code: 121007).

2.3. Study Setting and Population. The medical records of patients presenting with nontraumatic OHCA at the EDs of the CCH medical system during the study period were retrospectively reviewed. The CCH medical system is located in central Taiwan and covers a population of almost 2,500,000. Electronic medical records can be shared among these hospitals. Nontraumatic OHCA patients were not

included in this study if they had any one of the following characteristics: (1) cardiac arrest was caused by trauma, electric or burn injuries, or drowning; (2) age was <19 years (pediatric patients); (3) no resuscitation attempts were performed; (4) sustained ROSC was not achieved; and (5) medical records were incomplete (including a lack of laboratory data). Therefore, this study ultimately included 479 patients.

2.4. Study Protocol. Information regarding the prehospital resuscitation (including the period from scene to hospital) was obtained from public emergency medical system records or witness statements. Information on patient characteristics, demographics, clinical features, postresuscitation hemodynamics, outcome, and survival was obtained from medical charts (records made by physicians or nurses). All data were recorded according to the Utstein report system [32]. The resuscitation attempts for each patient in the EDs were based on advanced cardiopulmonary life support (ACLS), which is a standard resuscitation protocol supported by the American Heart Association (AHA). In this study, the possible etiologies that caused OHCA were classified as follows: (1) infection; (2) cardiovascular disease; (3) malignancy; (4) asphyxia; (5) electrolyte problem; (6) hypovolemia; and (7) other or unknown cause. The initial cardiac rhythm upon presentation to the ED was obtained using an electrocardiographic monitor. The rhythms were classified as asystole, pulseless electrical activity (PEA) or ventricular fibrillation (VF). In this study, VF also included pulseless ventricular tachycardia. The rhythm was obtained immediately upon arrival of the patient. In addition, inhospital CPR duration was defined as the CPR time in the ED. Once a patient achieved a sustained ROSC, the time after sustained ROSC was defined as the postresuscitation period. The initial laboratory data (the median data on blood gas, potassium levels, and creatinine clearance measured in the first 24 hours of the postresuscitation period) that might correlate with the chance of survival to discharge were analyzed.

The initial postresuscitation heart function measures and hemodynamics were recorded in the first hour of the postresuscitation period. Heart function-related factors were heart rate (tachycardia, normal, and bradycardia), blood pressure (hypertension, normal, and hypotension), and heart rhythm (sinus and nonsinus rhythms). The factor used to reflect kidney function and end organ perfusion was urine output (>1, <1 ml/kg/hr). The final factor used to reflect cell metabolism or gas exchange was blood pH level, which was recorded as acidosis (pH < 7.35) or nonacidosis (pH > 7.35). Since the values of these measurements can be very dynamic, the values were classified according to the mean value if more than one value was recorded during the first hour of the postresuscitation period. Finally, the associations between the above factors and different serum potassium levels (low: <3.5; normal: 3.5–5; high: >5 mmol/L) were analyzed. To understand the detailed dynamic changes in heart rates, the mean values of postresuscitation heart rates were recorded according to four different time periods (up to 15,

TABLE 1: Characteristics and clinical features of the patients.

	Patients who suffered nontraumatic OHCA and achieved sustained ROSC ($n = 479$)	
	Number	%
Personal information		
Age (mean ± SD) (y/o)	70.7 ± 12.4	
Male	271	56.6
Possible etiologies		
Infection	99	20.7
Cardiovascular disease	143	29.9
Malignancy	45	9.4
Asphyxia	114	23.8
Electrolyte problem	13	2.7
Hypovolemia	29	6.1
Other or unknown cause	36	7.5
Prehospital information		
Period from scene to hospital (mean ± SD) (min)	24.8 ± 10.8	
Inhospital resuscitation		
Initial cardiac rhythm		
Asystole	299	62.4
PEA	70	14.6
VF*	110	23.0
Inhospital CPR duration (mean ± SD) (min)	16.9 ± 9.4	
Outcome measurement		
Survival to discharge	132	25.3

*VF includes patients with pulseless VT.

16–30, 31–45, and 46–60 minutes). The correlation between the mean heart rates and different potassium levels was analyzed.

2.5. Data Analysis. Descriptive analyses were reported as percentage or mean ± standard deviation (SD) for independent variables (including age, sex, possible etiology, the period from the scene to the hospital, initial cardiac rhythm, inhospital CPR duration, and outcome). The associations between postresuscitation, heart function measures, and hemodynamics and different serum potassium levels were analyzed using the chi-square test. Furthermore, we used multinomial logistic regression analysis to determine the strongest effects at different potassium levels. A correlation analysis (Spearman's rank correlation test) between prehospital resuscitation duration and different serum potassium levels was performed. Laboratory data were also compared between survivors and nonsurvivors using the Mann–Whitney U test. The mean dynamic heart rates in patients with different potassium levels were analyzed by using one-way ANOVA. Finally, the relationships between the duration of survival and different potassium levels were analyzed using survival analyses (Kaplan–Meier curves).

3. Results

3.1. Patient Characteristics and Demographics. During the study period, a total of 479 patients were included. Their characteristics are shown in Table 1. Most of the patients presented with asystole (62.4%, $n = 299$) as their initial cardiac rhythm when arriving at the ED. Of all the patients, 25.3% ($n = 132$) survived to discharge from the hospital.

3.2. Serum Potassium Levels Influenced Hemodynamics. The associations between initial postresuscitation hemodynamics, heart function measures, and serum potassium levels are presented in Table 2. The heart rate, heart rhythm, urine output, and blood pH level differed significantly between the three levels of serum potassium (all $p < 0.05$). Comparing to the patients with low or normal levels of potassium, the patients with high potassium levels (hyperkalemia) were significantly more likely to suffer bradycardia (41.1%, $n = 118$), nonsinus rhythm (46.0%, $n = 132$), urine output < 1 ml/kg/hr (69.3%, $n = 199$), and acidosis (87.8%, $n = 252$). Finally, we also found that the longer prehospital resuscitation duration significantly associated with the higher levels of potassium ($r = 0.75$, $p < 0.05$).

3.3. Considering the Strength of Effects. Multinomial logistic regression analysis was performed to consider the strength of the effects on heart function and hemodynamic status (Table 3). Compared with a high level of potassium, a normal level was associated with a higher likelihood of presenting with a normal heart rate (OR: 2.97, 95% CI: 1.74–5.08) and sinus rhythm (OR: 2.28, 95% CI: 1.45–3.58). In addition, a low level of potassium was associated with a higher likelihood of presenting with tachycardia (OR: 3.54, 95% CI: 1.32–9.51), urine output > 1 ml/kg/hr (OR: 5.35, 95% CI: 2.58–11.10), and nonacidosis (blood pH > 7.35, OR: 7.74, 95% CI: 3.78–15.58).

3.4. Dynamic Heart Rates. The dynamic postresuscitation heart rates at different serum potassium levels are shown in Figure 1. Compared with low or normal potassium levels,

TABLE 2: Serum potassium levels influence the initial postresuscitation heart function and hemodynamic status.

| | Total patients ($n = 479$) | Serum potassium level | | | p value |
| | | Low ($K^+ < 3.5$, $n = 48$) | Normal ($K^+ = 3.5-5$, $n = 144$) | High ($K^+ > 5$, $n = 287$) | |
	Number (%)	Number (%)	Number (%)	Number (%)	
Heart rate*					<0.001
Tachycardia	121 (25.3)	17 (35.4)	43 (29.9)	61 (21.3)	
Normal	206 (43.0)	23 (47.9)	75 (52.1)	108 (37.6)	
Bradycardia	152 (31.7)	8 (16.7)	26 (18.0)	118 (41.1)	
Blood pressure					0.353
Hypertension	149 (31.1)	11 (22.9)	49 (34.0)	89 (31.0)	
Normal	142 (29.6)	16 (33.3)	47 (32.6)	79 (27.5)	
Hypotension	188 (39.3)	21 (43.8)	48 (33.4)	119 (41.5)	
Heart rhythm*					0.001
Sinus rhythm	291 (60.8)	33 (68.8)	103 (71.5)	155 (54.0)	
Nonsinus rhythm	188 (39.2)	15 (31.2)	41 (28.5)	132 (46.0)	
Urine output (median)*					<0.001
>1 (ml/kg/hr)	174 (36.3)	35 (72.9)	51 (35.4)	88 (30.7)	
<1 (ml/kg/hr)	305 (63.7)	13 (27.1)	93 (64.6)	199 (69.3)	
Blood pH*					
Acidosis (<7.35)	374 (78.1)	20 (41.7)	102 (70.8)	252 (87.8)	<0.001
Nonacidosis (>7.35)	105 (21.9)	28 (58.3)	42 (29.2)	35 (12.2)	

*Significant factors; the serum K^+ level is given in units of mmol/L.

TABLE 3: Multinomial logistic regression analysis for analyzing the strength of effects on heart function and hemodynamic status at different serum potassium levels.

| | Serum potassium level | | | | |
| | Low ($K^+ < 3.5$) | | Normal ($K^+ = 3.5-5$) | | High[†] ($K^+ > 5$) |
	OR	95% CI	OR	95% CI	—
Heart rate					
Tachycardia*	3.54	1.32–9.51	2.85	1.56–5.23	1
Normal*	2.68	1.05–6.82	2.97	1.74–5.08	1
Bradycardia[†]	—	—	—	—	—
Blood pressure					
Hypertension	0.47	0.19–1.12	1.05	0.62–1.76	1
Normal	1.01	0.45–2.28	1.33	0.79–2.26	1
Hypotension[†]	—	—	—	—	—
Heart rhythm					
Sinus rhythm*	2.05	0.99–4.24	2.28	1.45–3.58	1
Nonsinus rhythm[†]	—	—	—	—	—
Urine output (median)					
>1 (ml/kg/hr)*	5.35	2.58–11.1	1.17	0.75–1.85	1
<1 (ml/kg/hr)[†]	—	—	—	—	—
Blood pH					
Nonacidosis (>7.35)*	7.74	3.78–15.85	2.67	1.57–4.53	1
Acidosis (<7.35)[†]	—	—	—	—	—

[†]Reference group; *significant factors; OR: odds ratio; CI: confidence interval; the serum K^+ level is given in units of mmol/L.

a high level was clearly related to slower mean heart rates, especially in the first 30 minutes of the postresuscitation period (both $p < 0.05$).

3.5. Potassium Levels in Serum Were Associated with Survival.

Initial laboratory data (including blood gas levels, pH, and potassium levels in the first 24 hours of the postresuscitation period) were significantly associated with the chance of achieving survival to discharge (all $p < 0.05$). We found that the initial potassium levels of the patients who ultimately survived to discharge were clearly lower than those of the patients who died during the hospital stay. In addition, the survivors had higher levels of pH and PaO_2 than nonsurvivors. Initial creatinine clearance was also associated with survival to discharge (all $p < 0.001$, Table 4). Furthermore, survival analysis also showed that the patients with high levels of potassium had the shortest duration of survival (Figure 2).

4. Discussion

In this study, we found that a high level of serum potassium was a very common complication (59.9%, $n = 287$) during

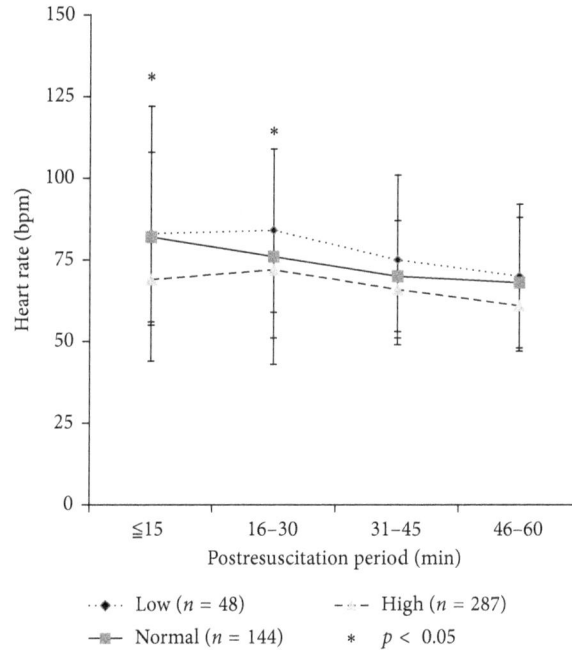

FIGURE 1: The dynamic postresuscitation heart rates at different serum potassium levels.

TABLE 4: Initial potassium levels in serum were associated with the chance of survival (the median laboratory data during the first 24 hours of the postresuscitation period).

	Total patients ($n = 479$)	Survival to discharge ($n = 132$)		
		Success	Failure	p value
Initial K level (median) (mmol/L)	4.4	4.2	5.1	<0.001
Initial pH (median)	7.10	7.14	7.08	<0.001
Initial PaO$_2$ level (median) (mmHg)	164.2	185.2	127.3	<0.001
Initial PaCO$_2$ level (median) (mmHg)	58.6	49.0	62.7	<0.001
Initial creatinine clearance (median) (mL/s)	1.2	0.9	1.4	<0.001

the postresuscitation period. Once OHCA occurred, several factors including hypoxic injury, ischemia/reperfusion injury, cell injury, and renal failure contribute to the rise in the potassium level [16, 17, 33]. Some previous studies have demonstrated that high level potassium might induce cardiac arrest and decrease the chance of successful CPR [34, 35]. However, only few studies have focused on OHCA patients. The OHCA patients with underlying kidney dysfunction were reported to be at a higher risk of hyperkalemia than patients without kidney disease. Moreover, kidney dysfunction might not be associated with survival [11, 36]. However, the impact of hyperkalemia on heart function was not further addressed. In this study, we aim to explore this association.

The postresuscitation heart rate was clearly decreased by a high potassium level. The action potential of heart rhythm cells has been clearly demonstrated to be more difficult to reach when the potassium level is high (a high level of extracellular potassium decreased the resting membrane potential and induced cardiac arrhythmia) [16, 20, 22, 37]. Since hypoxic injury and ischemia/reperfusion injury in cardiac arrest might include heart cell injury, we suspect that

the rate and rhythm of the damaged heart cells in the postresuscitation period might be more easily affected than those under normal conditions. However, this suspicion had never been clearly addressed. In this study, we found that bradycardia was more common in patients with high potassium (41.4%) levels than that in patients with low (16.7%) or normal (18.0%) potassium levels. Comparing with low or normal potassium levels, a high level of potassium was clearly related to slower mean heart rates, especially in the first 30 minutes of the postresuscitation period. After 30 minutes, the differences of heart rates presenting between low/normal/high levels were not significant (time points 31–45 and 46–60 minutes). This finding indicated that potassium levels might obviously influence the dynamic change of heart rate, especially in the first 30 minutes. Therefore, close observation and aggressive treatment for preventing hyperkalemia-related cardiac arrhythmia should be emphasized in this period.

Urine output was significantly lower in patients with high potassium levels. Some previous studies reported that decreased kidney function was one of the most common causes of hyperkalemia (because potassium excretion was

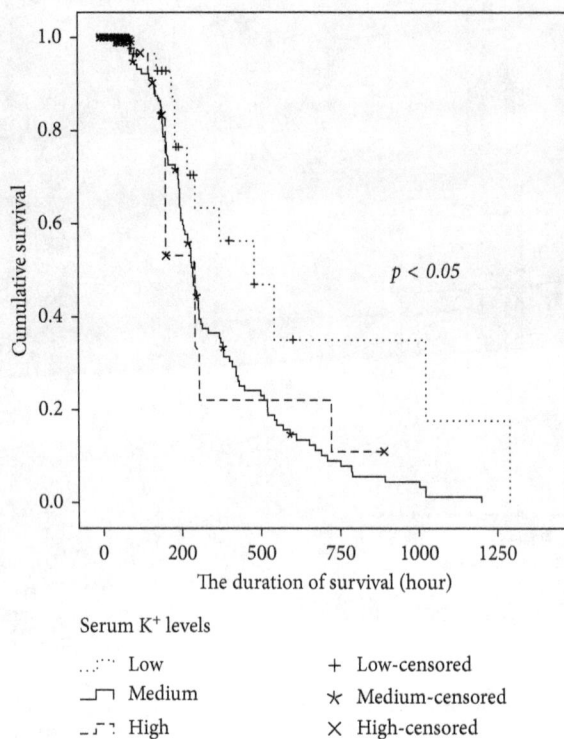

FIGURE 2: Time-related survival analysis. The serum potassium levels significantly influenced the duration of survival. The patients with high levels of potassium had the shortest duration of survival.

decreased) [16, 38, 39]. Moreover, the amount of urine output is clinically used to reflect the condition of glomerular filtration rate and tubular flow, which are important characteristics of decreased kidney function [40–42]. Therefore, close monitoring of the urine output in the early postresuscitation period might help ED physicians to recognize kidney dysfunction and consider the possibility of hyperkalemia. Metabolic acidosis, which is associated with an increased anion gap (infection, intoxication, and lactic acidosis), was also reported to be a common cause of hyperkalemia [40, 43–45]. The increased hydrogen ions in the cells might displace potassium from the cells (resulting in an increased serum potassium level) [46–48]. We suspect that routinely checking the blood pH during the early postresuscitation period would benefit evaluation of the potassium level. Finally, patients with normal or low potassium levels are likely to present with more stable postresuscitation function and hemodynamic status. Logistic regression analysis showed that patients with low levels of potassium were more likely to present with tachycardia, urine output > 1 ml/kg/hr, and nonacidosis than patients with normal levels of potassium. A possible reason could be massive fluid resuscitation. Increased intravascular fluid provided kidney perfusion and balanced the pH.

Overall, we found that a higher level of serum potassium predicted worse outcomes in OHCA patients. The initial potassium level was clearly lower in survivors than that in patients who died during the hospital stay. The duration of survival was also influenced by potassium levels. Therefore,

early treatment for hyperkalemia might be emphasized for patients entering the postresuscitation period.

4.1. Limitations. This retrospective study had some limitations. First, the association between hypothermia therapy and potassium levels was not evaluated in this study (not routinely used in each patient). Second, this study focused only on the early postresuscitation period (the first hour), whereas a high level of potassium might cause persistent complications. Finally, the patient outcomes in this study did not include neurological conditions (patient number is not enough to provide reliable evidence).

5. Conclusion

The impact of different potassium levels on early postresuscitation heart function was not clear until this study has more clearly stated. Early postresuscitation heart function and hemodynamics were associated with the serum potassium level. A high potassium level was more likely to be associated with bradycardia, nonsinus rhythm, urine output <1 (ml/kg/hr), and acidosis. The association between potassium levels and neurological outcomes should be analyzed by more patient number and analysis in the future. Most importantly, a high level of potassium decreased the duration of survival.

Authors' Contributions

Yan-Ren Lin drafted the manuscript and performed the data collection and data analysis. Chao-Jui Li supervised the study. Tsung-Han Lee managed the data quality control. Yan-Ren Lin, Chu-Chung Chou, and Chin-Fu Chang performed the statistical analysis.

Acknowledgments

The authors thank the Changhua Christian Hospital for financially supporting this research.

References

[1] J. S. Baekgaard, S. Viereck, T. P. Moller, A. K. Ersboll, F. Lippert, and F. Folke, "The effects of public access defibrillation on survival after out-of-hospital cardiac arrest: a systematic review of observational studies," *Circulation*, vol. 136, no. 10, pp. 954–965, 2017.

[2] C. F. Chang, C. J. Li, C. J. Ko et al., "The post-resuscitative urinalysis associate the survival of patients with nontraumatic out-of-hospital cardiac arrest," *PLoS One*, vol. 8, no. 10, article e75172, 2013.

[3] C. Y. Chen, Y. R. Lin, L. L. Zhao et al., "Epidemiology and outcome analysis of children with traumatic out-of-hospital cardiac arrest compared to nontraumatic cardiac arrest," *Pediatric Surgery International*, vol. 29, no. 5, pp. 471–477, 2013.

[4] G. Debaty, V. Babaz, M. Durand et al., "Prognostic factors for extracorporeal cardiopulmonary resuscitation recipients following out-of-hospital refractory cardiac arrest. A systematic review and meta-analysis," *Resuscitation*, vol. 112, pp. 1–10, 2017.

[5] G. Geri, J. Gilgan, W. Wu et al., "Does transport time of out-of-hospital cardiac arrest patients matter? A systematic review and meta-analysis," *Resuscitation*, vol. 115, pp. 96–101, 2017.

[6] C. P. Su, J. H. Wu, M. C. Yang et al., "Demographics and clinical features of postresuscitation comorbidities in long-term survivors of out-of-hospital cardiac arrest: a national follow-up study," *BioMed Research International*, vol. 2017, Article ID 9259182, 9 pages, 2017.

[7] S. Lin, C. W. Callaway, P. S. Shah et al., "Adrenaline for out-of-hospital cardiac arrest resuscitation: a systematic review and meta-analysis of randomized controlled trials," *Resuscitation*, vol. 85, no. 6, pp. 732–740, 2014.

[8] Y. R. Lin, C. J. Li, T. K. Wu et al., "Post-resuscitative clinical features in the first hour after achieving sustained ROSC predict the duration of survival in children with non-traumatic out-of-hospital cardiac arrest," *Resuscitation*, vol. 81, no. 4, pp. 410–417, 2010.

[9] A. Forrest, W. W. Butt, and S. P. Namachivayam, "Outcomes of children admitted to intensive care after out-of-hospital cardiac arrest in Victoria, Australia," *Critical Care and Resuscitation*, vol. 19, pp. 150–158, 2017.

[10] J. T. Niemann and C. B. Cairns, "Hyperkalemia and ionized hypocalcemia during cardiac arrest and resuscitation: possible culprits for postcountershock arrhythmias?," *Annals of Emergency Medicine*, vol. 34, no. 1, pp. 1–7, 1999.

[11] C. H. Lin, Y. F. Tu, W. C. Chiang, S. Y. Wu, Y. H. Chang, and C. H. Chi, "Electrolyte abnormalities and laboratory findings in patients with out-of-hospital cardiac arrest who have kidney disease," *American Journal of Emergency Medicine*, vol. 31, no. 3, pp. 487–493, 2013.

[12] N. Patel, N. J. Patel, C. J. Macon et al., "Trends and outcomes of coronary angiography and percutaneous coronary intervention after out-of-hospital cardiac arrest associated with ventricular fibrillation or pulseless ventricular tachycardia," *JAMA Cardiology*, vol. 1, no. 8, pp. 890–899, 2016.

[13] T. Tranberg, F. K. Lippert, E. F. Christensen et al., "Distance to invasive heart centre, performance of acute coronary angiography, and angioplasty and associated outcome in out-of-hospital cardiac arrest: a nationwide study," *European Heart Journal*, vol. 38, no. 21, pp. 1645–1652, 2017.

[14] Y. R. Lin, K. C. Ng, A. K. Exadaktylos, J. M. Ryan, and H. P. Wu, "Shock, cardiac arrest, and resuscitation," *BioMed Research International*, vol. 2017, Article ID 5743702, 2 pages, 2017.

[15] Y. R. Lin, H. P. Wu, C. Y. Huang, and C. F. Chang, "Predictors of successful resuscitation in non-trauma dead-on-arrival children," *Acta Paediatrica Taiwanica*, vol. 47, pp. 278–283, 2006.

[16] L. M. Kieneker, M. F. Eisenga, M. M. Joosten et al., "Plasma potassium, diuretic use and risk of developing chronic kidney disease in a predominantly White population," *PLoS One*, vol. 12, no. 3, article e0174686, 2017.

[17] G. Liamis, T. D. Filippatos, and M. S. Elisaf, "Electrolyte disorders associated with the use of anticancer drugs," *European Journal of Pharmacology*, vol. 777, pp. 78–87, 2016.

[18] P. A. McCullough, T. M. Beaver, E. Bennett-Guerrero et al., "Acute and chronic cardiovascular effects of hyperkalemia: new insights into prevention and clinical management," *Reviews in Cardiovascular Medicine*, vol. 15, pp. 11–23, 2014.

[19] H. Rosenberg, N. Pollock, A. Schiemann, T. Bulger, and K. Stowell, "Malignant hyperthermia: a review," *Orphanet Journal of Rare Diseases*, vol. 10, no. 1, p. 93, 2015.

[20] P. A. Sarafidis, P. I. Georgianos, and G. L. Bakris, "Advances in treatment of hyperkalemia in chronic kidney disease," *Expert Opinion on Pharmacotherapy*, vol. 16, no. 14, pp. 2205–2215, 2015.

[21] Z. Rafique, M. Kosiborod, C. L. Clark et al., "Study design of Real World Evidence for Treatment of Hyperkalemia in the Emergency Department (REVEAL-ED): a multicenter, prospective, observational study," *Clinical and Experimental Emergency Medicine*, vol. 4, no. 3, pp. 154–159, 2017.

[22] P. Rossignol, M. Legrand, M. Kosiborod et al., "Emergency management of severe hyperkalemia: guideline for best practice and opportunities for the future," *Pharmacological Research*, vol. 113, pp. 585–591, 2016.

[23] K. D. Patil, H. R. Halperin, and L. B. Becker, "Cardiac arrest: resuscitation and reperfusion," *Circulation Research*, vol. 116, no. 12, pp. 2041–2049, 2015.

[24] M. Zakkar, R. Ascione, A. F. James, G. D. Angelini, and M. S. Suleiman, "Inflammation, oxidative stress and postoperative atrial fibrillation in cardiac surgery," *Pharmacology & Therapeutics*, vol. 154, pp. 13–20, 2015.

[25] J. C. Jentzer, M. D. Chonde, and C. Dezfulian, "Myocardial dysfunction and shock after cardiac arrest," *BioMed Research International*, vol. 2015, Article ID 314796, 14 pages, 2015.

[26] C. E. Morris, P. F. Juranka, and B. Joos, "Perturbed voltage-gated channel activity in perturbed bilayers: implications for ectopic arrhythmias arising from damaged membrane," *Progress in Biophysics and Molecular Biology*, vol. 110, no. 2-3, pp. 245–256, 2012.

[27] N. Gaur, Y. Rudy, and L. Hool, "Contributions of ion channel currents to ventricular action potential changes and induction of early afterdepolarizations during acute hypoxia," *Circulation Research*, vol. 105, no. 12, pp. 1196–1203, 2009.

[28] X. C. Li, L. Wei, G. Q. Zhang et al., "Ca^{2+} cycling in heart cells from ground squirrels: adaptive strategies for intracellular Ca^{2+} homeostasis," *PLoS One*, vol. 6, no. 9, article e24787, 2011.

[29] F. O. Campos, A. J. Prassl, G. Seemann, R. Weber dos Santos, G. Plank, and E. Hofer, "Influence of ischemic core muscle fibers on surface depolarization potentials in superfused cardiac tissue preparations: a simulation study," *Medical & Biological Engineering & Computing*, vol. 50, no. 5, pp. 461–472, 2012.

[30] Y. R. Lin, Y. J. Syue, W. Buddhakosai et al., "Impact of different initial epinephrine treatment time points on the early postresuscitative hemodynamic status of children with traumatic out-of-hospital cardiac arrest," *Medicine*, vol. 95, no. 12, p. e3195, 2016.

[31] Y. R. Lin, H. P. Wu, W. L. Chen et al., "Predictors of survival and neurologic outcomes in children with traumatic out-of-hospital cardiac arrest during the early postresuscitative period," *Journal of Trauma and Acute Care Surgery*, vol. 75, no. 3, pp. 439–447, 2013.

[32] G. D. Perkins, I. G. Jacobs, V. M. Nadkarni et al., "Cardiac arrest and cardiopulmonary resuscitation outcome reports: update of the utstein resuscitation registry templates for out-of-hospital cardiac arrest," *Circulation*, vol. 132, no. 13, pp. 1286–1300, 2015.

[33] R. L. Chevalier, "The proximal tubule is the primary target of injury and progression of kidney disease: role of the glomerulotubular junction," *American Journal of Physiology-Renal Physiology*, vol. 311, no. 1, pp. F145–F161, 2016.

[34] M. A. Oliveira, A. C. Brandi, C. A. Santos, P. H. Botelho, J. L. Cortez, and D. M. Braile, "Modes of induced cardiac arrest: hyperkalemia and hypocalcemia–literature review," *Revista Brasileira de Cirurgia Cardiovascular*, vol. 29, pp. 432–436, 2014.

[35] H. Liu, Z. Yu, Y. Li et al., "Novel modification of potassium chloride induced cardiac arrest model for aged mice," *Aging and Disease*, vol. 9, no. 1, pp. 31–39, 2018.

[36] J. Yanta, F. X. Guyette, A. A. Doshi, C. W. Callaway, and J. C. Rittenberger, "Post cardiac arrest service. Renal dysfunction is common following resuscitation from out-of-hospital cardiac arrest," *Resuscitation*, vol. 84, no. 10, pp. 1371–1374, 2013.

[37] A. Q. Pham, J. Sexton, D. Wimer, I. Rana, and T. Nguyen, "Managing hyperkalemia: stepping into a new frontier," *Journal of Pharmacy Practice*, vol. 30, no. 5, pp. 557–561, 2017.

[38] H. J. Heerspink, P. Gao, D. de Zeeuw et al., "The effect of ramipril and telmisartan on serum potassium and its association with cardiovascular and renal events: results from the ONTARGET trial," *European Journal of Preventive Cardiology*, vol. 21, no. 3, pp. 299–309, 2014.

[39] B. C. Astor, K. Matsushita, R. T. Gansevoort et al., "Lower estimated glomerular filtration rate and higher albuminuria are associated with mortality and end-stage renal disease. A collaborative meta-analysis of kidney disease population cohorts," *Kidney International*, vol. 79, no. 12, pp. 1331–1340, 2011.

[40] J. Koeze, F. Keus, W. Dieperink, I. C. van de Hrorst, J. G. Zijlstra, and M. van Meurs, "Incidence, timing and outcome of AKI in critically ill patients varies with the definition used and the addition of urine output criteria," *BMC Nephrology*, vol. 18, no. 1, p. 70, 2017.

[41] J. A. Vassalotti, R. Centor, B. J. Turner et al., "Practical approach to detection and management of chronic kidney disease for the primary care clinician," *American Journal of Medicine*, vol. 129, no. 2, pp. 153.e7–162.e7, 2016.

[42] V. Mohsenin, "Practical approach to detection and management of acute kidney injury in critically ill patient," *Journal of Intensive Care*, vol. 5, no. 1, p. 57, 2017.

[43] D. Jammalamadaka and S. Raissi, "Ethylene glycol, methanol and isopropyl alcohol intoxication," *American Journal of the Medical Sciences*, vol. 339, no. 3, pp. 276–281, 2010.

[44] S. M. Bagshaw, M. Darmon, M. Ostermann et al., "Current state of the art for renal replacement therapy in critically ill patients with acute kidney injury," *Intensive Care Medicine*, vol. 43, no. 6, pp. 841–854, 2017.

[45] S. Gaudry, D. Hajage, F. Schortgen et al., "Initiation strategies for renal-replacement therapy in the intensive care unit," *New England Journal of Medicine*, vol. 375, no. 2, pp. 122–133, 2016.

[46] P. S. Aronson and G. Giebisch, "Effects of pH on potassium: new explanations for old observations," *Journal of the American Society of Nephrology*, vol. 22, no. 11, pp. 1981–1989, 2011.

[47] A. C. Van Slyke, Y. M. Cheng, P. Mafi et al., "Proton block of the pore underlies the inhibition of hERG cardiac K^+ channels during acidosis," *American Journal of Physiology-Cell Physiology*, vol. 302, no. 12, pp. C1797–C1806, 2012.

[48] L. Niu, Y. Liu, X. Hou et al., "Extracellular acidosis contracts coronary but neither renal nor mesenteric artery via modulation of H^+,K^+-ATPase, voltage-gated K^+ channels and L-type Ca^{2+} channels," *Experimental Physiology*, vol. 99, no. 7, pp. 995–1006, 2014.

12

Synthesis, Structural Characterization, and Antibacterial Activity of Novel Erbium(III) Complex Containing Antimony

Ting Liu⑩, Rong-Gui Yang, and Guo-Qing Zhong⑩

School of Material Science and Engineering, Southwest University of Science and Technology, Mianyang 621010, China

Correspondence should be addressed to Guo-Qing Zhong; zgq316@163.com

Academic Editor: Konstantinos Tsipis

The novel 3D edta-linked heterometallic complex $[Sb_2Er(edta)_2(H_2O)_4]NO_3 \cdot 4H_2O$ (H_4edta = ethylenediaminetetraacetic acid) was synthesized and characterized by elemental analyses, single-crystal X-ray diffraction, powder X-ray diffraction (XRD), Fourier transform infrared spectroscopy (FTIR), and thermal analysis. The complex crystallizes in the monoclinic system with space group *Pm*. In the complex, each erbium(III) ion is connected with antimony(III) ions bridging by four carboxylic oxygen atoms, and in each $[Sb(edta)]^-$ anion, the antimony(III) ion is hexacoordinated by two nitrogen atoms and four oxygen atoms from the $edta^{4-}$ ions, together with a lone electron pair at the equatorial position. The erbium(III) ion is octacoordinated by four oxygen atoms from four different $edta^{4-}$ ions and four oxygen atoms from the coordinated water molecules. The carboxylate bridges between antimony and erbium atoms form a planar array, parallel to the (1 0 0) plane. There is an obvious weak interaction between antimony atom and oxygen atom of the carboxyl group from the adjacent layer. The degradation of the complex proceeds in several steps and the water molecules and ligands are successively emitted, and the residues of the thermal decomposition are antimonous oxide and erbium(III) oxide. The complex was evaluated for its antimicrobial activities by agar diffusion method, and it has good activities against the test bacterial organisms.

1. Introduction

Much attention is currently focused on the rational design and controlled synthesis of metal-organic complexes with novel topological structure because of various potential applications of these complexes as function materials, catalysts, and medicaments [1–5]. Metal-based drugs continue to play a very important role in clinical medicine, and antimony-based metallotherapeutic drugs were used in medical applications very early in the past. Nowadays many of antimony (III) complexes have been clinically used because of their biological activities and drug efficacies [6–19], such as the treatment of a variety of microbial infections including leishmaniasis, parasitic diseases, diarrhea, peptic ulcers, helicobacter pylori, and so forth. More recently, the use of antimony complexes in cancer chemotherapy has become a topic of interest, and antimony(III) compounds have been tested *in vitro* for their cytotoxic effects on the proliferation of some leukemia and solid tumor cells [20–26].

The aminopolycarboxylate ligands can act as multidentate ligand, and their important characteristic bases on the bridging mode of the carboxylate groups [27–31]. Among the investigation of syntheses and structures of various aminopolycarboxylate complexes, heterometallic complexes are of great interest in view of their fascinating structural diversity and potential applications. Some edta-linked heterometallic complexes containing transition metals have been synthesized and structurally characterized (H_4edta = ethylenediaminetetraacetic acid) [32, 33]. However, less work on the main group elements participating in the heterometallic complexes due to the particularities of main group elements has been reported [34–40]. Antimony compounds are easy to be hydrolyzed in aqueous solutions, which makes difficult to synthesize their complexes [23], so the study of antimony complexes is much less than that of transition metal and rare earth metal complexes.

In continuation of our interest on the antimony(III) [41–43] and bismuth(III) [44, 45] complexes with aminopolycarboxylate ligands, we report herein a novel antimony-based hetero-metallic complex [Sb$_2$(edta)$_2$-μ_4-Er(H$_2$O)$_4$]NO$_3$·4H$_2$O; its composition and crystal structure have been characterized by elemental analyses, FTIR spectrum, single crystal X-ray diffraction, and thermal analysis. The complex has been evaluated for its antimicrobial activities by agar diffusion method. The synthesis method for the complexes of the antimony-transition metal and antimony-lanthanide with aminopolycarboxylic acid ligands is different. Significant knowledge about these complexes is very interesting due to their fascinatingly special structures and interesting properties, and antimony ion has weaker coordination ability than transition metal or lanthanide series ions leading to fewer reports about its complexes. The structural variety of antimony complexes is not similar to bismuth complexes. Bismuth(III) displays a marked propensity to form the complexes with high coordination number, such as the coordination number of 6–10 [32]. However, antimony(III) is generally hexacoordinated, and the stereochemistry of antimony(III) complexes is usually based on a distorted trigonal bipyramid with a pair of active lone electrons in one of the trigonal planar sites. The lone pair electrons located on antimony atom plays an important role in the final geometry obtained [23].

2. Experimental Section

2.1. Materials and Physical Measurements. All chemicals purchased in the experiments were of analytical reagent and used as received without further purification, and the solvents were also commercially available and further purified before use. The antimony trichloride, erbium nitrate hexahydrate, ethylenediaminetetraacetic acid, and ammonium bicarbonate were purchased from Sinopharm Chemical Reagent Co. Ltd. of Shanghai. The complex [Sb(Hedta)]·2H$_2$O was synthesized as described in the literature [46]. *Staphylococcus aureus, Escherichia coli, Salmonella typhi, Bacillus subtilis,* and *Staphylococcus epidermidis* were provided by the 404 hospital of Sichuan Mianyang.

Elemental analyses of C, H, and N were performed on an elemental analysis service of vario EL III elemental analyzer. Melting point was determined in capillary tubes on an X4 melting point apparatus. Molar conductance was measured by a DDS-11A conductometer. XRD pattern was recorded on a D/max-II X-ray diffractometer in the diffraction angle range of 5–80°. FTIR spectrum was measured with a KBr disk on a Nicolet 570 FT-IR system. Thermal gravimetric (TG) analysis was carried out on a STA 449C differential thermal balance in air, with a heating rate of 10°C·min^{-1} and α-Al$_2$O$_3$ reference.

2.2. Synthesis of [Sb$_2$Er(edta)$_2$(H$_2$O)$_4$]NO$_3$·4H$_2$O. 2 mmol (0.90 g) of [Sb(Hedta)]·2H$_2$O was dissolved in 60 mL hot distilled water, and the solution was heated to 95°C. Then, 2 mmol (0.16 g) NH$_4$HCO$_3$ was gradually added to the above solution, and the solution was stirred for about 30 min. After cooling the solution to room temperature, 2 mmol (0.92 g) Er(NO$_3$)$_3$·6H$_2$O was added to the above solution; in this

case, the transparent solution was obtained. The mixture solution was held for a week, and the pink block crystals were isolated from the solution. The yield was about 58%. m. p.: 192°C (decomposition). Anal. Calc. for the complex C$_{20}$H$_{40}$N$_5$O$_{27}$ErSb$_2$: C, 20.13; H, 3.38; N, 5.87%. Found: C, 20.01; H, 3.22; N, 5.51%. FTIR (KBr disk): 3426(s), 2986(w), 2956(w), 1654(s), 1593(s), 1508(w), 1469(m), 1448(m), 1402 (w), 1385(m), 1356(m), 1317(m), 1294(m), 1254(m), 1158(m), 1082(m), 1039(m), 1000(m), 948(m), 916(m), 864(m), 828 (m), 741(w), 710(m), 661(m), 619(w), 594(w), 562(m), 529 (m), 516(w), 460(m), 448(m), 434(m), 426(m), 420(m), and 407(m) cm^{-1}.

2.3. X-Ray Cystallography. All measurements were made on a Siemens P4 diffractometer at 289(2) K using graphite monochromated Mo K$_\alpha$ ($\lambda = 0.71073$ Å). A pink block with dimensions $0.48 \times 0.44 \times 0.20$ mm^3 was mounted on a glass fiber. Diffraction data were collected in ω mode in the range $1.84° < \theta < 26.00°$. Data were corrected for Lorentz and polarization effects, and an empirical absorption correction was applied. The structures were solved by the SHELXS-97 program and refined using full-matrix least squares on F^2 with the SHELXL-97 program [47]. For the complex, the hydrogen atoms attached to the oxygen atoms of water molecules were not located from the difference Fourier map due to the effect of heavy erbium and antimony atoms, while other nonhydrogen atoms were refined anisotropically, and hydrogen atoms were introduced at the calculated positions. CCDC 637089 contains the supplementary crystallographic data for the title complex. These data can be obtained free of charge via http://www.ccdc.cam.ac.uk/conts/retrieving.html or from the Cambridge Crystallographic Data Centre, 12 Union Road, Cambridge CB2 1EZ, UK.

3. Results and Discussion

The complex is stable in air and soluble in hot water and difficult to dissolve in most common organic solvents and slightly soluble in DMF. The molar conductance values of the complex in DMF and deionized water (10^{-3} mol·L^{-1} solution at 25°C) are 88.2 and 92.5 S·cm^2·mol^{-1}, respectively. The results show that the complex belongs to 1 : 1 electrolyte nature [48].

3.1. Crystal Structure Analysis. The molecular structure of the title complex with atomic labeling scheme is shown in Figure 1. Crystallographic data and structure refinement parameters of the complex are given in Table 1, and selected bond lengths and bond angles are given in Table 2. The asymmetric unit of the complex consists of a crystallographically independent heterometallic motif [Sb$_2$-μ_4-(edta)$_2$Er(H$_2$O)$_4$]$^+$, nitrate counterion, and four free water molecules. Each edta^{4-} ion consists of carboxylate groups adopting monodentate mode coordination to the antimony ion and bidentate bridging over one erbium ion and two different antimony ions. Each erbium(III) ion has a distorted trigonal dodecahedron environment with an O$_8$ donor atom

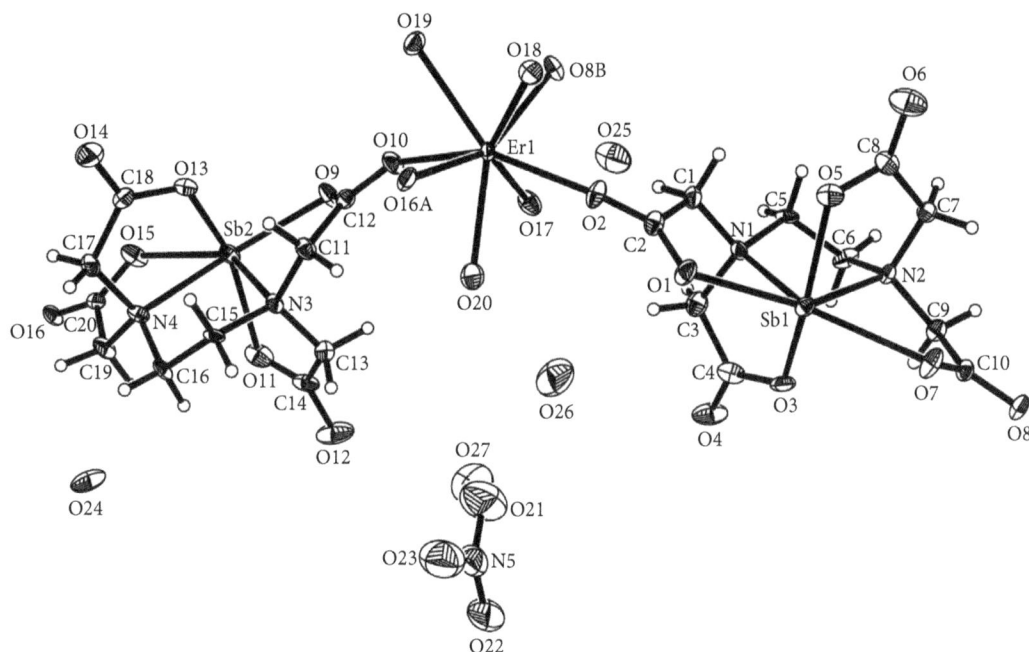

FIGURE 1: Thermal ellipsoid representation (at 50% probability) of molecular structure unit of the complex. All the H atoms are omitted for clarity. Symmetry codes: A $x - 1/2$, $-y$, $z - 1/2$; B $x + 1/2$, $-y + 1$, $z - 1/2$.

TABLE 1: Crystallographic data and structure refinement parameters of the title complex.

Empirical formula	$C_{20}H_{40}N_5O_{27}ErSb_2$	D_{calc} (g·cm^{-3})	2.278
Formula weight (g·mol^{-1})	1193.33	Absorption coefficient (mm^{-1})	4.043
T (K)	289(2)	F (000)	1162
Crystal system	Monoclinic	Crystal size (mm^3)	$0.48 \times 0.44 \times 0.20$
Space group	Pm	Theta range for data collection ($^\circ$)	1.84 to 26.00
a (Å)	7.3790(10)	Limiting indices	$-9 \leq h \leq 9$, $-27 \leq k \leq 27$, $-13 \leq l \leq 13$
b (Å)	22.116(5)	Reflections collected/unique	7847/6826 [R(int) = 0.0209]
c (Å)	10.661(3)	Goodness-of-fit (GOF) on F^2	1.002
β ($^\circ$)	90.55(2)	Final R indices [$I > 2\sigma(I)$]	$R_1 = 0.0392$, $wR_2 = 0.1051$
V (Å^3)	1739.7(7)	R indices (all data)	$R_1 = 0.0416$, $wR_2 = 0.1062$
Z	2	Largest diff. peak and hole (e·Å^{-3})	1.767 and -1.676

array: four bridged oxygen atoms [μ-O(2), μ-O(10), μ-O(8B), and μ-O(16A)] from four edta^{4-} ligands with Er–O bond distances ranging from 2.285(7) to 2.298(7) Å, four oxygen atoms from the four coordinated water molecules with Er–O bond distances ranging from 2.347(9) to 2.432(7) Å, and the coordination structure of the erbium(III) is shown in Figure 2. The antimony(III) ion is hexacoordinated by four oxygen atoms and two nitrogen atoms from the edta^{4-} ligand, and the lone pair electrons on the antimony atom cause the coordination geometry to be distorted octahedron with two oxygen atoms [O(3) and O(5)] at the axial sites. Two oxygen [O(1) and O(7)] and two nitrogen [N(1) and N(2)] atoms occupy the equatorial plane. The sum of the equatorial bond angles O(1)–Sb(1)–O(7), N(1)–Sb(1)–N(2), N(1)–Sb(1)–O(1), and N(2)–Sb(1)–O(7) is 361°, which shows that the O(1), O(7), N(1), N(2), and Sb atoms are almost located at one plane. The bond angle O(3)–Sb(1)–O(5) of 146.1° is almost twice as large as the bond angle N(1)–Sb(1)–N(2) (77.0°) or N(2)–Sb (1)–O(7) (67.0°), which may be due to the existence of a lone pair of electron in the diad direction [49]. The distances of

the Sb–O bonds are in the range of 2.127(8) to 2.541(8) Å, while the Sb–N bond lengths are in the range of 2.315(8) to 2.332(8) Å. The distances of the bidentate chelating bonds Sb(1)–O(1) (2.459 Å) and Sb(1)–O(7) (2.564 Å) are longer than the monodentate bond distances [Sb(1)–O(3) (2.215 Å) and Sb(1)–O(5) (2.127 Å)]. The O–Sb–O bond angles lie between 80.4(3)° and 148.3(3)°. These bonds and angles are slightly longer and wider, respectively, than those in the complex [CaSb$_2$(edta)(H$_2$O)$_8$]$_n$ [49]. These bond distances and angles (Table 2) are consistent with those of other edta-Sb compounds [50].

The carboxylate bridges between antimony and erbium atoms [O(1)–C(2)–O(2) and O(7)–C(10)–O(8)] form a planar array of metal atoms, with a maximum deviation of 0.416 Å, parallel to the (1 0 0) plane (Figure 3). Furthermore, there is an obvious weak interaction between antimony atom and oxygen atom of the carboxyl group from the adjacent layer. The interaction makes the layers extend to an infinite three-dimensional framework (Figure 4). Hydrogen bonds and short van der Waals force contact between oxygen atoms

TABLE 2: Selected bond lengths (Å) and angles (°) of the title complex.

Er(1)–O(16)#1	2.285(7)	Er(1)–O(10)	2.290(7)
Er(1)–O(2)	2.290(7)	Er(1)–O(8)#2	2.298(7)
Er(1)–O(20)	2.347(9)	Er(1)–O(17)	2.375(8)
Er(1)–O(19)	2.421(7)	Er(1)–O(18)	2.432(7)
Sb(1)–O(5)	2.127(8)	Sb(1)–O(3)	2.215(8)
Sb(1)–N(1)	2.315(8)	Sb(1)–N(2)	2.332(8)
Sb(1)–O(1)	2.459(7)	Sb(1)–O(7)	2.564(7)
Sb(2)–O(13)	2.124(8)	Sb(2)–O(11)	2.218(9)
Sb(2)–N(3)	2.318(8)	Sb(2)–N(4)	2.322(8)
Sb(2)–O(9)	2.438(7)	Sb(2)–O(15)	2.541(8)
O(16)#1–Er(1)–O(10)	85.6(3)	O(16)#1–Er(1)–O(2)	102.7(3)
O(10)–Er(1)–O(2)	150.5(3)	O(16)#1–Er(1)–O(8)#2	148.0(3)
O(10)–Er(1)–O(8)#2	102.5(3)	O(2)–Er(1)–O(8)#2	85.4(3)
O(16)#1–Er(1)–O(20)	70.8(3)	O(10)–Er(1)–O(20)	80.9(3)
O(2)–Er(1)–O(20)	75.5(3)	O(8)#2–Er(1)–O(20)	140.7(3)
O(16)#1–Er(1)–O(17)	138.7(3)	O(10)–Er(1)–O(17)	74.5(3)
O(2)–Er(1)–O(17)	81.1(3)	O(8)#2–Er(1)–O(17)	72.8(3)
O(20)–Er(1)–O(17)	70.5(3)	O(16)#1–Er(1)–O(19)	78.6(2)
O(10)–Er(1)–O(19)	69.4(3)	O(2)–Er(1)–O(19)	139.7(3)
O(8)#2–Er(1)–O(19)	75.6(3)	O(20)–Er(1)–O(19)	138.7(3)
O(17)–Er(1)–O(19)	124.6(3)	O(16)#1–Er(1)–O(18)	76.3(3)
O(10)–Er(1)–O(18)	139.0(3)	O(2)–Er(1)–O(18)	70.2(3)
O(8)#2–Er(1)–O(18)	77.6(3)	O(20)–Er(1)–O(18)	125.0(3)
O(17)–Er(1)–O(18)	140.0(3)	O(19)–Er(1)–O(18)	71.1(3)
O(5)–Sb(1)–O(3)	146.1(3)	O(5)–Sb(1)–N(1)	79.4(3)
O(3)–Sb(1)–N(1)	73.3(3)	O(5)–Sb(1)–N(2)	75.4(3)
O(3)–Sb(1)–N(2)	79.2(3)	N(1)–Sb(1)–N(2)	77.0(3)
O(5)–Sb(1)–O(1)	82.5(3)	O(3)–Sb(1)–O(1)	105.2(3)
N(1)–Sb(1)–O(1)	68.7(3)	N(2)–Sb(1)–O(1)	142.0(3)
O(5)–Sb(1)–O(7)	99.1(3)	O(3)–Sb(1)–O(7)	91.1(3)
N(1)–Sb(1)–O(7)	142.9(3)	N(2)–Sb(1)–O(7)	67.0(3)
O(1)–Sb(1)–O(7)	148.3(3)	O(13)–Sb(2)–O(11)	146.9(3)
O(13)–Sb(2)–N(3)	81.3(3)	O(11)–Sb(2)–N(3)	72.4(3)
O(13)–Sb(2)–N(4)	75.5(3)	O(11)–Sb(2)–N(4)	79.5(3)
N(3)–Sb(2)–N(4)	77.6(3)	O(13)–Sb(2)–O(9)	80.4(3)
O(11)–Sb(2)–O(9)	107.0(3)	N(3)–Sb(2)–O(9)	68.3(3)
N(4)–Sb(2)–O(9)	140.7(3)	O(13)–Sb(2)–O(15)	96.1(3)
O(11)–Sb(2)–O(15)	93.6(3)	N(3)–Sb(2)–O(15)	143.6(3)

Symmetry transformations used to generate equivalent atoms: #1 $x - 1/2, -y, z - 1/2$; #2 $x + 1/2, -y + 1, z - 1/2$; #3 $x - 1/2, -y + 1, z + 1/2$; #4 $x + 1/2, -y, z + 1/2$.

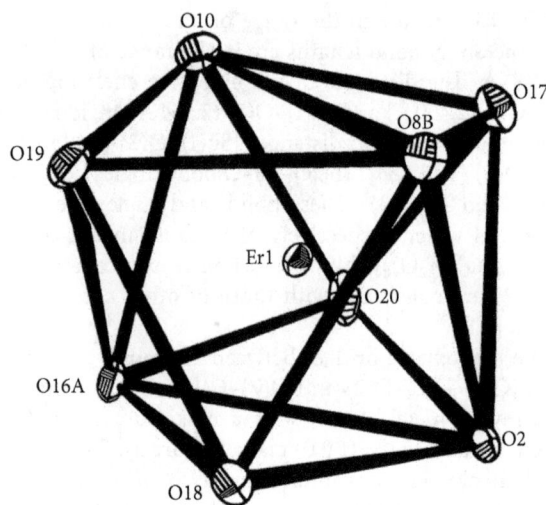

FIGURE 2: Coordination polyhedron structure of Er(III).

from carboxyl groups and water molecules and also between water molecules strengthen this three-dimensional arrangement.

3.2. FTIR Spectrum. The FTIR spectrum of the title complex is shown in Figure 5. The broad band at about 3426 cm^{-1} is due to $\nu(\text{OH})$ vibration of the water molecule. The frequency of the peak is higher than 3400 cm^{-1} showing that the oxygen atoms of the water molecule are coordinated to the metal ions [51]. The absorption peaks at 1593, 1402, and 1385 cm^{-1} may be from the asymmetric and symmetric stretching vibration in the carboxyl groups, respectively [52]. It is found that the absorption peak $\nu_{as}(\text{COO}^-)$ at 1690 cm^{-1} of $\text{Na}_2\text{H}_2\text{edta}$ is shifted red to 1593 cm^{-1} and the absorption peak $\nu_s(\text{COO}^-)$ at 1353 cm^{-1} of $\text{Na}_2\text{H}_2\text{edta}$ is shifted blue to 1402 and 1385 cm^{-1} in the complex. The difference values $[\Delta\nu(\nu_{as} - \nu_s) = 191 \text{ and } 208 \text{ cm}^{-1}]$ between the frequencies of the asymmetric and symmetric stretching vibration confirm that the oxygen atoms of carboxylic groups are coordinated

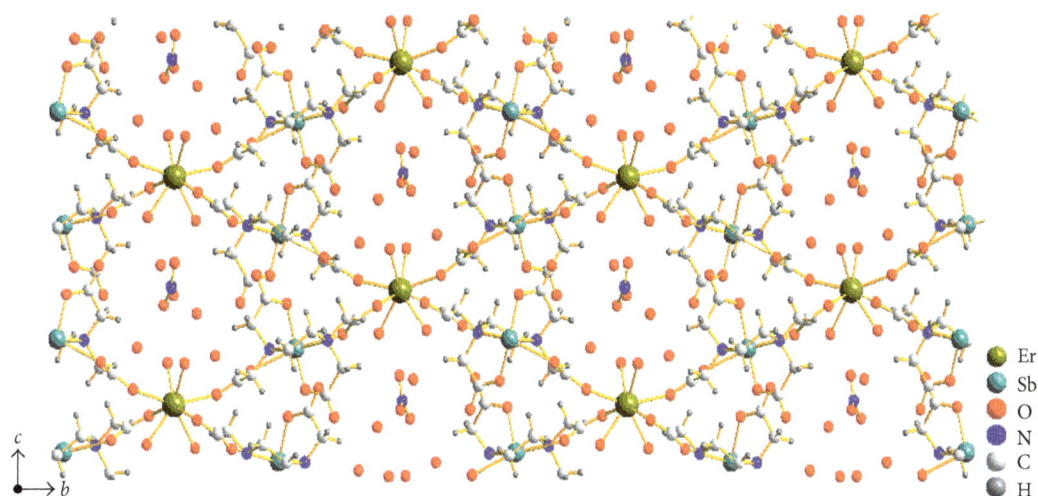

FIGURE 3: A view of the two-dimensional layer of the complex parallel to the (1 0 0) plane.

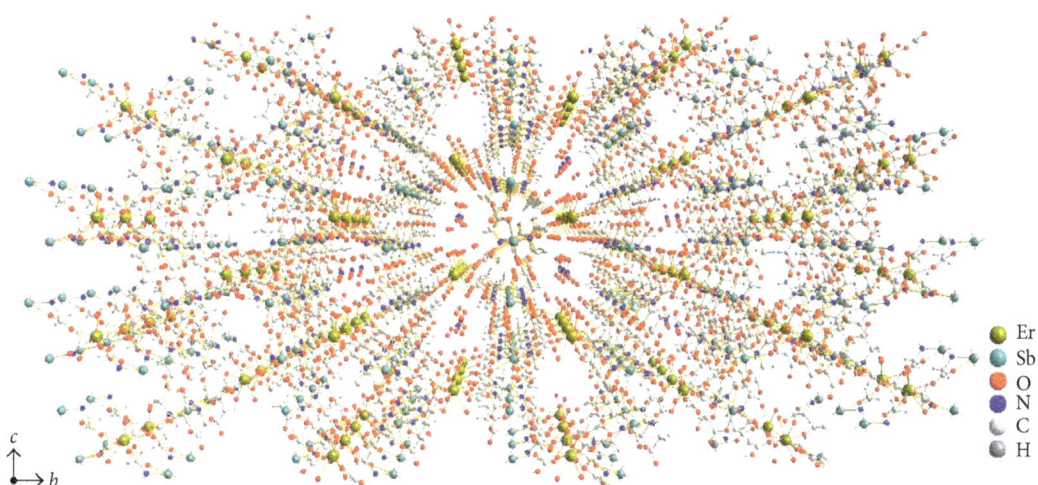

FIGURE 4: Packing diagram of the title complex viewed along the *a* axis.

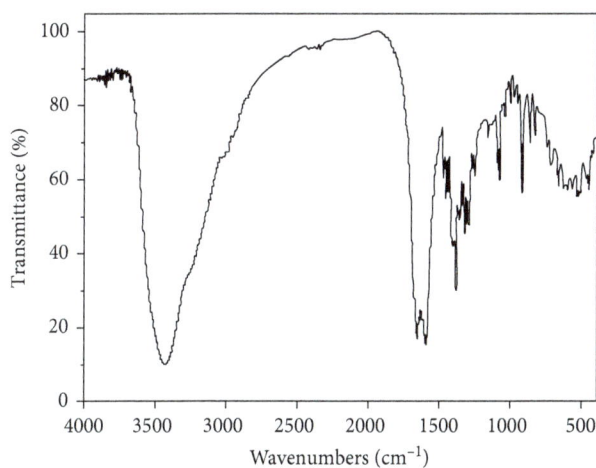

FIGURE 5: FTIR spectrum of the title complex.

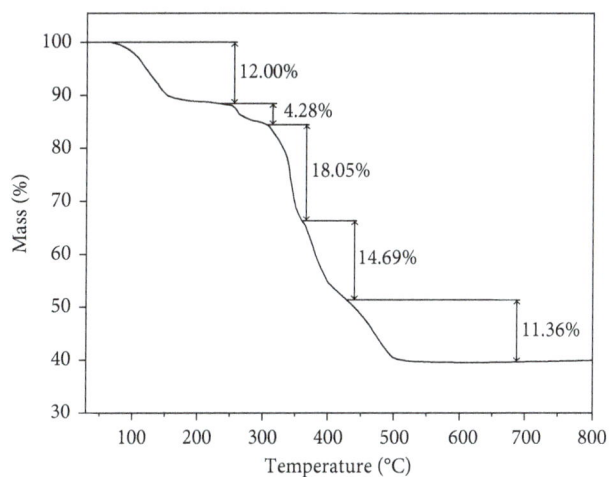

FIGURE 6: TG curve of the title complex.

to metallic ions by the monodentate mode and bidentate bridge mode in the complex [42], and it is in agreement with the crystal structure. The weaker absorption peaks at about 1356 and 828 cm^{-1} may be from the stretching vibrations in the free nitrate ion. This indicates that the nitrate ion is not coordinated to the metallic ions. The absorption peaks at

TABLE 3: Thermal decomposition data of the title complex.

Reaction	Temperature (°C)	Mass loss (%)	
		m_{exp}	m_{theor}
$[Sb_2(edta)_2$-μ_4-$Er(H_2O)_4]NO_3$·$4H_2O$			
↓ $-8H_2O$	70–220	12.00	12.08
$[Sb_2(edta)_2Er]NO_3$			
↓ $-NO_2$, $-1/4O_2$	220–310	4.28	4.53
$[Sb_2(edta)_2Er]O_{0.5}$			
↓ $-2(CH_2)_2NCH_2COO$, $-1/4O_2$	310–360	18.05	17.45
$(Sb–Sb)[N(CH_2COO)_3]_2(Er–Er)_{0.5}$			
↓ $-2N(CH_2)_3$, $-2CO$	360–430	14.69	14.09
$(OSbO–OSbO)(CO)_4(OErO–OErO)_{0.5}$			
↓ $-4CO$, $-3/4O_2$	430–510	11.36	11.40
$Sb_2O_3 + 0.5Er_2O_3$		39.62^a	40.45^b

[a]The experimental mass percent of the residue in the sample; [b]the calculated mass percent of the residue in the sample.

1082 and 1039 cm^{-1} may be from various stretching vibrations of the C–N and C–C bonds in the edta^{4-} ligand, respectively. In the far-infrared region, the frequency of the stretching vibration of the Sb–N bonds is 460 and 448 cm^{-1}, the frequency of the stretching vibration of the Sb–O bonds is 434 and 426 cm^{-1}, respectively. It may be reasonable to assign the peaks at 420 and 407 cm^{-1} to the stretching vibration of the Er–O bonds in the complex [31, 50].

3.3. Thermal Analysis. Studying the thermal decomposition process of complexes is helpful to the understanding of the coordination structure of these complexes [30, 43]. The TG curve of the complex in air atmosphere from room temperature to 800°C is shown in Figure 6, and the data of possible thermal decomposition processes are listed in Table 3. The first mass loss of 12.00% occurs between 70 and 220°C, corresponding to the gradual loss of the free water molecules and the coordinated water molecules (calculated as 12.08% for $8H_2O$). Then, the sample will gradually lose the free nitrate ion at between 220 and 310°C and the corresponding mass loss of 4.28% (calculated as 4.53%). Between 310 and 360°C, two $(CH_2)_2NCH_2COO$ groups in the complex are oxidized and decomposed, and meanwhile, one quarter oxygen molecules are lost, and the experimental mass loss (18.05%) is close to the calculated one (17.45%). The fourth step mass loss of the complex from 360 to 430°C is 14.69%, corresponding to the mass loss of two $N(CH_2)_3$ groups and two CO molecules (calculated as 14.09%) [30]. Upon further heating, the complex is decomposed completely between 430 and 510°C, and the mass loss of 11.36% in TG curve corresponds to lose the group of four CO molecules and three-fourths of oxygen molecules (calculated as 11.40%). The remaining mass is almost constant until 510°C, and the final residues of the thermal decomposition of the complex are the mixture of Sb_2O_3 and Er_2O_3, and the experimental result (39.62%) is in agreement with the result of theoretical calculation (40.45%).

To check the residue, a certain mass of the complex is placed in an alumina crucible and heated in a muffle furnace at 500°C for 2 h. Then the powder X-ray diffraction pattern of the pyrolysis products is recorded. As Figure 7 shows, its characteristic peaks are consistent with the mixture of Sb_2O_3

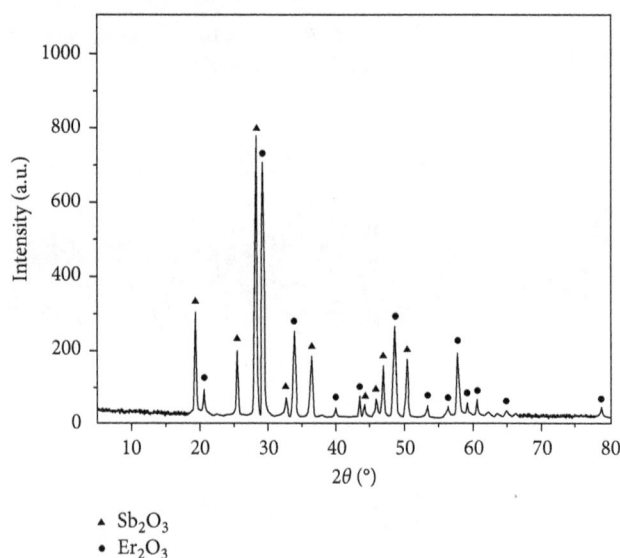

FIGURE 7: XRD pattern of the pyrolysis residue.

(JCPDS no. 71–0383) and Er_2O_3 (JCPDS no. 08–0050). Therefore, the pyrolysis residues must be the mixture of Sb_2O_3 and Er_2O_3.

3.4. Antimicrobial Activity. The culture maintenance and preparation of inoculum were referenced by the literature method [53]. The antimicrobial activities of these compounds were determined qualitatively by agar diffusion method [54]. The inhibition was labeled as the diameter of bacteriostatic circle. A lawn of microorganisms was prepared by pipetting and evenly spreading inoculums (10^6-10^7 CFU·cm^{-3}) onto agar set in petri dishes, using nutrient agar for the bacteria. Furacilinum was dissolved in DMSO, and penicillin, the title complex, and [Sb(Hedta)]·$2H_2O$ were dissolved in sterilized water. The Oxford cups were sticked on the previously inoculated agar surface and injected solution of the complex (0.15 mL) under sterile condition. The plates were incubated for 24 h at 37°C. The antimicrobial activity was indicated by the presence of clear inhibition zones around the discs.

TABLE 4: Antibacterial activities of the title complex.

Compound	Concentration (mg·mL^{-1})	Inhibition zone diameter (mm)				
		S. aureus	E. coli	S. typhi	B. subtilis	S. epidermidis
DMSO	—	—	—	—	—	—
[Sb(Hedta)]·2H$_2$O	1.0	14	17	13	14	12
[Sb$_2$(edta)$_2$-μ_4-Er(H$_2$O)$_4$]NO$_3$·4H$_2$O	1.0	17	26	16	22	16
Penicillin	1.0	15	18	17	19	18
Furacilinum	1.0	14	23	19	16	20

Preliminary screening for antimicrobial activities of the complex was performed qualitatively using the disc diffusion assay in Table 4. Each of the compounds was tested three times and the average data were recorded. DMF exhibited no effect on the organisms tested. Furacilinum and penicillin were used as standard drugs, and their activities had been compared with the activities of the title complex. The complex yielded clear inhibition zones around the discs. The results show that the complex has significant antibacterial activities against five tested bacteria, and the antibacterial activities of the sequence are *Escherichia coli*, *Bacillus subtilis*, *Staphylococcus aureus*, *Salmonella typhi*, and *Staphylococcus epidermidis*, respectively. The complex has good antibacterial activity against *Escherichia coli* and *Bacillus subtilis*, and the diameter of inhibition zone of the complex is 26 and 22 mm with the concentration of 1.0 mg·mL^{-1}. Meanwhile, the complex shows greater or equal activities against bacteria than the penicillin and furacilinum standard drugs.

4. Conclusions

The edta-linked heteronuclear complex [Sb$_2$(edta)$_2$-μ_4-Er (H$_2$O)$_4$]NO$_3$·4H$_2$O was synthesized with erbium nitrate and [Sb(Hedta)]·2H$_2$O as the raw materials, due to easy hydrolysis of antimony ion and its weaker coordination than that of erbium with edta^{4-} ion. The complex was characterized by elemental analyses, FTIR spectrum, X-ray diffraction analyses, and thermogravimetry analysis. The crystal structure of the complex belongs to the monoclinic system and space group *Pm* with cell parameters of $a = 7.3790(10)$ Å, $b = 22.116(5)$ Å, $c = 10.661(3)$ Å, $\beta = 90.55$ (2)°, and $Z = 2$. X-ray crystallography analysis reveals that the complex adopts 3-dimensional structures through the weak interactions of antimony and oxygen atoms. The bridging carboxylate-O,O' groups of edta^{4-} ions connect with erbium(III) ion and antimony(III) ions. In the complex, the carboxyl oxygen atoms participate in bridging to form diantimony entities and the entities are linked through the carbonyl oxygen atoms to form chains. The metal atoms occupy the space between the chains and are surrounded by the coordinated water molecules, which form hydrogen bonds with the other oxygen atoms of the structure. The complex displays strongly antimicrobial activities on the five tested bacteria.

Acknowledgments

This work was supported by the Scientific Research Funds of Education Department of Sichuan Province (no. 10ZA016) and the Longshan Academic Talent Research Supporting Program of SWUST (no. 17LZX414). The authors are very grateful to Professor Kai-Bei Yu from Chendu Branch of Chinese Academy of Science and Dr. Juan Shen from Southwest University of Science and Technology for the testing and analysis of the single crystal structure.

References

[1] G.-D. Zou, Z.-P. Wang, Y. Song, B. Hu, and X.-Y. Huang, "Syntheses, structures and photocatalytic properties of five new praseodymium–antimony oxochlorides: from discrete clusters to 3D inorganic–organic hybrid racemic compounds," *Dalton Transactions*, vol. 43, no. 26, pp. 10064–10073, 2014.

[2] H. P. S. Chauhan and J. Carpenter, "Synthesis, characterization and single crystal X-ray analysis of chlorobis(*N*, *N*-dimethyldithiocarbamato-*S,S'*)antimony(III)," *Journal of Saudi Chemical Society*, vol. 19, no. 4, pp. 417–422, 2015.

[3] H. P. S. Chauhan, J. Carpenter, and S. Joshi, "Mixed bis(morpholine-4-dithiocarbamato-*S,S'*)antimony(III) complexes: synthesis, characterization and biological studies," *Applied Organometallic Chemistry*, vol. 28, no. 8, pp. 605–613, 2014.

[4] M. Kořenková, M. Erben, R. Jambor, A. Růžička, and L. Dostál, "The reactivity of *N,C,N*-intramolecularly co-ordinated antimony(III) and bismuth(III) oxides with the sterically encumbered organoboronic acid 2,6-*i*-Pr$_2$C$_6$H$_3$B (OH)$_2$O," *Journal of Organometallic Chemistry*, vol. 772-773, pp. 287–291, 2014.

[5] H. P. S. Chauhan, S. Joshi, and J. Carpenter, "Synthetic, spectral, thermal and powder X-ray diffraction studies of bis(*O*-alkyldithiocarbonato-*S,S'*) antimony(III) dialkyldithiocarbamates," *Spectrochimica Acta Part A*, vol. 136, pp. 1626–1634, 2015.

[6] H. D. Yin and J. Zhai, "Synthesis, characterizations and crystal structures of antimony(III) complexes with nitrogen-containing ligands," *Inorganica Chimica Acta*, vol. 362, no. 2, pp. 339–345, 2009.

[7] D. C. Reis, M. C. X. Pinto, E. M. Souza-Fagundes et al., "Investigation on the pharmacological profile of antimony(III) complexes with hydroxyquinoline derivatives: anti-trypanosomal activity and cytotoxicity against human leukemia cell lines," *BioMetals*, vol. 24, no. 4, pp. 595–601, 2011.

[8] Z.-P. Zhang, G.-Q. Zhong, and Q.-Y. Jiang, "Biological activities of the complexes of arsenic, antimony and bismuth," *Progress in Chemistry*, vol. 20, no. 9, pp. 1315–1323, 2008.

[9] J. A. Lessa, D. C. Reis, I. C. Mendes et al., "Antimony(III) complexes with pyridine-derived thiosemicarbazones:

structural studies and investigation on the antitrypanosomal activity," *Polyhedron*, vol. 30, no. 2, pp. 372–380, 2011.

[10] E. H. Lizarazo-Jaimes, P. G. Reis, F. M. Bezerra et al., "Complexes of different nitrogen donor heterocyclic ligands with SbCl$_3$ and PhSbCl$_2$ as potential antileishmanial agents against Sb^{III}-sensitive and -resistant parasites," *Journal of Inorganic Biochemistry*, vol. 132, pp. 30–36, 2014.

[11] J. Shen, B. Jin, Q.-Y. Jiang, G.-Q. Zhong, Y.-M. Hu, and J.-C. Huo, "Synthesis, structures, luminescent and magnetic properties of heterometallic 5p-4f compounds with ethylenediaminetetraacetate," *Zeitschrift für anorganische und allgemeine Chemie*, vol. 639, no. 1, pp. 89–95, 2013.

[12] M. N. Rocha, P. M. Nogueira, C. Demicheli et al., "Cytotoxicity and *in vitro* antileishmanial activity of antimony(V), bismuth(V), and tin(IV) complexes of lapachol," *Bioinorganic Chemistry and Applications*, vol. 2013, Article ID 961783, 7 pages, 2013.

[13] L. Dawara and R. V. Singh, "Microwave-assisted synthesis, characterization, antimicrobial, and pesticidal activity of bismuth and antimony complexes with coumarin-based ligands," *Journal of Coordination Chemistry*, vol. 64, no. 6, pp. 931–941, 2011.

[14] I. I. Ozturk, S. Filimonova, S. K. Hadjikakou et al., "Structural motifs and biological studies of new antimony(III) iodide complexes with thiones," *Inorganic Chemistry*, vol. 49, no. 2, pp. 488–501, 2010.

[15] I. I. Ozturk, S. K. Hadjikakou, N. Hadjiliadis et al., "New antimony(III) bromide complexes with thioamides: synthesis, characterization, and cytostatic properties," *Inorganic Chemistry*, vol. 48, no. 5, pp. 2233–2245, 2009.

[16] H. P. S. Chauhan, J. Carpenter, and S. Joshi, "Synthetic aspects, spectral, thermal studies and antimicrobial screening on bis(*N*,*N*-dimethyldithiocarbamato-*S*,*S'*)antimony(III) complexes with oxo or thio donor ligands," *Spectrochimica Acta Part A*, vol. 130, pp. 230–237, 2014.

[17] E. D. L. Piló, A. A. Recio-Despaigne, J. G. D. Silva, I. P. Ferreira, J. A. Takahashi, and H. Beraldo, "Effect of coordination to antimony(III) on the antifungal activity of 2-acetylpyridine- and 2-benzoylpyridine-derived hydrazones," *Polyhedron*, vol. 97, pp. 30–38, 2015.

[18] T. Tunç, M. S. Karacan, H. Ertabaklar, M. Sarı, N. Karacan, and O. Büyükgüngör, "Antimony(III) complexes with 2-amino-4,6-dimethoxypyrimidines: synthesis, characterization and biological evaluation," *Journal of Photochemistry and Photobiology B*, vol. 153, pp. 206–214, 2015.

[19] H. P. S. Chauhan, A. Bakshi, and S. Bhatiya, "Synthesis, spectroscopic characterization and antibacterial activity of antimony(III) bis(dialkyldithiocarbamato)alkyldithiocarbonates," *Spectrochimica Acta Part A*, vol. 81, no. 1, pp. 417–423, 2011.

[20] O. S. Urgut, I. I. Ozturk, C. N. Banti et al., "Addition of tetraethylthiuram disulfide to antimony(III) iodide; synthesis, characterization and biological activity," *Inorganica Chimica Acta*, vol. 443, pp. 141–150, 2016.

[21] I. I. Ozturk, C. N. Banti, M. J. Manos et al., "Synthesis, characterization and biological studies of new antimony(III) halide complexes with ω-thiocaprolactam," *Journal of Inorganic Biochemistry*, vol. 109, pp. 57–65, 2012.

[22] D. C. Reis, M. C. X. Pinto, E. M. Souza-Fagundes, S. M. S. V. Wardell, J. L. Wardell, and H. Beraldo, "Antimony (III) complexes with 2-benzoylpyridine-derived thiosemicarbazones: cytotoxicity against human leukemia cell lines," *European Journal of Medicinal Chemistry*, vol. 45, no. 9, pp. 3904–3910, 2010.

[23] I. I. Ozturk, O. S. Urgut, C. N. Banti et al., "Synthesis, structural characterization and cytotoxicity of the antimony (III) chloride complex with *N*,*N*-dicyclohexyldithiooxamide," *Polyhedron*, vol. 52, pp. 1403–1410, 2013.

[24] I. I. Ozturk, C. N. Banti, N. Kourkoumelis et al., "Synthesis, characterization and biological activity of antimony(III) or bismuth(III) chloride complexes with dithiocarbamate ligands derived from thiuram degradation," *Polyhedron*, vol. 67, pp. 89–103, 2014.

[25] I. I. Ozturk, O. S. Urgut, C. N. Banti et al., "Synthesis, structural characterization and cytostatic properties of *N*,*N*-dicyclohexyldithiooxamide complexes of antimony(III) halides (SbX$_3$, X: Br or I)," *Polyhedron*, vol. 70, pp. 172–179, 2014.

[26] A. Han, I. I. Ozturk, C. N. Banti et al., "Antimony(III) halide compounds of thioureas: structures and biological activity," *Polyhedron*, vol. 79, pp. 151–160, 2014.

[27] J. J. Borras-Almenar and E. Coronado, "Single crystal EPR study of the bimetallic ferrimagnetic chain MnCu(EDTA)·6H$_2$O," *Inorganica Chimica Acta*, vol. 207, no. 1, pp. 105–109, 1993.

[28] E. Coronado, M. Drillon, P. R. Nugteren, L. J. Jongh, D. Beltran, and R. Georges, "Low-temperature investigation of the ferrimagnetic chains MnM′(EDTA)·6H$_2$O [M′ = cobalt, nickel, and copper(II)]: thermal and magnetic properties," *Journal of the American Chemical Society*, vol. 111, no. 11, pp. 3874–3880, 1989.

[29] F. Sapina, E. Coronado, D. Beltran, and R. Burriel, "From 1-D to 3-D ferrimagnets in the EDTA family: magnetic characterization of the tetrahydrate series MtM(M′EDTA)$_2$·4H$_2$O [Mt, M, M′ = cobalt(II), nickel(II), zinc(II)]," *Journal of the American Chemical Society*, vol. 113, no. 21, pp. 7940–7944, 1991.

[30] G. Q. Zhong, J. Shen, Q. Y. Jiang, Y. Q. Jia, M. J. Chen, and Z. P. Zhang, "Synthesis, characterization and thermal decomposition of Sb^{III}-M-Sb^{III} type trinuclear complexes of ethylenediamine-N,N,N′,N′-tetraacetate (M: Co(II), La(III), Nd(III), Dy(III))," *Journal of Thermal Analysis and Calorimetry*, vol. 92, no. 2, pp. 607–616, 2008.

[31] J. Shen, Q.-Y. Jiang, G.-Q. Zhong, Y.-Q. Jia, and K.-B. Yu, "Synthesis, crystal structure and thermal decomposition of a novel 3D heterometallic Sb(III)-Pr(III) complex [Sb$_2$-μ_4-(EDTA)$_2$Pr(H$_2$O)$_5$]NO$_3$·4H$_2$O," *Acta Chimica Sinica*, vol. 65, no. 16, pp. 1588–1592, 2007.

[32] V. Stavila, R. L. Davidovich, A. Gulea, and K. H. Whitmire, "Bismuth(III) complexes with animopolycarboxylate and polyamino-polycarboxylate ligands: chemistry and structure," *Coordination Chemistry Reviews*, vol. 250, no. 21-22, pp. 2782–2810, 2006.

[33] Q.-Y. Jiang, H.-Q. Deng, G.-Q. Zhong, P. He, and N.-H. Hu, "Synthesis, crystal structure and thermal stability of 3D heterometallic Bi(III)-Pr(III) polymer complex," *Chemical Journal of Chinese Universities*, vol. 29, no. 12, pp. 2521–2524, 2008.

[34] G. Q. Zhong, S. R. Luan, P. Wang, Y. C. Guo, Y. R. Chen, and Y. Q. Jia, "Synthesis, characterization and thermal decomposition of thiourea complexes of antimony and bismuth triiodide," *Journal of Thermal Analysis and Calorimetry*, vol. 86, no. 3, pp. 775–781, 2006.

[35] G.-Q. Zhong, Y.-C. Guo, Y.-R. Chen, X.-S. Zang, and S.-R. Luan, "Synthesis and crystal structure of the complex of thioglycollic acid and trivalent antimony ion," *Acta Chimica Sinica*, vol. 59, no. 10, pp. 1599–1603, 2001.

[36] S. Smola, N. Rusakova, and Y. U. Korovin, "New luminescent heteronuclear Ln(III)–Bi(III) complexes (Ln = Nd, Eu, Tb,

Yb) based on aminopolycarboxylic acids," *Journal of Coordination Chemistry*, vol. 64, no. 5, pp. 863–874, 2011.

[37] H. Ajaz, S. Hussain, M. Altaf et al., "Synthesis and characterization of antimony(III) complexes of thioamides, and crystal structure of $\{[Sb(Imt)_2Cl_2]_2(\mu_2\text{-}Imt)\}Cl_2$ (Imt = Imidazolidine-2-thione)," *Chinese Journal of Chemistry*, vol. 29, no. 2, pp. 254–258, 2011.

[38] Y.-Z. Li, R. Ganguly, and W. K. Leong, "Oxidative addition of halogen across an Os-Os or Os-Sb bond: formation of five-membered osmium-antimony carbonyl rings," *Journal of Organometallic Chemistry*, vol. 811, pp. 66–73, 2016.

[39] H. Choujaa, A. L. Johnson, G. Kociok-Köhn, and K. C. Molloy, "The synthesis of a novel heterobimetallic amidotungsten-antimony complex," *Polyhedron*, vol. 29, no. 6, pp. 1607–1611, 2010.

[40] Y.-Z. Li and W. K. Leong, "Raft-like osmium- and ruthenium-antimony carbonyl clusters," *Journal of Organometallic Chemistry*, vol. 812, pp. 217–225, 2016.

[41] J. Shen, B. Jin, Q.-Y. Jiang, G.-Q. Zhong, Y.-M. Hu, and J.-C. Huo, "Edta-linked 5p-4f trinuclear heterometallic complex: syntheses, X-ray structure and luminescent properties," *Journal of Coordination Chemistry*, vol. 65, no. 17, pp. 3040–3049, 2012.

[42] G.-Q. Zhong, J. Shen, Q.-Y. Jiang, and K.-B. Yu, "Synthesis and structural determination of a novel heterometallic complex $[Sb_2(edta)_2\text{-}\mu_4\text{-}Co(H_2O)_2]\cdot5.15H_2O$," *Chinese Journal of Chemistry*, vol. 29, no. 12, pp. 2650–2654, 2011.

[43] J. Shen, B. Jin, Q.-Y. Jiang, G.-Q. Zhong, Y.-M. Hu, and J.-C. Huo, "Synthesis, characterization, and magnetic properties of heterometallic trinuclear complex with Sb(III) and Ho(III)," *Inorganica Chimica Acta*, vol. 385, pp. 158–163, 2012.

[44] Q.-Y. Jiang, H.-Q. Deng, P. He, G.-Q. Zhong, and K.-B. Yu, "Sm(III)-Bi(III) heterometallic complexes with aminopolycarboxylate ligand: structure, thermal stability and spectral property," *Chinese Journal of Chemistry*, vol. 29, no. 12, pp. 2637–2642, 2011.

[45] Q.-Y. Jiang, H.-Q. Deng, Y.-M. Hu, J. Shen, G.-Q. Zhong, and N.-H. Hu, "Crystal structure and thermal decomposition of a 2D heterometallic coordination polymer $\{[NdBi(cydta)(NO_3)_2(H_2O)_4]\cdot2.5H_2O\}_n$," *Acta Chimica Sinica*, vol. 66, no. 12, pp. 1429–1434, 2008.

[46] D. Li and G.-Q. Zhong, "Synthesis and crystal structure of the bioinorganic complex $[Sb(Hedta)\cdot2H_2O]$," *Bioinorganic Chemistry and Applications*, vol. 2014, Article ID 461605, 7 pages, 2014.

[47] G. M. Sheldrick, "A short history of SHELX," *Acta Crystallographica Section A*, vol. 64, no. 1, pp. 112–122, 2008.

[48] W. J. Geary, "The use of conductivity measurements in organic solvents for the characterisation of coordination compounds," *Coordination Chemistry Reviews*, vol. 7, no. 1, pp. 81–122, 1971.

[49] B. Marrot, C. Brouca-Cabarrecq, and A. Mosset, "$[CaSb_2(EDTA)_2(H_2O)_8]_n$: synthesis, crystal structure, and thermal behavior," *Journal of Chemical Crystallography*, vol. 28, no. 6, pp. 447–452, 1998.

[50] J. Wang, X.-D. Zhang, Z.-R. Liu, and W.-G. Jia, "Synthesis and structural determination of binuclear nine-coordinate $(NH4)_4[Yb_2(dtpa)_2]\cdot9H_2O$," *Journal of Molecular Structure*, vol. 613, no. 1-3, pp. 189–193, 2002.

[51] K. Nakamoto, *Infrared and Raman Spectra of Inorganic and Coordination Compounds*, John Wiley & Sons Inc., New York, NY, USA, 6th edition, 2009.

[52] H.-P. Xiao, S. Aghabeygi, W.-B. Zhang et al., "A new Zn^{II} two-dimensional coordination polymer, $\{[Zn(\mu\text{-}4,4'\text{-}bipy)(1,4\text{-}ndc)(H_2O)_2]\cdot(H_2O)\}_n$ (4,4'-bipy = 4,4'-bipyridine and 1,4-ndc = 1,4-naphthalenedicarboxylate)," *Journal of Coordination Chemistry*, vol. 61, no. 22, pp. 3679–3686, 2008.

[53] Y.-C. Guo, Y.-Q. Feng, Z.-P. Qiao, S.-Y. Chen, and S.-Z. Huang, "Syntheses, crystal structures and antibacterial activities of dithiocarbamate complexes $[M(MeBnNCS_2)_3]$, M = Sb(III), Bi(III)," *Chinese Journal of Inorganic Chemistry*, vol. 30, no. 5, pp. 1031–1037, 2014.

[54] A. C. Ekennia, D. C. Onwudiwe, L. O. Olasunkanmi, A. A. Osowole, and E. E. Ebenso, "Synthesis, DFT calculation, and antimicrobial studies of novel Zn(II), Co(II), Cu(II), and Mn(II) heteroleptic complexes containing benzoylacetone and dithiocarbamate," *Bioinorganic Chemistry and Applications*, vol. 2015, Article ID 789063, 12 pages, 2015.

13

Synthesis and Characterization of *trans*-Dichlorotetrakis (imidazole) cobalt(III) Chloride: A New Cobalt(III) Coordination Complex with Potential Prodrug Properties

Kaila F. Hart,[1] Natalie S. Joe,[1] Rebecca M. Miller,[1] Hannah P. Nash,[1] David J. Blake,[2] and Aimee M. Morris (ID)[1]

[1]Department of Chemistry and Biochemistry, Fort Lewis College, 1000 Rim Dr., Durango, CO 81301, USA
[2]Department of Biology, Fort Lewis College, 1000 Rim Dr., Durango, CO 81301, USA

Correspondence should be addressed to Aimee M. Morris; ammorris@fortlewis.edu

Academic Editor: Giovanni Natile

Numerous therapies for the treatment of cancer have been explored with increasing evidence that the use of metal-containing compounds could prove advantageous as anticancer therapeutics. Previous works on Ru(III) complexes suggest that structurally similar Co(III) complexes may provide good alternative, low-cost, effective prodrugs. Herein, a new complex, *trans*-[Co(imidazole)$_4$Cl$_2$]Cl (**2**), has been synthesized in high yields utilizing ligand exchange under refluxing conditions. The structure of **2** has been characterized by elemental analysis, ^{1}H and ^{13}C·NMR, ESI-MS, CV, and UV-Vis. The ability of **2** to become reduced in the presence of ascorbic acid was probed demonstrating the likely reduction of the Co(III) metal center to Co(II). In addition, preliminary cell line testing on **2** shows a lack of cytotoxicity.

1. Introduction

The use of metals or metal-containing complexes for the treatment of cancer dates back to the sixteenth century but is still considered a relatively unexplored area of cancer research [1]. Increasing evidence suggests that the use of metal-containing compounds could prove to be advantageous as anticancer therapeutics [2, 3]. The use of metal complexes in anticancer therapies offers several advantages including the use of multiple oxidation states of metals, a larger range of geometries to be explored and utilized, and a so-called "tunability" of the thermodynamics and kinetics of ligand substitution [4–8].

Cisplatin targets DNA and has been successful for the treatment of many different cancers but has the disadvantage of being toxic to both healthy and affected tissues [9]. In a movement to develop less-toxic anticancer drugs, recent attention has been focused on the use of prodrugs. A prodrug remains inactive until it reaches diseased cells at which time it becomes activated and cytotoxic. To date, the most successful and studied metal-containing prodrugs are so-called functional models with Ru(III) centers, NAMI-A [10, 11] and KP1019 [12, 13] (Figure 1). Functional models refer to the "naked" metal center performing the desired therapeutic function as both have mounting evidence to strongly suggest loss of their original ligands and can complex to other Lewis bases *in vivo* [14–17]. While the mechanism of action for NAMI-A and KP1019 is still unknown [11, 18, 19], recent studies on NAMI-A and derivatives of KP1019 suggest that the ruthenium center can coordinate to nitrogen, oxygen, or sulfur amino acid donors on protein targets [15, 20–22].

It has been suggested that Co(III) complexes, with their octahedral geometries and low-spin d^6 configurations, may also represent an effective class of prodrugs for the treatment of cancer [2, 3, 20–23]. In addition, cobalt is ca. 30,000 times more abundant in the earth's crust and 1.2% of the current cost of ruthenium [24]. The octahedral, low-spin d^6 electron configurations give rise to diamagnetic complexes that can easily be assigned and followed using nuclear magnetic

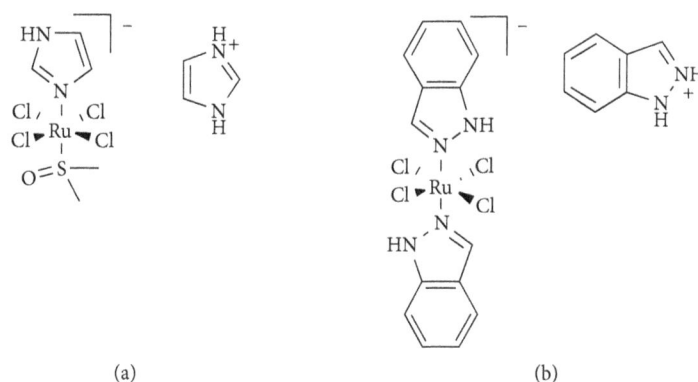

(a)   (b)

Figure 1: Structures of (a) NAMI-A and (b) KP1019.

resonance (NMR) spectroscopy. Substantial research has been published involving the synthesis and characterization of Co(III) complexes with one or more bidentate or polydentate ligands [8, 9, 25–30], having the potential disadvantage of the kinetic chelate effect if open sites on the metal center are needed to induce apoptosis. However, the synthesis of functional models of Co(III) complexes similar in structure to NAMI-A or KP1019 have not been reported and provide a relatively unexplored realm for anticancer prodrug therapies.

Herein, the synthesis of a new potential Co(III) functional prodrug, *trans*-dichlorotetrakis(imidazole)cobalt(III) chloride (2) is presented utilizing a ligand substitution reaction from a known Co(III) starting material, *trans*-dichlorotetrakis(pyridine)cobalt(III) chloride (1) [31, 32]. Characterization of 2 by ESI-MS, IR, ^{1}H and ^{13}C·NMR, UV-Vis, elemental analysis, and cyclic voltammetry all support the proposed structure. Furthermore, the characterization and preliminary biological studies of 2 suggest its potential as a prodrug.

2. Experimental

2.1. Materials and Methods. Chlorine (≥99.5%) and anhydrous pyridine (99.5 + %) were purchased from Sigma-Aldrich and Alfa Aesar, respectively, and used as received. Cobalt(II) chloride hexahydrate (98–102%; Alfa Aesar), imidazole (97 + %; Alfa Aesar), methyl isobutyl ketone (EMSURE), dimethyl sulfoxide (Fisher), and diethyl ether (Fisher) were of ACS grade and used as received without further purification.

Mass spectra were collected using a Finnigan LCQ Deca XP Plus ion trap with ions generated using an electrospray ionization (ESI) source with a spray voltage of 4.5 kV, heated capillary temperatures of 200–250°C, and a flow of 10 μL/min. The electrospray solution contained ~10^{-5} M of 2 in CH_3CN (anhydrous, 99.8%; Sigma-Aldrich). Elemental Analysis was performed by Galbraith Laboratories, Inc., Knoxville, TN. The ^{1}H and ^{13}C·NMR spectra were run on a JEOL ECX-400 spectrometer, operating at 400 MHz and 100 MHz, respectively. NMR spectra were run in DMSO-d_6 (Acros; 99.6%) or $CDCl_3$ (Acros; 99.8%) dried over activated molecular sieves, or D_2O (Cambridge Isotope Laboratories;

99.9%). NMR spectra were referenced to the residual solvent peak. FTIR in the solid state was taken on the Thermo Scientific Nicolet iS10 with Smart iTR from 600–4000 cm^{-1}.

2.2. Electrochemistry. Cyclic voltammetry (CV) was carried out with a BASi epsilon digital potentiostat at a scan rate of 100 mV/s using a glassy carbon working electrode, a platinum auxiliary electrode, and an Ag/AgCl reference electrode. The solution for analysis contained 1 mM of 2 in DMSO with 0.1 M tetraethylammonium perchlorate. The solution was degassed with nitrogen prior to running CV at room temperature, using a ferrocene internal standard (Fc/Fc$^+$ = +0.68 V in DMSO).

2.3. Synthesis of Cobalt(III) Starting Material, trans-Dichlorotetrakis(pyridine)cobalt(III) Chloride, 1. The synthesis of this starting material was achieved using a modified version of literature procedures [31, 32] (see Figure S1 of the Supplementary Materials for a detailed experimental setup). Specifically, 10.0 g (0.042 mol) of cobalt(II) chloride hexahydrate was solubilized in 40.0 mL of distilled water in a round-bottom flask with stirring and 30°C heating resulting in a magenta-colored solution. Next, 13.0 mL (0.161 mol) of pyridine was added dropwise (~1 drop/second) resulting in a final deep red/purple solution. The round-bottom flask was sealed with a septum and oxidized with vigorous bubbling of chlorine gas at 30°C for thirty minutes. The final dark brown solution with a visible solid green product present was then purged with N_2 for 20 minutes before being placed in the refrigerator for 1 week. The desired flaky green product was collected over a medium filter frit and washed 3 times with ~10 mL portions of ice-cold water, followed by ice-cold ethanol and ice-cold ether. The product was dried overnight on a high vacuum line. Yield: 5.0 g (25%). ^{1}H·NMR (400 MHz, CDCl$_3$, δ): 7.42 (t, J = 6.6 Hz, 2H), 8.08 (t, J = 7.2 Hz, 1H), and 8.42 (d, J = 4.8 Hz, 2H).

2.4. Synthesis of trans-Dichlorotetrakis(imidazole)cobalt(III) Chloride, 2. In a 50 mL round-bottom flask, 0.10 g (0.21 mmol) of 1 was partially solubilized in 10 mL of methyl isobutyl ketone with stirring and slight heating (40°C). In a beaker, 0.071 g (1.0 mmol) of imidazole was fully solubilized in 20 mL of methyl isobutyl ketone with stirring. The solubilized imidazole was added to the 50 mL round-bottom flask. The initial solution is light green with green particles visible in the solution. A water condenser was attached, and the solution was refluxed at 115°C for 2 hours. During the reflux, the starting material becomes fully solubilized. After the reflux period, a new green solid in blue solution is observed in the round-bottom flask. The solid was collected on a medium filter frit and washed three times with ~5 mL portions of ice-cold methyl isobutyl ketone followed by 3 × 5 mL portions of ice-cold diethyl ether. Yield: 0.070 g (77%). Note that this procedure has been reproducibly scaled up by 10 times maintaining reported yields with an increased reflux time of 24 hours.

2.5. Analytical Data for Complex 2.

$[Co(C_3H_4N_2)_4Cl_2]Cl$ (437.60 g/mol): calculated [found]: C, 32.94 [33.66]; H, 3.69 [4.06]; and N, 25.61 [24.70]. MS (ESI): m/z 401 $[M^+]$. FTIR (cm^{-1}): ν(N–H) 3135; ν(C–H) 3086, 2975, and 2871; ν(C=N/C=C) 1551, 1500, 1450, and 1334. ^{1}H·NMR (DMSO-d^6) δ: 6.86 (s, 1H), 7.18 (s, 1H), 7.71 (s, 1H), and 13.1 (s, NH); ^{13}C·NMR (DMSO-d^6) δ: 116.76, 130.47, and 140.79.

2.6. UV-Vis Studies of Complex 2.

Samples for UV-Vis were prepared in distilled water with 2 mM of complex 2. Studies using nitrogen-degassed and sealed cuvette samples as well as samples in air showed no experimental difference. The ability for the reduction of the Co(III) metal center of 2 was studied by adding 10 equivalents of ascorbic acid.

2.7. Cell Culture, Reagents, and Cytotoxicity Studies.

A549 cells (human epithelial lung carcinoma) were obtained from American Type Culture Collection (Manassas, VA). The cells were cultured at 37°C in a 5% CO_2 incubator in complete media containing modified Eagle's media with 10% fetal bovine serum and antibiotics (100 U/mL penicillin and 10 μg/mL streptomycin; Life Technologies). Complex 2 was dissolved in sterilized DMSO and diluted in cell culture media immediately before use.

Cell viability was assessed 24 h after exposure to varying concentrations of either 2 (10 nM–1 mM) or the corresponding concentration of the DMSO control through the CellTiter-Blue Cell Viability Assay (Promega). Fluorescence (560$_{EX}$/590$_{EM}$) was acquired using the Infinite® M200 Microplate reader (Tecan). Fluorescence at 590$_{EM}$ is proportional to the number of viable cells.

3. Results and Discussion

3.1. Synthesis and Characterization of Complex 2.

Complex 2 was synthesized in 77% yield in micro- and macroscale quantities according to the stoichiometric equation outlined in Scheme 1. Under refluxing conditions, the pyridine ligands can be displaced from 1 and replaced with the desired imidazole ligands to yield the new desired

product, 2, as a green powder. Synthetically, the highest yields of pure product were obtained by using a slight excess of the imidazole ligand: five equivalents versus the stoichiometric four equivalents. The bulk solid sample composition was verified by elemental analysis and FTIR and is consistent with the assigned formula, $[Co(C_3H_4N_2)_4Cl_2]Cl$. The FTIR of 2 is similar to free imidazole with the most significant change being the disappearance of hydrogen bonding vibrations that exist between 2800 and 2400 cm^{-1} in free imidazole (Figure S2 of the Supplementary Materials), as has been observed in other cobalt imidazole complexes [33, 34]. ESI-MS with collision-induced dissociation (CID) also supports the bulk composition proposed for complex 2 (Figure S3 of the Supplementary Materials).

The octahedral, low-spin, diamagnetic Co(III) metal center allowed for straightforward characterization of 2 by ^{1}H and ^{13}C·NMR, providing additional support of the assigned chemical formula and structure with trans-chloro ligands (Figure 2; full spectra are found in Figures S4 and S5 of the Supplementary Materials). Resonance and chemical equivalency for two of the free imidazole protons (H$_B$ and H$_C$) and carbon atoms C$_2$ and C$_3$ are lost upon binding to the Co(III) center. Assigned chemical shifts were determined by 1-D NOE difference and 2-D HSQC experiments (Figures S6 and S7 of the Supplementary Materials).

The UV-Vis of 2 in water (black line in Figure 3) displays two maxima at 390 and 520 nm with molar absorptivity values of 100 and 47 $M^{-1} \cdot cm^{-1}$, respectively. These are assigned as the expected tetragonally distorted d-d transitions $^1A_{1g}(D_{4h}) \rightarrow {}^1E_g(D_{4h})$ and $^1A_{1g}(D_{4h}) \rightarrow {}^1A_{2g}(D_{4h})$ of low-spin, octahedral Co(III) complexes [35]. Taken together, the ESI-MS, IR, elemental analysis, ^{1}H and ^{13}C·NMR, and UV-Vis of 2 all support the formation of a new Co(III) coordination complex, trans-dichlorotetrakis(imidazole) cobalt(III) chloride.

3.2. Potential Prodrug Capabilities of Complex 2.

The solubility properties of complex 2 are amenable to biological applications as 2 is soluble in polar solvents (e.g., H_2O, CH_3OH, and DMSO) at concentrations >10 mM. Furthermore, ligand lability is observed with complex 2 in H_2O and D_2O, demonstrating its potential as a functional Co(III) prodrug capable of displacing ligands.

Evidence of ligand lability in complex 2 was demonstrated by ^{1}H·NMR in D_2O. Over the course of an hour, the solution of 2 visibly changed from green to red. The time-lapsed ^{1}H·NMR spectra in D_2O (Figure S8) suggest lability of the chloro and also possibly imidazole ligands over time, demonstrating favorable characteristics for desired in vivo prodrug applications. The ligand lability in H_2O was further verified by UV-Vis. In the absence of a reducing agent, 2 shows a blue shift of the two λ_{max} values after 36 hours likely due to ligand exchange (Figure 3 and Figure S9 of the Supplementary Materials). Furthermore, the blue shifts are indicative of a ligand exchange involving a weak-field chloro ligand being replaced by a stronger-field aqua or imidazole ligand.

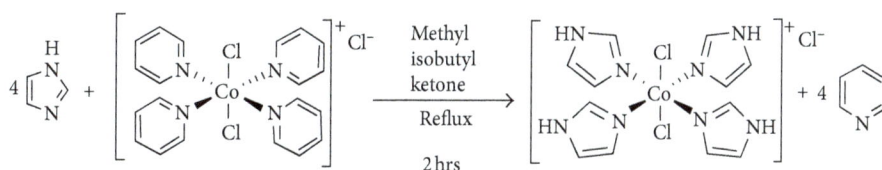

SCHEME 1: Stoichiometric equation for the formation of *trans*-[Co(imidazole)$_4$Cl$_2$]Cl (**2**).

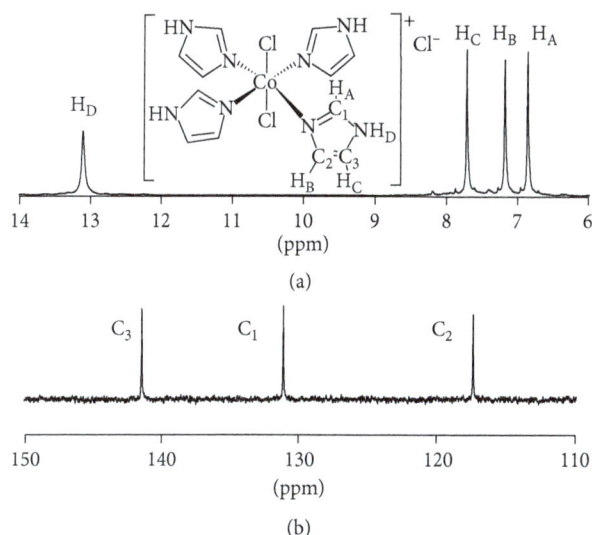

FIGURE 2: ^{1}H·NMR (a) and ^{13}C·NMR (b) of **2** in DMSO-d_6.

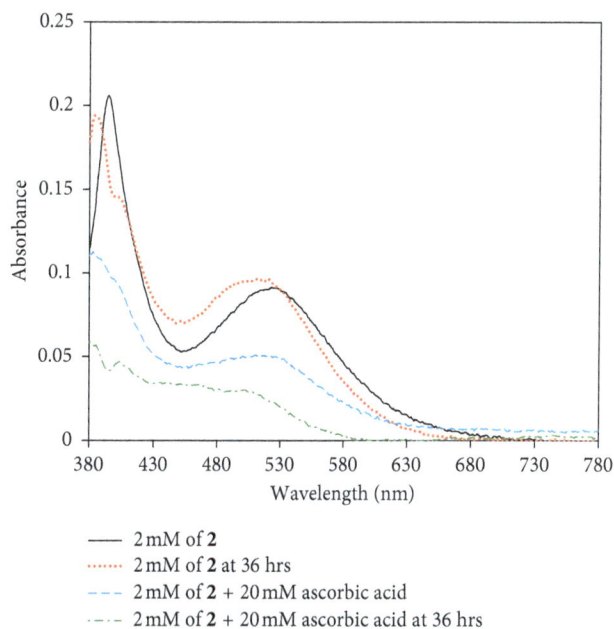

— 2 mM of **2**
········ 2 mM of **2** at 36 hrs
– – – 2 mM of **2** + 20 mM ascorbic acid
–··– 2 mM of **2** + 20 mM ascorbic acid at 36 hrs

FIGURE 3: UV-Vis absorption spectra of 2 mM of **2** in H$_2$O at time = 0 (black) and time = 36 hours (red). Immediately after the addition of 10 equivalents of the biological reducing agent, ascorbic acid (blue), the signature Co(III) *d-d* transitions decrease in intensity over time (after 36 hours is shown in green) suggesting that reduction of the cobalt center is observed.

While the "activation by reduction hypothesis" in Ru (III) prodrugs such as NAMI-A and KP1019 is still debated [11, 18, 19], exploration of biologically accessible reduction potentials and properties of potential prodrugs may provide important insights. For complex **2**, the ability of the Co(III) metal center to be reduced to Co(II) was also investigated. Previous studies indicate that a reduction potential between −0.1 V and −0.5 V versus NHE in water at pH 7 is required for successful redox cycling by common flavoproteins *in vivo* [36]. Complex **2** shows an irreversible reduction of the Co(III) metal center to Co(II) at −0.51 V versus NHE, indicating that reduction of this complex may be possible *in vivo* (Figure 4). The observed irreversibility is common in Co(III) prodrugs with multidentate ligands and is likely due to a chemical change in the structure of **2** that occurs after reduction to Co(II) [37–39].

The desired properties displayed by **2** along with the increasing interest and potential of transition metal complexes in prodrug applications prompted us to investigate a preliminary biological study utilizing a cancerous cell line. The *in vitro* cytotoxicity of **2** at concentrations ranging from 10 nM to 1 mM was evaluated against the human A549 cell line (Figure 5). It is observed that the cells exposed to **2** (up to 1 mM) retain similar metabolic activity levels to their DMSO controls. This suggests that **2** is not cytotoxic, a result that correlates with observations of other successful metastatic prodrugs such as NAMI-A [11, 40, 41].

3.3. Similarities and Differences of 2 to NAMI-A and KP1019.

Structural similarities of **2** to NAMI-A and KP1019 include octahedral geometries and low-spin configurations, and categorically, these are all considered kinetically inert complexes, although ligand lability is displayed in NAMI-A [11] and **2**. In addition, **2** and NAMI-A both have imidazole and chloro ligands and are very soluble in polar solvents. Admittedly, a major difference between **2** and NAMI-A and KP1019 is the overall charge on the coordination complexes; the Ru(III) complexes are overall anionic, while **2** is cationic. However, previous studies have shown that cationic (+1 and +2) octahedral Co(III) chaperone complexes are able to penetrate into hypoxic regions of tumor models [8, 35, 42] and cancerous cells [43]. Furthermore, NAMI-A has been described as "the ultimate prodrug" [11] since *in vivo* experimental evidence suggests the loss of its original ligands very quickly, suggesting the overall charge on the original coordination complex may be less important for a functional prodrug.

KP1019 displays cytotoxicity similar to traditional platinum chemotherapeutics [12, 13]. In contrast and similar

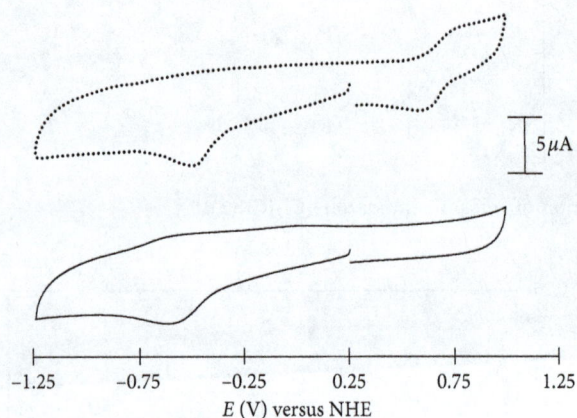

FIGURE 4: Cyclic voltammetry of 1 mM of **2** in DMSO with the internal Fc/Fc$^+$ standard (top) and without (bottom).

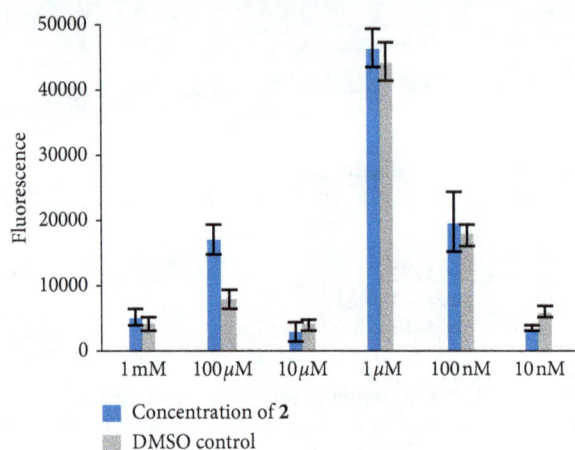

FIGURE 5: Viability of human epithelial cells is unaffected by **2**. Cells were exposed to **2** at different concentrations up to 1 mM for 24 h. Viability, which is correlated with fluorescence, was quantified using the CellTiter-Blue assay. Data presented are mean fluorescence $\pm$ SEM ($n = 3$–5 for each condition).

to our preliminary results on **2**, prodrugs such as NAMI-A lack cytotoxicity but display favorable pharmacological properties [11, 40, 41]. It has been suggested that *in vitro* cytotoxicity may provide a measure of metal uptake but cannot predict the potential *in vivo* anticancer activity [11]. In support of this idea, evidence is mounting that the selective antimetastatic activity and observed lack of cytotoxicity of NAMI-A are the result of Ru(III)-serum albumin adduct formation [15, 16, 20]. Hence, the presented characterization and lack of cytotoxicity displayed by the new Co (III)-containing complex **2** demonstrate similarities to one of the most successful metal-based prodrugs that warrants further investigation.

4. Conclusions

In summary, optimal reaction conditions to synthesize a new Co(III) coordination complex, **2**, in 77% yield were discovered. All physical measurements are consistent with the reported octahedral structure of **2** as *trans*-[Co(imidazole)$_4$Cl$_2$]Cl.

Characterization of the new complex was accomplished with ESI-MS, elemental analysis, IR, ^{1}H and ^{13}C·NMR, CV, and UV-Vis. Furthermore, initial evidence suggests that **2** may be suitable in prodrug applications due to its water solubility, ligand lability, ability to be reduced by biological reductants, and lack of cytotoxicity. Further biological studies on complex **2** and similar complexes are needed and represent an important new direction for future research.

Acknowledgments

The authors would like to thank Mr. Aaron R. Wegener and Dr. Callie A. Cole for the collection of the ESI-MS data. This work was supported by the National Institute of Health (grant no. 5T34GM092711-04), Fort Lewis College Foundation, and Fort Lewis College Faculty Development funding.

References

[1] B. Desoize, "Metals and metal compounds in cancer treatment," *Anticancer Research*, vol. 24, no. 3, pp. 1529–1544, 2004.

[2] T. W. Hambley, "Metal-based therapeutics," *Science*, vol. 318, no. 5855, pp. 1392-1393, 2007.

[3] P. Zhang and P. J. Sadler, "Redox-active metal complexes for anticancer therapy," *European Journal of Inorganic Chemistry*, vol. 2017, no. 12, pp. 1541–1548, 2017.

[4] C. X. Zhang and S. J. Lippard, "New metal complexes as potential therapeutics," *Current Opinion in Chemical Biology*, vol. 7, no. 4, pp. 481–489, 2003.

[5] T. W. Hambley, "Developing new metal-based therapeutics: challenges and opportunities," *Dalton Transactions*, vol. 43, pp. 4927–4937, 2007.

[6] M. Frezza, S. Hindo, D. Chen et al., "Novel metals and metal complexes as platforms for cancer therapy," *Current Pharmaceutical Design*, vol. 16, no. 16, pp. 1813–1825, 2010.

[7] T. Lazarevic, A. Rilak, and Z. D. Bugarcic, "Platinum, palladium, gold and ruthenium complexes as anticancer agents: current clinical uses, cytotoxicity studies and future perspectives," *European Journal of Medicinal Chemistry*, vol. 142, pp. 8–31, 2017.

[8] B. J. Kim, T. W. Hambley, and N. S. Bryce, "Visualising the hypoxia selectively of cobalt(III) prodrugs," *Chemical Science*, vol. 2, no. 11, pp. 2135–2142, 2011.

[9] P. C. A. Bruijnincx and P. J. Sadler, "New trends for metal complexes with anticancer activity," *Current Opinion in Chemical Biology*, vol. 12, no. 2, pp. 197–206, 2008.

[10] G. Mestroni, E. Alessio, and G. Sava, "New salts of anionic complexes of Ru(III) as antimetastatic and antineoplastic agents," *PCT International Applications*, article WO 98/00431, 1998.

[11] E. Alessio, "Thirty years of the drug candidate NAMI-A and the myths in the field of ruthenium anticancer compounds: a personal perspective," *European Journal of Inorganic Chemistry*, vol. 2017, no. 12, pp. 1549–1560, 2017.

[12] M. R. Berger, F. T. Garzon, B. K. Keppler, and D. Schmahl, "Efficacy of new ruthenium complexes against chemically induced autochthonous colorectal carcinoma in rats," *Anticancer Research*, vol. 9, no. 3, pp. 761–765, 1989.

[13] C. G. Hartinger, S. Zorbas-Seifried, M. A. Jakupec, B. Kynast, H. Zorbas, and B. K. Keppler, "From bench to bedside-preclinical and early clinical development of the anticancer agent indazolium trans-[tetrachlorobis(^{1}H-indazole)ruthenate (III)] (KP1019 or FFC14A)," Journal of Inorganic Biochemistry, vol. 100, no. 5-6, pp. 891–904, 2006.

[14] T. Gianferrara, I. Bratsos, and E. Alessio, "A catergorization of metal anticancer compounds based on their mode of action," Dalton Trans, vol. 37, pp. 7588–7598, 2009.

[15] A. Levina, A. Mitra, and P. A. Lay, "Recent developments in ruthenium anticancer drugs," Metallomics, vol. 1, no. 6, pp. 458–470, 2009.

[16] A. Levina, J. B. Aitken, Y. Y. Gwee et al., "Biotransformations of anticancer ruthenium(III) complexes: an X-ray absorption spectroscopic study," Chemistry–A European Journal, vol. 19, no. 11, pp. 3609–3619, 2013.

[17] A. Levina, D. C. Crans, and P. A. Lay, "Speciation of metal drugs, supplements and toxins in media and bodily fluids controls in vitro activities," Coordination Chemistry Reviews, vol. 352, pp. 473–498, 2017.

[18] A. Bergamo and G. Sava, "Ruthenium anticancer compounds: myths and realities of the emerging metal-based drugs," Dalton Trans, vol. 40, no. 31, pp. 7817–7823, 2011.

[19] A. R. Timerbaev, "Role of metallomic strategies in developing ruthenium anticancer drugs," Trends in Analytical Chemistry, vol. 80, pp. 547–554, 2016.

[20] M. Liu, Z. J. Lim, Y. Y. Gwee, A. Levina, and P. A. Lay, "Characterization of a ruthenium(III)/NAMI-A adduct with bovine serum albumin that exhibits a high anti-metastatic activity," Angewandte Chemie, vol. 49, no. 9, pp. 1661–1664, 2010.

[21] L. Messori and A. Merlino, "Ruthenium metalation of proteins: the X-ray structure of the complex formed between NAMI-A and hen egg white lysozyme," Dalton Transactions, vol. 43, no. 16, pp. 6128–6131, 2014.

[22] P.-S. Kuhn, S. M. Meier, K. K. Jovanovic et al., "Ruthenium carbonyl complexes with azole heterocycles–synthesis, X-ray diffraction structures, DFT calculations, solution behavior, and antiproliferative activity," European Journal of Inorganic Chemistry, vol. 2016, no. 10, pp. 1566–1576, 2016.

[23] S. H. Van Rijt and P. J. Sadler, "Current applications and future potential for bioinorganic chemistry in the development of anticancer drugs," Drug Discovery Today, vol. 14, no. 23-24, pp. 1089–1097, 2009.

[24] InfoMine, Inc., 2018, http://www.infomine.com/investment/metal-prices/.

[25] I. Ott and R. Gust, "Non platinum metal complexes as anti-cancer drugs," Archiv der Pharmazie, vol. 340, no. 3, pp. 117–126, 2007.

[26] L. Graf and S. J. Lippard, "Redox activation of metal-based prodrugs as a strategy for drug delivery," Advanced Drug Delivery, vol. 64, no. 11, pp. 993–1004, 2012.

[27] F. L. Bustamante, J. M. Metello, F. A. V. de Castro, C. B. Pinheiro, M. D. Pereira, and M. Lanznaster, "Lawsone dimerization in cobalt(III) complexes toward the design of new prototypes of bioreductive prodrugs," Inorganic Chemistry, vol. 52, no. 3, pp. 1167–1169, 2013.

[28] I. C. A. de Souza, L. V. Faro, C. B. Pinheiro et al., "Investigation of cobalt(III)-triazole systems as prototypes for hypoxia-activated drug delivery," Dalton Transactions, vol. 45, no. 35, pp. 13671–13674, 2016.

[29] B. M. Pires, L. C. Giacomin, F. A. V. Castro et al., "Azido- and chlorido-cobalt complex as carrier prototypes for antitumoral prodrugs," Journal of Inorganic Biochemistry, vol. 157, pp. 104–113, 2016.

[30] B. P. Green, A. K. Renfrew, A. Glenister, P. Turner, and T. W. Hambley, "The influence of the ancillary ligand on the potential of cobalt(III) complexes to act as chaperones for hydroxamic acid-based drugs," Dalton Transactions, vol. 46, no. 45, pp. 15897–15907, 2017.

[31] A. Werner and R. Feenstra, "Ueber dichlorotetrapyridinkobaltsalze," Berichte der Deutschen Chemischen Gesellschaft, vol. 39, no. 2, pp. 1538–1545, 1906.

[32] C. N. Elgy and C. F. Wells, "Kinetics of solvolysis of the trans-dichlorotetrapyridinecobalt(III) ion in water and in water + methanol," Journal of the Chemical Society, Dalton Transactions, vol. 12, pp. 2405–2409, 1980.

[33] W. J. Eilbeck, F. Holmes, and A. E. Underhill, "Cobalt(II), nickel(II), and copper(II) complexes of imidazole and thiazole," Journal of the Chemical Society A: Inorganic, Physical, Theoretical, pp. 757–761, 1967.

[34] J. Wisniewska and P. Kita, "Mechanistic study on the oxidation of promazine and chlorpromazine by hexaimidazolcobalt(III) in acidic aqueous media," Transition Metal Chemistry, vol. 31, no. 2, pp. 232–236, 2006.

[35] J. Springborg and C. E. Schäffer, "Tetrakis(pyridine)cobalt(III) complexes," Acta Chemica Scandinavica, vol. 27, pp. 3312–3322, 1973.

[36] E. Reisner, V. B. Arion, B. K. Keppler, and A. J. L. Pombeiro, "Electron-transfer activated metal-based anticancer drugs," Inorganica Chimica Acta, vol. 361, no. 6, pp. 1569–1583, 2008.

[37] A. K. Renfrew, N. S. Bryce, and T. Hambley, "Cobalt(III) chaperone complexes of curcumin: photoreduction, cellular accumulation and light-selective toxicity towards tumour cells," Chemistry–A European Journal, vol. 21, no. 43, pp. 15224–15234, 2015.

[38] D. C. Ware, P. J. Brothers, G. R. Clark, W. A. Denny, B. D. Palmer, and W. R. Wilson, "Synthesis, structures and hypoxia-selective cytotoxicity of cobalt(III) complexes containing tridentate amine and nitrogen mustard ligands," Journal of the Chemical Society, Dalton Transactions, vol. 6, pp. 925–932, 2000.

[39] E. T. Souza, L. C. Castro, F. A. V. Castro et al., "Synthesis, characterization and biological activities of mononuclear Co (III) complexes as potential bioreductively activated prodrugs," Journal of Inorganic Biochemistry, vol. 103, no. 10, pp. 1355–1365, 2009.

[40] P. Mura, M. Camalli, L. Messori, F. Piccioli, P. Zanello, and M. Corsini, "Synthesis, structural characterization, solution chemistry, and preliminary biological studies of the ruthenium(III) complexes [TzH][trans-RuCl$_4$(Tz)$_2$] and [TzH] [trans-RuCl$_4$(DMSO)(Tz)]·(DMSO), the thiazole analogues of antitumor ICR and NAMI-A," Inorganic Chemistry, vol. 43, no. 13, pp. 3863–3870, 2004.

[41] A. Bergamo and G. Sava, "Linking the future of anticancer metal-complexes to the therapy of tumour metastases," Chemical Society Reviews, vol. 44, no. 24, pp. 8818–8835, 2015.

[42] A. K. Renfrew, N. S. Bryce, and T. W. Hambley, "Delivery and release of curcumin by a hypoxia-activated cobalt chaperone: a XANES and FLIM study," Chemical Science, vol. 4, no. 9, pp. 3731–3739, 2013.

[43] N. Yamamoto, A. K. Renfrew, B. J. Kim, N. S. Bryce, and T. W. Hambley, "Dual targeting of hypoxic and acidic tumor environments with a cobalt(III) chaperone complex," Journal of Medicinal Chemistry, vol. 55, no. 24, pp. 11013–11021, 2012.

14

Metal Nanoparticles: Thermal Decomposition, Biomedicinal Applications to Cancer Treatment, and Future Perspectives

Ayodele Temidayo Odularu (iD)

Department of Chemistry, University of Fort Hare, Private Bag X1314, Alice 5700, South Africa

Correspondence should be addressed to Ayodele Temidayo Odularu; 201106223@ufh.ac.za

Academic Editor: Zhe-Sheng Chen

Monodispersed forms of metal nanoparticles are significant to overcome frightening threat of cancer. This review examined pragmatically thermal decomposition as one of the best ways to synthesize monodispersed metal nanoparticles which are stable and of small particle sizes. Controlled morphology for delivery of anticancer agent to specific cells can also be obtained with thermal decomposition. In addition to thermal decomposition, the study also looked into processes of characterization techniques, biological evaluation, toxicity of nanoparticles, and future perspectives.

1. Introduction

Cancer is a worldwide disease [1–3]. A report from the National Cancer Institute (NCI), titled "Statistics at a Glance: The Burden of Cancer Worldwide," said, "cancer was among the global leading fatality with 14.1 million cases which emerged in 2012 alongside with cancer-related issues of 8.2 million globally" [4]. It is the second cause of fatality after heart disease with 8.8 million lives affected in 2015, of which 70% came from low- and middle-income countries [5, 6]. World Health Organization (WHO) predicted the number of new cases of cancer disease to rise to 70% over the next twenty years [6]. In addition to this, alarming rates at which cancer subjects experienced resistance to antineoplastic drugs and challenges of its side effects when these drugs were administered, call for an alternative drug. A current approach to deal with the resistance and side effects can be drawn from nanotechnology. Nanotechnology is made up of small materials of magnitude between 1 and 100 nm, containing structures with arrangements of atoms, popularly referred to as nanoparticles [7]. It includes disciplines, such as chemistry, computing, electronics, energy, engineering, physics, and *biomedicine* [8]. In biomedicine, nanomedicine (medical application of nanotechnology) has a promising approach to detect and treat cancer [9, 10]. The focus of nanomedicine is to pave an effective way in the health sector to rid it of dangerous diseases, such as cancer [11]. Zainal et al. reported that one eminent way nanomedicine helped in going about this was through inorganic nanoparticles [12]. Nanoparticles enhance the delivery of promising anticancer agents on malignant cells [13]. It is also used as a anticancer therapeutic agent, as well as, to detect and diagnose cancer with magnetic resonance image (MRI) [9]. The goal of a better delivery system for drugs has made intense research to be delved in nanoparticles over the past ten years [12]. The greatest challenge in the research of nanotechnology is how to obtain a controlled nanometric size and shape. Nanoparticles with their small sizes when compared to bulk materials have helped to procure hopes with their edge advantages of *quantum size effect* and *high surface area to volume ratio* [13–15]. Uniqueness of inorganic nanoparticles due to the small size and morphology reflects optoelectronic characteristics [11, 12]. Monodispersed metal nanoparticles by thermal decomposition can be used to overcome resistance and side effects of the conventional drugs for cancer. The impact of monodispersed metal nanoparticles on cancer was considered in this study. This scientific communication also addressed synthesis, reaction method of coordination compounds, characterization, prevention of polydispersed, and toxicity of nanoparticles. The aim of this review was to look at the application of thermal decomposition as the most appropriate method of synthesis to obtain monodispersed forms of metal nanoparticles.

1.1. Reaction Methods prior to Synthesis of Metal Nanoparticles. Prepared coordination compounds are right precursors to synthesis of nanoparticles because they had been confirmed to be an effective path to give high-quality monodispersed metal nanoparticles [16]. Kelly et al. reported that coordination compounds and metallocenes were convenient precursors [17]. Further treatment of these coordination compounds before thermal decomposition involves the use of stabilizers and capping agents.

1.1.1. Reducing Agents (Stabilizers/Scavenging Agents). Reducing agents are inorganic and organic compounds used to prepare metal nanoparticles by decreasing oxidation states of metallic ions in coordination compounds or metallic salts to zero. They are also used to prevent agglomeration of metal nanoparticles [13, 18, 19]. Reducing agents are referred to as stabilizers or scavenging agents. Effective stabilizers can be natural polymers, (chitosan and oligochitosan) or artificial polymers (alginate, poly vinyl alcohol (PVA), and polyvinyl pyrrolidone (PVP)). Both forms of polymers have functional groups, such as amino ($-NH_2$), carboxylic acid (-COOH), hydroxyl (-OH), and thiols (SH) [18]. Natural polymers support green chemistry. They are sometimes used as capping agents for nanoparticles. The process of providing electrons or radicals to metallic ions allows them to be referred to as scavenging agents. They can be classified as weak and strong stabilizers. Weak reducing agents support slow reaction, thereby allowing particles to grow over a long period of time to give a faceted and less than one nanometer nanoparticles. Examples are sodium citrate and potassium bitartraate [19]. On the other hand, strong reducing agents, such as formamide and ortho-anisidine form bigger and spherical nanoparticles [19]. Other reducing agents which are very useful and not polymers are long chain organic molecules, sodium borohydride, and ethylene glycol.

1.1.2. Surfactants (Capping Agents) and Colloidal Nanoparticles. Surfactants are protective and surface acting agents which further prevent agglomeration by avoiding interaction of nanoparticles with one another [13]. They stabilize the nanoparticles formed by reducing agents, thereby allowing some researchers to refer to them as stabilizing agents. The qualities of soft-temperate model, capability to transform chemical kinetics, and easy maneuverability possessed by surfactants allow them to control morphologies of nanomaterials [20]. They are also referred to capping agents. During thermal decomposition, the reduced samples are injected inside the surfactants at a certain temperature, the colour of the sample changed to black after a while in an inert environment. Black solution is an indication of colloidal nanoparticles [21]. As it applies to stabilizers, polymers and functional groups of amines, carboxylic acids, hydroxyls, and

thiols are good capping agents for successful thermal decomposition [13].

1.1.3. Dual Stabilizing and Capping Agents. In some cases, researchers use neither stabilizing nor capping agents. In other cases, some compounds, such as sodium citrate can act as both a stabilizing and a capping agent [13].

1.2. Synthesis of Metal Nanoparticles. Two main approaches of synthesizing metal nanoparticles are top down and bottom up [22]. Both approaches have their advantages and disadvantages. Imperfection observed on the surfaces and damage of crystals of nanostructures is the biggest challenge in the top-down approach of synthesizing nanoparticles [22, 23]. This approach is still in use despite this challenge. Method used for synthesis of nanoparticles has a great impact on small particle size and morphology. Three different methods of synthesizing nanoparticles are physical (microwave irradiation, sonochemical, ultraviolet radiation, laser ablation, thermal decomposition (thermolytic), photochemical, or radical induced), chemical (supercritical fluid, coprecipitation, use of inorganic matrix as support, and organic solvents), and biological (use of algae, bacteria, fungi, or plants) [10–13, 24]. From the three methods highlighted, this review focused on the physical method of thermal decomposition because other methods are polluting, time consuming, forming agglomerated and aggregated nanoparticles with wide distributed sizes, and very expensive [15, 22]. Thermal decomposition is an innovative method to synthesize stable monodispersed nanoparticles product [21]. It is a research area that is fast developing, clearer, and economical when compared with conventional methods [25]. It is also one of the easiest and the most convenient way to synthesize monodispersed metal nanoparticles [26]. In addition, it answers the greatest challenge of obtaining a controlled nanometric size and shape in nanotechnology research [27]. Other factors to be considered to obtain controlled nanometric size are duration of synthesis, temperature, concentrations of the reactants, stabilizers, capping agents (surfactants), and types of surfactants [28]. Palacious-Hernández et al. and Kino et al. explained that the solventless method of thermal decomposition was an easy and moderate route which required no raw material [27, 28]. On a similar note, Tran et al. stated that thermal decomposition allowed large amount of nanoparticles to be produced once, unlike biological method which produced small amount of nanoparticles [26]. Tran et al. also emphasized the relevance of the precursor injection method, where the precursor was injected into hot solution of the surfactant [26]. This was to induce rapid nucleation of nanoparticles with small sizes which are either the same or similar and possess narrow size distributions [26]. These monodispersed nanoparticles could also be referred to as being homogeneous [22].

1.2.1. Thermal Decomposition. Factors such as nature of metallic ion and the force of reaction with the ligands in coordination compounds have effects on the temperature and pressure at which thermal decomposition takes place [17]. In other words, coordination compounds can be thermodynamically stable or kinetically stable. No particular stabilizing agent (stabilizer) is used for thermal decomposition. With regard to stabilizers, Rao et al. reported that capping agents, such as carboxylic acids and alkyl amines, influenced formation of monodispersed nanoparticles obtained from thermal decomposition [29]. The overall effect has an impact on achieving monodispersed nanoparticles [17].

1.3. Washing and Drying of Nanoparticles. After thermal decomposition, the reaction vessel is cooled at room temperature in a switched-on fume cupboard so as to lower the temperature [21, 28]. The switch is ensured to be put off after cooling. Nanoparticles can easily form precipitates in cold polar solvents, such as deionized water, ethanol, and methanol [21]. These solvents help to remove excess reaction materials (capping agents) [21]. The process of centrifugation as the next step separates the precipitates from colloidal nanoparticles. It also cleans the nanoparticles, but excess cleaning causes agglomeration of the nanoparticles. This process is completed when there is homogeneity in the precipitates. Several times of redispersion in nonpolar solvents like benzene, hexane, and toluene purify the precipitates and this signifies a nanoparticle surfactant core-shell structure [21, 28]. Drying of nanoparticles takes place by alcohol drying (chemical extraction), freeze drying (nonthermal), or in vacuum oven (thermal) for a temperature of less than 100°C and for a specific period of few hours or overnight after washing [18, 30–32].

1.4. Behaviour and Characterization of Nanoparticles. Most researchers studied the behaviour of nanoparticles using relevant techniques for appropriate characterizations. Wostek-Wojciechowska et al., Dallas et al., and Jung et al. stated that thermal decomposition could be performed as either thermolysis or pyrolysis [33–35]. Dallas et al. did the pyrolysis of his silver nanoparticles at a temperature of 300°C [34]. They also carried out thermolysis in both solid and liquid states [34]. In the case of Wostek-Wojciechowska et al., they reported that better nanoparticles were obtained when in solution than in the solid state [33]. In order to measure the level of thermal stabilities of thermolysed nanoparticles, thermogravimetric analysis is needed. Dallas et al and Khalil et al observed the temperature of nanoparticles with thermogravimetric analysis (TGA) and differential thermal analysis (DTA) [34, 36]. All the aforementioned authors reported temperatures below 300°C for the thermogravimetric analysis. They were all in line with Iravani et al., who reported that a temperature of an less than or equal to 300°C provides a broad range of reaction temperature and permits the effective control of nanoparticles by the variance in the heating temperature while the solvent is left constant [13].

Spectroscopic characterization techniques involve Fourier-Transform Infrared (FT-IR) Spectroscopy (FT-IR), Ultraviolet-Visible (UV-Vis) Spectroscopy, Florescence Spectroscopy (photoluminescence), and Raman Spectroscopy. The FT-IR functions to identify functional groups in the bulk materials and nanoparticles. For the UV-Vis spectroscopy, it supports

results from other spectroscopic characterization techniques, thereby giving the geometry and optical properties of the sample.

The optical property of metal nanoparticles requires the band gap [37]:

$$E_g \text{(Ev)} = \frac{1240}{k} \text{ (nm)}, \qquad (1)$$

where E_g is the band gap energy and k is the absorption edge.

Florescence spectroscopy often referred to as photoluminescence spectroscopy because the same instrument does the analyses of both techniques. Both provide the optical properties of the sample. Raman spectroscopy does the identification of the crystal structure and supplementary confirmation of phase purity of prepared samples of nanoparticles [38].

In addition to the aforementioned techniques, Mass Spectrometry (MS) and X-ray Photoelectron Spectroscopy (XPS) are important characterization techniques in Materials Chemistry. The MS measures the characteristics of charged particles of individual molecules by converting them into ions, while XPS analyzes surface and interface conditions of materials.

Other important techniques in Materials Chemistry are microscopic analyses, purity check, x-ray diffraction, magnetic characterization, and surface area. Microscopic analyses include Scanning Electron Microscopy (SEM), Transmittance Electron Microscopy (TEM), and Atomic Force Microscopy (AFM). The metal nanoparticles are always characterized for their morphologies using techniques of SEM and TEM. The SEM produces morphology of a sample based on scattered electrons to give images of three dimensional, such as cylinders. In the case of TEM, it produces morphology of a sample based on transmitted electrons to give images of two dimensional, such as thin sheets and thin wires. Atomic Force Microscope (AFM) observes the inside of materials directly.

Elemental Analysis (EA) and Electron Dispersive Spectroscopy (EDS) are used to check for the purity of bulk material and metal nanoparticles respectively.

X-ray diffraction can either be single crystal X-ray diffractometry or powder X-ray diffractometry (XRD). Single crystal X-ray diffractometer analyzes the complete structure of crystalline materials, from the range of simple inorganic solids to complex macromolecules.

The XRD is used to obtain data for the crystalline shape, crystallite size, and orientation in polycrystalline of powdered solid samples from the Scherrer equation:

$$D = \frac{0.9\lambda}{\beta \cos\theta_\beta} \theta, \qquad (2)$$

where D is the crystalline size of the metal nanoparticles; λ is the wavelength of X-ray radiation; β is the full width at half maximum (FWHM) of the diffraction peak; and θ_β, the Bragg diffraction angle [37].

The magnetic characterization includes Nuclear Magnetic Resonance (NMR), Electromagnetic Resonance (ESR) and Magnetic Susceptibility Sensors (MSS). The NMR as a scientific technique is employed in the study molecular physics, crystals, and non-crystalline materials, mostly for paired electrons. Nuclear magnetic resonance is used as diagnostic tool to confirm synthesized nanoparticles have specific moieties. The ESR, often called Electromagnetic Paramagnetic Resonance (EPR), provides information on the geometry of the radical and the orbital of the unpaired electron, while MSS provides quantitative measure to which a material may be magnetized in a magnetic field.

The surface area of nanoparticles can be determined using a mathematical expression called Brunauer-Emmett-Teller (BET) as shown in equations (3) and (4). The colloidal state of stable monodispersed metal nanoparticles can be determined using electrophoretic light scattering technique for its zeta potential [39]. Phase of *in vitro* testing prior to *in vivo* testing is a preliminary stage used to test the anticancer activities of the nanoparticles.

$$S_{\text{total}} = \left(\frac{v_m N s}{V}\right), \quad (3)$$

$$S_{\text{BET}} = \left(\frac{S_{\text{total}}}{a}\right), \quad (4)$$

where v_m is the unit of the monolayer volume of adsorbate gas, N is the Avogadro number, s is the adsorption cross section of the adsorbing species, V is the molar volume of the absorbing species, and a is the mass of the adsorbent.

1.5. Prevention of Polydispersed Nanoparticles. Four important steps in chemical synthesis of nanoparticles which aid particle size and uniformity are nucleation, growth of the colloidal particles, Ostwald ripening, and stabilization. During the stage of particle growth, it is possible to control the uniformity of the particle size. This is because once the reaction proceeds to Ostwald ripening, polydispersed nanoparticles are formed rather than monodispersed nanoparticles. Ostwald ripening is a slow diffusion-organized process, also called *second phase coarsening* [40]. It is defined as the dissolving of small nanoparticles and their redeposition on larger particles [41]. Apart from the dispersion of nanoparticles in liquid phase which promotes the formation of polydispersed nanoparticles, processes of drying in powdery form rather than in colloidal slurries form and sintering are also factors which contribute to polydispersity [30]. The process of sintering involves the heating of nanoparticles in powdery form (nanopowders) in order to make them solid [30]. It is high-temperature dependent [30] and, therefore, contrary to the temperature required for monodispersity. Schematic diagram for the synthesis, characterization, and *in vitro* anticancer testing of monodispersed metal nanoparticles is shown in Figure 1.

1.6. Toxicity of Nanoparticles. Small sizes of nanoparticles indicate that they are readily absorbed in human body than large sizes [42–44]. The absorption can be through ingestion, inhalation, injection, and transdermal delivery [44–47]. They are more toxic to the health of human beings than large sizes [48]. Research into toxicology of nanoparticles is still in its infant stage [44, 47]. The use of carbon nanotubes depending on the size, uses of silica, biodegradability or polymeric nanoparticles can reduce the toxicity of nanoparticles [48].

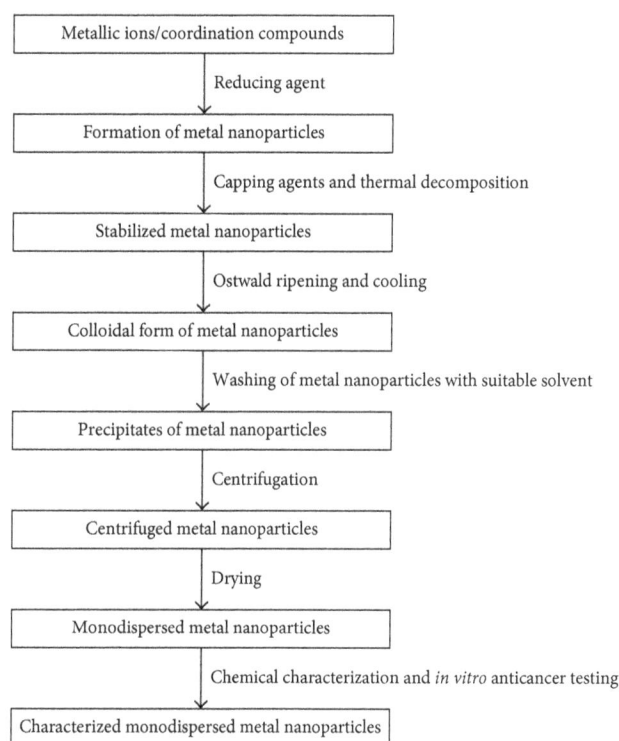

FIGURE 1: Schematic diagram for the synthesis of monodispersed metal nanoparticles.

There are a number of approaches used to assess the toxicity of nanoparticles [48]. The most cost-effective way which manages time approach is *in vitro* studies [48]. The generally assessed study is cell viability with examples such as biomarkers for apoptosis, cell membrane integrity with lactase dehydrogenase (LDH) assay, comet assay for genotoxicity, immunohistochemistry, and tetrazolium reduction assays [48]. Viable cells are detected using colorimetric assays such as 3-(4,5-dimethylthiazol-2-yl)-2,5-diphenyltetrazolium bromide (MTT), 2,3-bis-(2-methoxy-4-nitro-5-sulfophenyl)-2H-tetrazolium-5-carboxanilide (XTT), 3-(4,5-dimethylthiazol-2-yl)-5-(3-carboxymethoxyphenyl)-2-(4-sulfophenyl)-2H-tetrazolium, and water-soluble tetrazolium salts (WSTs) [48].

2. Conclusion

This mini review demonstrated thermal decomposition as the most suitable to synthesize stable monodispersed nanoparticles to order to achieve the goal of cancer treatment in the health sector. Thermal decomposition is easy and economical. Instrumental method of thermogravimetric analysis supported the measurement of the rate of decomposition while the stability as applied to biological application is done using zeta potential. Other relevant characterization techniques help in the identification. Assessments of the toxicity levels are done using *in vitro* assays.

3. Future Perspectives

Challenges of agglomeration of metal nanoparticles warrant ways to support thermal decomposition. Method of etching synthesis of metal nanoparticles will be encouraged to support thermal decomposition for the ease of preparing monodispersed nanoparticles. Melting points of samples of synthesized nanoparticles will be used to ascertain the temperature at which thermal decomposition will take place. Solubility test of nanoparticles will be done to detect toxicity and biomembral penetration. Green chemistry promotes good synthesis of nanoparticles in the areas of solvents, reducing agents, and capping agents; therefore, green compounds will be considered. Reference materials will also be considered to assess the efficiency of instruments and methodology relevant to characterization of nanoparticles generally used in preclinical biomedical research such as *in vitro* and interlaboratory proficiency testings. Results from zeta potential for colloidal stabilities of nanoparticles will introduce them to further testing of *in vivo* anticancer testing.

Acknowledgments

The author appreciates Govan Mbeki Research and Development Centre (GMRDC) for financial assistance.

References

[1] P. A. Lotufo, "Smoking and cancer: Brazil and the global burden of disease initiative," *Sao Paulo Medical Journal*, vol. 133, no. 5, pp. 385–387, 2015.

[2] A. Jemal, F. Bray, M. M. Center, J. Farlay, E. Ward, and D. Forman, "Global cancer statistics," *CA: A Cancer Journal for Clinicians*, vol. 61, no. 2, pp. 69–90, 2011.

[3] F. Fitzmaurice, D. Dicker, A. Pain et al., "The global burden of cancer 2013," *JAMA Oncology*, vol. 1, no. 4, pp. 505–527, 2015.

[4] L. A. Torre, F. Bray, R. L. Siegel, J. Ferlay, J. Lortet-Tieulent, and A. Jemal, "Global cancer statistics 2012," *CA: A Cancer Journal for Clinicians*, vol. 65, no. 2, pp. 87–108, 2015.

[5] D. İlem-Özdemir, E. Gündoğdu, M. Elinci, and M. Aşikoğlu, "Nanoparticles: from diagnosis to therapy," *International Journal of Medical Nano Research*, vol. 3, no. 1, pp. 1–5, 2016.

[6] World Health Organization, *Cancer Fact Sheet*, World Health Organization, Geneva, Switzerland, 2017.

[7] R. Bhattacharya and P. Mukherjee, "Biological properties of "naked" metal nanoparticles," *Advanced Drug Delivery Reviews*, vol. 60, no. 11, pp. 1289–1306, 2008.

[8] S. Singh and A. Singh, "Current status of nanomedicine and nanosurgery," *Anesthesia: Essays and Researches*, vol. 7, no. 2, pp. 237–242, 2013.

[9] P. Cherukuri, E. S. Glazer, and S. A. Curley, "Targeted hyperthermia using metal nanoparticles," *Advanced Drug Delivery Reviews*, vol. 62, no. 3, pp. 339–345, 2010.

[10] S. Chaturvedi, P. N. Dave, and N. K. Shah, "Applications of nano-catalyst in new era," *Journal of Saudi Chemical Society*, vol. 16, no. 3, pp. 307–325, 2012.

[11] S. Bhattacharyya, R. A. Kudgus, R. Bhattacharya, and P. Mukherjee, "Inorganic nanoparticles in cancer therapy," *Pharmaceutical Research*, vol. 28, no. 2, pp. 237–259, 2011.

[12] N. A. Zainal, S. R. A. AbdShukor, H. Azwana, and K. A. Rasak, "Study on the effect of synthesis parameters of silica nanoparticles entrapped with rifampicin," *Chemical Engineering Transactions*, vol. 32, pp. 2245–2250, 2013.

[13] S. Iravani, H. Korbekandi, S. V. Mirmohammadi, and B. Zolfaghari, "Synthesis of silver nanoparticles: chemical, physical and biological methods," *Research in Pharmaceutical Sciences*, vol. 9, no. 6, pp. 385–406, 2014.

[14] P. Logeswari, S. Silambarasan, and J. Abraham, "Ecofriendly synthesis of silver nanoparticles from commercially available plant powders and their antibacterial properties," *Scientia Iranica*, vol. 20, no. 3, pp. 1049–1054, 2013.

[15] A. A. D. Khalaji and F. Malekan, "Synthesis and characterization of nanowires hausmannite (Mn_3O_4) by solid-state thermal decomposition," *International Journal of Nano Dimension*, vol. 6, pp. 153–156, 2015.

[16] N. Moloto, N. Revaprasadu, M. J. Moloto, P. O. Brien, and J. N. Raftery, "N'-diisopropylthiourea and N,N'-dicyclohexylthiourea zinc(II) complexes as precursors for the synthesis of ZnS nanoparticles," *South African Journal of Science*, vol. 105, no. 7-8, pp. 258–263, 2009.

[17] C. H. M. Kelly and M. Lein, "Choosing the right precursor for thermal decomposition solution-phase synthesis of iron nanoparticles: tunable dissociation energies of ferrocene derivatives," *Physical Chemistry Chemical Physics*, vol. 18, no. 47, pp. 32448–32457, 2016.

[18] M. Muzami, N. Khalid, M. D. Aziz, and S. A. Abbas, "Synthesis of silver nanoparticles by silver salt reduction and its characterization," *IOP Conference Series: Materials Science and Engineering*, vol. 60, pp. 1–8, 2014.

[19] O. Masala and R. Seshadri, "Synthesis routes for large volumes of nanoparticles," *Annual Review of Materials Research*, vol. 34, no. 1, pp. 41–81, 2004.

[20] C. K. Latha, M. Raghasudha, Y. Aparna, R. M. D. Ravinder, J. K. P. Veerasomaiah, and D. Shridhar, "Effect of capping agent on the morphology, size and optical properties of In_2O_3 nanoparticles," *Materials Research*, vol. 20, no. 1, pp. 256–263, 2017.

[21] M. Salavati-Niasari, F. Davai, and M. Mazaheri, "Synthesis of Mn_3O_4 nanoparticles by thermal decomposition of a [bis (salicylidiminato)manganese(II)] complex," *Polyhedron*, vol. 27, no. 17, pp. 3467–3471, 2008.

[22] R. Betancourt-Galindo, P. Y. Retes-Rodriguez, B. A. Puente-Urbina et al., "Synthesis of copper nanoparticles by thermal decomposition," *Journal of Nanomaterials*, vol. 2014, Article ID 980545, 5 pages, 2014.

[23] Y. Wang and Y. Xia, "Bottom-up and top-down approaches to the synthesis of monodispersed effect of capping agent on the morphology, size and optical properties of In_2O_3 nanoparticles spherical colloids of low melting-point metals," *Nano Letters*, vol. 4, no. 10, pp. 2047–2050, 2004.

[24] J. Behari, "Principles of nanoscience: an overview," *Indian journal of experimental biology*, vol. 48, no. 10, pp. 1008–1019, 2010.

[25] D. Pomogalio and G. I. Dzhardimlieva, *Nanostructured Materials Preparation Via Condensation Ways*, Springer, Berlin, Germany, 2017, https://books.google.co.za/books.

[26] Q. H. Tran, V. Q. Nguyen, and A. Le, "Silver nanoparticles: synthesis, properties, toxicology, applications and perspectives," *Advances in Natural Sciences: Nanoscience and Nanotechnology*, vol. 4, no. 3, pp. 1–21, 2013.

[27] T. Palacious-Hernández, G. A. Hirata-Flores, O. E. Contreras-López et al., "Synthesis of Cu and Co metal oxide nanoparticles from thermal decomposition of tartrate complexes," *Inorganica Chimica Acta*, vol. 392, pp. 277–282, 2012.

[28] T. Kino, T. Kuzuya, K. Itoh, K. Sumiyama, T. Wakamatsu, and M. Ichidate, "Synthesis of chalcopyrite nanoparticles via thermal decomposition of metal-thiolate," *Materials Transactions*, vol. 49, no. 3, pp. 435–438, 2008.

[29] C. N. R. Rao, P. J. Thomas, and G. G. U. Kulkami, "Nanocrystals: synthesis, properties and applications," in *Springer Series in Materials Science*, vol. 95, pp. 1–182, Springer, Berlin, Germany, 2007.

[30] S. P. Yeap, "Permanent agglomerates in powdered nanoparticles: formation and future prospects," *Powder Technology*, vol. 323, pp. 51–59, 2018.

[31] C. P. Rezende, J. B. Da Silva, and N. D. S. Mohallem, "Influence of drying on the characteristics of zinc oxide nanoparticles," *Brazilian Journal of Physics*, vol. 39, no. 1, pp. 248–251, 2009.

[32] S. Sabir, M. Arshad, and S. K. Chaudhari, "Zinc oxide nanoparticles for revolutionizing agriculture: synthesis and applications," *Scientific World Journal*, vol. 2014, Article ID 925494, 8 pages, 2014.

[33] D. Wostek-Wojciechowska, J. K. Jeszka, P. Uznanski, C. Amiens, B. Chaudret, and P. Lecante, "Synthesis of gold nanoparticles in solid state by thermal decomposition of an organometallic precursor," *Materials Science Poland*, vol. 22, no. 4, pp. 407–413, 2004.

[34] P. Dallas, A. B. Bourlinos, P. Komninou, M. Karakassides, and D. Niarchos, "Silver nanoparticles and graphic carbon through thermal decomposition of a silver/-acetylenedicarboxylic salt," *Nanoscale Research Letters*, vol. 4, no. 11, pp. 1358–1364, 2009.

[35] Y. K. Jung, J. I. Kim, and J. Lee, "Thermal decomposition mechanism of single-molecule precursors forming metal sulfide nanoparticles," *Journal of the American Chemical Society*, vol. 132, no. 1, pp. 178–184, 2010.

[36] M. Khalil, M. M. Al-Qunaibitt, A. M. Al-zahem, and J. P. Labis, "Synthesis and characterization of ZnO nanoparticles by thermal decomposition of a curcumin zinc complex," *Arabian Journal of Chemistry*, vol. 7, no. 6, pp. 1178–1184, 2014.

[37] M. Y. Nassar, I. S. Ahmed, and I. Samir, "A novel synthetic route for magnesium aluminate ($MgAl_2O_4$) nanoparticles using sol-gel auto combustion method and their photocatalytic properties," *Spectrochimica Acta Part A: Molecular and Biomolecular Spectroscopy*, vol. 131, pp. 329–334, 2014.

[38] M. Krishnaiah, P. Bhargava, and M. Sudhanshu, "Low-temperature synthesis of Cu_2CoSnS_4 nanoparticles by thermal decomposition of metal precursors and the study of its structural, optical and electrical properties for photovoltaic applications," *RSC Advances*, vol. 5, no. 117, pp. 96928–96933, 2015.

[39] S. Singh, A. Bharti, and V. K. Meena, "Structural, thermal, zeta potential and electrical properties of disaccharide reduced silver nanoparticles," *Journal of Materials Science: Materials in Electronics*, vol. 25, no. 9, pp. 3747–3752, 2014.

[40] A. Baldan, "Review progress in Ostwald ripening theories and their applications to nickel-base superalloys. Part 1: Ostwald's ripening theories," *Journal of Materials Science*, vol. 37, pp. 2171–2202, 2002.

[41] Z. Zhang, Z. Wang, S. He, C. Wang, M. Jin, and Y. Yin, "Redox reaction induced Ostwald ripening for size-and shape-focusing of palladium nanocrystals," *Chemical Science*, vol. 6, no. 9, pp. 5197–5203, 2015.

[42] C. Buzea, I. I. P. Blandino, and K. Robbie, "Nanomaterials and nanoparticles: sources and toxicity," *Biointerphases*, vol. 2, no. 4, pp. MR17–MR172, 2007.

[43] N. Durán, P. D. Marcato, R. De Conti et al., "Potential use of silver nanoparticles on pathogenic bacteria, their toxicity and possible mechanisms of action," *Journal of the Brazilian Chemical Society*, vol. 21, no. 6, pp. 949–959, 2010.

[44] A. Elsaesser and C. V. Howard, "Toxicity of nanoparticles," *Advanced Drug Delivery Reviews*, vol. 64, no. 2, pp. 129–137, 2012.

[45] R. V. Ravishankar and B. A. Jamuna, "Nanoparticles and their potential application as antimicrobials," in *Formatex*, vol. 3, no. 1, pp. 197–209, Scientific Research Publishing, Wuhan, China, 2011.

[46] O. Kalantzi and G. Biskos, "Methods for assessing basic particle properties and cytotoxicity of engineered nanoparticles," *Toxics*, vol. 2, no. 1, pp. 79–91, 2014.

[47] S. A. Love, M. A. Maurer-Jones, J. W. Thompson, and Y. Lin, "Assessing nanoparticles toxicity," *Annual Review of Analytical Chemistry*, vol. 5, no. 1, pp. 181–205, 2012.

[48] H. Bahadar, F. Maqbool, K. Niaz, and M. Abdollahi, "Toxicity of nanoparticles and an overview of current experimental models," *Iranian Biomedical Journal*, vol. 20, no. 1, pp. 1–11, 2016.

β-Carboline Silver Compound Binding Studies with Human Serum Albumin: A Comprehensive Multispectroscopic Analysis and Molecular Modeling Study

Ali Alsalme ⓘ,[1] Rais Ahmad Khan ⓘ,[1] Arwa M. Alkathiri,[1] Mohd. Sajid Ali ⓘ,[2] Sartaj Tabassum,[2] Mohammed Jaafar,[1] and Hamad A. Al-Lohedan ⓘ[2]

[1]Department of Chemistry, College of Science, King Saud University, P.O. Box 2455, Riyadh 11451, Saudi Arabia
[2]Surfactant Research Chair, Department of Chemistry, College of Science, King Saud University, P.O. Box 2455, Riyadh 11451, Saudi Arabia

Correspondence should be addressed to Ali Alsalme; aalsalme@ksu.edu.sa and Rais Ahmad Khan; raischem@gmail.com

Academic Editor: Viktor Brabec

β-Carbolines (βCs) belong to the naturally occurring alkaloid family, derived from 9H-pyrido[3,4-b]indole, also known as norharmane (Hnor). Knowing the importance of the βCs alkaloid family in biological processes, a comprehensive binding study is reported of four Ag(I) compounds containing the ligand Hnor and having different counteranions, namely, NO_3^-, ClO_4^-, BF_4^-, and PF_6^-, with human serum albumin (HSA) as a model protein. Different approaches like UV-visible, fluorescence spectroscopy, circular dichroism (CD), and molecular docking studies have been used for this purpose. The fluorescence results establish that the phenomenon of binding of Ag(Hnor) complexes to HSA can be deduced from the static quenching mechanism. The results showed a significant binding propensity of the used Ag(I) compounds towards HSA. The role of the counteranion on the binding of Ag(I) compounds to HSA appeared to be remarkable. Compounds with (ClO_4^-) and (NO_3^-) were found to have the most efficient binding towards HSA as compared to BF_4^- and PF_6^-. Circular dichroism (CD) studies made clear that conformational changes in the secondary structure of HSA were induced by the presence of Ag(I) compounds. Also, the α-helical structure of HSA was found to get transformed into a β-sheeted structure. Interestingly, (ClO_4^-) and (NO_3^-) compounds were found to induce most substantial changes in the secondary structure of HSA. The outcome of this study may contribute to understanding the propensity of proteins involved in neurological diseases (such as Alzheimer's and Parkinson's diseases) to undergo a similar transition in the presence of Ag-β-carboline compounds.

1. Introduction

The history of silver applications started from its use in coins and jewelry. Women loved to decorate themselves with various trinkets of silver, but it is less well known that this metal can be an excellent metallotherapeutic agent. In the human body, silver (Ag) is not acting as an endogenous metal and exhibits relatively low toxicity. Ag(I) coordination compounds with a variety of ligands having nitrogen, phosphorus, and/or sulfur donor atoms have large applications in medicinal and analytical chemistry [1, 2]. Specifically, the antibacterial and antifungal actions of Ag(I) compounds are well known [3–7]. The properties of silver compounds,

like aqueous solubility, light stability, and biological activity, can be modified by varying the number and types of organic ligands [1–7]. Furthermore, in the more recent past, Ag(I) compounds have also been reported to display antitumor activity and have shown activities comparable to the clinical chemotherapeutic drug (cisplatin) [8–11].

Human serum albumin (HSA) is responsible for about 60% of the plasma protein in humans and is accountable for nearly 80% of the osmotic pressure of the blood, and it plays a prominent role in drug disposition and efficacy [12]. Various drugs bind reversibly to albumin and other serum components, which thereby function as carriers [13]. Serum albumin is known to increase the solubility of many drugs in plasma and modulates their delivery to cells *in vivo* and *in vitro* [14–16]. Hence, it is relevant to study the interactions of drug candidates with this protein.

Norharmane, 9H-pyrido[3,4-b]indole, is a rather unconventional ligand, belonging to an alkaloid family called β-carbolines (βCs). The molecule can act as a coordinating ligand, an H-bonding donor/acceptor, and may also act on π-π stacking; all these properties may act in synergy. A schematic chemical structure of norharmane is included in Figure 1.

β-Carbolines are present in many plants, arthropods, and insects. Endogenously, these alkaloids are synthesized by tryptophan or tryptophan-like indoleamines and also found in urine, plasma, platelets (~0.1 nM), in the case of mammals. Interestingly, after alcohol intake and smoking, β-carboline concentrations are found to increase to ~1.0 nM. It is also well known that some β-carboline derivates on photoexcitation induce chromosomal damages in mammal cells and may disarm viruses and bacteria [17]. Moreover, substituted aromatic β-carbolines may enter into the brain by crossing the blood-brain barrier (BBB) and may then be converted into methyl derivatives by specific enzymes, like methyltransferases. Some β-carbolines like 2,9-dimethyl-β-carbolines have exhibited mitochondrial damage leading to neurotoxicity, while 9-methyl-harmine has a neuroprotective effect. β-Carboline compounds are also known to reduce the expression of phosphorylated forms of the so-called tau protein, potently at multiple Alzheimer's disease-related sites [17–20].

In a previous study, some of us have shown that compounds of composition [Ag(Hnor)$_2$](anion) display significant anticancer activity for the anions, namely, NO$_3^-$, ClO$_4^-$, BF$_4^-$, and PF$_6^-$, with the ClO$_4^-$ compound being even comparable to cisplatin in two different cancer cell lines [21]. To explore the possible molecular mechanism of action, we have decided to undertake a study of protein binding of the four silver compounds (1–4) with Hnor varying the ionic sphere by changing the counteranions, namely, using NO$_3^-$, ClO$_4^-$, BF$_4^-$, and PF$_6$. HSA has been selected as a model for the protein binding studies. As spectroscopic techniques, UV–visible, fluorescence spectroscopy, and circular dichroism (CD) techniques have been chosen. Finally, the interactions of the Ag(I) compounds with the protein were studied using molecular modeling.

2. Experimental Section

2.1. Starting Materials and Syntheses. AgNO$_3$, AgClO$_4$, AgBF$_4$, AgPF$_6$, DL-tryptophan, and formaldehyde were purchased from Sigma-Aldrich. Human serum albumin (HSA; ≥99%, Sigma, USA) was essentially fatty-acid free and globulin free, purchased from Sigma, and used as received. We have synthesized the four Ag(I) compounds as described in before [21]. Other standard laboratory chemicals were used as available. Studies of protein folding have mainly been carried out in buffered dilute aqueous solutions to avoid loss of protein to the aggregation phenomenon. Stock solutions of HSA (300 μM) and Ag(I) compounds were prepared in a mixture of dimethyl sulfoxide (5%) DMSO and 95%

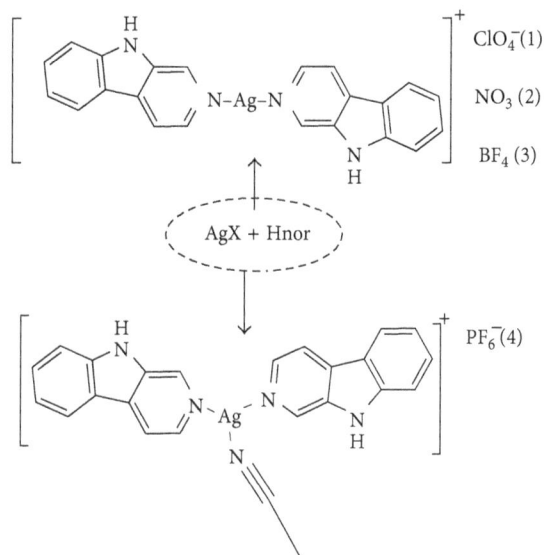

FIGURE 1: The structures of the ligand 9H-pyrido[3,4-b]indole (Hnor) and the four used Ag(I) compounds.

phosphate buffer (20 mM) of pH 7.4 (i.e., well above the isoelectric point of HSA, pH 4.7; hence, the protein possesses a net negative charge at this pH). Conductivity in solution was studied by using an Accumet AB30 Fisher Scientific conductometer at room temperature in MeOH solution of the compounds.

2.2. Protein Binding Studies. The samples of HSA were prepared in 20 mM phosphate buffer (pH 7.4), whereas silver complexes (1 mM) stock solution were prepared in DMSO and further diluted in 20 mM phosphate buffer to reach the desired concentration. In all the samples, the final concentration of DMSO was not more than 1%. The concentration of HSA was determined using the Beer–Lambert law with the molar extinction coefficient of 36500 M^{-1}·cm^{-1} at 280 nm. To study structural changes in HSA by the addition of [Ag(I)(Hnor)$_2$](anion) compounds, the UV absorption spectra were measured, by variation of the concentration of the [Ag(I)(Hnor)$_2$](anion) compounds, while keeping the concentration of HSA constant. UV absorption spectra, from 240 nm to 320 nm, were recorded on a Perkin–Elmer Lambda 45 spectrophotometer at 25°C. Quartz cuvettes of 1 cm path length were used for the measurements. Fluorescence measurements were performed on a Hitachi spectrofluorometer (Model F 7000) equipped with a PC. The fluorescence spectra were collected at 25°C with a path length cell of 1 cm. The slit width used was 5 nm with a protein concentration of 5 μM. The used excitation wavelength for the protein was 295 nm.

CD measurements were carried out on a Jasco spectropolarimeter (Model J-815) equipped with a microcomputer. The instrument was calibrated with D-10-camphorsulfonic acid. All the CD measurements were performed at 25°C with a thermostatically controlled cell holder, attached to a Neslab RTE-110 water bath with an accuracy of ±0.1°C. Spectra were collected with a scan speed of 0.2 nm/min and a response time of 1 s. Each spectrum was taken as the average of three scans.

The far-UV CD spectra were measured at a protein concentration of $20\,\mu M$ at a path length of 1 cm.

All the spectra were recorded after equilibration of the reaction mixture for 5 min.

2.3. Molecular Docking.

2.3. Molecular Docking. The rigid molecular docking studies were performed by using HEX 8.0.0 software [22], which is an interactive molecular graphics program for calculating and displaying possible docking modes of protein. The Hex 8.0.0 performs protein docking using spherical polar Fourier correlations [23]. Hex 8.0.0 necessitates the ligand and the receptor as input in PDB format. The parameters used for docking include the following: correlation type: shape only; FFT mode: 3D; grid dimension: 0.6; receptor range: 180; ligand range: 180; twist range: 360; and distance range: 40. The coordinates of compounds **1** and **2** were taken from its crystal structure as a .cif file and were converted to the PDB format using Mercury software. The crystal structure of the human serum albumin (PDB ID: 1h9z) was downloaded from the protein data bank (http://www.rcsb.org./pdb). Visualization of minimum energy favorable docked poses has been performed using Discovery Studio 4.1 [24] and PyMOL [25].

3. Results and Discussion

3.1. General Observations and Synthesis.

3.1. General Observations and Synthesis. Synthesis of the four Ag(I) compounds, namely, $[Ag(Hnor)_2](ClO_4)$ (**1**), $[Ag(Hnor)_2](NO_3)$ (**2**), $[Ag(Hnor)_2(MeCN)](PF_6)$ (**3**), and $[Ag(Hnor)_2](BF_4)$ (**4**), were done by dissolving the starting Ag salt and Hnor in acetonitrile (ratio 1 : 2); they were characterized according to the procedures reported by some of us previously [21]. Conductivity studies in MeOH solution (1–5 mM) have shown that the compounds behave as 1 : 1 electrolytes. Given the known kinetic lability of Ag(I), it is assumed that the Hnor ligand may dissociate from and associate with the Ag ion, in solution, but on average, they are largely coordinated. The schematic structures of the ligand and the used Ag(I) compounds are depicted in Figure 1.

3.2. HSA Binding Studies.

3.2. HSA Binding Studies. The interaction between the small bioactive molecules and protein receptors is a fundamental step in the drug discovery process. Obtaining a thorough idea of the interaction of the protein with chemical entities plays a vital role in the etiology of several diseases. The protein-drug intermediate products involved in governing various biochemical phenomena in both normal and diseased cells are known to play a significant role in metabolizing therapeutic compounds and their transport [26, 27].

3.2.1. UV Absorption Studies.

3.2.1. UV Absorption Studies. The binding propensity of the drug candidate with the biomolecule was first studied by using the UV technique. The cumulative absorption of three aromatic amino acid residues gives rise to an absorption peak at 280 nm for human serum albumins (HSA) [28]. Figure 2 displays the UV absorption spectra of HSA in the absence and presence of $[Ag(I)(Hnor)_2]ClO_4$. The behavior

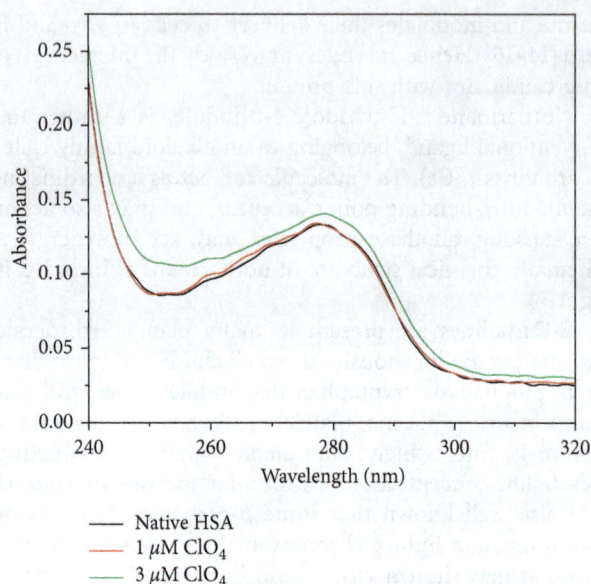

FIGURE 2: UV absorption difference spectra of HSA ($5\,\mu M$). These spectra represent the effect of compound 1 at $1\,\mu M$ and $3\,\mu M$ concentrations. (The difference spectrum was obtained by the HSA-Ag compound spectra minus Ag compound spectra).

of the other 3 compounds (compounds **2–4**) is quite similar, and these data are presented in Figure S1 (see Supplementary Materials).

With the concomitant increase in concentration of Ag(I) compounds, the absorbance of HSA increased, and shifts toward longer wavelengths were observed; the Ag(I) compounds give a definite pattern of the UV–Vis spectrum with weak absorbance at a higher concentration between 295 and 320 nm, ascribed to the ligand "Hnor." The profound enhancement of UV absorbance (hyperchromism) with a redshift (bathochromic effect) of 7 nm (**1**), 5 nm (**2**), 5 nm (**3**), and 6 nm (**4**) in the spectra is suggestive of the formation of adducts/intermediates between the Ag(I) compounds and HSA. HSA is known to act as an important extracellular antioxidant, and this antioxidant property resides in one free cysteine-derived redox-active thiol (-SH) group (i.e., Cys34), which can occur in either reduced or oxidized form. Ag(I) being a soft Lewis acid, its compounds are known to have a high affinity towards sulfur-ligand atoms and moderate affinity towards nitrogen donor atoms. However, binding of Ag(I) towards methionine (Met)/histidine(His) residues and disulfide bridges, and nitrogen atoms, such as deprotonated peptide nitrogen, imine, and indole nitrogen, also cannot be ruled out completely [29, 30]. However, the compounds **1–4** are quite stable in the solution.

The difference spectra of HSA have confirmed that the conformational changes to HSA are due to binding of the Ag (I) compounds (Figure 3 and also Figure S2 in Supplementary Materials). Nevertheless, the variation in binding extent/mode exhibited by the spectra of the compounds to HSA may be associated with the effect of the counteranions. The anions may have facilitated the microenvironmental changes of protein and exposed the targeted site of a subdomain of protein to assist or enhance the selectivity of the

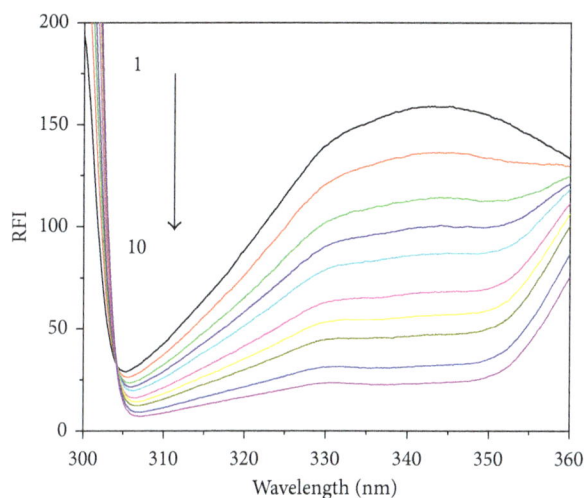

FIGURE 3: Fluorescence emission spectra of HSA ($5\,\mu M$) in the presence of various concentrations of $[Ag(I)(Hnor)_2]ClO_4$. Curves 1 to 10 correspond to compound concentrations of 0, 0.5, 1, 1.5, 2, 3, 4, 5, 7.5, and $10\,\mu M$, respectively, when excited at 295 nm.

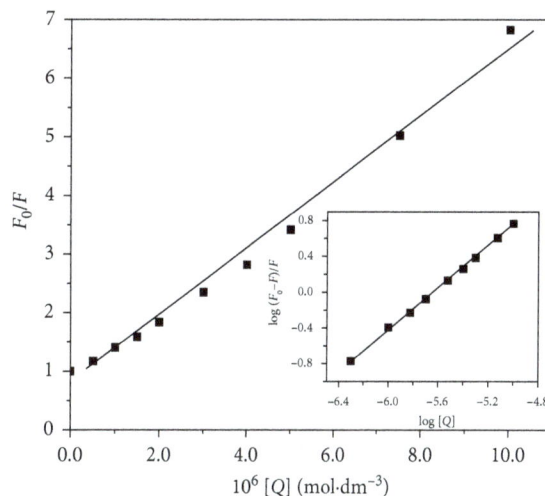

FIGURE 4: The Stern–Volmer plot of HSA fluorescence quenching by $[Ag(I)(Hnor)_2]ClO_4$ at 295 nm. Inset: plot of log $(F_0-F)/F$ as a function of log [complex].

metal center, that is, Ag(I). The effect of anions has been studied with the help of molecular docking to get further insight into their role. The possible role of anions has been discussed in computational studies section. However, the dissociation of the Ag(I) compound is a matter of main concern that we have also studied and found quite stable complex association of Ag(I)-Hnor. Shen et al. [31] have studied the interaction of Ag^+ alone with HSA. When we compared these results of Ag compounds with Ag^+ alone, silver compounds exhibited significantly worthy binding propensity compared to the Ag^+ alone. In the next stage, with the aim to gain more information about the mode of binding of the Ag(I) compounds with HSA, solution fluorescence studies were carried out.

3.2.2. Luminescence Studies. The interaction between metal compounds and proteins has been widely investigated by using fluorescence spectroscopies [32–34]. The luminescence response of HSA, upon addition of different concentrations of Ag(I) compounds, was studied in 20 mM phosphate buffer by using an emission titration experiment. Preliminary luminescence studies of the four starting Ag compounds **1–4** have been reported earlier by some of us [21]. The excitation at 290 nm for all four compound results in a strong luminescence band at around 370 nm. The fluorescence studies of the ligand "Hnor" as well as the free silver salt, "AgNO₃," were also studied under the same conditions to compare the effect of separate entities with a combination of the two. Upon excitation at 295 nm, fluorescence intensity around 340 nm is known to reflect the changes of the tryptophan residue microenvironment [35]. Encouragingly, the luminescence of HSA was found substantially decreased in the presence of increasing concentrations of $[Ag(I)(Hnor)_2]ClO_4$, as depicted in Figure 4. In Figure S3, similar data are presented for compounds **2–4**, with only marginal differences upon anion variation.

A gradual decrease in the luminescence of HSA was observed upon increasing concentration of Ag(I) compounds showed significant redshift at the maximal emission wavelength of tyrosine(Tyr) and tryptophan(Trp) residues of 7 nm (**1**), 5 nm (**2**), 4 nm (**3**), and 4 nm (**4**), which indicates the altered conformation of the HSA by decrease in the polarity around Tyr and Trp residues and increase in hydrophobicity. The continuous, gradual addition of Ag(I) compounds resulted in further decreases in the fluorescence intensity of HSA, which again is indicative of a strong interaction between the Ag(I) compounds and HSA. The factors linked to quenching of the fluorescence can be associated with, for example, excited-state reaction, energy transfer, molecular rearrangements, collisional quenching, and/or ground-state compound formation [36, 37]. Therefore, the consequence of these molecular interactions enforcing towards the static quenching that comprises the establishment of a ground state adduct between the fluorophore and the quencher. The involvement of static quenching in the binding process can be determined with the help of analyzed values of bimolecular quenching constant (K_q). A value of K_q higher than 2.0×10^{10} $M^{-1} \cdot s^{-1}$ reflects the quenching to be static [38–40].

The interaction of the Ag compounds with HSA was quantified; the Stern–Volmer equation has been employed [41]:

$$\frac{F_0}{F} = 1 + K_{SV}[Q] = 1 + K_q \tau_0 [Q], \qquad (1)$$

where F_0 and F are the steady-state fluorescence intensities in the absence and presence of quencher at 340 nm, respectively. K_{sv} is the Stern–Volmer quenching constant, K_q stands for bimolecular quenching constant, τ_0 is the lifetime of the fluorophore in the absence of quencher, and [Q] is the concentration of quencher (i.e., the Ag compound). By calculating the quenching rate constants, K_q, one can distinguish between the static quenching and the dynamic quenching, and this was evaluated by using the following equation:

(a)

(b)

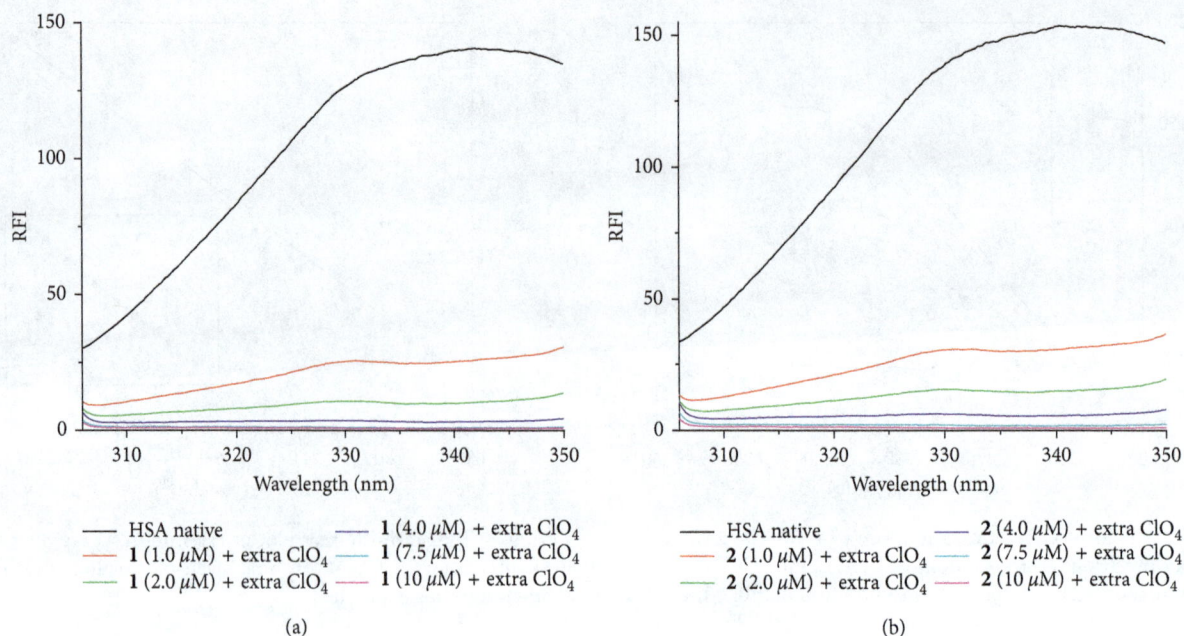

FIGURE 5: Fluorescence emission spectra of HSA ($5\,\mu M$) in the presence of various concentrations of (a) $[Ag(I)(Hnor)_2]ClO_4$ compound **1** + extra $KClO_4$, and (b) $[Ag(I)(Hnor)_2]NO_3$ compound **2** + extra $KClO_4$ corresponding to compound concentrations of 1.0, 2.0, 4.0, 7.5, and $10\,\mu M$, respectively, when excited at 295 nm.

TABLE 1: Stern–Volmer quenching constants and bimolecular quenching rate constant for the interaction of HSA with four Ag(I) Hnor compounds.

Complex	K_{SV} (10^5) (M^{-1})	K_q (10^{12}) ($M^{-1} \cdot s^{-1}$)	K (10^6) (M^{-1})	n	ΔG (kJ mol^{-1})
1 (ClO_4^-)	5.37	53.7	3.74	1.1	−37.50
2 (NO_3^-)	5.34	53.4	3.64	1.0	−37.44
3 (PF_6^-)	4.53	45.3	0.55	1.2	−32.73
4 (BF_4^-)	3.97	39.7	1.14	1.1	−34.55
Hnor	2.95	29.5	0.33	1.0	−31.40
AgNO$_3$	3.10	31.0	1.94	1.1	−35.88

$$K_q = \frac{K_{SV}}{\tau_0}. \qquad (2)$$

The value of tau(zero), τ_0, for biopolymers is known to be 10^{-8} s. [33] The Stern–Volmer quenching constant (K_{sv}) for the fluorometric titration of the Ag(I) compounds into HSA solution was calculated from the linear relationship between F_0/F and $[Ag(I)(Hnor)_2]ClO_4$ (Figure 5) and enlisted in Table 1. In Figure S4, similar data are presented for compounds **2–4**, confirming the above observations.

Thus, these data ascertain the static quenching in the interaction of Ag(I) compounds with HSA, by the calculated value of K_q. The equilibrium between free and bound molecules, when small molecules bind independently to a set of similar sites on a macromolecule, was ascertained by a modified Stern–Volmer equation [42]:

$$\log \frac{F_0 - F}{F} = \log K + n \log [Q], \qquad (3)$$

where K and n are the binding constant and the number of binding sites, respectively. A plot of $\log(F_0 - F/F)$ versus $\log[Q]$ was used to determine the value of K and n (inset

in Figure 5). Several forces, like electrostatic, hydrogen bonds, weak van der Waals, hydrophobic, and steric contacts, can be thought of being responsible for the interaction between the albumin and Ag compounds. The value of the binding constant, K, was used to calculate the standard free energy change $\Delta G°$ of the binding of the ligand to the HSA, by using the relationship [42]:

$$\Delta G = -2.303 \; RT \; \log K. \qquad (4)$$

The values of K, n, and ΔG (binding) are presented in Table 1. All estimated values of n are approximately 1, indicating the existence of just one major binding site in HSA for the present Ag(I) compounds.

The extent of interaction of Ag(I) compounds with HSA was found in the order **1** (ClO_4^-) > **2** (NO_3^-) > **4** (BF_4^-) > **3** (PF_6^-), which can be associated with the differences in the effect of counteranions. Thus, compounds **1** (ClO_4^-) and **2** (NO_3^-) exhibit significantly higher binding affinities towards HSA, compared to other two compounds.

Since it is known that HSA is a monomeric, three-domain, allosteric protein with only one free cysteine, Cys34, the Ag(I) ion could selectively bind at this site because it has a strong preference for S-donor atoms. Also, it can be assumed that the counteranions have facilitated the microenvironmental changes and may have contributed to the exposure of Cys34 from the subdomain of HSA and facilitated that the sulfur atom of cysteine will coordinate to the Ag(I) center of the compounds. Nevertheless, the known affinity of Ag(I) for Met residues and disulfide bridges and nitrogen atoms of HSA coordination on such sites cannot be ignored completely. This hypothesis is supported by the observation that the extent of microenvironmental changes of the protein is much higher in the presence of additional ClO_4^- and NO_3^- anions (see below), and which agrees with the trend in their binding parameter (Table 1); thus, a more profound effect on the Ag(I) compound and its HSA interaction is observed.

3.3. Effects of Addition of Additional Anions.

Given the fact that the highest anticancer activity and also the most substantial HSA interaction takes place in case of the perchlorate and nitrate salts, it was decided to add extra perchlorate (and also nitrate) for all cases and to study the effect on the binding affinity. So, we carried out two sets of experiments and studied the binding propensity of Ag(I) compounds with HSA. In one experiment, the extra ClO_4^- anion was added in a 1:4 ratio, compared to the Ag compound. In a second experiment, we used additional nitrate together with the Ag(I) compounds. Details are given in Figures 6 and 7, and the quite similar details for the other 3 compounds are presented in Figures S5 and S6 in Supplementary Materials.

To determine the role of only the anions (i.e., without Ag and Hnor), we also used $KClO_4$ and KNO_3 and studied their HSA binding by adding variable amounts of anions to the concentration maximum used in the experiments. We found that on the addition of $KClO_4$ and KNO_3 to HSA, only a negligible perturbance of the HSA structure has occurred. Hence, the effect of these anions (ClO_4^- and KNO_3^-) on HSA conformation can be neglected in comparison to the effect of Ag compounds and Ag compounds + additional anions on HSA conformation. When the extra anions + compounds were titrated, the results obtained did show an exponential increase in binding affinity of the Ag(I) compounds.

These findings can be attributed to the enhanced exposure of the cysteine sulfur atoms of HSA, most likely caused by the significant electrostatic effect of the additional anions. This behavior is indicative of the increased microenvironmental changes of the HSA, thereby allowing the more binding of the Ag(I) center to coordinate to the HSA binding site.

The binding strength was evaluated by calculating the Stern–Volmer constant (K_{sv}), a number of the binding sites (n) and Gibbs free energy (ΔG); (Table 2). It should be noted that the effect of ClO_4^- is quite high, whereas NO_3^- and BF_4^- mutually exhibit a nearly similar level of effect, while the PF_6^- anion comes at the end showing a smaller effect. However, no correlation with the size of these four anions is seen.

The effect of anions was also analyzed on the basis of binding constant (K) to study the binding affinity. The binding constant's value exhibited a vibrant increase in the presence of extra anions in particular with ClO_4^- and NO_3^-

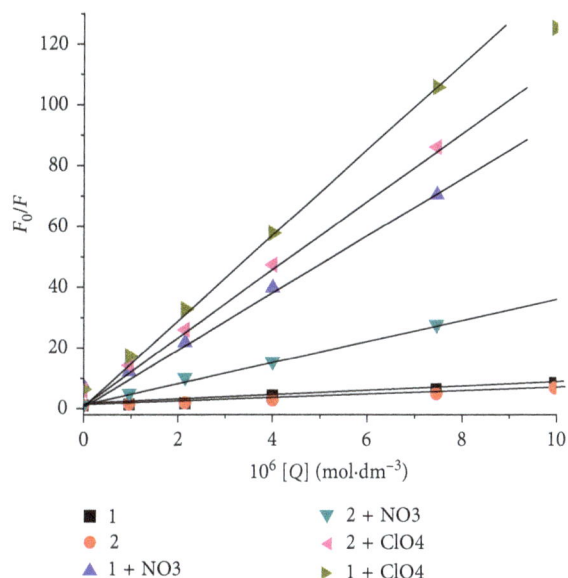

FIGURE 6: Stern–Volmer plot of HSA fluorescence quenching by (1) $[Ag(I)(Hnor)_2]ClO_4$ and (2) $[Ag(I)(Hnor)_2]NO_3$ with additional $KClO_4$ and KNO_3 at 295 nm.

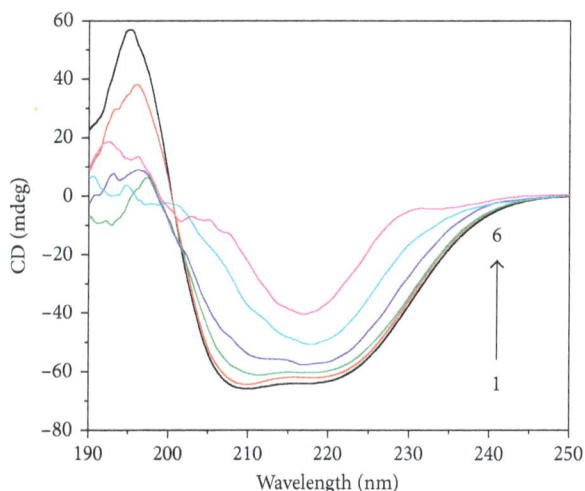

FIGURE 7: Far-UV CD spectra of HSA-compound system (HSA = 20 μM) in the presence of various concentrations of $[Ag(I)(Hnor)_2]ClO_4$ complex. Curves from 1 to 6 correspond to compound concentrations of 0, 20, 40, 60, 80, and 100 mM, respectively.

when compared with the results of complexes **1** and **2**. In complexes **1** and **2**, extra ClO_4^- showed significant increase in the binding affinity as compared to extra NO_3^- as evident from Table 2 (K values). This exponential increase of the magnitude of 10^4 says a lot about the role of anions in the stronger binding affinity of the silver compounds with HSA. The order of binding of Ag(I) compounds with HSA upon extra anion addition was found to be **1 > 2 > 4 > 3**. Thus, it is evident that the number of the binding sites (n) increases upon addition of extra anions.

3.4. Binding and Conformational Changes in Circular Dichroism.

CD spectroscopy is an ideal technique for

TABLE 2: Parameters for the interaction of HSA with Ag(I) compounds with the addition of extra anion, namely, ClO_4^- and NO_3^-.

Complex	K_{SV} (10^7) (M^{-1})	K (10^{10}) (M^{-1})	n	ΔG $(kJ\ mol^{-1})$
Set A				
1 (+extra ClO_4^-)	2.21	17.2	1.77	−64.12
2 (+extra ClO_4^-)	1.29	6.05	1.58	−57.48
3 (+extra ClO_4^-)	0.84	0.46	1.55	−55.13
4 (+extra ClO_4^-)	1.85	1.19	1.69	−61.52
Set B				
1 (+extra NO_3^-)	2.06	3.64	1.64	−60.26
2 (+extra NO_3^-)	1.24	0.66	1.53	−56.05
3 (+extra NO_3^-)	0.80	0.08	1.41	−51.01
4 (+extra NO_3^-)	1.38	0.38	1.48	−54.67

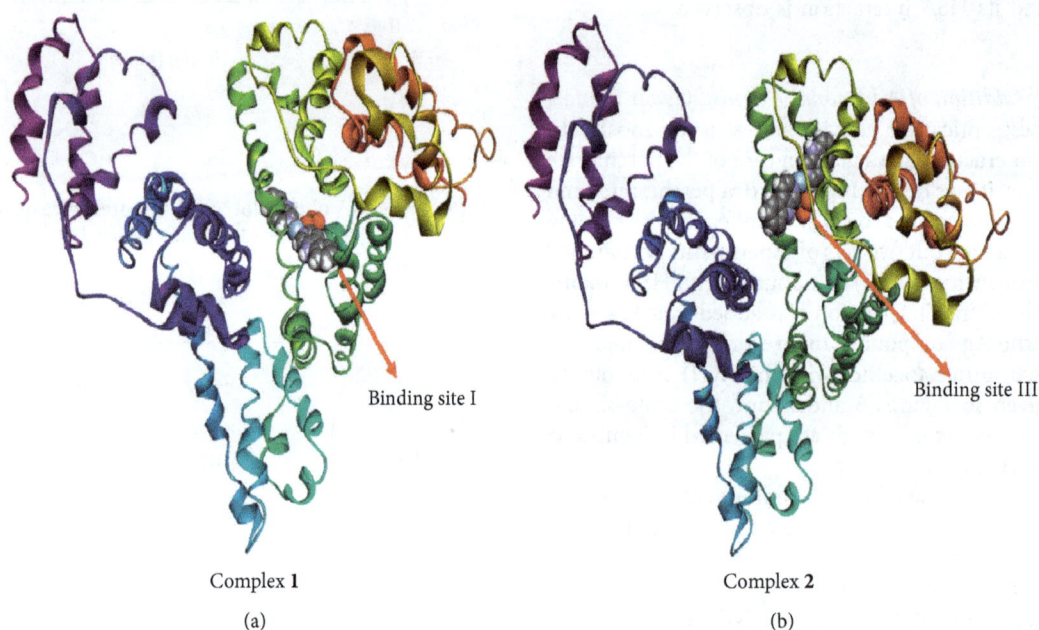

Binding site I

Complex **1**

(a)

Binding site III

Complex **2**

(b)

FIGURE 8: Binding of compounds (a) **1** and (b) **2** at different binding sites of HSA.

monitoring the conformational changes of proteins. As shown in Figure 8, the CD spectrum of free HSA (line 1) exhibits two negative bands at 208 nm and 222 nm in the ultraviolet region, attributed to n-π^* transfer for the peptide bond, with a positive ellipticity at 193 nm, which is the characteristic of an α-helix structure of HSA [43]. CD often allows obtaining information for almost all secondary structural variants, such as α-helices, β-sheets, β-turns, and random coil structures. All of these structures provide origins to bands of distinct forms and degrees in the far-UV region. Modifications of the ellipticity at 222 nm (−MRE222) are convenient methods for observing and quantifying changes in the α-helical content [44]. CD spectra recorded in the presence and absence of various concentrations of Ag(I) compounds are presented in Figure 8. Similar CD curves for the other three compounds are shown in Figure S8, as they show comparable behavior.

The CD signal, expressed in millidegree, obtained over the wavelength range of 190–250 nm, was converted to a mean residue ellipticity (MRE, θ), using the following conversion:

$$MRE = \frac{\theta_{obs}\ (mdeg)}{\left(10 \times n \times C_p \times l\right)}, \quad (5)$$

where θ_{obs} is the CD in millidegree, the number of amino acid residues (585) is given by n, "l" is the path length (cm), and C_p is the molarity. The unit of MRE is $deg\cdot cm^2\cdot dmol^{-1}$.

The α-helix contents of free and combined HSA were calculated from the MRE value at 222 nm, using the following equation [43, 45]:

$$\%\ \alpha-helix = \left[\frac{MRE_{222\,nm} - 2,340}{30,300}\right] \times 100. \quad (6)$$

With increasing concentrations of the Ag(I) compounds, the CD signal exhibits significant changes in ellipticity, and this change corroborates with the binding of the compounds with the HSA backbone (spectra 2–6, Figure 8). In fact with increasing concentration of the compounds the ellipticity decreases. However, at high concentration of silver compounds, a transition of the secondary structure (i.e., α-helix to β-sheet) was observed.

The results confirm that the compound binds to the amino acid residues of the primary polypeptide chain in HSA, which has distorted its hydrogen-bonding networks. Interestingly, the decrease in the α-helical content is indicative of substantial microenvironmental changes of the polypeptide chains of HSA, which increased the exposure of some hydrophobic regions that were previously submerged. The microenvironmental changes of the protein were found more pronounced in the case of **1** (ClO_4^-) and **2** (NO_3^-), as compared to other two compounds (Table 3). The results of all studies corroborate well with the findings of fluorescence and UV studies.

3.5. Molecular Docking. To provide further and deeper insight into the interactions of HSA with compounds **1** and **2**, a molecular docking technique was employed to learn more about the exact binding sites inside the molecular target HSA. From the 3D structure of crystalline albumin, it is known that HSA comprises three homologous domains that assemble to form a heart-shaped molecule, denoted I, II, and III: I (residues 1–195), II (196–383), and III (384–585). The principal region of compound **1** binding sites of HSA is located in hydrophobic cavities in the subdomains IIA and IIIA, corresponding to sites I and II, respectively, and the tryptophan residue (Trp-214) of HSA in the subdomain IIA. A large hydrophobic cavity in the subdomain IIA (a binding site at I) to accommodate compound **1** is present, while compound **2** preferentially appears to the bind at site III. The minimum energy docked pattern (Figure 9(a)) indicates that compound **1** is primarily located within the subdomain IIA of HSA, forming numerous hydrophobic contacts (π-σ, π-π stacked, and π-alkyl) with GLN196, HIS242, LYS199, LYS195, CYS200, CYS245, and CYS245 residues of the hydrophobic binding site IIA (Figure 10(a)). Furthermore, also a number of hydrogen bonds and specific electrostatic interaction formed by the compound **1** are observed (Table 4).

The resulting docked pattern (Figure 9(b)) indicates that compound **2** is located in the subdomains IA and IB, forming various noncovalent interactions like hydrogen bond, electrostatic, and hydrophobic within the binding cavity residues. A detailed description is given in Table 5 (Figure 10(b)). These noncovalent interactions formed by both the compounds **1** and **2** are dominated by hydrophobic contacts, with additional stabilization also assisted by the hydrogen bonding and electrostatic interaction with the polar residues of the binding site cavity. It has been previously observed [46, 47] that hydrogen bonding and electrostatic interaction decreased the hydrophilicity and increased the hydrophobicity to keep the compounds-HSA system stable. The relative binding energy of the docked structures was found to be −344.74 and −326.31 kJ/mol for **1** and **2**, respectively. The results obtained from the molecular docking studies revealed that the hydrophobic forces dominated the interaction of Ag compounds with the HSA. Notably, the binding of the compounds **1** and **2** to different sites can be largely associated with the difference of ionic sphere.

To find out the effect of counterion in the HSA binding, we also perform the docking of the individual counterion

TABLE 3: Effect of the Ag(I) compounds on the α-helical structure of HSA (model protein) expressed as percentage α-helix character.

Concentration (mM)	% α-helix			
	1 (ClO_4^-)	**2** (NO_3^-)	**3** (PF_6^-)	**4** (BF_4^-)
0	67.03	67.03	67.03	67.03
20	64.07	62.85	63.57	65.25
40	62.23	61.61	62.57	63.82
60	58.59	58.41	60.37	61.53
80	50.18	52.02	56.31	56.66
100	35.63	42.73	47.18	47.52

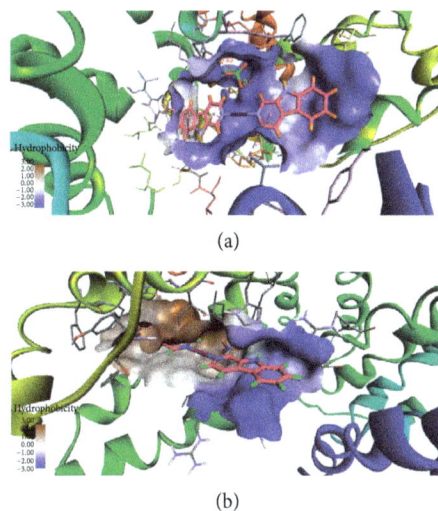

(a)

(b)

FIGURE 9: (a) Molecular docked model of compound **1** (stick representation) located within the hydrophobic pocket in the subdomain IIA of HSA; (b) molecular docked model of compound **2** (stick representation) located within the hydrophobic pocket in the subdomain IB of HSA.

(nitrate ion and perchlorate ion) with the HSA (Figure S9). In the minimum energy docked pose, the nitrate and perchlorate ion are found in the outer environment of the binding site's IIA and IIIA domains, respectively. These anions form electrostatic and other noncovalent interaction at these domains, which could be responsible for the microenvironmental changes or conformational changes of proteins. So that presence of the nitrate and perchlorate ions in the complex may have facilitated or enhanced the binding propensity (see Tables S1 and S2 in Supplementary Materials), which is also evident from the binding constant values in the presence of the anions. Thus, it plays an influential role in binding to the biomolecule. This observation confirms the significant role of the anions in the mode of interaction with the biomolecule.

4. Concluding Remarks

β-Carbolines motif crosses blood-brain membrane (BBM) and are known to inhibit the phosphorylated form of the tau protein [9]. These inhibitions are the essential features to study for any new compound to treat diseases such as Alzheimer's disease. Spectrophotometric methods have been

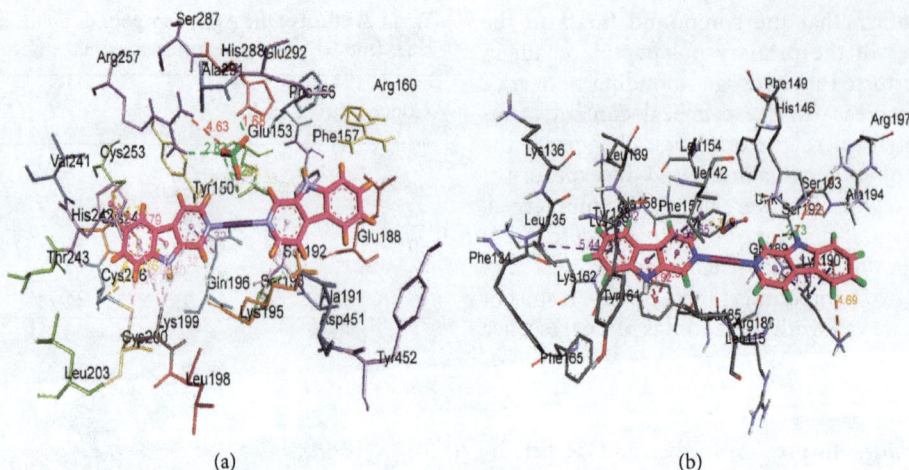

(a)   (b)

FIGURE 10: Noncovalent interactions of compounds (a) **1** and (b) **2** at the subdomains IIA and IB of HSA, respectively.

TABLE 4: Noncovalent interactions of compound **1** with the HSA binding site IIA.

Name	Distance (Å)	Category	Type
ARG257: H11-Complex **1**: O1′	2.81		
HIS288: H1-compound **1**: O3	1.68	Hydrogen bond	Conventional
Compound **1**: O1′-GLU153: O1	3.24		
Compound **1**: Cl1-GLU153: O1	3.63	Electrostatic	Attractive charge
GLN196-compound **1**	3.32		π-Sigma
HIS242-compound **1**	4.78		π-π Stacked
HIS242-compound **1**	3.92		π-π Stacked
Compound **1**-LYS199	5.38		π-Alkyl
Compound **1**-LYS199	4.19		π-Alkyl
Compound **1**-LYS195	4.75	Hydrophobic	π-Alkyl
Compound **1**-LYS199	5.03		π-Alkyl
Compound **1**-CYS200	5.25		π-Alkyl
Compound **1**-CYS245	4.23		π-Alkyl
Compound **1**-CYS246	4.92		π-Alkyl

TABLE 5: Noncovalent interactions of compound **2** with the HSA binding site III.

Name	Distance (Å)	Category	Type
Compound **2**: H4-GLY189:O	2.73	Hydrogen bond	Conventional
LYS190-compound **2**	4.88	Electrostatic	π-Cation
Compound **2**: O3-PHE157	3.39	Electrostatic	π-Anion
TYR161-compound **2**	4.29		π-π Stacked
TYR161-compound **2**	4.72		π-π Stacked
Compound **2**-A: TYR161	3.91		π-π Stacked
Compound **2**-A: ILE142	3.64		π-Alkyl
Compound **2**-A: ILE142	4.98		π-Alkyl
Compound **2**-A: LEU135	5.43	Hydrophobic	π-Alkyl
Compound **2**-A: ALA158	4.91		π-Alkyl
Compound **2**-A: LYS190	4.11		π-Alkyl
Compound **2**-A: LYS190	3.09		π-Alkyl
Compound **2**-A: LYS190	3.21		π-Alkyl

used to investigate the interaction between HSA and the present Ag(I) compounds. All Ag(I) compounds showed a strong affinity towards HSA. An extra effect of the perchlorate and nitrate anions was found in the binding studies, which ascertained the higher binding propensity of the Ag(I) compounds with counteranions ClO_4^- and NO_3^-. The moderate HSA binding in the case of BF_4^- and PF_6^- salts could be enhanced by the addition of extra anions, namely, ClO_4^- and NO_3^-. The results have shown a significant role of the anion in Ag-HSA binding and even showed an exponential increase in the binding propensity of the Ag(I) compounds. This affinity has been attributed to the increased exposure of the Ag-binding subdomains of HSA, as a result of the electrostatic and hydrophobic interactions. The presence of Hnor, despite being a labile ligand, appears to be important for the binding to HSA. The effect of anions was further validated by the molecular modeling studies which corroborate well with our findings of binding studies and ascertain that anions play an importance role in binding to HSA.

Abbreviations

HSA: Human serum albumin
CD: Circular dichroism
Cys: Cysteine
Met: Methionine
Hnor: 9H-pyrido[3,4-b]indole (norharmane).

Authors' Contributions

Ali Alsalme and Rais Ahmad Khan contributed equally to this work.

Acknowledgments

The authors extend their appreciation to the Deanship of Scientific Research at King Saud University for funding this work through the research group number RG-1438-006.

Supplementary Materials

Figure S1: UV absorption spectra of HSA in the absence and presence of compounds 1–4; concentration [HSA] = 5 μM. Figure S2: UV absorption difference of HSA (the difference UV absorption spectra were obtained by HSA-Ag compound spectra minus Ag compound spectra). Figure S3: fluorescence emission spectra of HSA (5 μM) in the presence of various concentrations of compounds 2, 3, and 4; curves from 1 to 10 correspond to compound concentrations of 0, 0.5, 1, 1.5, 2, 3, 4, 5, 7.5, and 10 μM, respectively, when excited at 295 nm. Figure S4: the Stern–Volmer plot for the quenching of the HSA fluorescence by compounds 2, 3, and 4 at 295 nm. *Inset*: plot of log $(F_0-F)/F$ as a function of log (complex). Figure S5: fluorescence emission spectra of HSA (5 μM) in the presence of various concentrations of compounds 1–4, corresponding to compound concentrations of 0, 1, 2, 4, 7.5, and 10 μM with the addition of extra KClO$_4$ in a 1 : 4 ratio, respectively, when excited at 295 nm. Figure S6: fluorescence emission spectra of HSA (5 μM) in the presence of various concentrations of compounds 1–4, corresponding to compound concentrations of 0, 1, 2, 4, 7.5, and 10 μM with the addition of extra KNO$_3$ in 1 : 4 ratio, respectively, when excited at 295 nm. Figure S7: CD spectra of the HSA-compound system (HSA = 20 μM) in the presence of various concentrations of compounds 2, 3, and 4; curves from 1 to 6 corresponding to compound concentrations of 0, 20, 40, 60, 80, and 100 mM, respectively. Figure S8: absorption spectra of Ag complexes only, 1: ClO$_4$ (red line) and 2: NO$_3$ (black line) of 1.0 mM solution. Figure S9:

molecular docked model of HSA in presence of anions. Table S1: noncovalent interactions of nitrate ion with the HSA. Table S2: noncovalent interactions of perchlorate ion with the HSA. (*Supplementary Materials*)

References

[1] M. C. Gimeno and A. Laguna, Comprehensive Coordination Chemistry II, J. A. McCleverty and T. J. Meyer, Eds., vol. 6, Elsevier Pergamon, Oxford, UK, 2nd edition, 2004.

[2] A. N. Khlobystov, A. J. Blake, N. R. Champness et al., "Supramolecular design of one-dimensional coordination polymers based on silver(I) complexes of aromatic nitrogen-donor ligands," *Coordination Chemistry Reviews*, vol. 222, pp. 155–192, 2001.

[3] A. B. G. Lansdown, *Silver in Healthcare: Its Antimicrobial Efficacy and Safety in Use*, Royal Society of Chemistry, London, UK, 2010.

[4] N. Farrell, Comprehensive Coordination Chemistry, J. A. McCleverty and T. J. Meyer, Eds., vol. 9, Elsevier Pergamon, Oxford, UK, 2nd edition, 2004.

[5] K. M. Fromm, "Give silver a shine," *Nature Chemistry*, vol. 3, p. 178, 2011.

[6] M. Rai, A. Yadav, and A. Gade, "Silver nanoparticles as a new generation of antimicrobials," *Biotechnology Advances*, vol. 27, no. 1, pp. 76–83, 2009.

[7] J. L. Clement and P. S. Jarret, "Antibacterial silver," *Metal-Based Drugs*, vol. 1, no. 5-6, pp. 467–482, 1994.

[8] S. Ray, R. Mohan, J. K. Singh et al., "Anticancer and antimicrobial metallopharmaceutical agents based on palladium, gold, and silver N-Heterocyclic carbene complexes," *Journal of the American Chemical Society*, vol. 129, no. 48, pp. 15042–15053, 2007.

[9] S. Medici, M. Peana, V. M. Nurchi, J. I. Lackowicz, G. Crisponi, and M. A. Zoroddu, "Noble metals in medicine: latest advance," *Coordination Chemistry Reviews*, vol. 284, pp. 329–350, 2015.

[10] S. Eckhardt, P. S. Brunetto, J. Gagnon, M. Priebe, B. Giese, and K. M. Fromm, "Nanobio silver: its interactions with peptides and bacteria, and its uses in medicine," *Chemical Reviews*, vol. 113, no. 7, pp. 4708–4754, 2013.

[11] V. Gandin, M. Pellei, M. Marinelli et al., "Synthesis and in vitro antitumor activity of water soluble sulfonate- and ester-functionalized silver(I) N-heterocyclic carbene complexes," *Journal of Inorganic Biochemistry*, vol. 129, pp. 135–144, 2013.

[12] D. C. Carter and J. X. Ho, "Structure of serum albumin," *Advances in Protein Chemistry*, vol. 45, pp. 153–203, 1994.

[13] R. E. Olson and D. D. Christ, "Chapter 33 plasma protein binding of drugs," *Annual Reports in Medicinal Chemistry*, vol. 31, pp. 327–337, 1996.

[14] T. Peters Jr., *All about Albumin: Biochemistry, Genetics and Medical Applications*, Academic Press, Cambridge, MA, USA, 1996.

[15] T. Peters Jr., "Serum albumin," *Advances in Protein Chemistry*, vol. 37, pp. 161–245, 1985.

[16] U. Anand and S. Mukherjee, "Binding, unfolding and refolding dynamics of serum albumin," *Biochimica et Biophysica Acta*, vol. 1830, pp. 5394–5404, 2013.

[17] M. Vignoni, F. A. O. Rasse-Suriani, K. Butzbach, R. Erra-Balsells, B. Epe, and F. M. Cabrerizo, "Mechanisms of DNA damage by photoexcited 9-methyl-β-carbolines," *Organic & Biomolecular Chemistry*, vol. 11, no. 32, pp. 5300–5309, 2013.

[18] M. A. Collins and E. J. Neafsey, Neurotoxic Factors in Parkinson's Disease and Related Disorders, A. Storch and

M. A. Collins, Eds., Kluwer Academic/Plenum Publishers, New York, NY, USA, 2000.

[19] J. Hamann, C. Wernicke, J. Lehmann, H. Reichmann, H. Rommelspacher, and G. Gille, "9-Methyl-β-carboline up-regulates the appearance of differentiated dopaminergic neurones in primary mesencephalic culture," *Neurochemistry International*, vol. 52, no. 4-5, pp. 688–700, 2008.

[20] D. Frost, B. Meechoovet, T. Wang et al., "β-Carboline compounds, including harmine, inhibit DYRK1A and tau phosphorylation at multiple Alzheimer's disease-related sites," *PLoS One*, vol. 6, no. 5, article e19264, 2011.

[21] R. A. Khan, K. Al-Farhan, A. de Almeida et al., "Light-stable bis(norharmane)silver(I) compounds: synthesis, characterization and antiproliferative effects in cancer cells," *Journal of Inorganic Biochemistry*, vol. 140, pp. 1–5, 2014.

[22] D. Mustard and D. W. Ritchie, "Docking essential dynamics eigenstructures," *Proteins: Structure, Function, and Bioinformatics*, vol. 60, no. 2, pp. 269–274, 2005.

[23] D. W. Ritchie and V. Venkatraman, "Ultra-fast FFT protein docking on graphics processors," *Bioinformatics*, vol. 26, no. 19, pp. 2398–2405, 2010.

[24] Accelrys Software Inc., *Discovery Studio Modeling Environment, Release 4.0*, Accelrys Software Inc., San Diego, CA, USA, 2013.

[25] *The PyMOL Molecular Graphics System, Version 1.5.0.4*, Schrödinger LLC, New York, NY, USA, 2013.

[26] Y. Hatanaka and Y. Sadakane, "Photoaffinity labeling in drug discovery and developments: chemical gateway for entering proteomic frontier," *Current Topics in Medicinal Chemistry*, vol. 2, no. 3, pp. 271–288, 2002.

[27] D. Robinette, N. Neamati, K. B. Tomer, and C. H Borchers, "Photoaffinity labeling combined with mass spectrometric approaches as a tool for structural proteomics," *Expert Review of Proteomics*, vol. 3, no. 4, pp. 399–408, 2006.

[28] H. Mach, D. B. Volkin, C. J. Burke, and C. R. Middaugh, "Ultraviolet absorption spectroscopy," in *Protein Stability and Folding: Theory and Practice, Methods in Molecular Biology*, B. A. Shirley, Ed., vol. 40, Humana Press, Totowa, NJ, USA, 1995.

[29] V. Chabert, M. Hologne, O. Sénèque, A. Crochet, O. Walker, and K. M. Fromm, "Model peptide studies of Ag+ binding sites from the silver resistance protein SilE," *Chemical Communications*, vol. 53, no. 45, pp. 6105–6108, 2017.

[30] S. Kracht, M. Messerer, M. Lang et al., "Electron transfer in peptides: on the formation of silver nanoparticles," *Angewandte Chemie International Edition*, vol. 54, no. 10, pp. 2912–2916, 2015.

[31] X.-C. Shen, H. Liang, J.-H. Guo, C. Song, X.-W. He, and Y.-Z. Yuan, "Studies on the interaction between Ag+ and human serum albumin," *Journal of Inorganic Biochemistry*, vol. 95, no. 2-3, pp. 124–130, 2003.

[32] V. D. Suryawanshi, P. V. Anbhule, A. H. Gore, S. R. Patil, and G. B. Kolekar, "Photoaffinity labeling combined with mass spectrometric approaches as a tool for structural proteomics," *Journal of Photochemistry and Photobiology B: Biology*, vol. 118, pp. 1–8, 2013.

[33] S. Tabassum, M. Zaki, M. Ahmad et al., "Synthesis and crystal structure determination of copper(II)-complex: in vitro DNA and HSA binding, pBR322 plasmid cleavage, cell imaging and cytotoxic studies," *European Journal of Medicinal Chemistry*, vol. 83, pp. 141–154, 2014.

[34] S. Tabassum, W. M. Al-Asbahy, M. Afzal, F. Arjmand, and R. H. Khan, "Interaction and photo-induced cleavage studies of a copper based chemotherapeutic drug with human serum albumin: spectroscopic and molecular docking study," *Molecular BioSystems*, vol. 8, no. 9, pp. 2424–2433, 2012.

[35] C. A. Royer and B. A. Shsirley, *Protein Stability, and Folding: Theory and Practice, Methods in Molecular Biology*, vol. 40, Humana Press, Totowa, NJ, USA, 1995.

[36] J. R. Lackowicz, *Principles of Fluorescence Spectroscopy*, Plenum Press, New York, NY, USA, 2nd edition, 1999.

[37] D. Romanini, G. Avalle, B. Farruggia, B. Nerli, and G. Pico, "Spectroscopy features of the binding of polyene antibiotics to human serum albumin," *Chemico-Biological Interactions*, vol. 115, no. 3, pp. 247–260, 1998.

[38] M. R. Eftink and T. G. Dewey, *Biophysical and Biochemical Aspects of Fluorescence Spectroscopy*, Plenum Press, New York, NY, USA, 1991.

[39] W. R. Ware, "Oxygen quenching of fluorescence in solution: an experimental study of the diffusion process," *Journal of Physical Chemistry*, vol. 66, no. 3, pp. 455–458, 1962.

[40] M. S. Ali and H. A. Al-Lohedan, "Interaction of human serum albumin with sulfadiazine," *Journal of Molecular Liquids*, vol. 197, pp. 124–130, 2014.

[41] D.M. Togashi, A.G. Ryder, and J. Fluoresc, "A fluorescence analysis of ANS bound to bovine serum albumin: binding properties revisited by using energy transfer," *Journal of Fluorescence*, vol. 18, no. 2, pp. 519–526, 2008.

[42] O. Stern and M. Volmer, "Über die abklingzeit der fluoreszenz," *Physikalische Zeitschrift*, vol. 20, pp. 183–188, 1919.

[43] Y. H. Chen, J. T. Yang, and H. M. Martinez, "Determination of the secondary structures of proteins by circular dichroism and optical rotatory dispersion," *Biochemistry*, vol. 11, no. 22, pp. 4120–4131, 1972.

[44] M. T. Rehman, H. Shamsi, and A. U. Khan, "Insight into the binding mechanism of imipenem to human serum albumin by spectroscopic and computational approaches," *Molecular Pharmaceutics*, vol. 11, no. 6, pp. 1785–1797, 2014.

[45] M. T. Rehman, S. Ahmed, and A. U. Khan, "Interaction of meropenem with "N" and "B" isoforms of human serum albumin: a spectroscopic and molecular docking study," *Journal of Biomolecular Structure and Dynamics*, vol. 34, no. 9, pp. 1894–1864, 2015.

[46] S. Tabassum, W. M. Al-Asbahy, M. Afzal, and F. Arjmand, "Synthesis, characterization and interaction studies of copper based drug with Human Serum Albumin (HSA): spectroscopic and molecular docking investigations," *Journal of Photochemistry and Photobiology B: Biology*, vol. 114, pp. 132–139, 2012.

[47] Y. Shua, W. Xueb, X. Xua et al., "Interaction of erucic acid with bovine serum albumin using a multi-spectroscopic method and molecular docking technique," *Food Chemistry*, vol. 173, pp. 31–37, 2015.

16

Anticancer Potential of Green Synthesized Silver Nanoparticles using Extract of *Nepeta deflersiana* against Human Cervical Cancer Cells (HeLA)

Ebtesam S. Al-Sheddi,[1] Nida N. Farshori [ID],[1] Mai M. Al-Oqail,[1] Shaza M. Al-Massarani [ID],[1] Quaiser Saquib,[2,3] Rizwan Wahab,[2,3] Javed Musarrat,[2,3] Abdulaziz A. Al-Khedhairy,[2] and Maqsood A. Siddiqui[2,3]

[1]Department of Pharmacognosy, College of Pharmacy, King Saud University, Riyadh, Saudi Arabia
[2]Zoology Department, College of Science, King Saud University, P.O. Box 2455, Riyadh 11451, Saudi Arabia
[3]Al-Jeraisy Chair for DNA Research, Zoology Department, College of Science, King Saud University, P.O. Box 2455, Riyadh 11451, Saudi Arabia

Correspondence should be addressed to Nida N. Farshori; nidachem@gmail.com

Guest Editor: Bon H. Koo

In this study, silver nanoparticles (AgNPs) were synthesized using aqueous extract of *Nepeta deflersiana* plant. The prepared AgNPs (ND-AgNPs) were examined by ultraviolet-visible spectroscopy, Fourier transform infrared (FTIR) spectroscopy, X-ray diffraction (XRD), transmission electron microscopy (TEM), scanning electron microscope (SEM), and energy dispersive spectroscopy (EDX). The results obtained from various characterizations revealed that average size of synthesized AgNPs was 33 nm and in face-centered-cubic structure. The anticancer potential of ND-AgNPs was investigated against human cervical cancer cells (HeLa). The cytotoxic response was assessed by 3-(4, 5-dimethylthiazol-2-yl)-2, 5-diphenyltetrazolium bromide (MTT), neutral red uptake (NRU) assays, and morphological changes. Further, the influence of cytotoxic concentrations of ND-AgNPs on oxidative stress markers, reactive oxygen species (ROS) generation, mitochondrial membrane potential (MMP), cell cycle arrest and apoptosis/necrosis was studied. The cytotoxic response observed was in a concentration-dependent manner. Furthermore, the results also showed a significant increase in ROS and lipid peroxidation (LPO), along with a decrease in MMP and glutathione (GSH) levels. The cell cycle analysis and apoptosis/necrosis assay data exhibited ND-AgNPs-induced SubG1 arrest and apoptotic/necrotic cell death. The biosynthesized AgNPs-induced cell death in HeLA cells suggested the anticancer potential of ND-AgNPs. Therefore, they may be used to treat the cervical cancer cells.

1. Introduction

Nobel metal nanoparticles have attracted the interest of scientific community due to their fascinating applications in the field of biology, material science, medicine, etc [1]. Silver nanoparticles specifically have gained attention due to their unusual physiochemical [2] (chemical stability and electrical conductivity) and biological activities such as antibacterial, antifungal, anti-inflammatory, antiviral, antiangiogenesis, anticancer, and antiplatelet activities [3–5]. In addition, silver nanoparticles have been used in clothing [6], room spray, laundry detergent, wall paint formulation [7, 8], sunscreens, and cosmetics [9]. Silver nanoparticles also inhibit HIV-1 virus from binding to the host cells *in vitro* [10]. Although a wide variety of metal nanoparticle preparation methods such as UV radiation, laser ablation, lithography, aerosol technologies, and photochemical reduction are available [11–13], the focus is shifting towards green synthesis of nanoparticles, using bacteria [14], yeast [15], fungi [16], and plants [17]. Green synthesis of nanoparticles reports to be clean, nontoxic, cost effective, and environmentally benign. Among the various biological methods available, the use of microbe-mediated synthesis has limited industrial use, as they require antiseptic conditions. On the contrary, the use of plant extract for the nanoparticles synthesis is valuable due to the ease of scale-up, less biohazardous nature, and avoiding the hideous procedure of maintaining the cell lines [18].

Cancer is a life threatening disease and leads the cases of deaths around the world [19]. According to the WHO, the annual cancer cases are to rise from 14 million in 2012 to 22 million in the next two decades [20]. Thus, the development of potent and effective antineoplastic drugs is one of the most persuaded goals. Among the various approaches, the exploitation of natural products is one of the most successful methods to identify novel hits and leads [21]. *Nepeta deflersiana* Schweinf. *(Labiatae)* is a medicinal plant growing in Saudi Arabia [22]. Traditionally *N. deflersiana* was used as a sedative; the leaf decoction was drunk with tea to release stomach and burn problems [23, 24]. The antimicrobial, anticancer, and antioxidant activities of *N. deflersiana* are documented [25]. Recently, we have reported the positive effects of *N. deflersiana* on human breast and lung cancer cell lines [26]. However, until the present, no published data are available on synthesis of nanoparticles using *N. deflersiana* plant. Herein, we report for the first time (i) the silver nanoparticles (ND-AgNPs) synthesis through a single-step silver ions reduction by *N. deflersiana* plant extract (Figure 1) and (ii) studied the anticancer activity of the biosynthesized silver nanoparticles against human cervical cancer (HeLa) cells.

2. Materials and Methods

2.1. Plant Material, Reagents, and Consumables.
Nepeta deflersiana (Lamiaceae) plants were collected from Shaza Mountains, Saudi Arabia. The identity of the plant was confirmed by Dr. Jakob Thomas, KSU, and a voucher specimen (#15797) was deposited in the herbarium. Cell culture medium, antibiotics-antimycotic solution, trypsin, and FBS were procured from Invitrogen, USA. Plastic wares and other consumables were obtained from Nunc, Denmark. Other chemicals/reagents used in this study were purchased from Sigma, USA.

2.2. Preparation of Plant Extract.
The aerial part of *N. deflersiana* was collected and washed several times with distilled water to remove dust and was dried under shade. The air-dried plant was cut into small pieces, macerated in distilled water, filtered under gravity, and the solvent evaporated under reduced pressure using a rotary evaporator. The dried extract was kept at 4°C (Figure 1).

2.3. Synthesis of Silver Nanoparticles.
The aqueous extract of *N. deflersiana* (500 mg) was dissolved in 100 ml distilled water. Further 10 ml of the above extract was added to 90 ml of 0.1 M $AgNO_3$ solution. After 24 h incubation, the solution

turned dark brown, which indicates the formation of AgNPs. The solution was then transferred into a round bottom flask and was heated with continuous stirring at 90°C. After 15 min, the centrifugation was done at room temperature and a speed of 9000 rpm. The black powder obtained after washing thrice with distilled water was dried overnight in an oven at 80°C.

2.4. Characterization of Synthesized Silver Nanoparticles.
The optical absorption of green synthesized silver nanoparticles was studied using FTIR (Shimadzu FT-IR Prestige 21) and UV-VIS (Shimadzu UV-VIS 2550, Japan) spectral analysis, respectively. Fourier transmission infrared (FTIR) spectra were recorded using KBr pellets in the range of 4000 to 400 cm^{-1}. The crystalline nature of green synthesized AgNPs was confirmed by XRD pattern. The XRD data were recorded using PANalytical X'Pert X-ray diffractometer using $Cu_{K\alpha}$ ($\lambda = 1.54056$ Å). Morphology, size, and electron diffraction pattern were examined by SEM (JSM-7600F, Japan) and TEM (JEM-2100F, Japan) at a voltage 200 kV, respectively. EDX analysis was used to confirm the presence of elemental silver in green synthesized AgNPs.

2.5. Cytotoxicity by MTT Assay.
Cytotoxicity of ND-AgNPs was examined by using MTT assay according to the method in [27]. In brief, HeLA cells obtained from American Type Culture Collection, USA, were plated in 96-well plates at a density of 1 x 10^4 cells/well. Cells were exposed to 1–100 μg/ml ND-AgNPs for 24 h. Following this, MTT was added in the wells, and plates were incubated for 4 h further. The reaction mixture was taken out and 200 μl/well DMSO was added and mixed several times by pipetting up and down. The absorbance of plates was measured at 550 nm. The results were expressed as percentage of control.

2.6. Cytotoxicity by Neutral Red Uptake (NRU) Assay.
Cytotoxicity by NRU assay was performed using the procedure [27]. Briefly, HeLA cells were treated with 1–100 μg/ml ND-AgNPs for 24 h. Then, cells were washed with PBS twice and incubated further in 50 μg/ml of neutral red containing medium for 3 h. The cells were washed off with a solution (1% $CaCl_2$ and 0.5% formaldehyde). The dye was extracted in a mixture of 1% acetic acid and 50% ethanol. The plates were measured at 550 nm. The results were expressed as percentage of control.

2.7. Morphological Analysis.
The changes in the morphology were observed under the microscope to determine the alterations induced by ND-AgNPs in HeLa cells treated with 1 μg/ml to 100 μg/ml of ND-AgNPs for 24 h. Images of the cells were grabbed at 20x by using the phase contrast inverted microscope (Olympus CKX 41, USA).

2.8. Glutathione (GSH) Level.
The depletion in GSH level was measured following the protocol [28]. In brief, HeLa

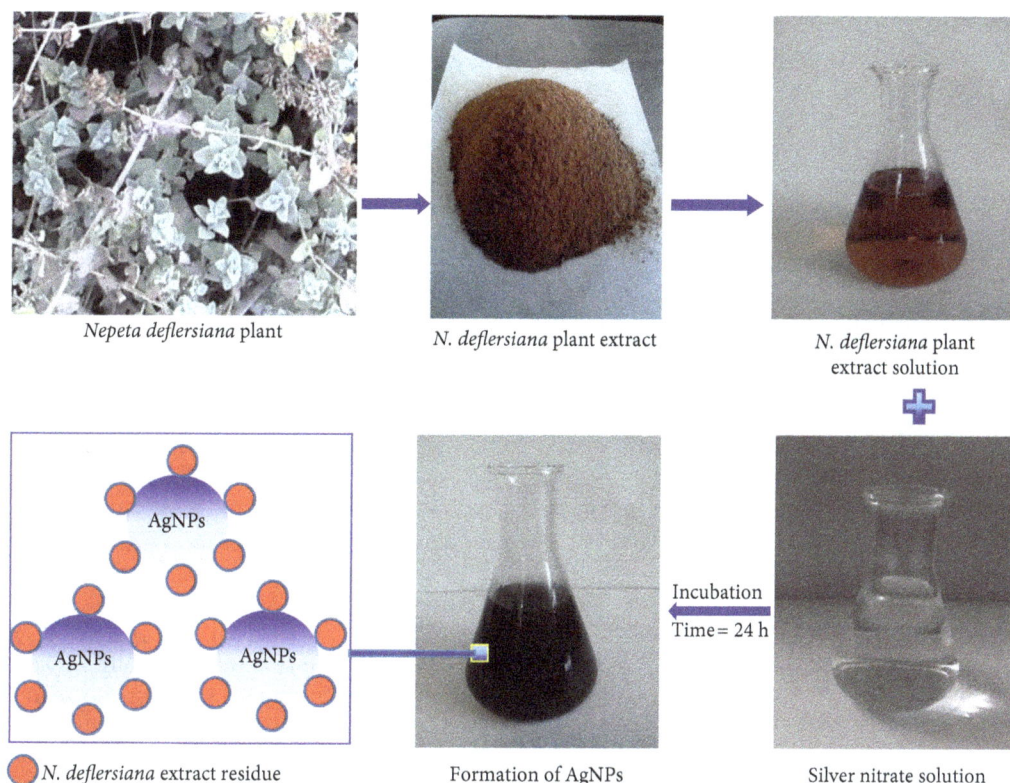

FIGURE 1: Schematic illustration of the green synthesis of silver nanoparticles (ND-AgNPs) using aqueous extract of the *Nepeta deflersiana* plant.

cells exposed to 5–25 µg/ml ND-AgNPs for 24 h were centrifuged, and cellular protein was precipitated in 10% TCA (1 ml). Following this, supernatant was taken by centrifugation at 3000 rpm for 10 min. Then, 2 ml Tris buffer (0.4 M) with EDTA (0.02 M) and 0.01 M 5, 5'-dithionitrobenzoic acid (DTNB) were added in the supernatant. The absorbance was measured at 412 nm after incubating for 10 min at 37°C.

2.9. Lipid Peroxidation (LPO). LPO in ND-AgNPs-exposed HeLA cells were measured following the method [28]. After respective treatment, cells were sonicated in chilled 1.15% potassium chloride solution. Following centrifugation, 1 ml of supernatant was added to 2 ml thiobarbituric acid solution (TCA (15%), TBA (0.7%), and 0.25 N HCl). The resulting solution was then boiled at 100°C for 15 min, and after the centrifugation for 10 min at 1000 × g, the absorbance was measured at 550 nm.

2.10. ROS Generation. The intracellular ROS generation was measured using 2, 7-dichlorodihydrofluorescein diacetate (DCFH-DA) dye [28]. In brief, HeLa cells were treated with different concentrations (10–50 µg/ml) of ND-AgNPs for 24 h. The cells were then with DCFH-DA (5 µM) at 37°C for 1 h. The cell pellet was collected in PBS (500 µl) by centrifugation at 3000 rpm for 5 min. Then, the cells were analysed using flow cytometer.

2.11. Mitochondrial Membrane Potential (MMP). The MMP level in HeLA cells was measured using the method defined by Zhang et al. [29]. In brief, HeLA cells were treated with 10 to 50 µg/ml of ND-AgNPs for 24 h. Then, treated and untreated cells were incubated with rhodamine-123 (5 µg/ml) for 1 h at 37°C in dark. Cells were washed twice, and finally, cell pellets were resuspended in PBS (500 µl). MMP was measured by using flow cytometer.

2.12. Cell Cycle Analysis. ND-AgNPs-induced changes in cell cycle were measured using the protocol [30]. In brief, HeLa cells were exposed for 24 h at 10–50 µg/ml ND-AgNPs. After the treatment, cells were fixed in chilled 70% ethanol for 1 h. Then, cells were washed twice by centrifugation, and cells were stained with propidium iodide for 60 min in dark. The stained cells were acquired by flow cytometer.

2.13. Apoptosis Assay. The apoptosis/necrosis induced by ND-AgNPs in HeLA cells were analysed using Annexin-V and 7-AAD Kit (Beckman Coulter) following the manufacturer's protocol. The amount of apoptosis/necrosis in the treated HeLa cells was analysed by flow cytometry following the protocol [31].

2.14. Statistical Analysis. Data were statistically analysed by ANOVA using the post hoc Dunnett's test. Value $p < 0.05$ was considered as a significant level between the exposed and control sets. The results are presented as mean ± standard deviation of three experiments.

3. Result and Discussion

3.1. Synthesis and Characterization of ND-AgNPs. Plant extract of *N. deflersiana* was used for the synthesis of ND-AgNPs under facile conditions. The colorless silver nitrate solution (Figure 1) turned dark brown indicating the formation of silver nanoparticles (AgNPs). The occurrence of brown color can be attributed to the surface plasmons [32], arising from the collective oscillations of valance electrons in the electromagnetic field of incident radiation. Figure 2(a) shows the UV-V is spectra of the synthesized AgNPs, giving the plasmon resonance at 400 nm. The characteristic λ max for AgNPs is in the range of 400–500 nm [33]. The position and shape of the surface plasmon absorption is dependent on the shape and size of particles formed, their interparticle distance, and the dielectric constant of the surrounding medium [34, 35]. Similar observations are reported earlier [32, 36]. FTIR measurements were carried out to identify the various functional groups in biomolecules responsible for the reduction of silver ions to AgNPs and capping/stabilization of AgNPs. The band intensities in different region of spectra for *N. deflersiana* extract (Figure 2(b)) and biosynthesized silver nanoparticles (Figure 2(c)) were analysed. The similarities between the two FTIR spectra, with some marginal shifts in peaks clearly indicate the plant extract is also acting as a capping agent. The *N. deflersiana* plant extract showed a number of peaks reflecting a complex nature of the plant extract. The shift in peaks at 3426 cm^{-1} corresponding to NH stretching of amide (II) band or C-O stretching or O-H stretching vibration implicated that their groups may be directly involved in the process of synthesis of AgNPs. Further, peak shifts from 1689 cm^{-1} to 1608 cm^{-1} indicated the possible involvement of C=O stretching or C-N bending in the amide group. Besides, the peak shifts from 1461 cm^{-1} to 1381 cm^{-1} suggest the involvement of C-H or O-H bending vibration of methyl, methylene, or alcoholic group in the reduction of Ag. Moreover, the observed peaks are more characteristic of flavonoids and terpenoids [37] that are present in the *Nepeta* species [25, 26]. It could be speculated that these secondary metabolites are responsible for the synthesis/stabilization of ND-AgNPs.

The crystalline structure of the green synthesized AgNPs was determined by XRD technique. Figure 2(d) displays the XRD pattern of synthesized AgNPs. The Bragg reflection with 2θ values of 37.89, 44.23, 64.26, and 77.24 corresponding to (111), (200), (220), and (311) sets of lattice planes, respectively, is observed. These can be indexed to the face centered cubic (fcc) structure of the synthesized AgNPs. The crystalline size of the AgNPs was determined by using Debye–Scherrer equation [38]:

$$D = \frac{0.9\lambda}{\beta \cos\theta}, \tag{1}$$

Where D is the grain size, λ is the wavelength of X-ray (1.54056 Å), and β is the full width at half maxima of the diffraction peak (in radians).

The average grain size determined by broadening of (111) reflection is estimated to be around 33 nm. Similar results have been reported earlier [39]. The absence of any reflection other than belonging to the silver lattice clearly indicates that the synthesized AgNPs lattice was unaffected by other molecules in the extract of plant. The scanning electron microscopy (SEM) and transmission electron microscopy (TEM) was employed to study the morphological and structural features of synthesized AgNPs. The SEM image (Figure 3(a)) shows that relatively spherical and uniform nanoparticles are formed. Some of the larger particles seen may be due to aggregation of nanoparticles induced by evaporation of solvent during sample preparation [40]. The TEM image (Figure 3(b)) revealed the nanoparticles formed have a narrow size distribution. The average size was about 33 nm, supporting the results of XRD further. Further, the energy-dispersion X-ray (EDX) spectroscopy study was employed to detect the existence of elemental silver. Figure 3(c) shows the EDX image of *N. deflersiana* synthesized AgNPs. The results clearly indicate an intense signal at approximately 2.98 KeV corresponding to the presence of metallic silver nanocrystals, occurring due to surface plasmon resonance (SPR) [41]. The other intense signal at around 0.0–0.5 Kev represents the characteristic absorption for oxygen and carbon. This indicates the presence of *N. deflersiana* plant extract as a capping ligand on the surface of AgNPs.

3.2. Cytotoxicity Assessments of ND-AgNPs by MTT and NRU Assays. The key results obtained by MTT and NRU assays in HeLA cells exposed to 1 μg/ml to 100 μg/ml for 24 h are summarized in Figures 4(a) and 4(b). The results exhibited a concentration dependent decrease in the viability of HeLa cells. The cell viability was recorded as 86% and 29% in ND-AgNPs at 2 μg/ml and 5 μg/ml concentrations, respectively; however, the maximum decrease in cell viability was measured as 9% each at 10, 25, 50, and 100 μg/ml of ND-AgNPs (Figure 4(a)). Like MTT assay, a concentration-dependent decrease in cell viability of HeLa cells exposed to ND-AgNPs was also observed by NRU assay. The cell viability was recorded as 87% and 43% in ND-AgNPs at 2 μg/ml and 5 μg/ml concentrations, respectively; however, the maximum decrease in cell viability was measured as 23% at 100 μg/ml of ND-AgNPs (Figure 4(b)). In this study, the cytotoxicity assessments were performed using two independent end points (MTT and NRU) assays [42]. The MTT, a colorimetric assay is based on the mitochondrial dehydrogenase enzyme of viable cells [43]; however, NRU assay is based on the lysosomal integrity of viable cells [44]. The cytotoxic responses of the ND-AgNPs, suggesting that biosynthesized AgNPs could contribute in search of alternative chemotherapeutic agent. Our results showed more than 50% of cell death even at 5 μg/ml of ND-AgNPs. The cytotoxic effects induced by ND-AgNPs at lower concentrations could be due to the plant components attached to the AgNPs [45]. The results obtained from this study are also very well supported with various evidences

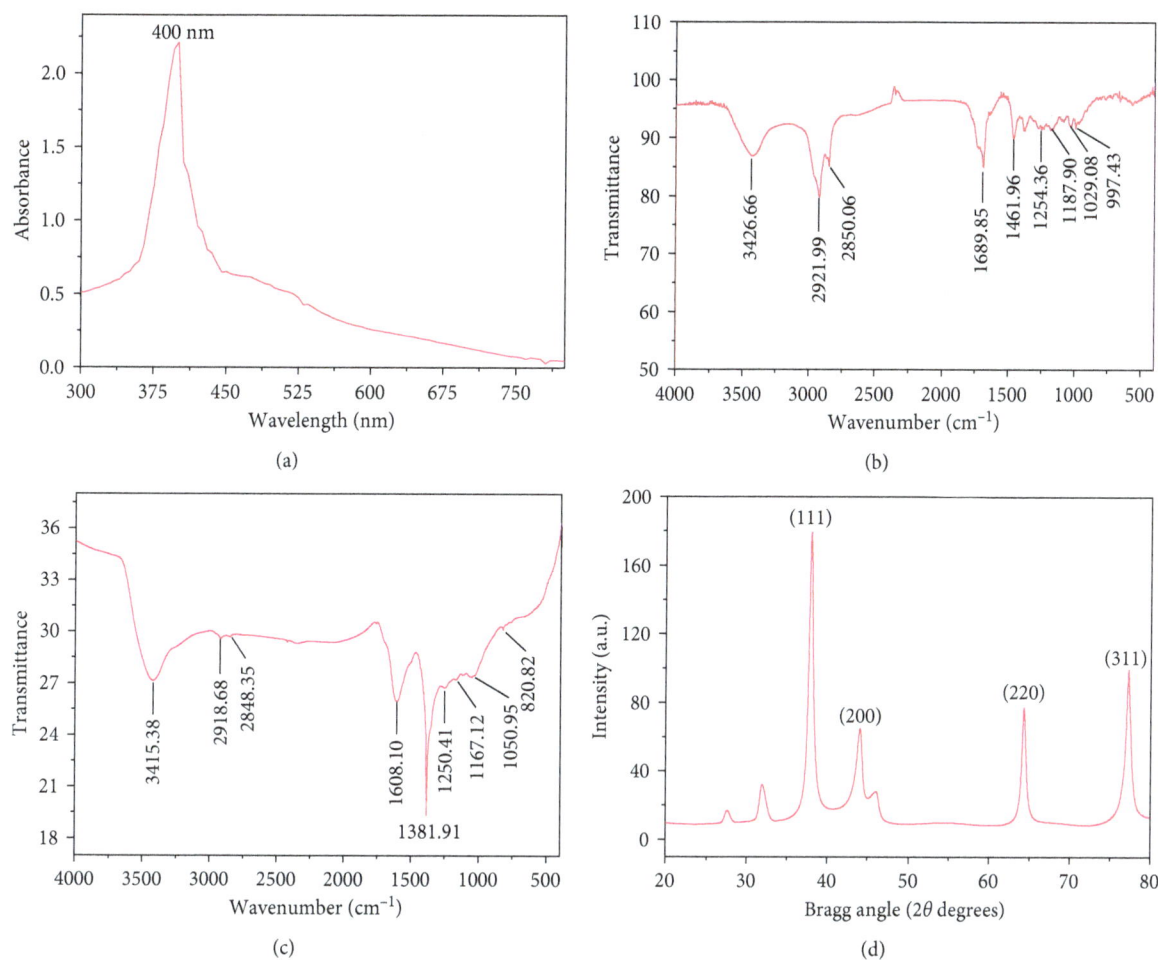

FIGURE 2: Characterization of green synthesized silver nanoparticles (ND-AgNPs) prepared using aqueous extract of the *Nepeta deflersiana* plant. (a) Ultraviolet-visible absorption spectra of synthesized silver nanoparticles (AgNPs). (b) Fourier-transform infrared spectra of *N. deflersiana* extract. (c) Fourier-transform infrared spectra of synthesized silver nanoparticles (ND-AgNPs). (d) X-ray powder diffraction pattern of synthesized silver nanoparticles (ND-AgNPs).

for the cytotoxic effect of biosynthesized AgNPs using *Annona squamosa* leaf extract against the breast cancer MCF-7 cell line [46], *Piper longum* leaf extracts against Hep-2 cancer cell line [47], and *Morinda citrifolia* against HeLa cell lines [48] *in vitro*.

3.3. Morphological Analysis under the Microscope.

The alterations observed in the morphology of HeLA cells treated with ND-AgNPs at 1–100 µg/ml for 24 h are presented in Figure 4(c). There was no significant change observed in the morphology of control HeLA cells. The control cells appeared in normal shape and were attached to the surface. However, the HeLA cells exposed to ND-AgNPs lost their typical shape and cell adhesion capacity, shrinked, and decreased the cell density. These kind of changes have also been reported using plant synthesized AgNPs in different cancer cell lines [46], suggesting that the cytotoxic effect of synthesized AgNPs may be due to the antineoplastic nature and their capability *via* numerous molecular mechanism to induce cell death [45].

3.4. Glutathione Depletion and Lipid Peroxidation Level.

Figures 5(a) and 5(b) summarize the decrease in glutathione level and increase in the lipid peroxidation in HeLa cells exposed to ND-AgNPs at 5–25 µg/ml concentrations for 24 h. The results indicate a concentration-dependent decrease in glutathione level. The depletion in the GSH was found to be 40%, 55%, and 69% at 5, 10, and 25 µg/ml, respectively, as compared to control (Figure 5(a)). The effect of ND-AgNPs-induced lipid peroxidation in HeLA cells exposed for 24 h is shown in Figure 5(b). A concentration-dependent statistically significant increase in the LPO level was also observed in HeLA cells. The increase in LPO level was observed as 25%, 56%, and 65% at 5, 10, and 25 µg/ml concentrations of ND-AgNPs, respectively (Figure 5(b)). Oxidative stress is known to be involved in the nanoparticles-induced cell death [49]. As observed in this study, the decrease in glutathione level and an increase in the level of lipid peroxidation suggest the role of oxidative stress in cell death in HeLa cell line exposed to ND-AgNPs. Our results are very well supported by previous report where a decrease in glutathione level and an increase in lipid

(a)

(b)

Element	(keV)	Mass(%)	Error(%)	At(%)	Compound	mass%	Cation K
O		6.90					
Ag L	2.983	93.10	3.33	100.00	Ag_2O	100/48.00	100.0100
Total		100.00		100.00		100/48.00	

(c)

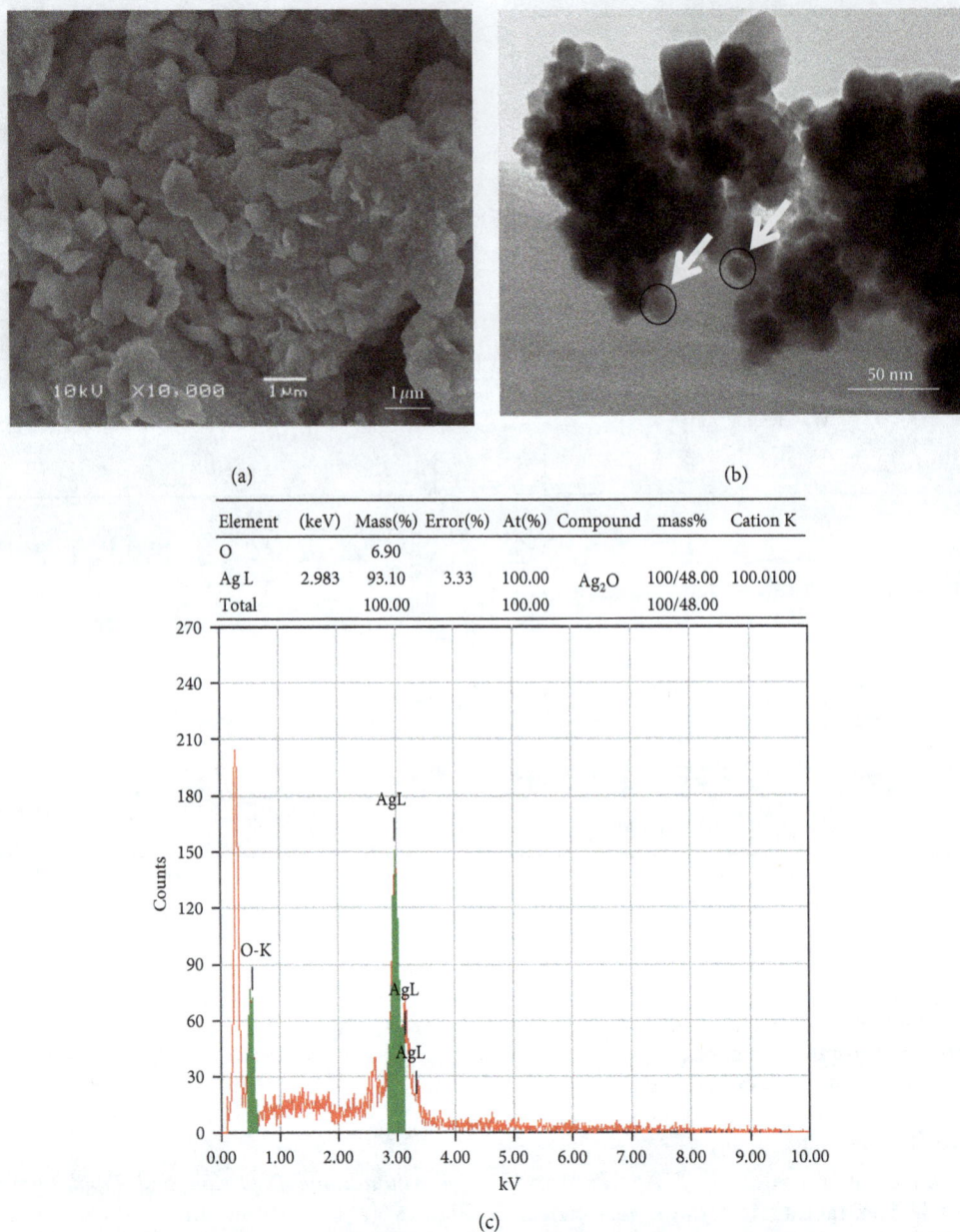

FIGURE 3: (a) SEM image of the green synthesized silver nanoparticles (AgNPs); (b) TEM image of green synthesized silver nanoparticles (ND-AgNPs) at 50 nm; (c) energy-dispersive X-ray spectrum of green synthesized silver nanoparticles (ND-AgNPs).

peroxidation level have been observed due to the exposure of nanoparticles in various cell lines [49, 50].

3.5. Determination of Intracellular Reactive Oxygen Species (ROS). The result obtained from ROS generation in HeLA cells exposed to ND-AgNPs for 24 h is shown in Figures 6(a) and 6(b). A statistically significant induction in ROS generation was measured in HeLA cells exposed to ND-AgNPs at 10, 25, and 50 μg/ml concentrations. As shown in Figures 6(a) and 6(b), an increase of 207%, 167%, and 160% was observed in ROS generation at 5, 10, and 25 μg/ml, respectively, as compared to untreated control. Nanoparticles are suggested to induce their toxicity through oxidative stress by generating reactive oxygen species (ROS) involved in a variety of different cellular processes ranging from apoptosis and necrosis to cell proliferation and carcinogenesis [51]. It have been reported that nanoparticles increase the ROS generation at cellular level. To investigate the potential role of ND-AgNPs in HeLA cell line, intracellular ROS generation was assessed by HDCF-DA dye using flow cytometer. An increase in the ROS level observed in this study established that AgNPs induced ROS generation, which leads to oxidative stress and cell death. Furthermore, consistent with previous reports that plant-synthesized AgNPs have capacity to induce ROS generation that can result in apoptotic cell death [52].

(a)

(b)

Control   1 μg/ml   2 μg/ml   5 μg/ml

10 μg/ml   25 μg/ml   50 μg/ml   100 μg/ml

(c)

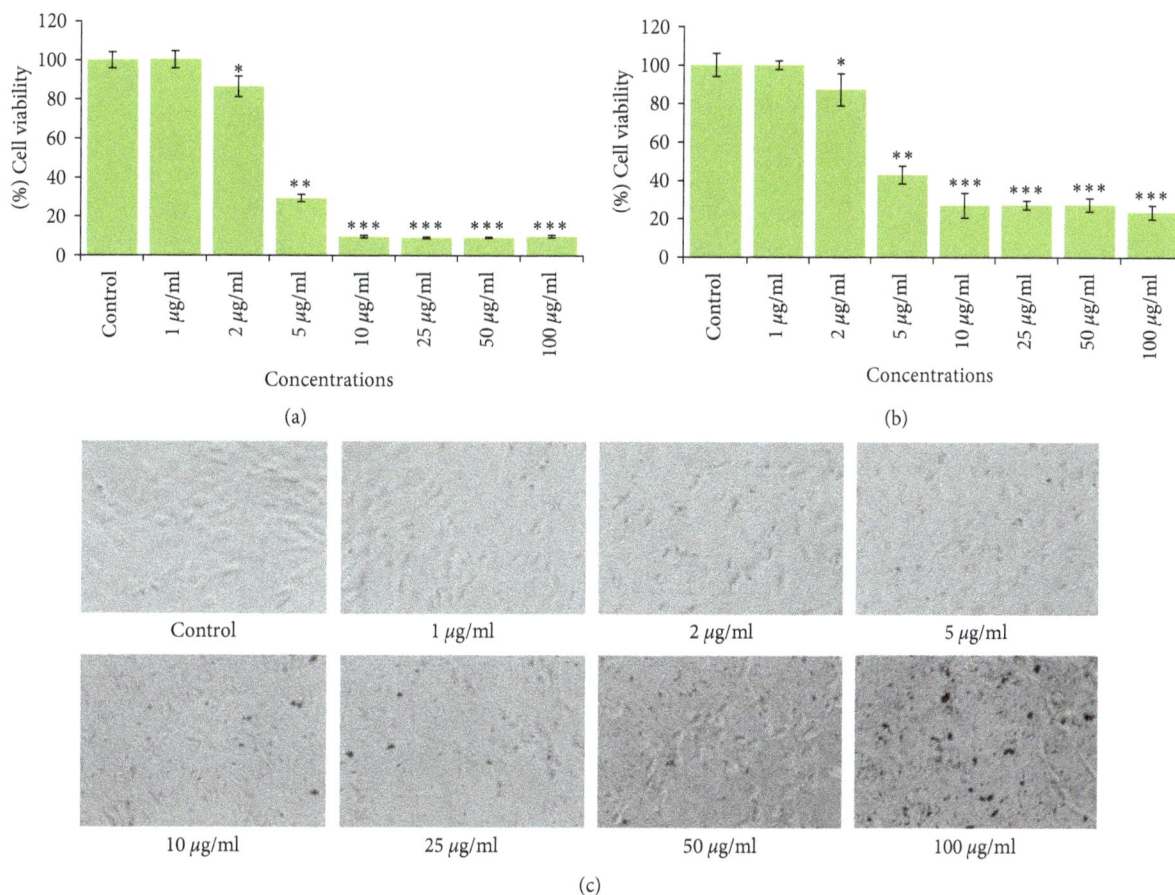

Figure 4: Cytotoxicity assessment in HeLA cells following the exposure of various concentrations of ND-AgNPs for 24 h: (a) MTT assay; (b) neutral red uptake assay. (c) Morphological changes. Images were taken using an inverted phase contrast microscope at 20x magnification. $^{*}p < 0.05,^{**}p < 0.01,^{***}p < 0.001$ vs control.

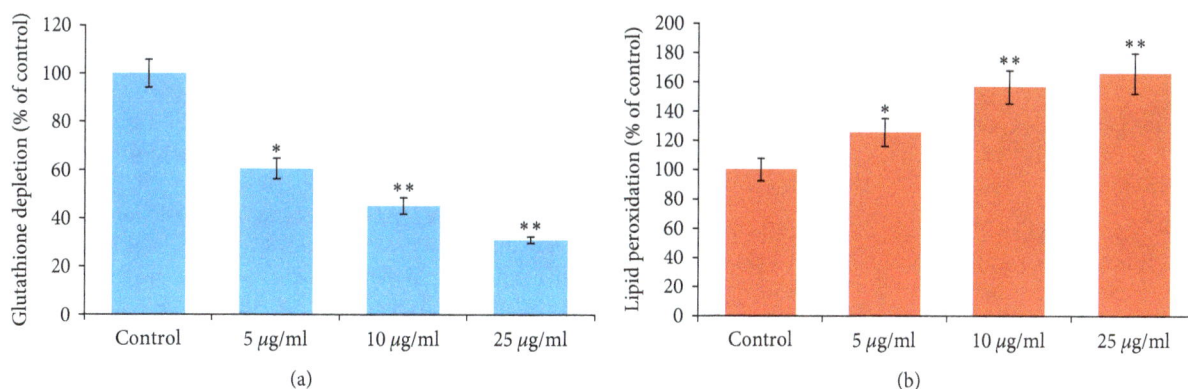

(a)

(b)

Figure 5: ND-AgNPs-induced oxidative stress in HeLA cells exposed for 24 h: (a) glutathione depletion; (b) lipid peroxidation. Results are expressed as the mean ± S.D. of three independent experiments. $^{*}p < 0.01,^{**}p < 0.001$ vs control.

3.6. Mitochondrial Membrane Potential (MMP).
Figures 6(c) and 6(d) illustrate the change in the MMP level. HeLA cells were treated for 24 h at 10–25 μg/ml of synthesized ND-AgNPs. A significant induction in MMP level was found in HeLA cells. The induction in MMP level was found to be 109%, 121%, and 114% at 5, 10, and 25 μg/ml, respectively, compared to control set (Figures 6(c) and 6(d)).

The results of this study suggested that the integrity of mitochondrial membrane might be involved in AgNPs-induced HeLA cell death. It is well documented that the ROS generation at high level can lead to cellular damage by resulting mitochondrial membrane damage, which can then induce toxicity [53, 54]. Based on cationic fluorescent probe Rh123 dye, the induction in MMP level indicated the role of

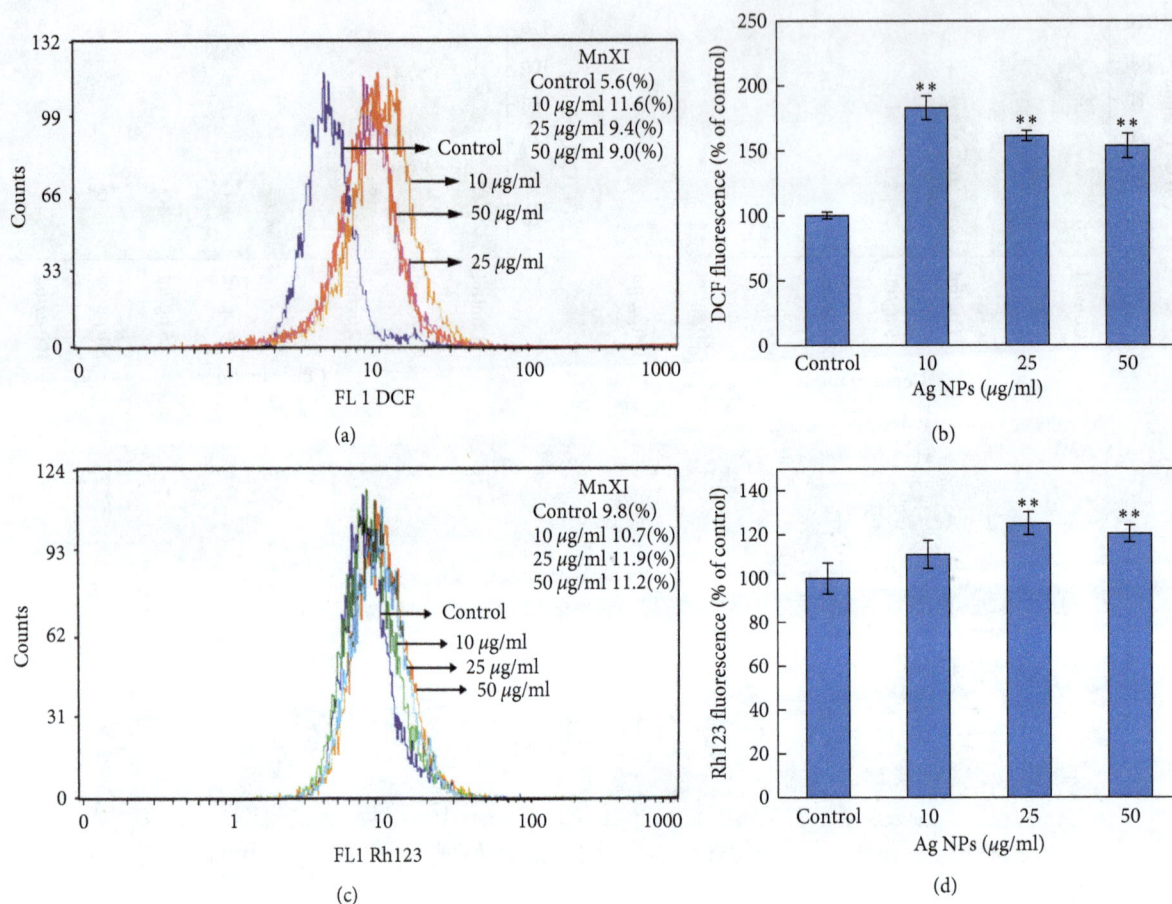

FIGURE 6: Flow cytometric analysis of intracellular ROS generation and mitochondrial membrane potential in HeLA cells exposed to ND-AgNPs for 24 h (a) Representative spectra of fluorescent DCF as a function of ND-AgNPs concentration. (b) Comparative analysis of the fluorescence enhancement of DCF with increasing concentrations of ND-AgNPs. (c) Representative spectra of fluorescence of Rh123 as a function of ND-AgNPs concentrations measured using a flow cytometer. (d) Comparative analysis of the fluorescence enhancement of Rh123 with increasing concentrations of ND-AgNPs. Each histogram represents mean ± S.D. values of DCF and Rh123 fluorescence obtained from three independent experiments. ** $p < 0.01$ versus control.

reactive oxygen species generation and oxidative stress in the AgNPs-induced HeLA cell death due to free radicals generation [55].

3.7. Cell Cycle Analysis. The results of cell cycle analysis in HeLA cell lines exposed to ND-AgNPs at 10–50 µg/ml for 24 h are represented in Figure 7. The flow cytometric measurement of propidium iodide-stained control and ND-AgNPs-treated HeLA cells showed an increase in apoptotic SubG1 peak. A significant increase in SubG1 arrest was observed at 50 µg/ml concentrations of ND-AgNPs-treated HeLA cells (Figure 7). The increase in the SubG1 (apoptotic) population found in this study suggests that ND-AgNPs-treated HeLA cells were not able to go through G2 checkpoint; therefore, G2/M transition was found to be affected. The apoptosis induction due to the presence of SubG1 peak in the process of cell cycle suggests the role of early and late apoptotic/necrotic pathway [56, 57].

3.8. Apoptosis/Necrosis Assessment Using Annexin V-PE and 7-AAD. The results obtained from the induction of

apoptosis/necrosis using flow cytometry are summarized in Figure 8. The flow cytometry data clearly showed that ND-AgNPs induced cell death in HeLA cells. Based on the Annexin V-PE/7-ADD staining, 94.2% of HeLA control cells were found alive with values of 0.56%, 3.31%, and 1.9% of cells, which are normal process for cells growing in cultures. The HeLA cells exposed to ND-AgNPs significantly increased the late apoptotic and necrotic cells as compared with untreated control cells. An increase in the percentage of apoptotic and necrotic cells was found with the values of 30.3–69.8% and 18.8–25.3% between 10 µg/ml and 50 µg/ml ND-AgNPs concentrations, respectively (Figure 8). Even at lower concentration, i.e., 10 µg/ml, ND-AgNPs were found to induce apoptotic and necrotic cell death. It is well known that high amount of ROS generation could lead to apoptotic and necrotic cell death [58]. The excessive ROS generation has been linked with the substantial DNA damage and apoptosis/necrosis [59]. Our results are in well accordance with the recent reports that have shown apoptosis cell death due to the exposure of nanoparticles [60], including the exposure of plant-synthesized silver nanoparticles [52].

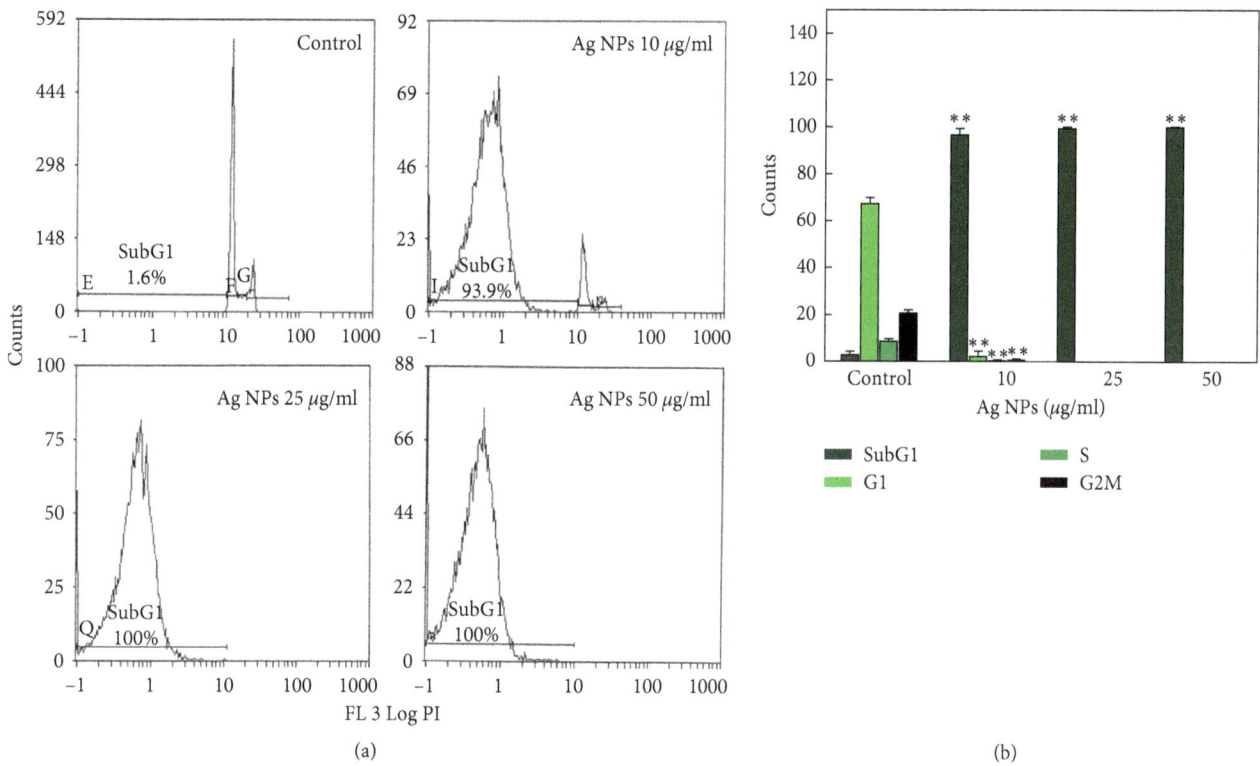

FIGURE 7: Cell cycle analysis in HeLA cells exposed to 10–50 μg/ml concentrations of ND-AgNPs for 24 h. (a) Representative flow cytometric image exhibiting changes in the progression of cell cycle. SubG1 in each micrograph represents the percentage of cells in the SubG1 phase. (b) Each histogram represents the percentage of cells arrested in different phases of cell cycle. Results are expressed as the mean ± S.D. of three independent experiments. **$p < 0.001$ vs control.

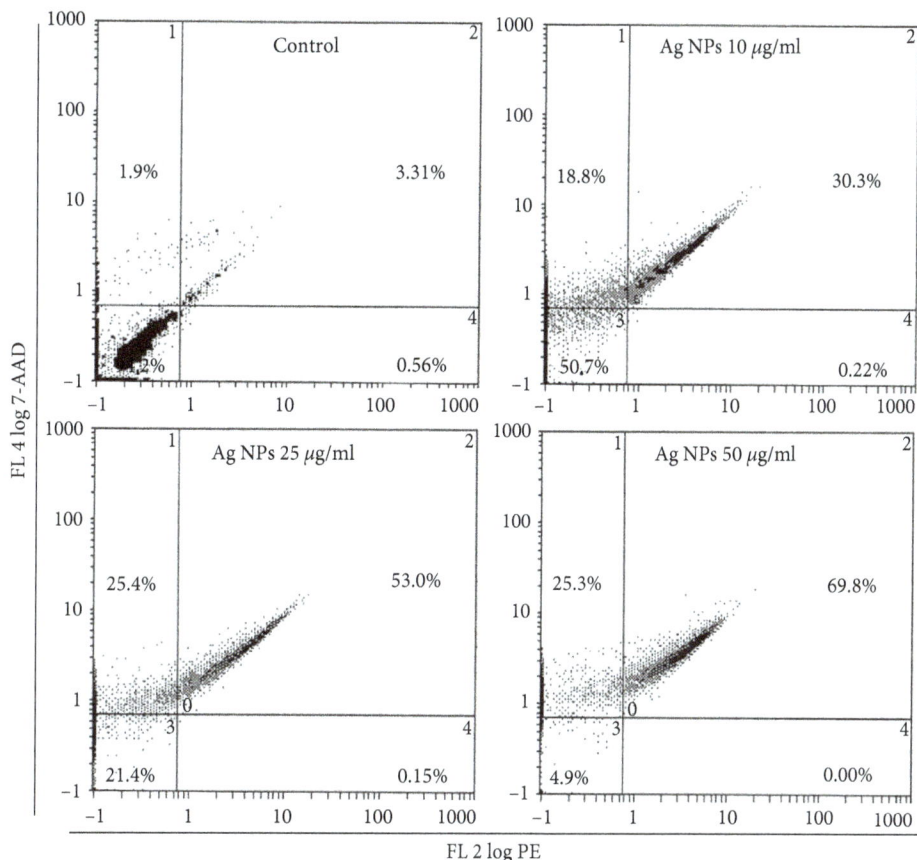

FIGURE 8: Annexin V-PE (phycoerythrin) and 7-AAD (7-amino actinomycin D assay. Bivariate flow cytometry analysis of HeLA cells treated with different concentrations of ND-Ag NPs. The scatter plots show early apoptotic, late apoptotic, and necrotic cells following 24 h treatment.

4. Conclusions

This investigation demonstrated the biosynthesis of silver nanoparticles (AgNPs) for the first time, via a single-step reduction of silver ions using *Nepeta deflersiana* plant and its anticancer potential against human cervical cancer (HeLa) cells. Our results showed that biosynthesized AgNPs (ND-AgNPs) induced a concentration-dependent cytotoxicity in HeLA cells. ND-AgNPs were also found to induce oxidative stress as observed by the increase in ROS and LPO level and the decrease in GSH level. The increase in the intracellular ROS generation was found eventually to trigger the development of mitochondrial membrane damage and cell cycle alterations. This study also showed that ND-AgNPs have the capacity of inducing apoptosis and necrosis cell death of HeLA cells through SubG1 cell cycle arrest. Thus, our findings suggest the anticancer potential of biosynthesized ND-AgNPs against human cervical cancer cells and could play an important role in the development of new therapeutic agent for the treatment of cancer.

Acknowledgments

This research project was supported by a grant from the "Research Centre of the Female Scientific and Medical Colleges", Deanship of Scientific Research, King Saud University.

References

[1] K. Yokohama and D. R. Welchons, "The conjugation of amyloid beta protein on the gold colloidal nanoparticles surfaces," *Nanotechnology*, vol. 18, no. 10, pp. 105101–105107, 2007.

[2] V. K. Sharma, R. A. Yngard, and Y. Lin, "Silver nanoparticles: green synthesis and their antimicrobial activities," *Advances in Colloid and Interface Science*, vol. 145, no. 1-2, pp. 83–96, 2009.

[3] K. K. Wong and X. Liu, "Silver nanoparticles—the real "silver bullet" in clinical medicine?," *MedChemComm*, vol. 1, no. 2, pp. 125–131, 2010.

[4] D. R. Monteiro, S. Silva, M. Negri et al., "Silver nanoparticles: influence of stabilizing agent and diameter on antifungal activity against Candida albicans and Candida glabrata biofilms," *Letters in Applied Microbiology*, vol. 54, no. 5, pp. 383–391, 2012.

[5] C. Krishnaraj, R. Ramachandran, K. Mohan, and P. T. Kalaichelvan, "Optimization for rapid synthesis of silver nanoparticles and its effect on phytopathogenic fungi," *Spectrochimica Acta Part A: Molecular and Biomolecular Spectroscopy*, vol. 93, pp. 95–99, 2012.

[6] F. Martínez-Gutierrez, E. P. Thi, J. M. Silverman et al., "Antibacterial activity, inflammatory response, coagulation and cytotoxicity effects of silver nanoparticles," *Nanomedicine: Nanotechnology, Biology and Medicine*, vol. 8, no. 3, pp. 328–336, 2012.

[7] R. Gottesman, S. Shukla, N. Perkas, L. A. Solovyov, Y. Nitzan, and A. Gedanken, "Sonochemical coating of paper by microbiocidal silver nanoparticles," *Langmuir*, vol. 27, no. 2, pp. 720–726, 2011.

[8] I. Osório, R. Igreja, R. Franco, and J. Cortez, "Incorporation of silver nanoparticles on textile materials by an aqueous procedure," *Materials Letters*, vol. 75, pp. 200–203, 2012.

[9] F. Martinez-Gutierrez, P. L. Olive, A. Banuelos et al., "Synthesis, characterization, and evaluation of antimicrobial and cytotoxic effect of silver and titanium nanoparticle," *Nanomedicine*, vol. 6, no. 5, pp. 681–688, 2010.

[10] J. Favero, P. Corbeau, M. Nicolas et al., "Inhibition of humanimmunodeficiency virus infection by the lectinjacalin and by a derived peptideshowing a sequence similarity with gp120," *European Journal of Immunology*, vol. 23, no. 1, pp. 179–185, 1993.

[11] K. Okitsu, A. Yue, S. Tanabe, H. Matsumoto, and Y. Yobiko, "formation of colloidal gold nanoparticles in an ultrasonic field: control of rate of gold (III) reduction and size of formed gold particles," *Langmuir*, vol. 17, no. 25, pp. 7717–7720, 2001.

[12] R. R. Naik, S. J. Stringer, G. Agarwal, S. E. Jones, and M. O. Stone, "Biomimetic synthesis and patterning of silver nanoparticles," *Nature Materials*, vol. 1, no. 3, pp. 169–172, 2002.

[13] K. B. Narayanan and N. Sakthivel, "Biological synthesis of metal nanoparticles by microbes," *Advances in Colloid and Interface Science*, vol. 156, no. 1-2, pp. 1–13, 2010.

[14] D. Mandal, M. E. Bolander, D. Mukhopadhyay, G. Sarkar, and P. Mukherjee, "The use of microorganisms for the formation of metal nanoparticles and their application," *Applied Microbiology and Biotechnology*, vol. 69, no. 5, pp. 485–492, 2006.

[15] M. Kowshik, S. Ashtaputre, S. Kharrazi et al., "Extracellular synthesis of silver nanoparticles by a silver-tolerantyeast strain MKY3," *Nanotechnology*, vol. 14, no. 1, pp. 95–100, 2002.

[16] P. Mukherjee, A. Ahmad, D. Mandal et al., "Fungus-mediated synthesis of silver nanoparticles and their immobilization in the mycelial matrix: a novel biological approach to nanoparticle synthesis," *Nano Letters*, vol. 1, no. 10, pp. 515–519, 2001.

[17] S. Iravani, "Green synthesis of metal nanoparticles using plants," *Green Chemistry*, vol. 13, no. 10, pp. 2638–2650, 2011.

[18] K. Kalishwaralal, V. Deepak, S. R. K. Pandian et al., "Biosynthesis of silver and gold nanoparticles using *Brevibacterium casei*," *Colloids and Surfaces B: Biointerfaces*, vol. 77, no. 2, pp. 257–262, 2010.

[19] X. Gao, Y. Lu, L. Fang et al., "Synthesis and anticancer activity of some novel 2-phenazinamine derivatives," *European Journal of Medicinal Chemistry*, vol. 69, pp. 1–9, 2013.

[20] S. McGuire, "World cancer report 2014. Geneva, Switzerland: world health organization, international agency for research on cancer, WHO press, 2015," *Advances in Nutrition*, vol. 7, no. 2, pp. 418-419, 2016.

[21] D. J. Newman and G. M. Cragg, "Natural products as sources of new drugs over the 30 years from 1981 to 2010," *Journal of Natural Products*, vol. 75, no. 3, pp. 311–335, 2012.

[22] M. A. Rahman, J. S. Mossa, M. S. Al-Said, and M. A. Al-Yahya, "Medicinal plant diversity in the flora of Saudi Arabia 1: a report on seven plant families," *Fitoterapia*, vol. 75, no. 2, pp. 149–161, 2004.

[23] J. S. Mossa, M. A. Al-Yahya, and I. A. Al-Meshal, *Medicinal Plants of Saudi Arabia*, King Saud University Press, Riyadh, Saudi Arabia, 2000.

[24] C. Hedge, "Studies in the flora of Arabia 2, some new and interesting species of labiatae," *Notes from the Royal botanic Garden Edinburgh*, vol. 40, no. 2, pp. 63–74, 1982.

[25] R. A. Mothana, R. Gruenert, P. J. Bednarski, and U. Lindequist, "Evaluation of the in vitro anticancer, antimicrobial and antioxidant activities of some Yemeni plants used in folk medicine," *De Pharmazie*, vol. 64, no. 4, pp. 260–268, 2009.

[26] M. M. Al-Oqail, E. S. Al-Sheddi, M. A. Siddiqui, J. Musarrat, A. A. Al-Khedhairy, and N. N. Farshori, "Anticancer activity of chloroform extract and subfractions of *Nepeta deflersiana* on human breast and lung cancer cells: an *in vitro* cytotoxicity assessment," *Pharmacognosy Magazine*, vol. 11, no. 44, pp. 598–605, 2015.

[27] M. A. Siddiqui, M. P. Kashyap, V. Kumar, A. A. Al-Khedhairy, J. Musarrat, and A. B. Pant, "Protective potential of transresveratrol against 4-hydroxynonenal induced damage in PC12 cells," *Toxicology in Vitro*, vol. 24, no. 6, pp. 1592–1598, 2010.

[28] M. A. Siddiqui, H. A. Alhadlaq, J. Ahmad, A. A. Al-Khedhairy, J. Musarrat, and M. Ahamed, "Copper oxide nanoparticles induced mitochondria mediated apoptosis in human hepatocarcinoma cells," *PLoS ONE*, vol. 8, no. 8, Article ID e69534, 2013.

[29] Y. Zhang, L. Jiang, L. Jiang et al., "Possible involvement of oxidative stress in potassium bromate-induced genotoxicity in human hepG2 cells," *Chemico-Biological Interactions*, vol. 189, no. 3, pp. 186–191, 2011.

[30] Q. Saquib, A. A. Al-Khedhairy, M. A. Siddiqui, F. M. Abou-Tarboush, A. Azam, and J. Musarrat, "Titanium dioxide nanoparticles induced cytotoxicity, oxidative stress and DNA damage in human amnion epithelial (WISH) cells," *Toxicology in Vitro*, vol. 26, no. 2, pp. 351–361, 2012.

[31] Q. Saquib, J. Musarrat, M. A. Siddiqui et al., "Cytotoxic and necrotic responses in human amniotic epithelial (WISH) cells exposed to organophosphate insecticide phorate," *Mutation Research/Genetic Toxicology and Environmental Mutagenesis*, vol. 744, no. 2, pp. 125–134, 2012.

[32] L. Mulfinger, S. D. Solomon, M. Bahadory et al., "Synthesis and study of silver nanoparticles," *Journal of Chemical Education*, vol. 84, no. 2, p. 322, 2007.

[33] M. Sastry, K. S. Mayyaa, and K. Bandyopadhyay, "pH dependent changes in theoptical properties of carboxylic acid derivatized silver colloid particles," *Colloids and Surfaces A: Physicochemical and Engineering Aspects*, vol. 127, no. 1–3, pp. 221–228, 1997.

[34] K. L. Kelly, E. Coronado, L. L. Zhao, and G. C. Schatz, "The optical properties of metal nano particles: the influence of size, shape and dielectric environment," *Journal of Physical Chemistry B*, vol. 107, no. 3, pp. 668–677, 2003.

[35] A. Rai, A. Singh, A. Ahmad, and M. Sastry, "Role of halide ions and temperature on the morphology of biologically synthesized gold nanotriangles," *Langmuir*, vol. 22, no. 2, pp. 736–741, 2006.

[36] N. E. El-Naggar, N. A. Abdelwahed, and O. M. Darwesh, "Fabrication of biogenic antimicrobial silver nanoparticles by streptomyces aegyptia NEAE 102 as eco-friendly nanofactory," *Journal of Microbiology and Biotechnology*, vol. 24, no. 4, pp. 453–464, 2014.

[37] A. K. Jha, K. Prasad, V. Kumar, and K. Prasad, "Biosynthesis of silver nanoparticles using Eclipta leaf," *Biotechnology Progress*, vol. 25, no. 5, pp. 1476–1479, 2009.

[38] H. Borchert, "Determination of nanocrystal sizes: a comparison of TEM, SAXS, and XRD studies of highly monodisperse CoPt3 particles," *Langmuir*, vol. 21, no. 5, pp. 1931–1936, 2005.

[39] K. D. Arunachalam, L. B. Arun, S. K. Annamalai, and A. M. Arunachalam, "Potential anticancer properties of bioactive compounds of Gymnema sylvestre and its biofunctionalized silver nanoparticle," *International Journal of Nanomedicine*, vol. 10, pp. 31–41, 2015.

[40] P. V. Kumar, S. V. N. Pammi, P. Kollu, K. V. V. Satyanarayana, and U. Shameem, "Green synthesis and characterization of silver nanoparticles using *Boerhaavia diffusa* plant extract and their anti bacterial activity," *Industrial Crops and Products*, vol. 52, pp. 562–566, 2014.

[41] K. Mallikarjuna, N. J. Sushma, G. Narasimha, L. Manoj, and B. D. P. Raju, "Phytochemical fabrication and characterization of silver nanoparticles by using pepper leaf broth," *Arabian Journal of Chemistry*, vol. 7, no. 6, pp. 1099–1103, 2014.

[42] D. R. Nogueira, M. Mitjans, M. R. Infante, and M. P. Vinardell, "Comparative sensitivity of tumor and nontumor cell lines as a reliable approach for *in vitro* cytotoxicity screening of lysine-based surfactants with potential pharmaceutical applications," *International Journal of Pharmaceutics*, vol. 420, no. 1, pp. 51–58, 2011.

[43] T. Mosmann, "Rapid colorimetric assay for cellular growth and survival: application to proliferation and cytotoxicity assays," *Journal of Immunological Methods*, vol. 65, no. 1-2, pp. 55–63, 1983.

[44] E. Borenfreund and J. A. Puerner, "Short-term quantitative in vitro cytotoxicity assay involving an S-9 activating system," *Cancer Letters*, vol. 34, no. 3, pp. 243–248, 1987.

[45] M. A. Farah, M. A. Ali, S. M. Chen et al., "Silver nanoparticles synthesized from adenium obesum leaf extract induced DNA damage, apoptosis and autophagy via generation of reactive oxygen species," *Colloids and Surfaces B: Biointerfaces*, vol. 141, pp. 158–169, 2016.

[46] R. Vivek, R. Thangam, K. Muthuchelian, P. Gunasekaran, K. Kaveri, and S. Kannan, "Green biosynthesis of silver nanoparticles from *Annona squamosa* leaf extract and its *in vitro* cytotoxic effect on MCF-7 cells," *Process Biochemistry*, vol. 47, no. 12, pp. 2405–2410, 2012.

[47] S. J. P. Jacob, J. S. Finub, and A. Narayanan, "Synthesis of silver nanoparticles using piper longum leaf extracts and its cytotoxic activity against Hep-2 cell line," *Colloids and Surfaces B: Biointerfaces*, vol. 91, pp. 212–214, 2012.

[48] T. Y. Suman, S. R. Rajasree, A. Kanchana, and S. B. Elizabeth, "Biosynthesis, characterization and cytotoxic effect of plant mediated silver nanoparticles using *Morinda citrifolia* root extract," *Colloids and Surfaces B: Biointerfaces*, vol. 106, pp. 74–78, 2013.

[49] M. A. Siddiqui, Q. Saquib, M. Ahamed et al., "Molybdenum nanoparticles-induced cytotoxicity, oxidative stress, G2/M arrest, and DNA damage in mouse skin fibroblast cells (L929)," *Colloids and Surfaces B: Biointerfaces*, vol. 125, pp. 73–81, 2015.

[50] S. M. El-Sonbaty, "Fungus-mediated synthesis of silver nanoparticles and evaluation of antitumor activity," *Cancer Nanotechnology*, vol. 4, no. 4-5, pp. 73–79, 2013.

[51] S. J. Soenen and M. De Cuyper, "Assessing cytotoxicity of (iron oxide-based) nanoparticles: an overview of different methods exemplified with cationic magnetoliposomes," *Contrast Media and Molecular Imaging*, vol. 4, no. 5, pp. 207–219, 2009.

[52] A. Stroh, C. Zimmer, C. Gutzeit et al., "Iron oxide particles for

molecular magnetic resonance imaging cause transient oxidative stress in rat macrophages," *Free Radical Biology and Medicine*, vol. 36, no. 8, pp. 976–984, 2004.

[53] S. Dwivedi, M. A. Siddiqui, N. N. Farshori, M. Ahamed, J. Musarrat, and A. A. Al-Khedhairy, "Synthesis, characterization and toxicological evaluation of iron oxide nanoparticles in human lung alveolar epithelial cells," *Colloids and Surfaces B: Biointerfaces*, vol. 122, pp. 209–215, 2014.

[54] I. Nicoletti, G. Migliorati, M. C. Pagliacci, F. Grignani, and C. Riccardi, "A rapid and simple method for measuring thymocyte apoptosis by propidium iodide staining and flow cytometry," *Journal of Immunological Methods*, vol. 139, no. 2, pp. 271–279, 1991.

[55] S. Ravi, K. K. Chiruvella, K. Rajesh, V. Prabhu, and S. C. Raghavan, "5-isopropylidene-3-ethylrhodanine induce growth inhibition followed by apoptosis in leukemia cells," *European Journal of Medicinal Chemistry*, vol. 45, no. 7, pp. 2748–2752, 2010.

[56] P. V. Asharani, M. P. Hande, and S. Valiyaveettil, "Antiproliferative activity of silver nanoparticles," *BMC Cell Biology*, vol. 10, no. 1, pp. 65–79, 2009.

[57] H. Ciftci, M. Turk, U. Tamer, S. Karahan, and Y. Menemen, "Silver nanoparticles: cytotoxic, apoptotic, and necrotic effects on MCF-7 cells," *Turkish Journal of Biology*, vol. 37, pp. 573–581, 2013.

[58] R. Foldbjerg, P. Olesen, M. Hougaard, D. A. Dang, H. J. Hoffmann, and H. Autrup, "PVP-coated silver nanoparticles and silver ions induce reactive oxygen species, apoptosis, and necrosis in THP-1 monocytes," *Toxicology Letters*, vol. 190, no. 2, pp. 156–162, 2009.

[59] Y. H. Hsin, C. F. Chen, S. Huang, T. S. Shih, P. S. Lai, and P. J. Chueh, "The apoptotic effect of nanosilver is mediated by a ROS-and JNK-dependent mechanism involving the mitochondrial pathway in NIH3T3 cells," *Toxicology Letters*, vol. 179, no. 3, pp. 130–139, 2008.

[60] Y. Pan, S. Neuss, A. Leifert et al., "Size-dependent cytotoxicity of gold nanoparticles," *Small*, vol. 3, no. 11, pp. 1941–1949, 2007.

Permissions

The contributors of this book come from diverse backgrounds, making this book a truly international effort. This book will bring forth new frontiers with its revolutionizing research information and detailed analysis of the nascent developments around the world.

We would like to thank all the contributing authors for lending their expertise to make the book truly unique. They have played a crucial role in the development of this book. Without their invaluable contributions this book wouldn't have been possible. They have made vital efforts to compile up to date information on the varied aspects of this subject to make this book a valuable addition to the collection of many professionals and students.

This book was conceptualized with the vision of imparting up-to-date information and advanced data in this field. To ensure the same, a matchless editorial board was set up. Every individual on the board went through rigorous rounds of assessment to prove their worth. After which they invested a large part of their time researching and compiling the most relevant data for our readers.

The editorial board has been involved in producing this book since its inception. They have spent rigorous hours researching and exploring the diverse topics which have resulted in the successful publishing of this book. They have passed on their knowledge of decades through this book. To expedite this challenging task, the publisher supported the team at every step. A small team of assistant editors was also appointed to further simplify the editing procedure and attain best results for the readers.

Apart from the editorial board, the designing team has also invested a significant amount of their time in understanding the subject and creating the most relevant covers. They scrutinized every image to scout for the most suitable representation of the subject and create an appropriate cover for the book.

The publishing team has been an ardent support to the editorial, designing and production team. Their endless efforts to recruit the best for this project, has resulted in the accomplishment of this book. They are a veteran in the field of academics and their pool of knowledge is as vast as their experience in printing. Their expertise and guidance has proved useful at every step. Their uncompromising quality standards have made this book an exceptional effort. Their encouragement from time to time has been an inspiration for everyone.

The publisher and the editorial board hope that this book will prove to be a valuable piece of knowledge for researchers, students, practitioners and scholars across the globe.

List of Contributors

Samuel Treviño, Alfonso Díaz, Víctor Enrique Sarmiento-Ortega, Jоśe Ángel Flores-Hern ández, Eduardo Brambila and Francisco J. Meléndez
Facultad de Ciencias Químicas, Benemérita Universidad Autónoma de Puebla, 14 Sury Av. San Claudio, Col. San Manuel, 72570 Puebla, PUE, Mexico

Eduardo Sánchez-Lara and Enrique González-Vergara
Centro de Química, ICUAP, Benemérita Universidad Autónoma de Puebla, 14 Sur y Av. San Claudio, Col. San Manuel, 72570 Puebla, PUE, Mexico

Leilei Xie, Lifang Liu, Wenming Wang, Zhiou Ma, Liqun Xu and Hongfei Wang
Key Laboratory of Chemical Biology and Molecular Engineering of Education Ministry, Institute of Molecular Science, Shanxi University, Taiyuan 030006, China

Leilei Xie, Lifang Liu, Wenming Wang and Hongfei Wang
Key Laboratory of Energy Conversion and Storage Materials of Shanxi Province, Institute of Molecular Science, Shanxi University, Taiyuan 030006, China

Xuan Zhao
Department of Chemistry, University of Memphis, Memphis, TN 38152, USA

Shalaka S. Rokade
Department of Microbiology, Modern College of Arts, Science and Commerce, Ganeshkhind, Pune 411016, India

Komal A. Joshi, Ketakee Mahajan, Saniya Patil and Geetanjali Tomar
Institute of Bioinformatics and Biotechnology, Savitribai Phule Pune University, Pune 411007, India

Dnyanesh S. Dubal
Indian Institute of Science, Education and Research, Pashan, Pune 411008, India

Vijay Singh Parihar
Department of Biomedical Sciences and Engineering, BioMediTech, Tampere University of Technology, Korkeakoulunkatu 10, 33720 Tampere, Finland

Rohini Kitture
Department of Applied Physics, Defense Institute of Advanced Technology, Girinagar, Pune 411025, India

Jayesh R. Bellare
Department of Chemical Engineering, Indian Institute of Technology Bombay, Powai, Mumbai 400076, India

Sougata Ghosh
Department of Microbiology, School of Science, RK University, Kasturbadham, Rajkot 360020, India

María del Carmen Sánchez-Navarro
Facultad de Estomatología, Universidad Autónoma de San Luis Potosí (UASLP), Avenida Manuel Nava 2, Zona Universitaria, 78290 San Luis Potosí, Mexico

laudio Adrian Ruiz-Torres, Nereyda Niño-Martínez, Gabriel Alejandro Martínez-Castañón, I. DeAlba-Montero and Facundo Ruiz
Facultad de Ciencias, Universidad Autónoma de San Luis Potosí (UASLP), Avenida Manuel, Nava 6, Zona Universitaria, 78290 San Luis Potosí, Mexico

Roberto Sánchez-Sánchez
Instituto Nacional de Rehabilitación LGII, CENIAQ, Calzada México Xochimilco No. 289, Colonia Arenal de Guadalupe, Delegación Tlalpan, 14389 Ciudad de México, Mexico

Vasiliki D. Papakonstantinou and Constantinos A. Demopoulos
Laboratory of Biochemistry, Faculty of Chemistry, National and Kapodistrian University of Athens, 15771 Athens, Greece

Nefeli Lagopati and Effie C. Tsilibary
Institute of Biosciences & Applications, NCSR Demokritos, 15310 Agia Paraskevi, Greece

Athanassios I. Philippopoulos
Laboratory of Inorganic Chemistry, Department of Chemistry, National and Kapodistrian University of Athens, Panepistimiopolis Zografou, 15771 Athens, Greece

Zbigniew Dutkiewicz
Department of Chemical Technology of Drugs, Poznán University of Medical Sciences, Grunwaldzka 6, 60-780 Poznán, Poland

Renata Mikstacka
Department of Inorganic and Analytical Chemistry, Ludwik Rydygier Collegium Medicum, Nicolaus Copernicus University in Torún, Dr A. Jurasza 2, 85-089 Bydgoszcz, Poland

Gamil A. Al-Hazmia and Khlood S. Abou-Melha
Chemistry Department, Faculty of Science, King Khalid University, Abha, Saudi Arabia

Gamil A. Al-Hazmi
Chemistry Department, Faculty of Applied Sciences, Taiz University, Taiz, Yemen

Nashwa M. El-Metwaly
Chemistry Department, College of Applied Sciences, Umm Al-Qura University, Makkah, Saudi Arabia
Chemistry Department, Faculty of Science, Mansoura University, Mansoura, Egypt

Kamel A. Saleh
Biology Department, Faculty of Science, King Khalid University, Abha, Saudi Arabia

Narendra Kumar Chaudhary and Parashuram Mishra
Bio-Inorganic and Materials Chemistry Research Laboratory, Tribhuvan University, M. M. A. M. Campus, Biratnagar, Nepal

Abdullah-Al Masum, Yosuke Hisamatsu, Kenta Yokoi and Shin Aoki
Faculty of Pharmaceutical Sciences, Tokyo University of Science, 2641 Yamazaki, Noda, Chiba 278-8510, Japan

Shin Aoki
Imaging Frontier Center, Tokyo University of Science, 2641 Yamazaki, Noda, Chiba 278-8510, Japan

Adebayo A. Adeniyi and Peter A. Ajibade
School of Chemistry and Physics, University of KwaZulu-Natal, Private Bag X01, Scottsville, Pietermaritzburg 3201, South Africa

Jiaqi Mai, Yunlan Li, Xiaozhi Qiao, Xiaoqing Ji and Qingshan Li
School of Pharmaceutical Science, Shanxi Medical University, Taiyuan 030001, China

Yunlan Li and Qingshan Li
Shanxi University of Traditional Chinese Medicine, Jinzhong 030619, China

Xuejiao Li and Hongqi Tian
Tianjin Key Laboratory of Radiation Medicine and Molecular Nuclear Medicine, Institute of Radiation Medicine, Chinese Academy of Medical Sciences and Peking Union Medical College, Tianjin 300192, China

Yahong Liu
Tianjin Binjiang Pharma, Inc., Tianjin 300192, China

Yan-Ren Lin, Tsung-Han Lee, Chu-Chung Chou and Chin-Fu Chang
Department of Emergency Medicine, Changhua Christian Hospital, Changhua, Taiwan

Yan-Ren Lin
School of Medicine, Kaohsiung Medical University, Kaohsiung, Taiwan

Yan-Ren Lin and Chu-Chung Chou
School of Medicine, Chung Shan Medical University, Taichung, Taiwan

Yuan-Jhen Syue
Department of Anaesthesiology, Kaohsiung Chang Gung Memorial Hospital, Chang Gung University College of Medicine, Kaohsiung, Taiwan

Chao-Jui Li
Department of Emergency Medicine, Kaohsiung Chang Gung Memorial Hospital, Chang Gung University College of Medicine, Kaohsiung, Taiwan
Department of Public Health, College of Health Science, Kaohsiung Medical University, Kaohsiung, Taiwan

Ting Liu, Rong-Gui Yang and Guo-Qing Zhong
School of Material Science and Engineering, Southwest University of Science and Technology, Mianyang 621010, China

M. Manjushree and Hosakere D. Revanasiddappa
Department of Chemistry, University of Mysore, Manasagangothri, Mysuru, Karnataka 570 006, India

Kaila F. Hart, Natalie S. Joe, Rebecca M. Miller, Hannah P. Nash and Aimee M. Morris
Department of Chemistry and Biochemistry, Fort Lewis College, 1000 Rim Dr., Durango, CO 81301, USA

David J. Blake
Department of Biology, Fort Lewis College, 1000 Rim Dr., Durango, CO 81301, USA

Ayodele Temidayo Odularu
Department of Chemistry, University of Fort Hare, Private Bag X1314, Alice 5700, South Africa

Ali Alsalme, Rais Ahmad Khan, Arwa M. Alkathiri and Mohammed Jaafar
Department of Chemistry, College of Science, King Saud University, Riyadh 11451, Saudi Arabia

Mohd. Sajid Ali, Sartaj Tabassum and Hamad A. Al-Lohedan
Surfactant Research Chair, Department of Chemistry, College of Science, King Saud University, Riyadh 11451, Saudi Arabia

Ebtesam S. Al-Sheddi, Nida N. Farshori, Mai M. Al-Oqail and Shaza M. Al-Massarani Department of Pharmacognosy, College of Pharmacy, King Saud University, Riyadh, Saudi Arabia

Quaiser Saquib, Rizwan Wahab, Javed Musarrat, Abdulaziz A. Al-Khedhairy and Maqsood A. Siddiqui Zoology Department, College of Science, King Saud University, Riyadh 11451, Saudi Arabia

Quaiser Saquib, Rizwan Wahab, Javed Musarrat and Maqsood A. Siddiqui Al-Jeraisy Chair for DNA Research, Zoology Department, College of Science, King Saud University, Riyadh 11451, Saudi Arabia

Index

www.ingramcontent.com/pod-product-compliance
Lightning Source LLC
Chambersburg PA
CBHW050445200326
41458CB00014B/5068